D1607041

Laurent Schwartz

# A Mathematician Grappling with His Century

Translated from the French by Leila Schneps

Birkhäuser Verlag
Basel · Boston · Berlin

Author
Laurent Schwartz
37, Rue Pierre Nicole
75005 Paris
France

The original French version was published in 1997 under the title "Un mathématicien aux prises avec le siècle" by Editions Odile Jacob.


Published with the help of the French Ministry of Culture – CNL.

2000 Mathematical Subject Classification 01A70

A CIP catalogue record for this book is available from the Library of Congress, Washington D.C., USA

Deutsche Bibliothek Cataloging-in-Publication Data
Schwartz, Laurent:
A mathematician grappling with his century / Laurent Schwartz. Transl. from the French by Leila Schneps. - Basel ; Boston ; Berlin : Birkhäuser, 2001
Einheitssacht.: Un mathématicien aux prises avec le siècle <eng.>
ISBN 3-7643-6052-6

ISBN 3-7643-6052-6 Birkhäuser Verlag, Basel – Boston – Berlin


Member of the BertelsmannSpringer Publishing Group
Cover design: Micha Lotrovsky, Therwil, Switzerland
Printed on acid-free paper produced of chlorine-free pulp. TCF ∞
Printed in Germany
ISBN 3-7643-6052-6

9 8 7 6 5 4 3 2 1

# Contents

# Foreword

I am a mathematician. Mathematics have filled my life; a passion for research and teaching, as a professor both in the university and at the Ecole Polytechnique. I have thought about the role of mathematics, research, and teaching, in my life in and the lives of others; I have pondered on the mental processes of research, and for decades I have devoted myself to urgently necessary reforms within the university and the Ecole Polytechnique. Some of my reflections are contained in this book, as well as a description of the course of my life. However, I do not discuss the university reforms, since I have written many articles and books on the subject. Inevitably, mathematics appear in this book; one cannot conceive of an autobiography of a mathematician which contains no mathematics. I have written about them in a historical form which should be accessible to large non-specialist sections of the scientific public; readers impervious to their charm may simply skip them. They concern only about fifteen per cent of the volume.

I have been involved in many activities besides mathematics, sometimes to such a point that my research suffered. I devoted a large part of my time to struggling for the oppressed, for human rights and the rights of peoples, first as a Trotskyist, and later outside of any party. It was normal that I would wish to describe these activities, so they will be remembered in the future. Furthermore, I have had another great passion, almost as intense as that for mathematics, but which I did not describe here at great length: entomology, and particularly butterfly collecting, but more generally the whole field of biology. It would have been too long to speak of that here, except for some anecdotes; I hope some day to write a book about my butterfly-hunting adventures during more than thirty trips to the tropics.

There are many people I wish to thank for their help. I showed several chapters to various colleagues, and to my wife and daughter; their advice was very precious to me. I am grateful to Odile Jacob for having accepted, with typical enthusiasm and dynamism, to publish a book written by a mathematician. I thank Mireille de Maistre for taking care of my scientific administration, thus providing me with the time necessary to write. I took great pleasure in my truly agreeable three years' collaboration with Isabelle Rozenbaumas. She read all the chapters with the competence of a translator, which enabled her to help me polish the style, without losing anything of my thought or my personality, so that everyone who knows me recognizes me in the book and feels, as certain readers have told me, that they can "hear the sound of my voice".

# Introduction
# The Garden of Eden

My mother always wanted a country house. As a little girl, she dreamed of being a shepherdess and adored nature, gardens, animals and flowers. When my parents bought the property of Autouillet, in September 1926 – "for a mouthful of bread"[1], they always said –, they were already reasonably well-off: my father was a surgeon. I was just getting better from polio, which I had caught in July. It left few traces, but it did leave me physically weak for my whole life, not really able to walk a long way or to run and jump. I was eleven, and my brothers Daniel and Bertrand were nine and seven. My parents had searched for years, but they had missed a lot of opportunities because of my father's professional conscience, which trembled for the life of his patients after each operation. In those days there were no antibiotics and postoperational infections were frequent.

Les Closeaux (the closed waters) covered two and a half hectares and included an eleven-room house and a garden with two meadows at the end of it. My mother loved bric-a-brac and antiques. She decorated the interior with furniture, objets d'art and pictures, all of which disappeared after a number of robberies, doubtless committed with the cooperation of various antique dealers. Now we have installed an alarm system, which cost almost as much as whatever remained to steal.

My mother went to antique fairs and made numerous acquisitions. The antique dealers knew her and would whisper: "Madame Schwartz, there's nothing for you here. You know that we're the ones who buy antiques, and you have to buy them from us. We'll raise the auction prices as much as we have to to make sure you don't get a thing." But she was a real collector and ended up wearing them down with sheer persistence.

In the living room there are two large paintings showing my father at around forty-five years old and my mother at twenty-five, just before my birth. I like the portrait of my father a lot; my mother's is enchanting. It is just exactly as she was when I was a child. Every time we go to Autouillet and I enter the

---

1) A typical French expression meaning "for a song"

living room, I look at her with emotion: she is so sweet, so beautiful. In spite of the alteration brought on by her many years (she died at the age of eighty-four), I never stopped seeing her as she looked in her flowering youth.

The house is located at the entrance to the property. If, from there, you look towards the garden, it falls into several grand sections. To the left lies a greenhouse which we never kept up; it only has one pane now, but for a long time it was covered with a thick and very perfumed honeysuckle. Also to the left, a bit farther on, there is a vegetable garden of about 50 meters by 30, whose paths are edged by boxtrees probably over a hundred years old. The fruit trees are mainly apple and pear trees. At the bottom, you discover a little marvel: the outbuildings of an old eighteenth century castle, with precious tile rooves. The castle itself has disappeared, carried off stone by stone, apparently, by an American, and rebuilt somewhere in the US. But the outbuildings remain a splendid antique: we lodged the gardener Charles and his wife Alphonsine there.

Beyond the greenhouse, a large lawn spreads more than fifty meters towards the outbuildings. It used to be covered by high grasses. It was a memorable delight for my brothers and me. I used to roll and wallow in the untamed grasses and admire all the beautiful flowers, particularly the big wild daisies. But they were soon replaced by a neat French garden. It was very charming though it didn't correspond to my taste as much as the wild lawn. As children, we used to lie on our backs after dinner to watch for falling stars. A juniper bush which we planted at the bottom of the garden when I was young, and which was about two meters high back then, now measures a good twelve meters. It's a truly remarkable tree which, I believe, ought not to survive at this latitude. Behind the juniper, a little stone wall full of gaps separates the house and garden from the old outbuildings. Beyond the little wall there is a vast area which used to be covered with very high trees – walnut trees, ash trees and lime trees – many of which had to be cut down because they were threatening the delicate tile rooves of the outbuildings.

From the center, the view is majestic. A vast green corridor in sections stretches away for more than a hundred meters. First comes a beautiful square garden edged with very old and splendid trees which are our pride and joy. Every year they are inhabited by one or two adorable couples of squirrels. In front, to the left and the right, stand two chestnut trees – one with white, the other with red flowers, both with monumentally thick round trunks, and about twelve meters high – they're probably a century and a half old. Generations of children have benefited from the yearly delight of gathering up heaps of chestnuts which they carried back to Paris for no purpose whatsoever. This square prairie ends, on the left and the right, with two enormous and very strange trees. The right-hand one is the largest chestnut I have ever seen. I measured its height this year, by a simple method of mathematical similitude using a stick stuck in the ground. I found it to be twenty-eight meters high, which is really impressive for a chestnut tree. It is amazingly ample and seems to touch the sky – it must be nearly two centuries old. I can't stop myself from contemplating it several times a day. The left-hand corner of the prairie is occupied by an elm and an ash grown together so closely that they've become glued together in at least five places. I never could figure out if even their sap mixed, like grafted trees. We should call that old married, intertwined couple Philemon and Baucis. The elm is actually dead, the ash alive. At the top, there is a strange grouping of five or six meters of skeletal and sinister but very picturesque elm branches, offering a perfect perch for numerous birds. At sunset, when the sun is sending its last rays along the ground and everything else is already in shadow, the square is at its most beautiful.

It gives onto a stagnant pond whose level fluctuates according to the level of the ground water at the bottom of the village. We put fish into it, mostly roaches, carp and rainbow trout, although these last usually need clear, oxygenated water. There wasn't enough for them to eat, so that the carps multiplied by thousands without ever growing. We should have introduced a predator, a pike for example, in the springtime, which we could have fished out again each autumn. One very cold year however, the pond froze over, to our great delight since we could skate on it for a whole day. But every single fish died from lack of oxygen, and we never replaced them. One of the goldfish weighed three pounds. When I was young, I used to fish there for tadpoles, toads, newts and salamanders, and dragonfly larvae, which we then raised.

Once out of the water, the young alyte toads used to come and nest in little holes right under the entrance to the house. You could hardly see them except with a powerful flashlight: all night long, starting at dusk, the males emitted a series of little monosyllabic cries with varying notes and periods. We liked to listen to them at length; it was a real concert. My brother Bertrand, who whistles admirably, used to amuse himself by answering them, which sometimes disconcerted them (he also claims to have talked with cuckoos, persuading them to change the order of their two syllables, which I can hardly believe). If a female is attracted by this song, the male fertilizes her and she lays large eggs which he carries away, all sticky, on his back towards the pond where the tadpoles will come out. I could never understand how they could sing and sing in vain night after night, from the spring until the end of summer, until one day I saw a male emerging from his hole, his back covered with eggs. A careful examination of the cracks between the stones revealed that under the stairs lurked a palace from the Thousand and One Nights, with constant arrival of females and incessant departure of males loaded with eggs. This ecological niche between the stairs and the pond (about thirty meters away) prospered from the 1960's until 1985, when the mason, on his own initiative and because that was his usual way of working, filled in all the holes. All the toads died. That's how civilizations sometimes come to an end. But in 1993, new alytes appeared, dispersed along a hedge, and a new niche was created.

So once again there are tadpoles in the pond. But I don't think there's a tremendous lot of life there, except maybe for microbes. Generations of rats also lived there – we used to hear "splatch!" whenever we approached. They're neither mean nor dangerous, and don't carry the plague. They're muskrats, well-known in Canada and in the US, but which have also become quite well-adapted to pond life in France. They are enormous. We actually caught one once and cooked it. But it wasn't as good as hare or rabbit, and we didn't feel like doing it again. In the US and Canada, they are eaten, though – a friend sent me a recipe for muskrat fricassee with beans. When I told our Museum friends what we had done, they were disgusted. We had eaten rat!

Not far from the pond, to the right, is the lollipop tree, where once a year we hung sweets for children (who were firmly convinced that they had grown there) to come and pick. Beyond that, another little garden leads to a low wall. A remarkable tree grows there: a mulberry tree, whose leaves feed silkworms. It's a rare tree around Paris. It's covered with mulberries, not white ones like you see in Provence, but red, and then almost black when they're ripe. They are quite sour but very refreshing; they're delicious and our guests are always surprised but happy to taste them. I'm always surprised by the vitality of this mulberry tree. It's been knocked down several times and broken in many places by storms: all the pieces remained firmly attached to the trunk. It reminds me of a banyan, whose huge branches plunge into the ground and come out again to form runners, which gives it fresh sap. This mulberry tree simply refuses to die. The only defect of the fruit is that its juice is bloodred, and children who gather the mulberries come home looking covered with blood.

On the right is an orchard, followed by two almost unused meadows; one of them does contain a tennis court where almost nobody plays except for Bertrand and his family. The second meadow is occupied by a pond built by my parents, the only remaining trace of a complete failure. The idea was to build a natural swimming pool; unfortunately the ground consists of muddy clay. The pool was to be fed by a spring which is still there. By now, it has become filled with mud up to the surface, and completely overgrown with aquatic and terrestrial vegetation, brambles, epilobes and a great deal of watercress. We got there by crossing a little bridge made of tree trunks to a small island whose trace has completely disappeared. Nothing is left except for an impenetrable marsh which we don't even dare approach any more for fear of the quicksand. The pool was a failure. However, I did use to fish there a lot. I always adored fishing.

At the bottom of the second field there used to be half-a-dozen beehives. But since they were located in the shade of a forest, which is very bad, they never gave more than about fifteen kilos of honey per year.

Of all the fruit trees, the one I specially prefer is the fig tree. Its fruit is unsurpassable and possesses a perfume of voluptuousness. Eating a ripe, fleshy, sweet green fig is one of the ineffable sweetnesses of existence. These trees were planted by my father at the end of his life, and I take care of them now. They're not made for the Parisian climate. One can easily imagine that a fig, not an apple, was the forbidden fruit. The fig tree is the first tree named in the Old Testament: Adam and Eve "sewed fig leaves together, to make themselves aprons" (Genesis 3.7). A fig tree reproduces by layering, and hardly multiplies. Of course, it must multiply sometimes, for the evolution of the species. But its reproductory system is incredibly complicated. It can only reproduce with the aid of a certain parasite, a species of little wasp called blastophage There are about seven hundred species of fig trees and exactly the same number of species of blastophage. Each and every species of blastophage is associated to exactly one species of fig tree and vice versa. The fig is not even actually a fruit; it's a receptacle containing non-synchronous male and female flowers, and it welcomes also non-synchronous male and female wasps. By an incredible sequence of genetically determined actions, the fig tree ensures the reproduction of the wasps and the wasps that of the fig tree.

What an imagination! There's definitely an exotic perfume in this property – figs near Paris, mulberries from the mulberry tree and since the creation of the Thoiry zoo, you can even hear lions roaring at the end of a summer afternoon.

. . .

My parents freely indulged in their hobbies. My mother supervised the flowers. They used to be abundant starting in springtime: tulips, irises, daffodils, primroses and cowslips, roses and enormous peonies, hawthorn, "snowball trees", forsythias, a profusion of purple and white lilacs, syringas, goldenrods and so many others; in the autumn, we had asters. My mother went each day from one end of the garden to the other with a basket and secators. The gardener Charles had planted magnificent rosebushes with varied, perfumed roses all around the French garden. My mother was enchanted to go around it and delicately choose the most beautiful roses, which she distributed in the bedrooms or carried off to Paris to decorate the house there or offer to friends.

My father was the director of the fruits. Since he knew nothing about them, he bought books on the right way of pruning, grafting and caring for fruit trees. He let the apples take care of themselves, but on the other hand gave the tenderest care to the pears over which he waxed enthusiastic. He surveyed them day after day, knew each one individually, protected them from wasps with little bags while they ripened. Every summer we had a considerable quantity of pears of all the most beautiful varieties: magnificent beurres, doyennes du comice, Williams. In the orchard there was also a peach tree which gave only half-a-dozen or at most ten peaches a year, but they were absolutely extraordinary for taste and size: the biggest weighed eight hundred grams! The last year of the peach tree was a glorious swan song. According to my father, the tree produced ten thousand inedible peaches that year, and then died. We also had cherry trees, but the birds swarmed down on the fruit before we even had time to raise our heads. Later, our gardener Charles succeeded in growing a quantity of melons in the greenhouse. The orchard gave my father an enormous amount of work which kept him busy throughout most of the summer.

He exercised such a monopoly over the fruits that we didn't even have the right to pick a pear from the tree. We were allowed to pick apples since they were so abundant, but the precious pears were forbidden. And yet – shouldn't a country house symbolise the freedom to eat the fruits directly from the tree? Out of the question! The pear played the role of the forbidden fruit. We were divided between regret and admiration for these treasures.

As adults, my brothers and I never became really interested in the fruits, and the beautiful orchard has slowly fallen into decay. We haven't planted new trees, nor taken care of those that are there, and in fact we hardly even pick the fruits. Sad decadence. The death of my father signaled the end of the fruits, except for the figs, the mulberries, the currants and the raspberries, which Bertrand takes care of. There are enormous numbers of them. In 1996, we had nearly a hundred kilos of currants, we were offering them to everyone!

On the right, next to the house, an old building contained a billiard table which used to be in my parents' Parisian apartment. It was unusable and the room it was in not fit to inhabit, but that was where I satisfied my passion for butterflies. When I was four and a half, my mother initiated me to butterflies; I started a collection and knew the names of several species. My mother bought me a glass-topped box to put them in, a board to prepare them, special pins, pincers and most importantly, a book by Berce on the diurnal and nocturnal butterflies of France, with their Latin and popular names, their locations and seasons of capture, descriptions and beautiful photos. I devoured this book which fed my dreams and of which certain passages have remained inscribed in my memory: "This satyr lives in July in dry and rocky parts of the Southern Alps." I went there later – and I found it.

By a strange chance, the first butterfly I ever caught in Autouillet, simply sitting on the ground in front of the billiard room, was a female *Apatura iris*, which is quite rare. I spent a a fair portion of my time in that billiard room.

I raised my caterpillars in boxes specially made for the purpose. I particularly remember one big disappointment: I had found a magnificent caterpillar which I'd been looking for for a long while. In fact, it was just a common hawkmoth, the privet hawkmoth, *Sphinx ligustri*, but at that time, it was a rarity for me. I raised it carefully, feeding it every day. One morning, coming into the room, I found the cat sitting on the crushed box. I still feel sorry when I think of it!

My mother had also introduced me to moths (night butterflies), which immediately and forever acquired a particular prestige in my eyes. Everyone knows they are attracted by light. I hunted them enormously in Autouillet and later in the tropics. Every summer for several nights, I left a big lamp (two hundred and fifty or even five hundred watts) lit in the second floor bathroom. Later, we actually discovered that a mercury vapor lamp which has green and ultraviolet radiation exercises an even greater attraction on moths, and it sufficed to use a fifty or one hundred watt bulb. I only learned this in Brazil in 1952. Quantities of big and little moths entered the bathroom and in the morning, I came to harvest them while they were still sleeping. Nobody except me dared enter that bathroom in the night, for fear of being surrounded by a whirlwind of moths. Later, in the tropics, I sometimes attracted thousands of moths with mercury lamps.

Butterfly hunting was my main occupation. Every day I went around the garden, and then around the village, to look for specimens. But most of my hunting took place in what we called "the Labouise meadow"; it wasn't a meadow and didn't belong to Mr. Labouise. In fact it was the waterway, a corridor covering the Avre aqueduct, a subterranean river bringing potable water to Paris. It was a paradise for butterflies, but also for plants. Every year I was filled with admiration for the ecological niches you could find there. In one place, there was always ivy on the ground; in another, hundreds of pimpernels; at the beginning, a clump of sweetpeas which are much more commonly found cultivated than wild; here, half-a-dozen orchids, there, abundant quantities of a kind of caterpillar (*Philudoria potatoria*) which interested me. I raised a dozen of them, and went back there every year to get some; I almost never found any anywhere else. After I abandoned my collection at the age of eighteen, I returned there fifteen years later, and I found the same caterpillars.

A hedge of blackthorn still lies along the edge of this path on the left, over quite a long distance. Sloes grow on blackthorn; several times I've amused myself by tasting them. Sloe gin is delicious, but the fruits are really sour and dry in the mouth. The woods running along the way are practically impenetrable. There are a lot of wild cherry trees. Clematitis stretches its giant vines everywhere; sometimes we carried home vines fifteen meters long. The little masses of violets still grow in the same places.

Every morning, I would go around the village to see if there were some moths on the lanterns or under the copings of the walls, because that's where moths often sleep during the daytime. My sight was very practiced, and I could make out a grey moth on a grey wall. But the main part of my hunting took place in the Labouise meadow. I went there around the middle of the morning and stayed till lunchtime, alone with my net and my box. I captured some of my very best butterflies there; the field was full of them. I also let myself dream all kinds of daydreams, about the future, about the butterflies I wanted to catch there or later, about the other animals I wanted to meet (pheasants, does, weasels, rabbits, hares, partridges, which I sometimes caught sight of). And starting at a certain age, probably twelve or thirteen, my imagination went towards certain little or young girls, in whom I became interested – purely abstractly since I suffered from desperate shyness. I fell in love with Marie-Hélène at the age of sixteen, when she was taking the preparatory special mathematics course at the Lycée Janson-de-Sailly[2]; she was a year ahead of me, I was just going into my last year of high school. I dreamed about her a lot in the Labouise meadow. Alas – I had to wait seven years before marrying her!

Sometimes I used to recount historical events to myself, which I modified to make good triumph over evil. I made Athens beat Sparta, Vercingetorix Caesar, Greece Rome, Jeanne d'Arc the English. I loved to just lie on the ground or sit around and contemplate and marvel over everything around me: the sky, the clouds, the sun. Sometimes I read. Eden belonged to me. The waterway stretched over many kilometers, which I explored sometimes quite far. These were my only expeditions outside of the actual village. The sunset is ravishing when it lights up the immensity of the umbellifers almost horizontally. Sage, scabious, rennet, quantities of pretty yellow bird's-foot trefoil still flower in the same places.

2) Lycée is the French word for high school.

At other times, I went into the forest which separates Autouillet from Thoiry. It's very well kept and belongs to the count of La Panouse. Besides numerous ferns which grow to a height of as much as two meters, and the omnipresent heather, it's mainly interesting for its magnificent chestnut trees, its oaks and its beeches. Some of these trees, especially the oaks, are very large, but none of them are as big as the biggest ones on our property. Even the one known as "the oak of Saint-Sentin", where an annual ball takes place in September, has a gigantic trunk but is not all that high. It must be two or three hundred years old. I often hunted butterflies in the forest, particularly *Aglia tau* in springtime. The forest walk was also like paradise. The edge was decorated with brambles, but also with forest honeysuckle, and there are a lot of *Limenitis sibilla* which procured me particular pleasure. Those times are over – there are no more butterflies in Autouillet. Pesticides have caused the disappearance of butterflies in the countryside. Ecologists have never really become interested in the disappearance of the butterflies. But a world without butterflies would be a sad place. Yet that is exactly what is happening. The only idea that ecologists and legislators have had to protect them is to forbid butterfly hunting. Of course, there have been abuses, almost to the point of driving certain rare species to extinction, but it's still a ridiculous idea! Real entomologists know the sites of the insects and how to protect them. It's the agriculturists who don't know and put them in danger. I know that it is necessary to extend and protect agriculture, but if we do it at the expense of the biotopes which shelter the most important insects, we're just preparing catastrophes for nature and for humans. We must learn how to slip into the narrow path between systematic destruction of butterflies for the profit of agriculture and destruction of agriculture for the profit of butterflies.

I didn't have a watch, when I was young, and when I got far away from the house, I had to get back on time for lunch. Along one wall of the house was an enormous bell, almost a little church bell. Ten minutes before each meal, the cook rang this bell loudly, and the entire village became aware that the Schwartz family was about to sit down to table. The chain of the bell is broken today, but now there is a cow bell which plays the same role for smaller distances, and we all have watches. Daniel had a particular job to do. We adored very cool water, and refrigerators didn't exist yet. He would run to the spring at the bottom of the second field, carrying bottles full of water, set them to soak in the spring and carry back the bottles placed there at the previous meal (the spring water itself wasn't potable, nor was the water from the taps in the village).

At the entrance to the forest lies the Autouillet cemetery. It's one of the very beautiful little gardens in the area, sweet and serene. We buried my father there when he died in 1957, then my mother, my uncle Jacques Debré, and our son

Marc-André in 1971. That is where our family tomb lies. It doesn't displease me that our graves remain all together, in a dream cemetery. We go there together, Marie-Helene and I, once or twice in the year, to think and remember our lost ones.

There are no stores in the village of Autouillet any more; it's just a little hamlet with one hundred and eighty inhabitants, but back then there was a grocery store run by Madame Sixe. She sold common articles, and also lollipops, much desired by the children and which cost two pennies for one and five pennies for two. We never understood this arithmetic. Was it stupid or astute? Did she not have enough lollipops? did she want to prevent children from buying too much candy? The children were never silly enough to buy two lollipops – they came back several times a day if they wanted more than one. . . .

Just at the exit of Autouillet, along the road to La Haye-Frogeay, where Marie-Helene and I often go at the end of the afternoon, lies the Lark field. Whatever day you come there in springtime, you're sure to see the male larks making their nuptial flight. It's still an ecological niche. Above a wheat field, you see the male suddenly rise to a very high altitude, sometimes as much as a hundred meters; we twist our necks trying to watch him. Then he comes down in a circular gliding motion, and finally dives in one swoop next to the female; if he lands a little too far, she throws herself at him. If you stay the whole afternoon, you can see them just about every three to five minutes, never more than two at a time. Even if there are several dozen larks, which I doubt, it still means a lot of flights for each male in a single day; a gigantic continuous effort. Probably the female really likes it. They follow each other with their eyes for the whole flight. Once we went there with friends; I always take them there in the evening. "Oh, I wouldn't mind if my husband did that for me!" declared one young lady of the company, which made us burst out laughing because her husband is famous, young and vigorous!

Listening to the bird songs has always ravished me; particularly, at dusk, one of the most beautiful ones, the marvelous song of the blackbirds. Little by little, I became somewhat deaf and lost the world of birds. But for a few years now I've worn an auricular implant which makes me particularly sensitive to high sounds. The world of birds suddenly came alive for me again. However, I don't hear English any better than I did before – although this was one of the goals of the implant – so I have trouble following mathematical seminars in that language and often go to sleep. I never suffer from insomnia during a seminar. But there are birds!

When we were young, we often saw wrens, warblers, and titmice. They disappeared completely decades ago. It's a real pity; they had become so familiar.

Warblers are not timid and didn't hesitate to approach us. A couple of wood pigeons used to fly around the garden in front of the house; we clearly heard the song of the male. The baby male learns the song of the species in the nest, by listening to his father. If he's born in the spring, he won't begin to sing before the following year, and then he'll seek out the song from somewhere in his memory. His first tries are somewhat deformed, and it's only after some weeks, progressively, that he will have completely reconstituted the song as it should be. In the spring, I've often heard these first tries of the wood pigeon. The young female also hears her father's song, and she'll know how to identify companions of her species.

However – birds can sometimes be fooled! If you let a couple of goldfinches bring up a young chaffinch, it will sing the goldfinch song as an adult. If you then let that bird bring up young chaffinches, it will teach them goldfinch and so on. It seems as though birds, like humans, are genetically predestined to learn a language, but not any particular language. Actually, things aren't really that simple. Successive generations of chaffinches brought up as goldfinches sing a song which is progressively less goldfinch and more chaffinch. Furthermore, in nature these birds are surrounded by other chaffinches and hear their song. After several generations, their song becomes pure chaffinch. Anyway, there exist several different chaffinch dialects from different regions. And like humans, even two chaffinches from the same place don't have the same voice. Among the innumerable chaffinches living in a given place, each female can immediately identify her male, and to mark their territories, they must be able to distinguish each other by voice. There is more. All humans on earth don't speak the same language. But chaffinches living over a region of thousands of square kilometers, from north to south and east to west, sing the same song apart from differences of dialect. This has been going on for tens of thousands of years. It's hardly possible to imagine such immobility without some genetic predetermination. The song of each species of bird, even if it is genetically fixed, can still be modified by special circumstances.

The male city pigeon, the rock pigeon, emits only a little rolling one-syllable coo, whereas the wood pigeon develops a very interesting song on five syllables. It understands three or four sequences of five syllables of the following form: too-too-too-too / too-too-too-too-too / too-too-too-too-too / too.

Only the middle word has five syllables, the first one has four and the last one only one. From the mathematical point of view, they ought perhaps to be organised differently: too-too-too-too – too / too-too-too-too – too / too-too-too-too – too.

I always wondered with the greatest surprise how the syrinx, the vocal organ of the birds, is brought – almost forced – to make this separation between the syllables, which doesn't seem at all natural.

I have no scientific ideas about the precise evolution of these two species, but I would believe that the rock pigeon appeared later than the wood pigeon, the city bird appearing later than the wild one, with a certain degeneration of the song. I love to listen to the rolling coo of the wood pigeon. But they are being chased away by Turkish turtledoves, which were not known in Europe before 1950. Little by little, these birds have penetrated Europe from Turkey and Greece, and they've multiplied, ceaselessly conquering new terrain and particularly from the wood pigeons. Two or three years ago I witnessed a battle between a turtledove and a wood pigeon in one of the chestnut trees in front of the house: a great many wing blows in the tree, and the wood pigeon flew away, the turtledove remained, and his victory appeared definitive. Yet the turtledove is much smaller than the pigeon and much more delicate. Now I only rarely hear the call of a wood pigeon, in the distance, while dozens of turtledoves people the area. They sing the whole day long. There are several couples in the garden which never cease to sing their little three-note melody. As soon as I hear one, I immediately locate it on the telegraph pole, or on a high branch or on the roof, or sometimes even on the ground. They always distract me from my work, but it is a song, and after all the alternation of mathematics and turtledoves is quite pleasant! Since 1992, a couple of wood turtledoves, which are much rarer than the Turkish variety, has nested in the elm-ash at the edge of the pond. Now there are several of those in the village as well.

Woodpeckers are also frequently to be met with, and I've often heard a great spotted woodpecker; the rapid series of blows they give with their beak in a tree to signal their presence is very impressive. One day I watched a woodpecker feasting in an old acacia which seemed to be dying yet which still prospers today, and whose trunk was full of hollows. About ten meters above my head, he extracted from the trunk five enormous Coleopterae larvae. You could feel the voluptuous pleasure with which he sat down to table. Just recently I also discovered the nuthatch in Autouillet; now there are many in the garden. I've even seen several couples simultaneously on the two grass plots. I hear the male almost all day long. His song is very characteristic, with a strong and repetitive part followed by several little moans. Like the woodpecker, this bird comes down a tree like a gymnast, along the trunk. It's a beautiful blue bird which I always love to see. One year, when they were being particularly noisy in the garden, the nuthatches actually followed me. I went to Delphi and visited parts of Greece. Imagine my surprise when I heard a nuthatch singing on the large tombs of the Mycenean kings! Then I traveled to Quebec and the nuthatch's song was the first thing I heard upon arriving. They actually belong to three different species: *Sitta europea* here, *Sitta niemeyer* in Greece and *Sitta canadensis* in Canada. I talked about them so much that my family thought it was turning into an obsession.

We heard a nightingale only once, at the beginning of one year. Chaffinches are very numerous from April to the end of July, and they sing their song imperturbably every few minutes throughout the day. Tits, which used to be very common, have become rare. As for daytime birds of prey, they've almost completely disappeared. Years ago, with a little luck I'd see a sparrow hawk every summer, hovering over the garden. Now there are no more in the area. At the Ecole Polytechnique in Palaiseau, you used to see one near the entrance during the first years after the school moved there, in 1976. But it's gone too. As for buzzards, in which I used to be so interested, I haven't seen any here for ages now. But there are still nocturnal birds of prey, in particular the tawny owl with its long, rather sinister ululation. I love to hear it; it seems like a symbol of life in the night. No summer passes without us hearing it. In autumn, you can hear them in the gardens of the Val-de-Grâce. I remember one autumn, when I was young, there was even a nest of them in Autouillet. If you went out at night and remained still, a strangely silent movement might pass above you sometimes. Now they're much rarer, but still there. There are some bats, but much less than before.

In my youth I was passionately interested in the mourning cloak *Vanessa antiopa*; it's the most splendid butterfly to be found in France. It's colored a dark chocolate brown, brilliant and velvety, and edged with a yellow border which frames minusculous blue spots; the whole is perfectly harmonious. I was six and a half when I saw my first mourning cloak, on vacation in Pralognan. "It's the most beautiful of the French butterflies" said my uncle Jacques when he showed it to me, and I pretty much agree. Except for certain places in the Alps, it's quite a rare butterfly. All collectors now have several though, since you can actually raise them – once I myself came across half-a-dozen caterpillars which I raised into perfect butterflies. But in Autouillet, where the mourning cloak appears only very exceptionally, I captured about one every two years, and such a capture always provoked general enthusiasm. I would see one somewhere, perhaps in a field, sitting on the trunk of a little birch tree. I'd try to catch it, miss it, and tell the whole family that there was a mourning cloak on the property. Nymphalids often return to the same places again and again. After following the butterfly from place to place around the garden, I'd finally capture it on the budleia. So I did have a few in my collection. There are never any more now, of course. It's really a very special butterfly, whose psychology I'd like to understand. By its form and its flight, it's infinitely noble: a long gliding flight, then some wing beats. It flies mostly quite high, seeking the summits of willows or poplars where the female lays her eggs. You can attract them to the ground and capture them with fermented bananas. Their flight is regal; a prince of rank, clad in his best, moving majestically. I have the vague feeling that the mourning cloak knows perfectly well that he is of noble family, that he is a king or a grand lady dressed in black silk, advancing slowly, surrounded by a court of admirers. Yet here there is no court and no admirers, except maybe sometimes someone of the other sex who can recognize a partner from afar by its flight. Judging by the majesty of its flight, I believe the butterfly must have a certain consciousness of its nobility. In contrast, the flight of the cabbage white (Pieris rapae) or the brown satyr is just plain vulgar. What is this division of butterfly society into social classes? I'd tend to interpret it this way. Vulgar butterflies, whose flight involves jumping, bounding, zigzagging, do all those manoeuvres to avoid birds, flying almost along the ground, because they're incapable of a long flight at a high altitude. But the mourning cloak rises, sure of himself, slowly and straight up into the heights, without detours. If he sees from afar a bird diving at him, he swerves away with a single, sudden move. Anyway, I never saw a bird chasing a mourning cloak. And even entomologists, who easily capture flying cabbage whites or satyrs with their nets, almost never capture a flying mourning cloak. I still think they must be conscious of that incredible strength, that speed, that agility and flexibility.

Daniel and I strongly believe in the intelligence of animals, which forms an inexhaustible source of conversation between us. Of course, purely instinctive, genetically determined behavior plays a much more fundamental role for animals than for humans, particularly in alimentary and sexual aspects. But we fool ourselves if we believe that animals are incapable of most appropriate strategic behavior in many situations. When an animal sees a predator coming, in which direction should it flee? In general their choices are amazingly astute, judging by the impressive number of butterflies who have succeeded in escaping my net. Here are three eloquent examples of animal intelligence. When I was very young, I was taken to the zoo of the Jardin d'Acclimatation. There was a crowd that day in front of the elephants: everyone was feeding them, stretching their hands towards the elephants to give them bread or vegetables. I was a bit nervous, but I stretched out my hand with a piece of bread. When an elephant's giant trunk approached me, I was scared and pulled my hand away. Then I took courage and tried again, but again I was scared and pulled my hand away. The elephant turned its back to me and I forgot it. But it didn't forget me! It went and filled its trunk in the elephants' pool – an elephant's trunk contains a lot of water – and in the most tranquil way, stepped in front of me and showered me, soaking me from head to toe.

The squirrels in Autouillet display the same type of discernment. One of them takes a charming little walk, its tail harmoniously lifted. Suddenly, it sees me; quick as a wink, it dashes to hide near the trunk of a tree, on the side where I can't see it. I can wait or not wait; I won't see it any more. Of course, it could climb quickly up to the top of the tree and flee from one tree to another. But sometimes it's just near a little round yew bush, with no trees nearby. There's just the flat green prairie. But I don't see it anywhere, neither to the right, nor to the left. Yet when I reach the bush, it isn't there either. There's only one way for the squirrel to flee without my seeing it: in a straight line rigorously extending the line from me to the bush. The yew bush is too small for it to be hidden inside. My two brothers and I have observed this dozens of times and we never can understand how it can happen. Let me end with the story of Crocro, a crow which was caught and hurt in a barbed wire fence. A man from the village freed it and brought it home to take care of it. Having discovered that humans could be kind and helpful, the bird became attached to the inhabitants of the village, and particularly to our granddaughter Lydia, who adores animals (she's raised a lot of large and small animals in her Grenoble apartment, in particular a big Picard dog called Vrac). Crocro quickly understood that he had a friend in Lydia. Almost every day, he'd come gliding over the garden from afar, and delicately settle on her shoulder or her head. We were always a bit worried, but nothing ever happened. Once, he came into the house, right into the second floor office where I was working on mathematics, and landed on my notebook. I ushered him out politely. The dog was always furious with him and chased him away, but I think Crocro was smarter than Vrac. When the dog chased him, he would lead Vrac, flying just high enough not to be caught, behind a hedge where the dog couldn't see us. From there he'd spurt up into the air so quickly the dog would lose sight of him completely, and even lose the trace of his smell. Then the bird would calmly return to Lydia, while the dog hunted everywhere else. It always reminded me of that silly sentence that adults sometimes toss at children: "Go look somewhere else to see if I'm there." Crocro died shortly afterwards, probably killed by some exasperated villager.

Hunting the hawkmoth (*Agrius convolvuli*, nocturnal, Sphingidae family) also became a family activity. It's very common, but you still have to capture it. It's one of the biggest French butterflies, measuring a good twelve centimeters or so. Like all Sphingidae, it has long narrow wings, rather like a jet, gray and not very beautiful, but its long thick body is prettily ringed in black and pink. Its flight is very rapid, and it's impossible to catch one in the countryside. It doesn't settle on flowers to suck their nectar; its heavy body would destroy them, but it hovers over them, beating its wings with a little sound like a motor, at a rhythm of twenty to fifty beats a second, which is a record difficult to imagine. While it is hovering, you can capture it if you sneak up very quietly, as it's a very prudent beast. All this occurs at dusk, or even later, so you don't see very clearly. The game consists in spying the slightest movements above the flowers and cocking your ear. When you spot one at a small distance, you sneak up on it very softly and catch it while it's hovering. It takes a certain agility. Unlike other Sphingidae which fly in June, this one flies in September. You can only start hunting it around 7:30 in the evening. It is particularly attracted, over large distances, by the perfume of petunias. Now, at that time we had rows of petunias around the French garden and the outbuildings. They smelled so good that we'd have liked to taste them ourselves. While hovering, the hawkmoth sucks a drop of nectar, with an enormous suction tube. This tube is about twelve centimeters long, and very thin; usually the butterfly keeps it rolled around in front of its head, but it unrolls it in less than a second to plunge it into a flower, then sucks the nectar for two to four seconds, rolls up the tube in less than a second and moves to another flower. It looks like he sucks up nectar the way we drink orange juice through a straw, using a vacuum. But probably that would be too slow for such a long, thin tube. Probably the nectar actually flows up it by capillary action, helped by a little extra sucking.

The whole family would move around silently in front of the row of petunias, including my father, who didn't usually share in our activities; but only the children had nets. As soon as one of us spotted one, he told everyone, and the hawkmoth usually ended up captured.

Between the middle of September (when we returned from Blonville) and the end of the month, we'd catch about ten. The Sweet Potato Hawkmoth (Herse Convolvuli) and the death's-head Hawkmoth (Acherontia Atropos) usually live in North Africa. A large number cross the Mediterranean in a single flight (at a speed which can attain 50 kilometers an hour) and arrive in France or even farther north in Europe. They don't stop to feed there, but immediately reproduce. The caterpillar (also enormous) lives in July, and the fresh butterfly emerges in September. But this generation is sterile; the abdomen of the females is empty. That's why these two types of hawkmoth fly in September while all their close relations which don't migrate fly in June. How can one explain this migration, without perpetuation of the species, from the viewpoint of natural selection? Perhaps it is a residue from some former period when the hawkmoth actually resided in France.

Schwartz as a boy with a butterfly net

We didn't purposely restrict ourselves to our own garden, but we didn't get out of it all that much. My parents were worried about leaving us a lot of geographical independence. My father was a surgeon and had all possible accidents perpetually in mind. We were allowed to go biking as long as we wore gloves and thick leggings and rode around within a limited radius, for fear of cars. For the same reason, we never went skiing. We weren't particularly attracted by hiking. Personally I was no good at it on account of my bout with polio. I liked fishing, but it never occurred to me (and I wouldn't have been allowed anyway) to go fishing in the Holland ponds, though they weren't a great distance away. There was never any question of allowing me to go traveling to the Alps or to Provence, though they are French paradises for butterflies.

Games occupied a special place in our lives. Given the size of the garden, games of cops and robbers were filled with adventures. Everyone wanted to be the robbers. We also played interminable games of croquet in the space between the house and the grass plots. We prepared the terrain on purpose to be uneven. For instance, we had a "terrible arc" placed at the top of a little rise. You needed treasures of agility to get through it; you'd have to approach it from behind, go upwards towards it and then actually past it, so as to go through it descending and backwards. It stopped every player for a long time! Naturally, the corsairs knocked away the balls of their adversaries. I became a specialist at this. I didn't do it sideways but placed the ball in front of me between my feet which pressed on it; then I'd lift my mallet high in the air and despite personal risk, I'd whack as hard as I could in my own direction. Sometimes I'd succeed in sending the adversary's ball ten or fifteen meters away. Because of the uneven terrain, a ball flew off one day and lodged in a fork of a lilac bush a meter and a half above the ground. Another time, a ball rolled towards the cellar, jumped over a step, and then rolled all the way down the staircase just when the cellar happened to be flooded by ground water. That poor player had to get in the water and hit the ball around in it for quite some time before getting out again. These events occasionally led to real quarrels, with rising voices. Then we called each other "sir". Yes sir, no sir. But harmony was always quickly reestablished; we never stopped getting along well and in fact we represent for each other the most solid links in the chain of friendship and solidarity.

Of course, we also had plenty of indoor activities, cards and chess. I rarely played chess and although the game is quite mathematical, I was far from brilliant; I couldn't get up enough interest to really want to win. For the same reason, I don't play well at ping-pong. When my mother was already elderly, we were content, all three of us, to spend the evening with her in the living room, solving the crossword puzzles in *Le Monde*; they're always quite tricky and perverse. My mother ended up becoming very good at them and it was usually she who found the solutions. We enjoyed ourselves a lot with them. We always liked to laugh, and we've kept beautiful memories of those evenings in front of the fire, in the living room, on cold days. My mother was always very talkative. She was the real conversationalist in Autouillet, especially after my father's death. We used to laugh at her for it. I'm a little bit less talkative than she is, and my daughter Claudine a bit less than me, but we are all pretty loquacious! One day, she had a short, benign attack. Afterwards she was mute for s short time, but she quickly became able to talk again, and said to Bertrand: "There, I can talk!" He immediately answered: "Oh! too bad" which made her laugh and relax.

In our youth, we used to come to Autouillet not only in summer but also every weekend. In general, we left on Sunday morning and came back on Sunday evening. We always left between eight and eight-thirty. First we'd go to the Maison de Santé, a clinic in Paris where my father performed a lot of his operations. He cared so much about his patients that he never accepted to leave Paris without having checked on them all personally. It always amused us to see with how much respect he was treated in the hospital. We waited in the car, and we could see, as soon as he stepped over the threshold of the Maison de Santé, that they rang a bell to let everyone know, as a sign of respect and also of prudence, probably. Then he came back to the car and we drove off to Autouillet. I've taken that road hundreds, if not thousands of times in my life, particularly these last years. It's true that I've always been driven by other people, certainly, sometimes even by taxis, since I don't drive myself. But I'm so hopeless in topography that I could simply never assimilate the route from Paris to Autouillet, except starting from Jouars-Pontchartrain, just before the end. Recently, a taxi driver asked me to indicate the beginning of the route to him. I had to say I didn't know. "But haven't you ever been there before?" he asked? I had to lie a bit: "Well yes, a few times, but not enough to know the way." I never looked on a map; I don't know where Autouillet is. My ignorance in topography is not a subject of foolish pride, but a real bother. Even around my own house, I can't orient myself. My parents were the same, but to a lesser degree. My mother boasted of not knowing whether San Francisco was on the East or West coast of the United States. Fortunately I know a little more than she did, but all my efforts at serious improvement have failed miserably. The left-hand side of my brain is pre-eminent – and what about the right-hand side? If after my death they do an autopsy on me, they'll probably find that it's just plain empty.

The moment we reached Autouillet, we entered into its enchantment. Sometimes one of us had too much work; then he brought his homework to Autouillet (ah, those revisions, those compositions in history and geography!) The Sunday evening trips back were spectacular. We had to leave around five o'clock to reach Paris around seven thirty, even though the distance is only about forty kilometers. In the spring, the car would be absolutely filled with flowers, fruits, vegetables, entire branches of perfumed lilac.

During that time, the gardener Charles and his wife Alphonsine lived in the old outbuildings and took care of the garden. Charles was an old, very capable Alsatian with a strong accent. He not only cultivated the melons, but it was he who planted a magnificent rose garden around the edge of the French garden. Come spring and summertime, every variety of rose was there for our pleasure. Later we planted petunias which exhaled their sweet perfume throughout the garden. Of all that, nothing is left except a rose here and there.

Alphonsine raised fowls and even some pigeons. The famous myth about the faithfulness of pigeons was seriously threatened by some of hers. A couple of pigeons lived together in the same pigeonhole for two whole months, never laying any eggs – we finally realized they were both male.

Every Sunday, we brought home two chickens which we ate hot or cold during the week. A that time, chicken was a rare and much appreciated meat. Naturally, we had an abundance of fresh eggs. I had the right to swallow them raw, after poking a little hole at each end of the shell. Sometimes we had so many, when no one else in the village had any, that Alphonsine would say "There's only us laying right now!" The return to Paris was slowed down by having to pass the city toll in Paris. Not only were there customs barriers between countries then, but also at the entrances to all large cities. We had to declare everything we brought into Paris. Very honestly, my father would declare his two chickens and pay a small fee. It was the great chemist Lavoisier who originated the strange idea of establishing city tolls at the gates to all big cities. We always arrived home a little numb and almost asphyxiated by the perfume of the flowers.

We also went to Autouillet during Easter vacation, at Pentecost and even during the winter vacations, when it was extremely cold. It was a bit sinister. But our parents desired it. The house was not heated. We shivered in our rooms when we got up in the morning. One had to wash a bit, but there were wood-burning stoves in some of the rooms. Thanks to a large fireplace, we had a fire in the living room as soon as the early evening began. We tossed in yew branches which burst into fireworks. There, we played together, told stories and read. Reading occupied a fair part of our time, though I must admit it was never the favorite activity of any of us, though I don't know why. That changed later. We read a lot of books, but it always seemed difficult to settle down to reading. So much so, in fact, that at one point our parents used to oblige us to sit down to an hour's reading after lunch. That's when I discovered some of the major classics which unfortunately I've never reread since then. If I compare my reading of that time with my friends', I find it consisted in about the same things: the scholar Cosine, Alexandre Dumas, Jules Verne and the classics. I read the stories which interested me without always appreciating them from a literary viewpoint. So my youth left me with a lack of literary culture and an even worse lack of artistic culture, since Sundays in the country deprived us of museums. On the other hand, I read a lot of scientific books in Autouillet, especially on insects. A large portion of my time was employed in devouring books on entomology, especially the remarkable writings of Jean Henri Fabre. I also read during the 1930's a whole series of books by Jean Rostand on chromosomes. Butterflies introduced me to biology. A lot of people imagine that the natural sciences are primitive amusements which can't lead to serious biology. But in my case, they stimulated my curiosity which was far too scientific to be restrained. I often hesitated between mathematics and biology, but I really lack the basic practical sense an experimentalist needs. I once was so impregnated with my reading of Jean Rostand that I had a very strange dream. I had read that he had obtained the parthenogenesis of a male toad. Using hormones, he transformed a male into a female, without of course modifying the gametes. The toad then laid eggs, Jean Rostand dissected them and caused them to multiply although they had not been fertilized. In this way, he obtained parthenogenetic toads from a male. After this amazing reading, I dreamed that someone came and announced to me: "We've just performed an amazing experiment: we succeeded in fertilizing the egg of a grape with the sperm of a general of bees. What do you think the result is?" I reflected and answered: "Obviously nothing. It can't possibly be viable." "Yes it is," persisted my interlocutor; not only is it viable but do you know what it is?" "No, tell me." I admire these situations where I interrogate someone in a dream and they answer me, even though it's just me dreaming the whole time! "Well, it's a general!" was the answer. And I burst out laughing. "And now, what happens if you fertilize the egg of a general of bees with the sperm of a grape?" I still didn't know. "A drunkard," continued the dream, and I was further informed that "This explains the existence of generals and of drunkards."

After the war, my father retired. He remained expert surgeon for Social Security, but he never operated again. My parents were very well-organised; they spent half the week in Paris, the other half in Autouillet, at least during the beautiful season. But we, the children, were no longer the same. Daniel and I were both married, and Bertrand was soon to be. Marc-André, our son, was born during the war in 1943, and Claudine was born in 1947. She first came to Autouillet at the age of two weeks, in summer. Daniel had a son, Maxime, born on the first of June 1940 (excellent timing!) and later another son, Yves and a daughter, Irene. Bertrand and his wife Antoinette Gutmann had four children, Olivier, Alain, Isabelle and Jean-Luc, all born after 1953. They've all spent summers in Autouillet, more or less together. We had professions and social responsabilities. I was a professor at the Scientific University of Nancy, Bertrand was director of the Ecole des Mines in Nancy and Daniel was a tobacco engineer, actually a researcher. So we were very busy and couldn't spend every weekend in Autouillet. Daniel bought a house, Bois-Fleuri, in Autouillet-le-Haut. It wasn't the Closeaux, but it was still in Autouillet, to which he was attached. Our uncle Jacques also bought a property there, the Florelles. Daniel's sister-in-law did the same, settling in Pacyphore, next to Bois-Fleuri. So Autouillet slowly filled with "Schwartzes" and their clan, always happy to get together. In summer the house in Closeaux was occupied my parents, the "Bertrands" and the "Laurents" with their descendants. So every year we had magnificent family reunions.

Yet my life had changed considerably. I used to be a young man passionately interested in my studies and in entomology, and I have kept these interests. But whereas in my youth I knew only Paris, Autouillet and Blonville, suddenly I began to travel all over the world. Invitations to foreign universities allowed me to visit more than fifty countries, giving mathematical lectures and hunting butterflies in tropical and equatorial third world countries. I spent several summers in foreign countries and have been around the world four times. For all of us, Autouillet paled somewhat. Its landscapes interested me less, eclipsed by the immense stretches of tropical and equatorial forests. There were new civilizations to discover. Art, literature, the political life of foreign countries, scientific communication occupied the major portion of my existence. I became detached from Autouillet. And then, I had become a Trotskyist in 1936, and remained very militant until 1947. I found it somewhat incongruent, with my political ideas, to own such a grand and magnificent property, and twice (in 1946 and 1947, I believe), I actually spent part of the summer in Paris militating. Nowadays, I no longer believe that forcing oneself to the level of deprived peoples is a duty or even a good way of improving the future of humanity. I'm not troubled by such scruples any more. Even my family now spent half the vacations in the mountains and the other half in Autouillet, when they weren't actually in a foreign country: even in France our horizons were widening.

Marc-André became a poet and a writer. He wrote a lot of poetry and also a novel inspired by Autouillet. I initiated him to butterfly collecting when he was tiny, just as my mother did me at four and a half. He adored it just as much as I did. I had stopped my collection at the start of the war and I didn't start it up again immediately, but Marc-André took it up. I still remember the day in the mountains when he came running to tell me: "I caught a machaon, I'm wild with joy!" And to an old lady who asked him what that orange butterfly he was holding was, he answered right off the bat, to her amazement: "This is a female *Argynnis paphia*." He was just six, it was at Notre-Dame-de-Bellecombe, in the Alps, in 1949. Starting in the summer of 1952, I got excited by the butterflies I'd seen in Brazil, and I went back to the collection as well; it became a joint project. Later, we even hunted together on certain trips (he was much more agile than I was!) and he used to mount and label the butterflies with me.

In 1971, Claudine married Raoul Robert. He came from a family with roots in a family property located in Sisteron. So Claudine divided her vacations between Sisteron and Autouillet, and partly because of the climate, a certain rivalry sprang up between the two places. When we were all together in Autouillet, my parents were surrounded by big tables of joyful people: Bertrand and me, our children and grandchildren – Claudine has two children, Magali and Lydia

– who continually burst out laughing without always even knowing why. I remember once how one of the little girls served herself some jam. Slowly and tranquilly, she poured the entire pot out onto her dish. Then she glanced at us, a little embarrassed, and seeing our astonished and amused faces, suddenly burst into tears. That was our Autouillet, as it has always been. Our children had a happy childhood there. And if from time to time they are a little bored, it's not comparable to how bored Daniel used to feel, because now there's always a horde of children around, playing at all kinds of games, battling out great soccer matches. Lydia is passionately interested in horses. She's read every book she can find about them and can recognize one or two hundred different races. When she comes to Autouillet, she spends most of her time in the equestrian center where she's in charge of several horses which she rides and grooms and cares for and feeds. She adores all animals and in Autouillet, she raises rabbits, chickens and cats which she carries around in her arms. Magali also likes Autouillet, at least if she can bring a friend along. When Claudine comes, Olivier Schwartz, Bertrand's oldest son, often comes too and brings his daughter Emmanuelle. She's three years older than Lydia and they get along perfectly. Without their friends, perhaps Lydia and Magali would be just as bored in Autouillet as Daniel used to be. It's all part of a normal and natural childhood.

Autouillet was my garden of Eden. It wasn't so for either of my brothers, and particularly not for Daniel. There weren't any distractions. I had butterflies, but they didn't. Croquet games were rare. No carpentry, no gardening; the gardening was done by the gardener, the flowers by my mother, the fruits by my father. No walks, no bike rides. The swimming pool in the field ended up a failure. The tennis court was built only after the war. The piano was terrible and anyway, Daniel didn't like music. We almost never left Autouillet or Blonville until our adulthood. In fact, like many parents at that time, our parents had a very restricted conception of the desires of young people. Real emancipation came only much later. Yet a kind of ideal remained with them from their own childhood, because as adults, Autouillet also appeared to them as a garden of Eden. Surely this paradise must have been linked to their own impressions of childhood. Autouillet always played a fundamental psychological role in our family, as the anchor of the family and a source of childhood memories. The various apartments we lived in in Paris could never replace Autouillet for these things.

Our parents have been dead now for many years. My father died in 1957 at the age of eighty-five, and my mother in 1972, at eighty-four. My father died in a quarter of an hour, in Paris. It was a beautiful death, but brutally quick, and came as a great shock. For months, I wasn't able to enjoy anything. My mother, though she suffered from serious heart trouble, survived him for fifteen years. Her cerebral capacity remained absolutely intact, but she became more and more fragile physically. When we left Autouillet in September 1972, she seemed so frail to me that she put me in mind of an autumn leaf, ready to fall at the least puff of wind. She died of the flu two weeks later. She had become the "matriarch" of a large tribe of children, grandchildren, nephews, nieces, grand-nephews and grand-nieces; she was beloved and respected by all. Seventy-five people called her grandmother or Aunt Claire. She was at the heart of an intense epistolary activity, letting everyone know what the others were doing, a real information center. Dozens of individual and group photos of the whole family hung above her bed in Autouillet.

My father was born in Balbronn, near Westhoffen in Alsace. He lost his father very young and lived alone with his mother, his brother Isaac who became a rabbi and died of cancer quite young, and his sister Delphine. Their mother sold flour. She never really seized the rise in the social scale of her son Anselme Schwartz, my father. To the end, she persisted in saying to him: "You know, if you had stayed with me in Balbronn selling flour, the business would be very good now." They say she was very beautiful. She died of the flu in the middle of the thirties, apparently eighty years old. But according to her birth certificate, she was actually ninety! Nobody had ever noticed her coquettish untruth. Very young, my father decided to leave Balbronn and go to France. Life was quite agreeable there and he was apparently an astute little peasant. Anyway, Balbronn was a fairly tolerant village since a church, a temple and a synagogue shared the worship of a small number of inhabitants. My father was brought up in the Jewish religion, but he became an atheist. He brought up his own children in total atheism. We were never involved with any kind of religious questions.

Born in 1872, just after the annexation of Alsace-Lorraine by Germany, his desire to live in France was motivated by patriotism. He absolutely didn't want to be German. He must have been a very stubborn young man since at only fourteen, he locked himself in the attic and declared that he would not eat or drink until his mother accepted that he leave for France. His sister Delphine "secretly" brought him a little sustenance, with their mother's agreement, naturally. But she had to give in. He packed his bags and left for Paris where he lived for several years in the Israelite seminary, rue Vauquelin. That is where, little by little, he lost his faith, but nobody in the seminary ever asked him about it. He didn't know a word of French when he arrived, but he had to pass his baccalaureate[3]. Very quickly, he decided to become a surgeon, and

3) French high school (lycée) ends with a stiff national examination, the baccalaureate.

as at that time the medical profession was open only to those who passed the baccalaureate A (Latin-Greek), sciences being considered useless, his memories of mathematics were not very brilliant. Once he said to me: "You'll learn later that $ax^2 + bx + c$ is always equal to zero." Nevertheless, my father passed his exams brilliantly, studied medicine (externat, internat, prosectorat) and passed his thesis. When he reached the age at which he could have become a hospital surgeon, anti-Semitism was still quite virulent in medical circles: his professors warned him frankly that he would never be a hospital surgeon because he was Jewish. But with the stubborn insistence that characterized him, he had decided that he would succeed in spite of all. His talents were so impressive that at one point his colleagues simply had no other alternative than to nominate him. That is how, in 1907, at the age of thirty-five, he finally became the first Jewish hospital surgeon in Paris. It was a remarkable promotion. He always spoke of his professors, Quénu and Reclus, with emotion, but his youth had been so difficult that he never spoke to us about it.

The same year, he married my mother who was his first cousin and sixteen years younger than he was. He had seen her as a newborn and bounced her on his knee as a baby.

Naturally, I didn't see my father very often. He was extremely able and had written a book on thoracic surgery which remained a classic for a long time. His strong personality showed itself through a truly exception moral courage and sense of honesty. At the beginning of the 1914–1918 war, the French army was completely disorganized. Many of the officers and non-commissioned officers were not worth much, the army was ill-equipped, and the army doctors were mediocre. My father, and his professor Quénu were mobilized as ... stretcher-bearers! Their competence as hospital surgeons was considered superfluous, as sufficient science was to be found in the hands of the military surgeons and doctors. My father suffered to feel so useless. His unit was directed by a captain whose knowledge was non-existent. No precautions were taken against asepsis. My father was furious and complained to the captain who ceaselessly repeated to him that he and no other was the captain of the ship. However, one day the captain was absent for a whole day, and my father took advantage of it to completely reorganize the hospital. Naturally, he expected a storm when the captain returned. Indeed, the captain immediately set to searching for the author of all the upheavals. "It was I," announced my father. "Who is the commander here, then?" And my father answered firmly and tranquilly: "Starting now, it is myself, sir." The captain was so amazed at this response that he gave in, and my father took over the organization of the hospital unit. He had taken a big risk.

His battle against dichotomy gave us another proof of his upright nature and his moral involvement. Dichotomy was a well established custom between doctors and surgeons. When a doctor decided that one of his patients needed to be operated, he sent him to a surgeon who gave the doctor one half of the fee. Certain surgeons even gave the doctor the whole fee for the first operation. Thus, doctors were led to choose, not the best surgeons, but those who could command the highest fees. My father rejected this system, and this limited his clientele. However, his clientele was extremely faithful to him because his exceptional competence was admitted even by his peers. The majority of doctors practiced dichotomy, so they never sent their patients to my father – but they sent him any member of their own family who needed to be operated! In a word, dishonesty festering with corruption reigned in the medical profession.

My father was so determined that he founded a League against dichotomy. I don't think he ever succeeded in recruiting more than one or two hundred members, often practicing Catholics and political right-wingers but morally intransigent. The liberal doctors, more or less left-wingers, were more tolerant about these dishonest practices. At the height of the struggle, my father took me aside and solemnly explained to me: "If, in a given circumstance, you find that you are alone with your opinion against everybody else, try to listen to them, because maybe they are right and you are wrong. But if, after having thought it out, you still find yourself alone with your opinion, then you should say it and shout it and let everybody hear it." That is what he did with the League against dichotomy. His explanation remained engraven inside me, and guided me in all my political activities in my adult life. Some people, reading Freud, would say that I have been dominated for my whole life by the gigantic stature of my father. This judgement doesn't give me any complexes. Throughout my years in the Ecole Normale, I had deep conflicts with my parents, and rejected some of their deep convictions. But I kept their fundamental values and whatever I found right and good in them, and these are the best things about me. The end of the dichotomy story is quite surprising. My father had a student that he was particularly fond of, whom he had trained entirely himself, and who had also become a hospital surgeon, while remaining a close and devoted collaborator of my father. But he was a convinced Vichyste, and accepted to become Minister of Health under Pétain. And from this position, he induced Pétain to suppress the practice of dichotomy. It ended up being forbidden by a decree signed by Pétain! This decree was not annulled after the victory of 1944. The ways of Providence are strange.

After seven years of marriage, my parents still had no children, and my father was forty-two years old when I was born. This difference in age meant that a certain distance always remained between us. He was very affectionate and always had conversations with us, especially at meals, but that was about it. Except for an occasional game of chess, he never played with us. I don't even think he had leisure to do so. He was very serious, we almost never, perhaps never actually saw him laugh. He didn't really bother with our studies, either. Only our lessons in natural history were actually taught by him. He specialized in anatomy and his pedagogy was really extraordinary, his explanations limpid. Without a doubt, he trained the three of us as pedagogues.

My mother was always very sweet and indulgent with her children. Our studies, which she followed closely, were essentially directed by her tender ministrations. She knew loads of games and told us innumerable stories. Her father, Simon Debré, was the Chief Rabbi of Neuilly. He suffered from serious heart trouble and died rather young. He taught us the sacred history, with images, like a series of beautiful legends. Naturally, he was a practicing Jew, but judging from the way he taught us sacred history, I always suspected him of not being a real believer, in spite of his profession.

My very first memory of childhood is of a lie. Once I peed on the floor. My brother Daniel was next to me. My father burst in and said: "Hey, who peed on the floor?" "Not me!" I hastily replied. Well, obviously, it could only have been me. My father gave me a good spanking, put me in the corner and said I would be deprived of dessert for eight days. At that time, putting a child in the corner was a common punishment. Naturally I cried and wailed. Then he explained to me that one should never, ever lie. In fact, I got my dessert that very evening, as my parents had forgotten all about the whole story. But I never forgot it. I was about three years old and I remembered that lesson my whole life. Clearly, you have to lie sometimes. But when I'm really obliged to avoid telling the truth, I always tell it to myself, inside. In other words, I know exactly what I think and if I am obliged to deform it for others, I make sure that my own thoughts retain their integrity.

At the death of my mother, we divided the property in two. The house, the garden and the fields down to the wall came to me, the outbuildings and their surroundings to Bertrand. It's very intimate, being neighbors in this way. The wall separating us from the outbuildings is not really continuous, and we get through it all the time, and meet just about every day. Daniel and I visit each other frequently too.

My mother communicated her passionate love of nature to all three of us. However, we never really got to exercise this passion as children. Daniel liked flowers, but it was my mother, with the help of the gardener, who reigned over them as queen in our garden. Daniel had nothing to say about it. The vegetable garden was the domain of the gardener, the fruits that of our father. Bertrand, who often comes down on weekends, immediately metamorphoses into an agricultural worker. He puts on his overalls, grabs his tools and cultivates the vegetable garden from which he gets all kinds of vegetables: salads, herbs, currants, raspberries and strawberries. It takes up all his Sundays except in winter, and every summer day, from morning to evening. Then he gives out his fresh fruits to all his friends. Daniel cultivates his garden, too. It's a magical place where he grows a large number of rare flowers. He has multiple varieties of tulips, azaleas, rhododendrons and many others. He buys the bulbs or seeds in France or in foreign countries and knows a lot of professional horticulturists. His garden is actually classified and he receives flower lovers there for tours. In the months of May and June, his garden is absolutely enchanting; you can easily spend two hours just visiting it. In summer, I often join him in that paradise where we chat at length about his horticultural triumphs and difficulties and about lots of other subjects, medicine, probability, psychology.

Mathematical research still procures me my greatest pleasure. But mathematics in Autouillet has more charm than mathematics at my desk in Paris. Being disturbed in my research by the phone ringing in Paris or by the birds chirping in Autouillet is not at all the same. Here, I have a sense of total solitude and liberty, happy in the fresh air, as long as I cover myself well and, in case of wind, put something heavy on my papers so they don't go flying all over the place. I proved many of my most interesting theorems in Autouillet. Most of my course notes for the Ecole Polytechnique were written up there. I settle in front of the left-hand chestnut tree, in front of the house, next to the French garden. The children playing round about don't bother me at all. I work all morning and in the afternoon until six, stopping only for lunch and a one-hour nap. I'm eighty-two years old and feel like my whole life is still before me. Nothing makes me happy like nature.

I was a sort of aristocrat in Autouillet. It's not very recommendable, but I've always been "kept". Marie-Hélène takes care of all contacts with the carpenters, the plumbers, the roofers and masons. In the garden, Bertrand and his wife take care of everything. I simply enjoy the place. I don't say it's right – it's very egotistic of me. I'll try to do better in another life. But for now, I'm going to tell the story of mine.

# Chapter I
# The Revelation of Mathematics

## Classical beginnings

I always possessed the spirit of research. And it is a fact that I became a researcher. It was predictable from my earliest childhood. As far back as I can remember, I asked a multitude of easy and difficult questions, and was never satisfied by evasive answers. Of course, all children do this; there are ages and moments where they constantly ask "Why?" But most of the time, they are happy with simple answers. When I asked why, I needed a complete explanation, but I rarely received it. Some of my questions lasted for as long as ten or fifteen years. Often it was impossible to answer me properly because the true answer would have been too difficult for me to understand. For example, in my bath, I had of course tried the broken stick experiment: you thrust a stick into the water and it looks broken. I had also remarked how deformed my hand looked if I stuck it in water, either horizontally or vertically. I believe my mother had some recollections of the idea of refraction, but not enough to satisfy me. I reflected at length on the subject of floating bodies, just like Archimedes, but without arriving at the same result as him! I believe my mother explained to me that light bodies floated and heavy bodies sank because "the water pushed". This answer did not suffice. Of course, I was also bothered by the caustics (curves due to light refraction) which I saw in her bowl of café au lait in the mornings. Who could have explained them to me? Why, at the same temperature, does air seem warmer or cooler than water? Why is this also true for wood and metal? It's a question of heat conduction. Why does a spoonful of water never empty completely when you turn it over; one drop always remains sticking to the spoon? It's a question of capillarity. I'm too warm in my bed, my arm is warm. If I move it to a new place it cools off, but then becomes warm again. Who heats it? Animal heat. I didn't understand my father's explanation that hemoglobin oxydizes in the lungs, becoming oxyhemoglobin, then arriving to the muscles and delivering the oxygen to them, it absorbs carbon dioxide instead, becoming carbohemoglobin, which is then discharged upon its return to the lungs. What did it prefer? Did it want oxygen or carbon dioxide? Could it prefer anything? Did it have a conscience? My father didn't know enough about physics and chemistry to give me full answers. He simply said "You silly idiot, if it wasn't like that, how could we breathe?" My life as a child and a schoolboy was loaded with questions of this kind. When I asked them, I was often told that I asked absurd questions. So after a while, I stopped asking.

After this period rich in questions, I was overtaken by a passion for Latin and Greek. These two languages, along with early notions of linguistics and philology, were my favorite subjects from sixth to eleventh grade. Starting in seventh or eighth grade, I studied them intensively on my own, for pleasure. I loved Latin and Greek and studied them whenever I could, independently of my schoolwork. If we had to explain twenty lines of Virgil in class, I couldn't resist going on on my own and preparing sixty or eighty. I read the Iliad and the Odyssey for pleasure, as well as Herodotus and of course Livy, which is quite easy. Naturally, I always needed a dictionary, but I managed quite well. What I call "reading" doesn't just mean reading through them, but learning them inside out. Nothing remained foreign to me: I mastered genders and cases, declinations and syntax. I didn't mean to dominate the class or surpass the teacher, but I did want to dominate Latin and Greek. I adored studying. What counted was not my success but the fact that studying made me completely happy, in the most personal way. I was incredibly happy with everything we learned. I remember how happy I was thinking about all the questions which naturally arose from the different subjects we studied. In eleventh grade, I actually started to write a Greek grammar of my own! I had a lot of good friends with whom I could discuss the things which interested us. I'm not at all certain that the kind of atmosphere I remember still exists in high schools.

In spite of my success, I was always deeply uncertain about my own intellectual capacity; I thought I was unintelligent. And it is true that I was, and still am, rather slow. I need time to seize things because I always need to understand them fully. Even when I was the first to answer the teacher's questions, I knew it was because they happened to be questions to which I already knew the answer. But if a new question arose, usually students who weren't as good as I was answered before me. Towards the end of eleventh grade, I secretly thought of myself as stupid. I worried about this for a long time. Not only did I believe I was stupid, but I couldn't understand the contradiction between this stupidity and my good grades. I never talked about this to anyone, but I always felt convinced that my imposture would someday be revealed: the whole world and myself would finally see that what looked like intelligence was really just an illusion. If this ever happened, apparently no one noticed it, and I'm still just as slow. When a teacher dictated something to us, I had real trouble taking notes; it's still difficult for me to follow a seminar.

At the end of eleventh grade, I took the measure of the situation, and came to the conclusion that rapidity doesn't have a precise relation to intelligence. What is important is to deeply understand things and their relations to each other. This is where intelligence lies. The fact of being quick or slow isn't really relevant. Naturally, it's helpful to be quick, like it is to have a good memory. But it's neither necessary or sufficient for intellectual success. The laurels I won in the Concours Général[4)] liberated me definitively from my anguish. I won first prize in Latin theme and first accessit in Latin version; I was no longer merely a brilliant high school student, I acquired national fame. The Concours Général counted a lot in my life, by helping me to get rid of a terrible complex. Of course, I was not instantly metamorphosed, and I've always had to confront the same problems; it's just that since that day I know that these obstacles are not unsurmountable and that in spite of delicate and even painful moments, they will not block my way to accomplishment, which research represents for me. Fortunately, I had an excellent memory. For instance, in twelfth grade, in math, I believe that at the end of the year I remembered every single thing I had learned, without ever have written anything down. At that point, I knew my limits but I had a solid feeling of confidence in my possibilities of success.

This type of competition is an excellent thing. Many young people feel self-doubt, for one reason or another. The refusal of any kind of comparison which reigns in our classrooms as a concession to egalitarianism, is all too often quite destructive; it prevents the young people who doubt their own capacities, and particularly those from modest backgrounds, from acquiring real confidence in themselves. But self-confidence is a condition of success. Of course, one must be modest, and every intellectual needs to recall this. I'm perfectly conscious of the immensity of my ignorance compared with what I know. It's enough to meet other intellectuals to see that my knowledge is just a drop of water in an ocean. Every intellectual needs to be capable of considering himself relatively, and measuring the immensity of his ignorance. But he must also have confidence in himself and in his possibilities of succeeding, through the constant and tenacious search for truth.

---

4) The Concours Général, "General Competition", is a set of examinations offered in many different subjects to high school students. Only the best students usually make the attempt, and generally they choose to try only one or two of the exams. There are six national winners for each examination; first, second and third prize, first, second and third accessit.

## The Seduction of Geometry

It would have been natural for me to enter the philosophy class in October 1931, and then to enter khagne[5] to prepare the entrance competition in literature for the ENS, the Ecole Normale Supérieure[6]. However, one could also enter the mathematics class in the last year of high school, and pass the mathematics and philosophy baccalaureates the same year. Should I try this? It was a way of leaving doors open for my future choices. But it would mean abandoning Latin and Greek almost completely for a whole year. My eleventh grade teacher did not want me to do it, because according to him I was destined to become a Latinist or a Hellenist. In mathematics, it is true that I was always the best in the class, but it was a Latin-Greek class and the level of the students and professors in mathematics was not very high. Guy Iliovici, a cousin of my mother's and a professor of higher mathematics (who later died in deportation with his wife and daughter) tested my mathematical abilities and concluded that I was a good student but nothing more. However, I remember my seventh grade literature teacher, whose name was Thoridenet, who said to my mother "I never had a student like him in Latin, but he's not that good in French literature and his reasoning is that of a mathematician and a linguist. Don't make a mistake: no matter what they'll tell you later, he ought to become a mathematician." At a time when I wasn't particularly good in mathematics, this teacher showed himself to be much more lucid than the ones I had later. When she was faced with this kind of dilemma, my mother used to turn to her brother, the pediatrician Robert Debré. He had a remarkable knowledge of the psychology of children and young people. He said he could more or less predict the intellectual capacities of a child of two or three. He said it was absolutely independent of the rapidity of the child in acquiring language. He had always told my mother, who of course was only too willing to be convinced, that I would be an exceptional adult. She used to remind me of this to encourage me, even as far back as fifth grade when I had not yet shown my aptitude for schoolwork in any precise way, and of course she repeated it to everyone else as well. My uncle did not change his mind in spite of my occasional manifestations of naivete; he said without hesitation that I should enter the mathematics class and pass the two baccalaureates. He thought I would become a mathematician and wanted to make sure I left the possibility open. Without the advice of Thoridenet and my uncle, I probably would have become a linguist or a Hellenist, and I'd probably have been perfectly happy, but *a posteriori* I think I did end up following my

5) see footnote p. 41

6) ENS: see footnote p. 39 and following chapter

own true path. I have studied some linguistics and admired the comparative properties of languages, but deep down I'm just more of a mathematician than a linguist.

I entered the class of elementary mathematics. My philosophy professor had a solid reputation even outside of the Lycée Janson-de-Sailly where I went. But he turned out to be catastrophic; he dictated his lessons in a monotonous voice and didn't require even the slightest effort at reflection from his students. He never discussed with us and didn't inspire me with any new ideas. For many of my friends and colleagues, their philosophy classes represented a new opening to the world, and these classes are frequently considered as the crowning glory of high school education. I knew that and I'm still convinced of the considerable role played by philosophy in the history of humanity. Numerous philosophers of yesterday and today have worked hard to conserve this dignity. But there is undeniably a large number of philosophers of doubtful quality.

My great-uncle Jacques Hadamard, one of the greatest mathematicians of his time (1865–1963 – he was my great-uncle by marriage, having married a sister of my maternal grandmother) – used to say rather uncharitable things about them, such as "Those people study problems that anyone can solve and problems that no one will ever solve." Or even worse, "Philosophers are people who are searching in a dark room for a black cat which isn't there, and finding it." Quite recently, I heard a talk at the Paris philosophy seminar given by a professor of the epistemology of biology from the University of Quebec in Montreal, in which he said incredibly stupid things. He explained to us that the fundamental problem of biology today is to explain the difference between individuism and classism. Classism consists in considering that the set of animals is a set in the mathematical sense, and the set of lions is an equivalence class (the equivalence relation being that two animals are equivalent if they belong to the same species). This is undeniable. Individuism consists in considering the set of lions as a single individual, of which the different lions form the various limbs, legs, arms etc. I forced myself to listen to the talk for the full hour, all of it was in the same style, with long discourses on the relative advantages and disadvantages of the two systems. It reminded me of the following caricature (due to Sartre) of the type of exam question usually given in philosophy: "The soul and the body, similarities and differences, advantages and disadvantages." I feel the same way about a certain type of modern painting. When Duchamp exhibits a urinal or a perfectly white painting, and experts discuss whether it's a white square on a white background or a white disk on a white background, I think someone is thumbing his nose at the world. I remember at a sale whose profits were to go to the battle against Greek colonels, a perfectly just cause, a "pile of razors" being sold for a high price; just a bunch of old electric razors piled into a glass box. It's plain cheating. Paintings using even very simple forms and colors can produce an agreeable decorative effect, but that doesn't mean they express any talent whatsoever.

So, I entered the class of elementary mathematics. For the first time, I had a really exciting schoolteacher in mathematics, called Julien. He had a certain fondness for the bottle, but nobody saw anything wrong about it. He explained the most marvelous problems of geometry with joy and simplicity. I discovered a mathematical universe which I knew nothing about, which I had barely glimpsed in my eigth-grade geometry class. After just two or three weeks, I had decided to become a mathematician. I realize now that this tendency must have been in me the whole time, but restrained by the mediocrity of my teachers; now, it exploded suddenly. Because of this, the value of a teacher for the future of his students has always struck me as fundamental.

In this new class, I had plenty of competition. The students came from various different eleventh grade classes, and some of them were really better than I was. It was a challenge. For years, I just "had" to be the first of the class in composition. Now, I "had" to love mathematics. It was natural and indispensable to me to study mathematics for pleasure, just as I had studied Greek and Latin before. But now there was a new dimension: my future career was as yet completely undetermined. If, in spite of the great attraction that mathematics held for me, I turned out as Iliovici had said to be just a medium to good student, given the talent I had shown for Greek and Latin, I would almost be forced to return to them and to give up mathematics. I was ambitious and refused to be merely a mediocre mathematician, if I could be a first-class linguist or Hellenist. So I accomplished an exceptional effort, probably the greatest one of my existence, finding in myself a capacity for work and innovation which perhaps I never really reached again.

From the beginning, Julien taught us Simson's line and Euler's circle, the nine-point circle, which fascinated me. I understood things more or less in class, but when I got home I had to go over everything again, and it wasn't always easy. I read the few great books of elementary geometry which are written for students interested in mathematics: Guichard, Petersen, Duporcq and the two magnificent books on elementary geometry written by Jacques Hadamard. That year, I solved practically every question in those books, meaning I probably solved about a thousand problems. I particularly liked the geometry of the triangle (the theorems of Menelaus and Ceva, and all their consequences), and pushed it quite far. The study of the geometry of the triangle has now been more or less abandoned as not sufficiently fruitful; it no longer contains theorems which are really interesting for modern mathematics or other sciences. The different domains of a given science do often have such periods of success and decline; at any given moment each science has domains which are on the rise and others which are being abandoned and forgotten, sometimes only to be rediscovered later on. Nobody is interested in the geometry of the triangle nowadays, but I liked it and studied it eagerly, helping myself by reading Guichard.

Ceva's theorem gave conditions for three lines issuing from the three vertices of a triangle to intersect in a single point. Guichard generalized the problem to the "quadruple hyperboloid": under what conditions do four lines issuing from the four vertices of a tetrahedron form a "quadruple hyperboloid", i.e. are the four generators of the same system of a quadric? The conditions turn out to be very simple: the three heights of a triangle always intersect in a single point, but the four heights of a tetrahedron form a quadruple hyperboloid; if they do meet in a point, the tetrahedron is said to be orthocentric. These things are fairly advanced compared to the school syllabus which contained only quadrics. I did study quadrics and their double systems of generators at length. I also liked the transformations of the plane and three-dimensional space, which play a fundamental role in many parts of mathematics. I thought about the group generated by inversions and discovered for myself the formula $STS^{-1}$ for the transformation of $T$ by $S$. I didn't know about groups, so I didn't know that the integers form a group, but I did know about transformation groups. I understood that the important thing about an inversion was not so much the center or the power of the inversion as the circle or sphere of inversion, and that the sphere of inversion of $STS^{-1}$ was the transform by $S$ of the sphere of inversion of $T$. I can't even think how many properties I amused myself by transforming under inversion. I understood what was meant by the anallagmatic plane (or space) having a single point at infinity; a circle passing through this point was a line. By stereographic projection, it became a sphere in dimension 2. I didn't realize I had the right to do the same in three dimensions, since the sphere would then have become a three-dimensional sphere embedded in a four-dimensional space, which was unknown to me. I was enchanted by sheaves and lattices of circles and spheres. If you took two sheaves of conjugate circles and one of the two base points of the first one, and you sent the other to infinity by an inversion, it became the sheaf of lines passing through a fixed point. The Poncelet points of the conjugate sheaf were given by that point and the point at infinity. It was just a sheaf of concentric circles. I was initiated to the paratactic circles of André Bloch by Jacques Hadamard; they are systems of two circles whose planes are orthogonal. The intersection of the two planes is a common diameter of the two circles, and is cut by them according to a harmonic division. Then any sphere containing one of them is orthogonal to any sphere containing the other. Consequently, parataxy is invariant under inversion, and every inversion with respect to a point situated on one of them transforms them into a circle and its axis.

The manuscript containing this discovery had been posted to Jacques Hadamard, and contained a return address: 57 Grand-Rue, Saint-Maurice. Hadamard was always interested in elementary geometry, and he enthusiastically wrote to the author to ask him to dinner. The author replied that this was impossible, but invited Hadamard to visit him. Hadamard went and was stupefied to find that he lived in the Charenton insane asylum. André Bloch spent the rest of his life there. In a moment of insanity, he had killed one of his uncles, an aunt and one of his brothers. For years afterwards, he was a perfectly peaceful man, deeply interested in mathematics. He discovered quite important things in the theory of functions of one complex variable, and there is even a Bloch constant. Hadamard went to see him, and they talked at length. Later they exchanged letters, and Hadamard included Bloch's paratactic circles in his big book on elementary geometry. Bloch was protected from deportation during the Occupation. In 1953, we happened to meet his surviving brother, in Mexico; it was deeply moving for all of us. Later, I tried to work with Jacques Hadamard. He was the author of an immense quantity of mathematics, and he was one of the greatest mathematicians of his century. His face always showed the gentleness characteristic of great sages. His culture in mathematics, physics and biology was vast. However, he was interested in many other things and even collected ferns, which he loved as much as I loved butterflies. At the age of eighty-five, he was invited to be the president of honor of the International Congress of Mathematicians in 1950, and he requested to be taken on horseback to a certain forest in Mexico where a rare fern grew which was not represented in his collection. His wife, our aunt Lou (for Louise) pleaded with Mandelbrojt and me to dissuade the organizers of the conference from acceding to his request. Finally we came to a compromise and took him on horseback to a nearby forest, which also contained interesting ferns which overjoyed him. Some of my colleagues reproached me a bit about this. Should I have restricted his freedom? After all, he was old; if he had died, what of it? But I was also worried that he could get hurt. And it was his wife who asked us.

Hadamard had flung himself bodily into the Dreyfus affair – he was related to Dreyfus through his wife – and worked actively for the League for Human Rights (for a long time, he actually lived rue Jean-Dolent, next to the League's offices, which made it particularly easy for him to belong to their central committee). For years, he had been a teacher at the Lycée Buffon, while writing his thesis (at that time, there was no equivalent of the CNRS). His books on elementary geometry date back to this period. He worked in extremely varied domains: analytic number theory (he gave the first proof of the distribution of primes using the zeta-function $\zeta(s)$, in a memoir which received a prize from the Academy of Sciences in 1892), the analytic continuation of Taylor series, the orders of entire functions, the geometry of surfaces with constant negative curvature, and above all partial differential equations, one of the most magnificent parts of his work. Although he had been a high school teacher, his pedagogy was entirely unsuited to my needs. Once, he was astonished that I had never heard of the function $\zeta(s)$. "I find you very ignorant," he said. But later, we were talking about solutions to quadratic equations, and when I mentioned the usual formula for the two roots, he congratulated me with an admiring look: "Ah, I see you really know a lot!" I soon understood that working with him would be useless, in spite of our excellent beginning with paratactic circles. I only learned recently that he was a terrible teacher at the Lycée Buffon, who had no idea how to get on the same level as the students, and that after a huge number of parental complaints, he had almost been penalized. Nevertheless, in my eyes he remained a role model, haloed with prestige, and even if I didn't actually study with him much, my ideas of mathematical research came from him. However, I had no idea what rings and fields were, and never guessed that similarities with centers at the origin in the Euclidean plane formed a field (the field of complex numbers, which I had never heard of). I lived with a delicious feeling, like a fish in water, amongst circles, spheres and inversions.

I also adored transformations by reciprocal polars, which Julien had taught us with his usual communicative enthusiasm. With harmonic divisions and the anharmonic proportion, called the cross-ratio I passed to the projective plane and then to projective space, then to homographies and involutions, and to correlations in these spaces. At that time, the projective plane was defined as a plane with a line at infinity, so that I could clearly envision the difference between the projective plane and the anallagmatic plane. The modern definitions given today, where the projective plane is the set of lines passing through the origin in a three-dimensional vector space, would have been beyond my comprehension at the time. But it didn't matter; I handled all these notions with ease.

What I used to call my competitive spirit was confirmed in the mathematics class. Obviously, I wanted to be number one. However, I never actually worked for that precise goal, any more than I had worked on Latin in order to win the Concours Général. I sat at my table, opened my books and solved problems in a kind of euphoric state. Naturally, this was very helpful in becoming first of the class, yet I did it for my own personal pleasure. I was also very interested by some subjects which had no effect on my class rank or competition results. I simply found mathematics beautiful – extraordinarily beautiful – and geometry in particularly incomparably elegant. I still believe it ought to be taught more in high school, because a high school student cannot easily make discoveries in analysis or in arithmetic (although it occasionally happens). But the study of geometry is a source of perpetual research, comparable to what a university researcher does, though at a lower level. It wasn't enough for me to solve the problems in the books; I gave myself problems and even managed to solve some of them, which gave a whole new dimension to my work. I remember feeling a happiness so intense during the whole of that year, that just talking about it makes me feel it surge forth with the same freshness.

The geometry books I nourished myself with all had one particular surprising feature: they contained no trigonometry or cartesian coordinates, although these formed part of the official school program. There are generally two ways of attacking a problem in geometry: by algebraic computations with coordinates, or without them, by pure geometry. I never considered it dishonorable to use a mixture of the two – why not use all the methods you have at your disposal? I do understand the reason behind the dichotomy; usually one method is much better than the other for understanding the real nature of the problem. For instance, in tensor theory or topological vector spaces, or in differential geometry, the intrinsic method is usually preferable to computational methods. However, in a given problem, it may become indispensable to use actual coordinates, as the purely theoretical method may become desperately abstract. For instance, it's impossible to avoid computations when working with partial differential equations. Every mathematician should know the trace of an operator in finite dimension: it is an element of the contracted tensor product of the space with its dual, but its also simply the sum of the diagonal elements of the matrix of the operator in any basis. Even high school students are taught to use coordinates, usually working within a suitable basis. The authors I mentioned before use exclusively geometric methods in their books. This means they sometimes need recourse to astonishing acrobatics. For example, one wants to construct the complexification of ordinary 3-dimensional real space. Without giving any formal definition, Jacques Hadamard simply says that if an involution on a line has no double point, then we decide to state that it has two complex conjugate points. These imaginary points are introduced right in the middle of a paragraph, with no warning, so that when they are used later in the book, it's practically impossible to find where they were defined. They aren't in the table of contents, either. Later however, he uses them to deduce that if two circles in the plane do not intersect each other, then we can decide that they have two complex conjugate points of intersection, and he explains how to determine when two pairs of non-intersecting circles in the same plane have the same pair of complex conjugate intersection points: the answer is, that it happens exactly when the four circles all belong to a given sheaf of circles. The cyclic points of a plane are the intersections of its circles with the line at infinity, giving rise to isotropic lines, and in space, to isotropic cones or spheres of radius zero, isotropic planes, and the imaginary marvel of the umbilical conic. I juggled with all this without too much difficulty, but I never realized how much simpler everything could be made. Indeed, I learned about computing with coordinates only the following year; you fix a basis and prove what you want, and then you show that everything is independent of the choice of basis.

Today, complex numbers are taught in twelfth grade (and so are exponentials and logarithms, which in my time were only taught in the first year of Classes Préparatoires[7]); I didn't know about them in twelfth grade, I was advanced only in the subject of pure geometry. I knew nothing of $z = x + iy$ or $r = e^{i\theta}$. In his book, Hadamard barely talks about quadratic equations, possibly having two complex conjugate solutions: the number $i$ does not exist. At any rate, the proper definition of the complexification of the real vector space $E$, given by $E + i\tilde{E}$ where $E$ is the affine space and $\tilde{E}$ is the associated vector space, is not given in Hadamard, nor in Guichard, nor in Petersen or Duporcq, nor even in the Classes Préparatoires in my time, since the notions of affine space and vector space (of any dimension) did not exist, even at university. Even when I passed the agrégation[8], I had never seen these notions; I was capable of considering *space*, but not *a space* of given dimension $n$. Life is strange. Indeed, in geometry, one doesn't imagine an affine complex line (for example when working on Ceva's theorem in a triangle) at all like the field of complex numbers $x + iy$. When I think about it, the imaginary points in geometry are grey, the real points are black and the intersection of two complex conjugate grey lines is a real, black point. The beautiful umbilical conic is silvery, isotropic cones and lines are more pink. As for the mixture of projective and Euclidean structures, it is well understood in modern language. In a projective space, the complement of a hyperplane, considered as a plane at infinity, has a canonical

7) The Classes Préparatoires, preparatory classes, are two-year advanced courses taught in lycée after the baccalaureate, which both replace the first two years of university study and serve as intensive preparation for the entrance competitions to the so-called Grandes Ecoles (Big Schools) such as the ENS (Ecole Normale Supérieure), in which students are paid a small stipend for attending university.

8) The agrégation is an examination which follows undergraduate university studies, qualifying students to become high school teachers.

affine structure, and there is no difficulty adapting this to a Euclidean structure. For an inversion of negative power, the inversion sphere is imaginary.

Conics, defined by their foci, their axes and their equations in cartesian coordinates, interested me greatly, but because they had to be studied within projective geometry, I was led to define them as second-degree curves and to extend their study to quadrics, surfaces of degree 2, with their real and imaginary points, at finite or infinite distances, differentiating the different types of quadrics in read affine Euclidean geometry. I didn't have any trouble defining convergence in projective space (topology); I didn't deal with the "differential structure". No one else seemed to either. Progressively, I became able to swim effortlessly, voluptuously, in the ocean of these objects. I sometimes became so excited that I used to rush feverishly up and down my room, talking out loud, very loud, and sometimes even leaning out of my window, over the courtyard of our house, to shout something like "Sheaves of orthogonal circles!" The neighbors obviously thought I was a bit crazy, but I didn't care at all. My parents restricted themselves to some mild observations. Via complex projective maps, I had understood that all non-degenerate conics could be transformed projectively into each other. Thus, for example, it was possible by a projective transformation to bring two conics to two circles, by sending two of their intersection points to cyclic points. I knew about bitangential conics, i.e. conics which are tangent to each other at two points: if one sent these two points to cyclic points, the conics became two concentric circles. The transformation by reciprocal polars was not related to the circle and to orthogonality. The same polarity exists with respect to any conic; the polar of a point is simply the line joining the two points of contact of the tangents to the conic coming out of this point, real or imaginary. I also knew the conchoid of a curve and its podary, Pascal's limaçon and the cardioid. These were curves we had studied with Julien. He had also taught us the strophoid, Diocles' cissoid, hypocycloids and epicycloids, as well as billiard problems. I particularly admired the hypocycloids with three or four cusps. Julien had also had us study Cassini's oval, the locus of points the products of whose distances from two given points is a constant; Bernoulli's lemniscate is a special case.

Seeing me studying trigonometry, my uncle Jacques Debré taught me the series expansions of sine and cosine, which excited me greatly. For me, the sine and cosine of an angle were abstract concepts; I didn't think of actually proceeding to compute their values, other than by the very vaguest of approximations. But my uncle explained to me that the series converged very rapidly, and giving me the explicit expansions of the sine and cosine of an angle measured in radians, he showed me how to calculate close approximations, which finally made the notions of sine and cosine become really concrete in my eyes. With all this, I began to have a non-negligible baggage of mathematical culture. I strolled among the figures just as I used to stroll in the garden of Autouillet, digging in corners and dreaming as though I was in the Labouise meadow. I admired the beauty of geometry, but also the classification of mathematical structures. Well before becoming aware of the existence of Bourbaki, I had vaguely understood the idea of mathematical structure, since the projective and anallagmatic structures were so different from each other (they are even structures on different sets). From this point of view, it's probably too bad that I abandoned geometry to study classical analysis at the ENS. Today, algebraic geometry is one of the essential branches of mathematics. Of course, it is not really based on high-school elementary geometry, but still, using a certain type of "algebrization" which I learned later from Bourbaki, one may pass from one to the other. Indeed, I learned in hypotaupe and taupe[9] to work with multiple points and intersections. I think the local definition used in modern analytic geometry of an analytic space (not necessarily reduced), via germs of holomorphic functions, is something which I would have understood very well and which would have greatly interested me, if I had seen it a little later. But that is not the road I ended up taking; I almost completely abandoned geometry after entering the ENS, and devoted myself to the differential geometry of curves and surfaces, which was very elementary at that time. Today, projective spaces and generally the whole of projective geometry, as well as part of differential geometry, have been removed from the high school mathematics program. Today's mathematics teaching in France is not based on geometry, even though it is an essential subject for engineers and physicists, and has become one of the fundamental branches of modern mathematics. It's terribly sad, even disastrous. On the other hand, the whole subject of "projective-Euclidean" geometry, the umbilical conic, sheaves of conics and quadrics, the theorems of Apollonius and other Greeks, has perhaps now become as obsolete as the geometry of the triangle; I admit I don't know. At any rate, I myself never used it, and have rather forgotten most of it, along with Latin and Greek. Not without a certain feeling of nostalgia.

It's natural to ask why mathematics is beautiful. A finished mathematical theory is a well-constructed edifice whose foundations rest on virgin terrain and which rises forth as a majestic palace, like a time-defying monument. Like a cathedral, minus the religious value, but with the addition of perfect logic. Read-

9) The word "taupe", a mole, is used familiarly to describe the second year of Classes Préparatoires in mathematics; "hypotaupe" refers to the first year. The words "khagne" and "hypokhagne" are used correspondingly for the literary classes.

ing, discovering a beautiful theory or an astute proof, is like mountaineering; the going is sometimes rough, but from the top one can contemplate the whole of the path one has just followed. A beautiful mathematical argument is comparable to a concerto by Bach or a harmonious ballet. Of course, mathematics is beautiful because we find it beautiful. And how is our sense of æsthetics formed? Natural selection leads us to adapt ourselves to the world we live in. At a certain point, man, given the gift of reason, became capable of solving certain complex problems in order to survive, using the logic of our universe. And to be able to do this, he must have found this logic beautiful, and felt pleasure in solving these problems. If man finds beauty in the logic of nature, then mathematics, the culmination of abstract reasoning, must be particularly beautiful. For similar reasons, we like nature, flowers, leaves, the colors green, red, blue, and all the other colors of the spectrum, birds, butterflies and all sciences. By a kind of necessity, man must be in harmony with his surroundings. Generally speaking, it's normal to appreciate what is around you. And if certain things are really ugly, one can imagine that there is a good reason for them to disgust us. The odor of excrement or of corpses repels us for good reasons, whereas these same odors are supremely agreeable to many kinds of animals, mammals, birds and insects. For our hygiene, it's healthy that we loathe these odors. I admit that this concept is very finalist, but it follows from the notion of natural selection. I think animals function the same way as humans. Females are desirable to males, and their odor is attractive. But vultures are repulsive, not only to humans, but also to lions and tigers – Egyptians admired flying vultures, and so do we, but not the vultures sitting next to cadavers. When a lion devours a corpse, he protects it jealously and tries to keep it for himself as long as possible. But the meat of the corpse slowly rots and becomes unhealthy for him. At that point, it must smell bad to him. Then the vultures arrive and pester him until he gives up and abandons the remains to them. At that point, it's much better for the vultures to be repulsive to the lion, even frightening if there are many of them, and this is surely why they must appear ugly to the lion's sense of æsthetics. The situation of the bearded vulture is quite different. It eats the bones of mammals, after the vultures have finished their work. It flies very high and looks out for piles of bones with no vultures near them. Then it swoops down, picks up the bones, rises in the air and drops them onto a flat rocky surface to break them, after which it descends and consumes the marrow. It does the same with living turtles that normal carnivores cannot eat. This is quite a convincing proof of intelligence. Legend has it that Aeschylus was killed when a turtle fell on his head, which a bearded vulture had taken for a smooth rock. Unlike the vulture, the bearded vulture has no reason to appear ugly to lions, and indeed it's a beautiful bird. Thus we have proved that mathematics is beautiful because vultures are ugly.

I was also very excited by twelfth grade physics. We studied everything concerning the law $F = m\gamma$, falling bodies, acceleration and weight, the difference between weight and mass, the motion of a pendulum, and all sorts of vibrations, and resonances. Alternating currents, with condensers and induction phenomena also fascinated me, and particularly the existence of high frequency currents. Waves entered my life never to depart, since my research always remained in the region of Fourier transforms. There is such a multitude of different types of vibration; vibrating strings and tubes, and all musical instruments. Leon Brillouin, an excellent physicist and a family friend, brought me to his laboratory and showed me many things which I had never heard of in high school, such as protons and electrons, and electric current as a displacement of electrons in the opposite direction to the current. But physics never led me as mathematics did to attempting my own efforts at research, or reading books for my own pleasure. I satisfied myself with learning the school program. And this program was extremely old-fashioned. We still learned about the "atomic hypothesis", even though Niels Bohr was running all over Europe lecturing on the atomic model which he had discovered in the twenties.

I also enjoyed natural sciences immensely. I studied them with my father, as I did every year. When he explained the double circulation of blood, I found it a magnificent architectural structure. I studied natural sciences for my pleasure. It's too bad that after twelfth grade, natural sciences stopped; they were not taught in hypotaupe and taupe. Of course, the human anatomy we studied contained no sexual organs. Only sexual reproduction of plants was considered. We did learn about the antherozoid and the oosphere, but there were no spermatozoids, no eggs. Our souls were saved and only plants reproduced.

I was very excited at first about organic chemistry, which was a completely new subject for me, but eventually I became disappointed. Everything we learned was interesting, but I wanted to learn about the protids, lipids, and glucids we studied in natural sciences, which I thought of as being part of organic chemistry. But we studied no such thing. I waited in vain until the end of the year. Later, I studied organic chemistry all alone, during my military service.

At the end of the year, I passed the Concours Général in mathematics and received a second accessit; I also passed the baccalauréat with the result "Très Bien"[10). I don't remember the problem we had to solve at the Concours Général, I only remember a certain second degree cone. I introduced the focals of the

10) According to the score one receives, baccalauréat results are called "Très Bien", "Bien",

cone, i.e. the lines passing through the vertex such that the tangent planes containing these lines were isotropic; this was not in our twelfth grade program. I said that if the cone was cut by a plane perpendicular to these focals, the intersection points of the focals with the plane were foci of the section of the cone in the plane. I wrote all this in excellent mathematical style. Julien always worked on improving my style, and by the end of the year I wrote in a style so elegant and correct it was almost literary. He wasn't surprised that I received an accessit, even though I didn't completely finish the problem.

My written exam for the baccalauréat was catastrophic. The formula $a^2 = b^2 + c^2 - 2bc\cos A$ for the sides and angles of a triangle was constantly needed, and o horror, I wrote $\sin A$ instead of $\cos A$. For this mishap due to nerves, I received a 0 on that half of the exam. On the other half, I got a perfect score, but the average of the two was not very glorious. I came to the oral exam feeling a bit nervous. I don't remember the problem I was given at all, but it had to do with conics tangent to four given lines, and a certain property of their centers. I immediately set about describing tangential sheaves of conics, the locus of the poles of a line which also formed a line, and thus the locus of the centers, poles of the line at infinity, and thus also a line. The direction of this line was the same as the axis of the parabola of the sheaf; in particular, considering the degenerate (tangential) conics of the sheaf, one saw that each of them was composed of two points, one the intersection of two of the lines and the other the intersection of the other two lines, so that one found three degenerate conics. The center of a tangential conic formed by two points was the midpoint between them, so the three centers of the three tangential conics were colinear. This way, one reproved the theorem of Pappus on four given lines. I seem to remember explaining all this as easily as though it were seven times eight equals fifty-six. The examiner stared at me in disbelief (I'm not at all sure he understood everything I was talking about, because the properties of tangential sheaves, degenerate tangential conics, the parabola tangent to the line at infinity and so on, were not in the program, and he may not have been familiar with these concepts at all). I can't imagine what the original problem I was supposed to solve might have been. It's still a mystery to me. However, he said with a satisfied look "I guess there's no need for me to continue!" and gave me 39.5 out of 40. Thanks to this, I got my result "Très Bien" in spite of my failure at the written test.

Looking back, the mathematics I learned that year appear particularly æs-

"Assez Bien" or merely "Passable", the three honors corresponding more or less to summa cum laude, magna cum laude, cum laude.

thetic to me. Parts of it became obsolete, such as the geometry of the triangle. Inversion, homographic transformations are today considered as linear transformations in a vector space, giving rise to a projective space, and transformation by reciprocal polars is now considered as a correlation between the points and the hyperplanes of projetive space, or as a non-Euclidean orthogonality in a vector space with respect to a non-degenerate symmetric bilinear form. Apart from the change of language, these notions are still frequently used. At any rate, I learned a quantity of things which still surprises me. I knew much more than Julien by the end of the year. For instance, once he spoke to me of "the" cyclic point, whereas in fact there are two cyclic points in the plane. My perfect understanding of geometry was not so extraordinary in itself; one was supposed to acquire this kind of understanding in hypotaupe and taupe. But I learned it all myself, without any course notes, thanks to key lines here and there in books I read, particularly those of Hadamard, which led me to a whole personal creative theory. Starting from odd, vague and acrobatic definitions of imaginary points, I built a whole beautiful coherent inner palace, as I later called it. Perhaps never again in my life did I work so well, and with so much happiness.

My comprehension of geometry has one quite unusual feature. I already mentioned my "empty right hemisphere" and my inability to orient myself in space. I visualized almost nothing when studying geometry, not even figures made of lines and circles in the plane. If I drew a figure, I could see something, but immediately forgot it. In studying geometry, I worked only with the geometric properties which somehow can be expressed in a simple, local manner. This was quite enough to attack all the problems I've mentioned above. For the study of inversion, projective space, the umbilical conic, there's no need for real geometric vision. But it is a peculiar way of doing geometry, and later, I was more and more troubled by my complete absence of vision. For instance, I couldn't come to a real understanding of the standard high school theorem stating that the volume of a tetrahedron is the third of the product of the height by the area of the base. The decomposition of the parallelepiped into three tetrahedra of equal volume was invisible to me. I had an enormous memory and remembered everything I had ever read and every problem I had ever solved. I never studied geometry in more than 3 dimensions, and wasn't even sure that it was meaningful to speak of 4-dimensional space. I asked Julien if, as the ellipse was the orthogonal projection of the circle, an ellipsoid of revolution around its small axis was the orthogonal projection of a sphere in 4-dimensional Euclidean space. He found the suggestion a bit suspicious and advised me to avoid such properties; I heard this again the following year. I didn't persevere, since equations $x' = ax$, $y' = by$, $z' = cz$ could be used to obtain any ellipsoid from some sphere without changing dimensions. But I'm still surprised, when I think back about how much I had studied and learned, that I had no idea about $n$-dimensional spaces for $n$ greater than 3. They were never mentioned in any of the books I read, and I didn't discover them for myself. And I never heard of them even in the following years, including at the ENS, except for Leray's course at the Collège de France, in which he mentioned infinite-dimensional Banach spaces. But I never saw simple Euclidean space of dimension greater than 3. It's strange.

The contrast between my love for geometry and my total absence of geometric vision is really mysterious. If I consider what I know now, I see that I only know very non-figurative mathematics: analytic geometry (complex varieties), differential geometry and algebraic topology at a moderate level, but not differential topology, nor Thom's catastrophe theory, nor knot theory. I like geometry which is close to algebra or analysis, not visual geometry. It's because of this lack of spatial vision that I never learned to drive. I have no topographic vision whatsoever. Even in Paris, after two or three turns, I no longer have the slightest idea how to find the street I started from. I know by heart the list of metro or bus stops on certain lines, because the name of each station engenders the next one, but I have no idea of their actual situations. I know that if you orthogonally project a point of a circle circumscribing a triangle onto its three sides, the projections lie on a line, called Simson's line of the point, but I don't imagine the figure in my head. However, I can still prove Simson's theorem. I knew how to prove that the hull of the Simson's lines of a triangle formed a hypocycloid with three cusps, but I didn't visualize it. I also can prove the theorem of Pappus, but don't visualize any figure; the proof I gave at my oral exam, which is certainly not the most elementary one possible, doesn't need any. If I don't have a family tree, I have to think a long time in order to represent to myself my exact genealogical position with respect to some distant relation. I could never play a game of chess without looking at the chessboard, but I guess I'm not alone in this. My education and my passion for butterfly-hunting should have cured me of this topographical cretinism, but they never did succeed in curing me of this strong genetic trait, which also afflicts the other members of my family, though somewhat less powerfully. Quite the opposite – an unexpected event during my last year of high school considerably worsened it.

## The music of love

The courtyard of our school was also used by a few girls who were in hypotaupe and taupe. They formed an inexhaustible subject of conversation for us boys.

Among them was Marie-Hélène Lévy, the daughter of Paul Lévy, one of the best mathematicians of the time. She wasn't altogether unknown to me since I had seen her once or twice before, and our families had known each other for three generations, but I knew her sister Denise, who was in my solfège class, a little better. The first time I had met Marie-Hélène, several years earlier, I had really liked her. She was a year older than me and entered hypotaupe when I entered twelfth grade. In spite of – maybe because of – the pleasure of her going to school in the same building as me, I almost never spoke to her throughout the year, but quite soon I was madly in love with her. When I met her, I spoke to her incredibly timidly, while becoming more and more smothered by emotional distress. As soon as I entered the school courtyard, a quick circular glance allowed me to determine exactly where she was. I then avoided her carefully. Thanks to this technique, I only exchanged words with her two or three times during the year. I had only brothers; the lack of sisters made my relations with girls extremely difficult. Sexual oppression and the careful hiding of any sign of love on the part of our parents made our generation totally ignorant. I had almost never met young girls. The idea of going out with a girl never occurred to me, and if I had done it, my parents would have been deeply worried. Once I remember staying in my classroom a little longer than usual, and suddenly a group of hypotaupe students entered; Marie-Hélène was among them. They began trying to solve a hypotaupe problem of projective geometry. While they got stuck, I quickly solved the problem in my head. I was absolutely dying to go over and show them the solution, to impress my Dulcinea. But I was too shy, and for a long time I never ceased to regret the lost opportunity.

The more I loved, the more I was covered with confusion. I tried to hide it from my family, but I couldn't prevent them from noticing it after a while. Every time a word was spoken at table which had anything to do with Marie-Hélène, even remotely, for instance "Marie", "Hélène", "Lévy", or her father's first name "Paul", I felt faint. First I became brick red, than white, then became horribly pale and nearly fainted. Sometimes this happened in front of guests. I knew exactly what was happening to me; the whole cycle took several minutes. Of course, everybody noticed it. My mother always asked me if I was all right, and I always (untruthfully) said yes. My mother and Daniel finally sat together and went through every time this had happened to me, noting the words which had provoked it; by intersecting them, they hit on the truth. My mother used to come see us in our rooms every evening towards ten or eleven o'clock, to see if we were sleeping well. In and after the last year of high school, we'd often still be studying at that hour, and she would just ask if we didn't mean to go to bed too late. We never did keep very late hours. One evening, she came in when I was already in bed, and I gathered up all my courage and blurted out "I'm terribly in love with Marie-Hélène Lévy". She asked me some discreet questions, which I answered, feeling more relaxed. During our little conversation, she let me know that it was a universal truth that one's first love was not the true one. I could go right ahead and love Marie-Hélène, but I should keep in mind that probably she would not be my wife. I would have opportunities to meet other girls, and to have a few love affairs, before finding Miss Right. For myself, I was totally convinced that I had found the love of my life. And indeed, Marie-Hélène did become my wife, although I had to wait seven years. Unlike me, she has a solid sense of spatial geometry and topography. In an unknown town, on country roads, in the middle of any trip, she always knows where she is and never loses her sense of direction. She's also a mathematician, but she sees where I am blind, complex geometric figures, stratified spaces of all dimensions. This is why love made my own cretinism even worse. Why should I furnish incredible efforts to orient myself when she always knows long before I do where to go? I just follow her unthinkingly. I've never even really been able to understand her mathematical work – it's just too visual for me!

My relation with music also climaxed at the beginning of my love affair. In our youth, Daniel played violin, Bertrand cello and I played piano. My piano teacher was Germaine Hinstin, the sister of Jacques Hadamard; she was a remarkable teacher and I have excellent memories of her. But after sixth grade I stopped taking lessons. Impelled by my love for Marie-Hélène, I started piano again in twelfth grade; without a teacher, I slaved over scales and arpeggios, and tried to play all kinds of pieces, preferably Beethoven sonatas. I had perfect pitch, so that I could hear exactly I was playing. But my technical abilities were only sufficient to allow me to play the andantes; the other movements were too fast. I think I played every single andante. I felt as though I was interpreting them with all the sentiment that a Beethoven andante should contain, alternating fortes and pianos. I also played a little Mozart, Bach, Vivaldi and Haydn, but that was about it. My musical culture has remained minuscule. I never went to concerts, quite simply because my parents didn't go, and we never went out alone. My father had adored Wagner in his youth, and I'd also heard that he danced extremely well, but when I was a child, I never saw any signs of this. I did hear a few Beethoven symphonies, which powerfully impressed me, especially the Pastorale and the Ninth Symphony, with its chorus which I found absolutely sublime. But I didn't understand them as I did the sonatas, since I never played them myself. And for good reason! The lack of functioning of the right hemisphere is once again visible here. If someone plays me a little sequence of say 12 notes, I'm practically unable to repeat it, even after several times. However, the left side of my brain analyzes it and figures out what the notes are, and when I know them, then of course I can reproduce the tune and in those days I could even sing it. So, for instance, I could never sing the folk song *En passant par la Lorraine* until the day I figured out all the notes. If I listen to a concert, I can more or less identify the notes, but I don't have time to then put them together synthetically. So I just give up and let myself float along in the sonorous ambiance created by the music, appreciating the melodies without understanding them, and capturing a handful of notes here and there. But it's not completely satisfying to me. I lose my footing rather easily, and after a few pieces of a certain length, I end up falling asleep because I don't have time to analyze what I'm hearing.

Anyway, in twelfth grade, and then in hypotaupe and taupe, I played piano pretty frequently, and then I stopped again when I entered the ENS. I could have gone on practicing – many students did – but it was materially too complicated. The piano was shared by many students, and to use it you had to sign up on a sheet several hours ahead of time. It was totally incompatible with my taste for freedom. So I stopped playing piano at the ENS, and because of ulterior events, I never took it up again. I regret it a lot; I can imagine it would have been a real pleasure to improve, to add to my playing what time would have brought me of culture and humanity, and to play throughout my professional life. My musical life is very poor compared to many of my mathematical colleagues.

In high school, I was lucky to encounter excellent teachers, to study well-balanced programs and to have plenty of free time. I gained a solid basis in geometry that year. And for whatever reasons, it was one of the happiest years of my existence.

## Classical analysis

The two following years were less brilliant, and in my memory they do not leave anything like the same luminous trace. As I've said before, the first year of scientific Classes Préparatoires is called "hypotaupe" and the second year "taupe"; perhaps these names have to do with the darksome activity of the mole, who digs under the earth without ever seeing the light. I entered hypotaupe in October 1932. Our professor, Coissard, was perhaps even more remarkable than Julien, and my memories of him are marvelous. He said "jometry" instead of "geometry", and he preferred it, just as I did; he frequently said "Ah, when jometry takes to being elegant ..." and the sentence stopped there. We used to love his verbal tics: "999 chances out of 1,000,000" and "What is this curve? Pouou!! It's a parabola!" His beginning on the very first day was rather abrupt.

He walked into the classroom, put his coat down and his pipe on it – he always did this, and his coat suffered rather over the course of the year – and started in: "I'm dictating the lecture: Combinatorial analysis." That was his tone. He was friendly and not at all severe, but sometimes he could be abrupt, and he expected hard work. Complex numbers, exponentials, also complex exponentials with Moivre's formula, logarithms, series, computation of integrals, differential equations; all highly interesting subjects. There were determinants – difficult – and solving $m$ equations with $n$ unknowns with Rouché's and Cramer's theorems, not always fun. His explanations were very clear and always purely algebraic, using neither matrices nor vector spaces, which we had not yet studied in dimension greater than 3, whereas here $n$ was arbitrary. Geometry, which I had completely mastered the previous year, did not hold my attention as much as analysis, which I bit like a fish a hook; I became an analyst for the rest of my life. There was almost no analysis in high school, and Julien had mentioned this to me, saying "You'll see analysis next year, and I'm sure you'll love it."

But I had some difficulties. Unlike my experience with geometry, I didn't completely master the program in analysis. At the end of the year, I didn't know enough to feel really at ease. Series were very interesting, but they were taught formally. To see if a series was convergent, we first applied d'Alembert's criterion, then Cauchy's criterion, then the $n^\alpha u_n$ criterion, which is never used in this form. If the series was not absolutely convergent, we then had to apply the unique theorem we had at our disposal on this subject, the alternating series theorem, to see if it was semi-convergent. All this was in the nature of a mechanical procedure. Coissard never told us clearly that for absolute convergence, the only real question is the rapidity with which coefficients decrease, and that all our criterion were based on this. In his memoirs, Paul Lévy says that he understood this immediately, but I certainly didn't. Naturally, we also studied partial sums and everything that can be deduced from them about the construction of curves. I quickly assimilated the concepts of limits and successive derivatives. The competition in this class was more vehement than it had been the previous year. The two-year program aimed at preparing the entrance competition for the Ecole Normale Supérieure (ENS) and the Ecole Polytechnique[11)], so it had to be at the highest possible level. There were two particularly talented students in the class, Neyman and Martinelli, who came from the same high school as I did and who had also gotten ahead of the program by working alone, but

11) These two schools are the most prestigious of the French Grandes Ecoles. Entering them affords the happy student a scholarship, a dormitory room and specially taught courses to obtain his university degrees.

their advance was more in analysis than geometry. In particular, they already understood differential equations. I managed to conserve my position as first of the class, but only at the price of hard labor and terrible nerves during exams. We used to endlessly discuss mathematics and physics and we were excellent friends. My third friend, slightly less close than the other two, was Hue de La Colombe. Neyman wanted to be a physicist. He was a Communist, but didn't succeed in getting me interested in politics.

Coissard was completely at ease in geometry, but not in analysis. When he talked to us about the upper bound of a subset of the line, he gave us the following theorem: "Either the upper bound is in the set, in which case it is called a maximum, or it is not in the set. In the second case, we call it $B$, and then we have the following result on $B$: for any real number $\epsilon > 0$, there is *at least one* point of the subset between $B - \epsilon$ and $B$." He always emphasized *at least one* very strongly. I was quite impressed by this, since it seemed to me that there must be an infinity. But he always invariably said *at least one!* I shared my perplexity with Neyman, who after some reflection answered me "Imagine the set containing a single element, which is the largest possible number strictly less than 3." Of course, there is no such number. But he insisted – "For physicists, this number exists! Then the set contains a single element, and it's clear that for any $\epsilon$, there's only a single element of the set between $3-\epsilon$ and 3." Our controversy on this subject lasted for a long time, but I think I finally convinced him. When I talked it over with Coissard, he became uneasy, in a typical way for a professor who's been teaching hypotaupe for ten or fifteen years when suddenly faced with the difficulties of analysis. "Pouou! watch out for algebra!" he said. Algebra meant analysis, and he meant that I should be careful of following my intuition, and that it could probably happen that only a single element of the set could lie between $B - \epsilon$ and $B$. I had no trouble proving to him that there existed an infinity. But he only replied "Pouou!! watch out for algebra. I'll think about it." The next day he did, however, toss out "You were quite right", and from then on, he never failed to state "Between $B - \epsilon$ and $B$, there exists an *infinite number* of elements of the set". Julien and Coissard both taught us geometry with true maestria, which they transmitted to me. But obviously, for analysis, the effect was exactly the opposite. Coissard was not sure of himself and presented everything in a timorous manner, taking infinite precautions. Thus, he not only gave us the impression that analysis was so complicated as to be absolutely inaccessible, but also that one should beware of it as of the plague. "Pouou! watch out for algebra." I've often noted that a professor who has to teach something he's not too familiar with fills his students with a sense of discomfort which they have incredible difficulty ridding themselves of later on, even if the professor's actual lecture is perfect. But if he knows his subject inside out, then the students will inherit something of this mastery from him.

Neyman and Martinelli both suffered tragic destinies during the war. Neyman was in the Resistance in Nantes. After invading France, the English, French and American troops continued their attack towards Germany, leaving a few pockets in France still under Nazi control. Nantes was such a pocket and Neyman was caught there and shot. As for Martinelli, who had entered the Ecole Polytechnique (ranked seventh) and had brilliantly succeeded there, he was mobilized and made a prisoner of war. He was not too badly treated in his prisoners-of-war camp, but died of an abscess which did not receive proper medical treatment. I lost a third friend, Raymond Croland, who was a student at the ENS at the same time as I was, but in biology; like me, he collected butterflies and was a political left-winger. He was in the Resistance and died in deportation. After the war, I contacted Martinelli's mother and we invited her over several times (I also contacted the parents of Neyman and Croland). She had lost her husband during the first World War and her only son during the second; she ended her life alone, and died after an existence filled with sadness.

The physics we learned in hypotaupe and taupe did not interest me very much. The part called "heat" was a long series of considerations about faucets and Regnault's experiments. The errors of optical instruments did not make for particularly interesting problems. Chemistry was horrible; we studied only the chemistry of metalloids. Of course, I paid attention to Mendeleiev's classification theory, but it was the only glimpse we got of atomic theory. Our professors were very vague on the subject. I did hear "the atomic hypothesis" mentioned towards the end of my taupe, in 1934. And in 1945, the atomic bomb exploded.

While I was still in hypotaupe with Coissard, I made the astute discovery of the following theorem: There is a point in the plane through which every real line in the plane passes. This theorem is obviously wrong, but it is quite extraordinary. When I proved it, the error did not leap to my eyes. It was during one of my periods of personal research. This was the proof. Let $D$ and $D'$ be two complex conjugate lines. Their intersection point $O$ is real. If $M$ and $M'$ are distinct complex conjugate numbers on $D$ and $D'$ respectively, then the line $MM'$ is real. Conversely, every real line cuts $D$ and $D'$ in mutually complex conjugate points. Thus, we have an algebraic bijection between the points of the two lines $D$ and $D'$. So by a well-known theorem, it is homographic. But if there is a homographic correspondence $(M, M')$ between the points of two lines, and if the point $O$ of intersection of the two lines corresponds to itself, which is the case since it is real, then the line $MM'$ passes through a fixed point. So, all real lines passed through some fixed point. I had a lot of trouble finding the error, but when I did, I was really thrilled – I dashed into class and proved my result for Neyman and Martinelli. They both agreed that the correspondence was algebraic, and obviously bijective .... I asked Coissard and he also agreed that the correspondence was algebraic ("c'est évident") and also a bijection, so .... Martinelli and Neyman couldn't find the mistake, but Coissard said "I don't see it, but I'll think about it and tell you tomorrow." And indeed, on the morrow, he had discovered that the correspondence between the complex conjugate points of $M$ and $M'$ was absolutely not algebraic. I proudly showed my discovery to Paul Lévy and Jacques Hadamard, but they immediately said "That correspondence isn't algebraic". Jacques Hadamard was delighted by the whole story, and considered that it showed that the theorem "an algebraic and bijective correspondence is homographic" ought not to be taught to hypotaupe students, who would do better to learn how to show directly when a correspondence is homographic, using results about the stability of the correspondence. This would be more reasonable than studying algebraic correspondences which they didn't understand. He actually went so far as to campaign for the exclusion of this theorem from the program, and to support his case, he decided to publish my theorem, together with its "proof". The theorem was published in the *Journal de mathématiques spéciales*. This voluntary publication of a false theorem was my very first publication. I'm quite proud of it!

I won't speak much about my year of taupe. Within a month, I came to feel the easy comfort with analysis which the whole of the previous year had been unable to give me. I knew already that analysis would become an integral part of my thought. I stopped working on the given program in mathematics at that point, since it was basically just a revision of hypotaupe, and this was absolutely uninteresting to me; furthermore the professor was mediocre. I started studying the program for the Licence[12)], using course notes by Paul Lévy, Marie-Hélène's father, and Hadamard. These courses notes were really well written and precise, and it gave me a lot of pleasure to read and work on them.

Thus, when I entered the ENS, I already knew everything in the analysis and differential geometry syllabi of the Licence in mathematics. I was plain hopeless in descriptive geometry, not to speak of graphic design, for which I

12) The Licence is the third-year university degree. The first two years at university are crowned with the degree DEUG (diplôme des études universitaires générales, diploma of general university studies) or replaced by the two years of Classes Préparatoires, the third with the degree of Licence and the fourth with the degree of Maîtrise. Students then usually spend a year preparing the agrégation. Students at the Grandes Ecoles often combine these five years into three or four.

was just as inept as for ordinary drawing, always due to the same dysfunction of my right hemisphere – if I tried to draw a dog, it was indistinguishable from a potato. I'm happy that at least my descendents have not inherited this trait from me. For descriptive geometry, you had to be able to visualize in space, for instance the intersection of two cones. But I didn't. It happened that some students who were not all that good in mathematics were be extremely good in descriptive geometry, able to visualize and draw complicated intersections with visible and hidden parts. Often they were the same ones who were very good at drawing. But I couldn't do it at all. My homework in descriptive geometry was not very brilliant. In fact, my grade at the descriptive geometry examination for the entrance competition to the ENS was 2 out of 20. We had to find the intersection of two surfaces. I proved geometrically that it was a circle, and I computed its center and its radius. I handed in this paper, in which instead of constructing each point of this curve, I simply drew the center, and then used my compass to draw the whole circumference.

The French system of Classes Préparatoires is unique in the world. Their main virtue resides in the fact that students receive an intensive training for hard, deep and prolonged work. Whoever has been in taupe knows how to compute (except for myself). Even in the third cycle of university, you can always recognize a student who has been in taupe. But at the university, students are not always given the training necessary to attack mechanical computations. Naturally, when I say I can't compute, I'm not lying, but I do always know the basis of the problem I'm considering; I know where I'm going, I don't get my inequalities backwards. I might find $4x^2$ instead of $3x^2$, but I could never accidentally find $4/x^2$. Unfortunately, there are a lot of computations where the main thing is not to make mistakes in the signs $\pm 1$. The mediocre student in Licence, the one who hasn't been in taupe, can obtain the most shocking results without being disturbed, for instance $10^3$ instead of $10^{-3}$, or seconds instead of meters. Taupe gives solid habits of hard work, deep reflection and the readiness to attack boring things with energy and stubborness. Every student is trained to be rigorous, to reason with solid logic, and to write well.

The worst defect of taupe, from which I myself was protected thanks to my curiosity and my taste for personal, untrammeled work, is that it is kind of unique mold, which is supposed to produce all the future scientists of value. It's not possible to treat a whole generation in a unique and single manner, eliminating the ones who are not adapted to this mold. You can see, for instance, that the students who just barely scrape into the Ecole Polytechnique, sometimes after having repeated their year of taupe, don't really have the level to do good work there. Now, if you look for the three or four hundred best young scientists in France, you'll probably find a very high and quite homogeneous level among them; in each age class, France has at least one or two thousand young scientists of a very high level, even if they're not all Nobel Prizes. But if you look for three hundred students who are all good in sciences and able to endure the constraints of taupe, you'll see the level descend from the first to the third hundredth. It's even true if you just consider the Lycée Louis-le-Grand. Louis-le-Grand selects the students which are allowed to follow its Classes Préparatoires. I am in no way opposed to the principle of selection, which is a good thing. But if it is practiced according to one unique criterion, then it can only generate perverse effects. A large proportion of the students who go to the Ecole Polytechnique come from the Lycée Louis-le-Grand, so that they have all been selected and trained according to a uniform method, with no room left for individual originality or efforts at personal scientific research. It's even worse in the humanities, where more than one-third of each class comes from the Lycée Henri IV.

Before passing my oral entrance exam at the ENS, I listened to the exam of the candidate before me. He was asked which algebraic relation had to be satisfied by two numbers $x$ and $y$ in order for some property to be satisfied. He rapidly gave the correct answer. Then he stepped back, reflected an instant and added: "It's interesting – that relation actually means that $x$ and $y$ are harmonic conjugates with respect to the roots of the equation $az^2 + bz + c = 0$," then after a short pause, he remarked: "It should be possible to prove that by a direct geometric method." He then gave the geometric method. He received an excellent grade for his oral exam and was admitted to the ENS. When I left the examination room with him, I congratulated him for his prompt and precise answer to the problem, which was really a small masterpiece. He answered simply: "You know, it's the third time I've been asked that question." This anecdote is absolutely representative of the force-feeding methods used in these classes. The professors of taupe are very good. They are judged by the results of their students at the competitions; if a professor is not good, his students' results will not be good, and it will have consequences for him. The country needs to take advantage of this potential. However, it would be a good thing if these professors had some research experience in their training, like the ones teaching even in the first years of university do. They should have written a Ph.D. thesis or at least a Masters', above and beyond the agrégation. Then there would be a little aroma of research in their teaching. I understand that a professor of taupe has no time to do research, but everything which could be done in the direction of research would be welcome. I've said that I myself did not fall into the typical mold. When I entered the ENS, I had rejected the typical force-feeding and was still a free researcher. Fortunately, most of the people who later end up doing research are quite original people, who are more likely to attend the ENS than the other Grandes Ecoles.

The taupe classes are literally exhausting. Most of the students who graduate from them and go to one of the Grandes Ecoles are so exhausted that they spend the entire first year resting. This naturally diminishes the level of education they receive there. In all the Grandes Ecoles and many smaller ones as well, there are students who work very little, if at all, throughout their entire stay, and yet they receive the diploma at the end, taking advantage of the unique fact of having passed the entrance competition. This is particularly flagrant at the Ecole Polytechnique. The universities of Oxford and Cambridge recruit their students directly after high school, and they select them according to certain criteria of scientific value. The students recruited in this way have absolutely no possibility of not working once they enter the university; they would be excluded immediately. And even if they were not, no industry would be interested in a poor or failed Oxford student with barely more than high school knowledge, whereas French industries hasten to employ graduates from the Ecole Polytechnique and to give them important positions of responsibility, knowing that at the very least, they got through taupe and passed the difficult entrance competition. And if they didn't do a stroke of work afterwards – too bad! In this way, France has trained a number of very mediocre engineers.

I certainly am not suggesting that the taupe classes be brutally suppressed. They work well, which is not the case of the first years of university, and the result would simply demolish the whole system. But I believe they should undergo a certain number of modifications. The Grandes Ecoles should have several recruitment methods, one by the usual competition, and others, for instance, by individual application together with an interview, or any other method which would attract talented students. Diversity of recruitment is a vital necessity. The physicist Alfred Kastler put forth a proposition which nobody followed. He suggested that the ENS should propose to every winner of the Concours Général to enter the ENS immediately, without having to attend the Classes Préparatoires or pass the entrance competition, and that once there, these special students could follow an adapted five-year program instead of the usual three-year program, beginning with the contents of the preparatory classes or the first two years of university, but taught entirely by university professors. I know that the fact of winning a prize at the Concours Général is not always significant, however in mathematics and in the humanities it is certainly somewhat indicative of unusual talent, though this is much less true in the experimental sciences such as physics, chemistry or biology, since the exam is purely theoretical. Kastler's proposition should be modified, but it is a good idea. Moreover, it is interesting to note that girls tend to succeed as well or better than boys in the humanities, but less in the sciences and particularly badly in mathematics where their results are catastrophic. Other classes besides the taupe, together with a variety of recruitment methods, would certainly allow them to show their talents better. The whole question deserves to be studied with entirely fresh ideas.

# Chapter II
# A Student at the Ecole Normale Supérieure in Love

The Ecole Normale Supérieure in the rue d'Ulm was founded by the Convention in 1794. It is different from the other "Grandes Ecoles", which are essentially greenhouses for growing young engineers. Originally intended as a school for the training of high school teachers, its vocation evolved little by little towards the training of researchers, particularly in the sciences. It is a remarkable institution, grouping together young students of a very high level in science and the humanities, and for the space of three years on the average allows them to debate grand problems in a first-class intellectual atmostphere. The Ecole Normale Supérieure in the rue d'Ulm used to be for boys only, while the girls had a separate school in Sèvres, and later in the boulevard Jourdan. Since then, new Ecoles Normales have been founded, in Fontenay-aux-Roses for the humanities, in Lyons and in Cachan for science. So there are a lot more "Ecole Normale students" than there used to be.

The role of the ENS, which turns out a multitude of scholars in various domains, has always been of considerable importance in France. One came out of the ENS as a mathematician, a physicist, a chemist or a biologist, with the mathematicians forming the largest group at the time I attended the school. Until about thirty years ago, almost all French mathematicians came from the ENS. That isn't the case any longer, except for those of the very highest level: all those who have obtained Fields Medals are from the Ecole Normale, except Grothendieck whose studies were irregular because of Nazism and the war. However, the Ecole Polytechnique and the universities now turn out a sizeable group of research mathematicians. At the Academy of Sciences, there are nine former Ecole Normale students among the fourteen members of the mathematics section, ten among the twenty-four members of the physics section, seven among the twenty-four in mechanics and eight among the seventeen of science of the universe. In chemistry however, only two of the thirteen members are alumni of the ENS; very few of the biologists are, and no one in the medical section. This enumeration is not in itself interesting; I never actually try to find out whether a given colleague is or is not an alumnus of the ENS. Its value lies in the indication it gives of the level of the school.

In Preparatory Class, I had decided to attempt only the entrance competition for the ENS. At that time, they admitted twenty students in literature and twenty in science each year. There were about four hundred candidates in science. The competition was thus significantly tougher than at the Ecole Polytechnique, where they accepted about two hundred out of two thousand candidates. I wasn't at all certain of succeeding, but still I didn't try any other competition because I didn't want to be an engineer. The situation is different today since you can very well become a researcher even if you go to the Ecole Polytechnique. So I was taking a risk. I didn't find my second year classes very interesting, so I decided that if I failed the competition, I wouldn't redo the year, I would go to the university. My destiny would probably have been somewhat different, but there I would have been led to frequent the students of the ENS, then later I could have attended their seminars, and I'd have probably ended up much the same way.

I was always weaker in written tests than in oral ones. I like to be able to think while moving around, breathing and oxygenating myself. Sitting for four hours amongst hundreds of other candidates gives me a feeling of imprisonment. Still I succeeded very well in the first exam, with a grade of 18 2/3 out of 20 which put me in second place after Queysanne. There was an error in the statement of the problem. In the last question, we were supposed to prove that a certain curve was a conic; I showed easily that in fact it was a Pascal limaçon. But the series of four-hour tests tired me out more and more and my results showed it. By the end of the written exams, I was down to seventy-second place. Fortunately, I didn't know it! Coissard said he knew the results, and to give me confidence he told me I was third. I was surprised, but prudently chose to believe him. I needed to pass an excellent oral exam to improve the low ranking which had been so charitably hidden from me. But Coissard knew I was good at oral exams. In fact, I succeeded so well that I got a first place in the oral exam. Coissard then went to see one of the members of the jury. "I've never been to see you before to ask for special consideration for any student. But this time I'm doing it. Laurent Schwartz is a first-class student; the best I have ever had, and it is necessary, I mean absolutely necessary to admit him to the ENS." They apparently heeded his message. I was accepted twentieth out of twenty-one (Marie-Helene was accepted too – but girls were not classed in the list of twenty-one). Probably the jury augmented my grade at the oral exam a little in order to accept me. That year, only one single student refused his admission to the ENS in order to attend the Ecole Polytechnique. I just barely scraped in! Strictly speaking, from the logical point of view of the competition, there may even have been some injustice. However, the rigid system applied at the

Ecole Polytechnique, where they simply add up your points without thinking, isn't satisfying either. The vocation of the ENS is to train the best scientific researchers in the nation, and they do care about whether the candidates have the right qualities to fulfill this role. Collectively, the jury does take these factors into consideration as well as the number of points obtained at the exams; the slight distorsions of the results produced in this way are all for the good of the school and of the country. My second place in the first written exam sufficed to cause one of these distorsions, and Coissard's intervention clinched it. It's interesting that it was Coissard who intervened, rather than the professor of my second-year class, who wasn't interested in my results at all.

There are equivalents of the ENS in foreign countries: "Oxbridge" in England, and some of the great American universities: Yale, Harvard, Princeton, MIT, the Courant Institute and Columbia in New York, and the University of California at Berkeley, Stanford and Caltech, etc. The Ivy League is the union of the best universities on the East Coast. I'm probably forgetting some. The atmosphere in these places is comparable to that of the ENS. Students are accepted, not by competition but by careful examination of applications consisting of the entire scholastic record of the student, as well as interviews. Personally, I find this system, which doesn't try to give a linear classification of the applicants but tries to distinguish those who are best adapted, as better than the competition system. A competition necessarily forces the students into a kind of stereotype. In the US, there are about seven thousand students at MIT alone, and there are about the same number in the Polytechnic School in Switzerland. These universities can be proud of their level, largely because of their diverse approach to accepting students.

In any case, one's entrance rank at the ENS or the Ecole Polytechnique doesn't count much afterwards. It's the result of too many chances which don't really reflect reality. Everyone knows about the role played in classes by certain students who act as "locomotives", the best students who pull everybody else upwards. In a class which is too homogeneous, many talents never really bloom. There were locomotives in the ENS, too, although in fact all the students seem to pull each other ahead. In my year, three particular students emerged: Gustave Choquet, Raymond Marrot and myself. This was quickly noticeable to everyone. Choquet had won first prize in the Concours Général in mathematics, and he entered the ENS as "bizuth inté", which means that he had skipped the first year of the Classes Préparatoires and gone directly into the second. Marrot had been seriously handicapped by polio. He was incredibly energetic. His knowledge of analysis was enormous, much greater than mine and Choquet's, already when he entered the ENS, and even more so afterwards. He really knew how to read and assimilate things very well. He initiated me to the theory of classes of quasi-analytic functions, of which I knew nothing. After his thesis, he obtained a position in the Scientific University of Bordeaux. Some time later, he died tragically: the gas jet remained on in his apartment, and he and his mother were both asphyxiated during the night. We were close friends. I was also very close to him politically, and we often had interminable discussions; I was extremely sad when he died. An investigation showed that his death was the responsibility of an employee in the Gas Company. The Company was sued and the case was won; it was defended by the lawyer Pierre Stibbe, a very talented and honest man who became famous later on because of his role in the Algerian war. There were also outstanding students in physics (in particular André Blanc-Lapierre and Noël Félici, who had been first at the entrance competition). It was a first-class group.

The competitive spirit, which had motivated me to try to be first in my high school class, disappeared when I entered the ENS. Never again did I feel tempted to adopt that attitude or to establish individual comparisons. I knew I belonged to the best group, but I didn't really care about wondering who else was in it, or asking myself if other mathematicians were better than I was. The great mathematician Norbert Wiener once abruptly asked his colleague Mandelbrojt: "In your opinion, who is the better mathematician, Hardy or myself?" Such questions appear to me painfully foolish and fortunately not very representative. Competition in high school was useful to me to overcome my doubts about myself; I was complexed about my slowness. But once at the ENS, I didn't need to reassure myself any more and turned my back on what I considered a primitive and warlike state, already over and done with in Classes Préparatoires. It seemed necessary to me to suppress a too fierce competition. It's only beneficial in the softened form of "healthy emulation", which leads everyone to give the best of himself. It becomes frankly negative and one should avoid it like the plague when it actually becomes an obstacle to scientific collaboration. A scientist who publishes his results gives them to the community and allows others to go further. Science gives rise to its own type of particularly interesting habits of collaboration. However, even if the competition which reigned in the high schools of my youth was especially exaggerated, to the point of poor taste, I do find that things have now gone too far the other way in elementary and high schools. Everything is dominated by a small-minded egalitarianism, forbidding any student to stand out amongst the others. Yes to modest competition, no to ferocious competition, definitely no to no competition at all. Studies by Bourdieu have shown that children from the working classes suffer from a handicap which has two origins: the low cultural level of their circle and *the almost complete absence of ambition.* The latter is also frequently encountered in the upper classes of society. Personally, I believe that this factor is more decisive than the cultural one. A student from a modest social background vitally needs to feel a certain type of emulation, and the encouragement of his professors, to assert himself and to succeed. France today is training large numbers of young people without any kind of higher goal – in fact the aspiration towards a fictitious equality even leads us to cut off any heads which stand out above the others. The existence of higher levels is a locomotive which pushes everything upwards, whereas egalitarianism levels things out from below. In the Greece of Antiquity, the Olympic Games were a remarkable motor for intellectual development. They were not limited to sports as they are today, but included also poetry readings, and victory was not rewarded by money, but by a laurel wreath. Similarly, in Asia under Chinese influence, the competition for mandarins, while only selecting a few winners in each village, incited everyone to study.

The students at the ENS recognized each other in the courses at the Sorbonne, by the tiny yellow notebooks given out by the school. Many students actually never went to the courses for their Licence, and just worked by themselves at the school. We read the big books by Goursat, Picard, etc. I must mention the books of the "Borel collection" written on all kinds of subjects. They were written in a simple, intuitive, easy style, more sentimental than rigorous – they were hardly even divided into chapters. You could find intriguing sentences beginning "We proved earlier ..." but where? You had to search for the proof, do a little astute detective work – sometimes, it must be said, in vain. You also might read: "It is true that the theorem proven earlier is not identical to the one we are using here, but the reader will easily adapt the proof." Not always!

At the Sorbonne, we took courses which would be called "third cycle courses" today. Among the professors, who were all very good, were Maurice Fréchet, Paul Montel, Emile Borel, Arnaud Denjoy, Gaston Julia, Elie Cartan and at the Collège de France, Henri Lebesgue and the Hadamard seminar. The Hadamard seminar replaced his course, and mathematicians from the whole world, mostly already professors, came there. I only went at the beginning; I was discouraged by a very recent article by Polya which Hadamard had given me to lecture on. I was used to books, to complete course notes, not to this form of publication, however indispensable to a scientist, and with which of course I eventually came to feel at ease. I wasn't used to articles which condense a subject into a dozen pages or so, and tried in vain to understand why it was interesting before giving up. I shouldn't have stopped there; I ought to have gone back to the seminar in my second or third year.

We also had "seminars" at the ENS, given by the professors at the scientific universities of Paris, sent to the ENS temporarily. It's a tradition that the ENS has no professors of its own (and it's a good thing). These seminars were reserved to students at the ENS and a few auditors, and their subjects were outside of regular course work and sometimes really difficult. I profited from the seminars of Georges Valiron, René Garnier, Joseph Pérès, Francis Perrin and Georges Darmois.

Quite soon, Choquet and I thought of having seminars for our class. Mine was organized around the theory of harmonic functions, potentials, Dirichlet's problem and Lebesgue integrals, research themes which I've never abandoned since. Choquet organized his around Baire classes and the fine theory of sets, in which he's remained interested ever since as well. Another great subject of enthusiasm was the theory of entire functions, which we studied in Valiron's seminar. We particularly studied results on the order of entire functions; there is a whole set of extremely stimulating theorems on which the decreasing of the coefficients of their Taylor series, the growth of the maximum of the module on concentric circles, the distribution of zeros, etc. Poincaré and Hadamard collaborated in this, but I owed the best proofs to Paul Lévy. The canonical product forms part of this theory, et Hadamard's famous theorem of the minimum, which roughly speaking says that there exists an infinity of circles centered at the origin, with radii tending towards infinity, on which the minimum of the module is nearly the inverse of the maximum: an undoubtedly magnificent result.

And isn't Picard's theorem, stating that a non-constant entire function takes every value (except at most one) one of the most beautiful theorems of mathematics? Its proof using the modular function takes only a few lines, and made Picard world-famous overnight. He was still alive in those days; he died in 1941. I learned this theorem in the gardens of the ENS from Dufresnoy, a mathematician of the class of 1933, who died quite young.

Montel, Picard, Hadamard, Lebesgue, Riesz, Hilbert: we were actually learning theorems of contemporary mathematicians! That had never happened to us before. This wasn't due to the glory of higher education; rather it was a serious weakness of secondary education, which has disappeared today. The role played by students of older classes in the education of the younger ones is not the least of the virtues of the system of Grandes Ecoles; the phenomenon could never exist within the heavy administrative machine of the university, at least not until graduate level.

The life of the ENS was a marvel for a young person of my temperament. In one blow, the field of mathematics became infinitely wide. Nevertheless, my impression of my own progress while at the ENS was that it was much more laborious and less fabulous than it had been in high school and in the Classes Préparatoires.

## An invention: finite parts

I did however make one discovery which counted a lot for my future: the finite parts of divergent integrals. If you consider an integral like $\int_a^b (x-a)^\alpha g(x)\,dx$ where $g$ is a regular function, and if $\alpha \leq -1$, in general it diverges. But you can find the principal part as $\epsilon$ tends to 0 of the integral from $a+\epsilon$ to $b$; it's a monomial in $\epsilon$ (or a term in $\log \epsilon$), together with a finite part of the expansion of the integral in powers of $\epsilon$ (with possibly a logarithmic term). The beginning of the expansion has terms with negative powers in $\epsilon$ and a term in $\log \epsilon$, then there's a constant term, and then positive powers of $\epsilon$. The constant term is called the finite part of the integral, written Pf $\int_a^b (x-a)^\alpha g(x)\,dx$. This finite part has remarkable properties with respect to change of variables, convergence, analytic continuation. We only learned a little bit about it, for instance as Cauchy's principal value vp $\int$.

Now, Jacques Hadamard, to whom I showed my discoveries, had actually discovered these things himself, and even in several spatial dimensions, and had used it quite remarkably in his solution of hyperbolic partial differential equations, which he was working on nearly alone (apart from some foreign mathematicians such as Florent Bureau in Belgium and particularly Marcel Riesz in Sweden). So, to my great disappointment, I hadn't produced anything absolutely new. I still had to learn that unlike previous centuries, the early discoveries of young people now most frequently already belong to the enormous mass of existing knowledge. I didn't sulk however, and felt some pleasure in having rediscovered important ideas of Hadamard in this promising start of my research activities. Given the close relations between the future theory of distributions and the theory of partial differential equations, it's not surprising that finite parts played an important role throughout the theory of distributions; analogous notions are used in renormalization in quantum physics and more recently, in probability. The concept can be naturally extended to the finite part of the sum of a divergent series, using the Euler-Maclaurin expansion, where it gives some beautiful results. For instance, if you consider the Riemann $\zeta$ function $\zeta(s) = \sum_{n=1}^{\infty} 1/n^s$, this form is only valid for $\mathrm{Re}\, s > 1$, but the same series makes sense for all values of $s$ if you consider Pf $\sum$, and it gives the analytic continuation of $\zeta$ to all complex $s$ except $s = 1$. At $s = 1$, $\zeta$ has a pole of order 1 and Pf $\sum_{n=1}^{\infty} 1/n^s$ is equal to the Euler constant $C$, so not to

$\zeta(1)$ which has no meaning, but to the limit as $s$ tends to 1 of $\zeta(s) - 1/(s-1)$, which we can call $\mathrm{Pf}\lim_{s\to 1}\zeta(s)$ or even $\mathrm{Pf}\,\zeta(1)$, so that $\mathrm{Pf}\,\zeta(1) = \mathrm{Pf}\sum 1/n$.

In working with this, I wasn't dealing with partial differential equations, which I knew very little, but with analytic continuation. I had talked all of this over with Marrot before discussing it with Hadamard, and I remember he said to me: "Oh, then you already have a thesis!" I remember thinking he was really exaggerating!

By the end of the year, I had completely lost the feeling of superiority over my professors that I had always felt in high school. On the contrary, they seemed to be much better than I was, and this feeling persisted throughout my time at the ENS. However, I didn't feel that in dealing with them, I was dealing with unapproachable genius. Between these two realities, I felt a great sense of lack, of frustration. French science seemed to have gone to sleep after the First World War. A generation of young people had been decimated, and students of the ENS had not been spared. Entire classes had been sent to the front lines where half of their number was killed, and France was having great difficulties recovering from their loss.

Among the professors at the Sorbonne, apart from Elie Cartan, only Denjoy really emerged as extraordinary, but he was as anti-pedagogical as it was possible to be, speaking into his beard, always turned towards the blackboard, and totally uninterested in knowing whether he was being comprehensible. They used to say about him: "he thinks A, he says B, he writes C, and in fact he's talking about D." Without really being able to express it, I think we all felt a certain frustration. Our professors were good, even very good, but they were not the giants of the nineteenth century or even those of the preceding decade, or those of today. It wasn't that they lacked time for research, those privileged professors at the Sorbonne. They only taught one semester a year, two hours a week, didn't have any administrative work, didn't even direct assistants to take care of the students, who were left completely to themselves. They simply didn't seem to see what research they could do. Mathematics appeared to be a finished subject. We were emerging from a period in which great progress had been accomplished, and it seemed that there wasn't much left to do. And we didn't even have an idea of that little bit. We were ignorant of entire branches of mathematics, because there were no locomotives among the mathematical scholars around us to give us information about them. The train was moving in slow motion. I felt that I could become a researcher, write a thesis, create something in mathematics, but nothing really exalting suggested itself. The future seemed closed. I never suspected the existence of the Bourbaki group, which had started working in 1935, but I can say that what was lacking for me was Bourbaki. I needed their inspiration to really become a mathematician. I clearly perceived something lacking in everything we learned; the absence of some unifying thread. All this changed in my second year, thanks to my meeting with Paul Lévy.

In contrast to French science, which if not absolutely mediocre was progressing rather slowly, Germany experienced a prodigious scientific development after World War I. During the war, Germany kept back its scientists, who were set to work for the military. After its defeat, it showed much more dynamism than France. Between 1930 and 1932, German science was at its zenith. The best mathematicians in the world were working in Göttingen. The French not only didn't realize this, but were entirely uninterested in everything going on on the other side of the Rhine. The scientists themselves suffered from a case of general chauvinism which continued long after the war. The separation between France and Germany was deep. At the first International Congress of Mathematicians which took place in 1920, and then again in 1924, the French opposed the invitation of German mathematicians to the Congress. Only in 1928 were mathematicians from other countries able to overcome French chauvinism. Hilbert was one of the greatest mathematicians who ever lived, yet his work was almost unknown in France. The remarkably modern book *Modern Algebra* by van der Waerden was not available in France. In 1907, Hilbert showed that the space $L^2$ was complete; the generalization to $L^p$ by Fischer-Riesz came the following year. During a Hadamard seminar, in 1924, we realized that not a single mathematician there, even the most famous ones, knew if $L^2$ was complete or not. It's not possible to do good analysis – it's not possible to walk in the street not knowing that! The funny thing is that the young Polish mathematician Stephan Banach was there; a few years later he would publish his famous book on linear operations, and he knew all that very well, but he didn't say a word: he must have been intimidated. The admirable book *Die Idee der Riemannschen Fläche* (1913) by Hermann Weyl describes in detail the notion of a differential variety by charts, well-known to van der Waerden but unknown in France, and which Bourbaki only picked up during the war. Far ahead of France, England or the United States, Germany captured most of the Nobel Prizes. In the twenties, atomic theory was developed in Denmark. Niels Bohr travelled throughout Europe to talk about his new revolutionary theory of the atom. Dirac's discoveries, in England, of spin and the electron date from 1928. Heisenberg and Schrödinger worked in Germany. Most of wave and quantum mechanics was also developed during the twenties. Relativity dates back to 1905. The French general public was told that it was so complicated that only Einstein could understand it. It never occurred to me to try to learn it at the ENS. In fact, it was only taught several years after the war – today it can be found (I've taught it myself) in the third year university program. It's true that France saw the so-called "discovery of fire", the wave mechanics of Louis de Broglie, who discovered the dual wave-particle nature of matter (1924; he received the Nobel Prize for this in 1929). But in his courses and lectures, he was so condensed that it was difficult to follow. This doesn't lessen his greatness as a scientist. Unfortunately, he let himself be surrounded by a group of courtiers of no real value and didn't continue in a fruitful direction – another aspect of French torpor. Scientific work was more rigorously written in Germany than in France, as well. In their books, you didn't read about "the plane" or "space", but about vector spaces of arbitrary dimension.

German science collapsed brutally under the effect of Nazism. The flight of the liberals who couldn't bear to support the regime worsened the effects of the exile of Jewish scientists. This disaster is partially the origin of American science. Receiving the flow of German scholars, America soon became the most advanced country in the world. Many universities in Canada and Latin America also benefited from the arrival of so many German and European mathematicians.

However, a renewal was beginning in France already before the war. Artificial radioactivity, discovered by Frédéric and Irène Joliot-Curie (1934) earned them the Nobel Prize in 1935. Jean Perrin had received it for computing the Avogadro number for Brownian motion in 1926. In mathematics, André Weil, Henri Cartan, Claude Chevalley, Jean Dieudonné, Jean Delsarte, René de Possel, Jean Leray and Szolem Mandelbrojt had begun a formidable job of renewal. I entered the ENS in October 1934, knowing absolutely nothing of these mathematicians, except for de Possel and Leray, since I took their "Peccot courses" in the Collège de France. I believe that they were doing exactly what I was hoping for without realizing it. Furthermore, these mathematicians profited from the last remains of German science thanks to André Weil who had spent a few years across the Rhine around 1926, when chauvinism was still virulent. Amazed by the level of mathematics as it was practiced there, he had returned to France with a large harvest of new ideas. I think Bourbaki would never have seen the light of day if André Weil hadn't spent those years in Germany.

## Life at the ENS

Independently of a certain general dissatisfaction which I felt all through my first year, certain non-negligible circumstances shed light on my problems. When I entered the ENS, I went through a period of confusion which lasted at least four months. I felt lost. I'd always been well-protected in the cocoon of my family, and suddenly here I was tossed into a very particular style of student life, among people I didn't know. The rules were different from those of high school. The complete absence of any weekly schedule forced us to organize our work by ourselves. I started by deciding to fix a strict schedule for myself, day by day, hour by hour. That was obviously a total failure. Then I tried another method consisting of studying a different subject every week. Very soon, like most of the other students, I stopped taking the courses at the Sorbonne because they were too easy. We had to read books and organize ourselves. In high school and Classes Préparatoires, I had been fairly autonomous, but within a strict framework, not left completely to myself.

I was allergic to dormitory life. We slept in one big dormitory which didn't consist of individual rooms, but cubicles opening onto a big corridor, from which we were isolated only by a curtain. Life in the ENS was very free, the doors were only locked between one and two in the morning. Students wearing heavy boots were coming in and going out all night long. You could get up when you liked, though not too late if you cared about having breakfast. The meals were all right, often even quite good, but the bread was terrible, a thick grey bread with doughy insides, not at all like the delicious fresh, crusty bread I was used to at home. The discontented students shouted "Death to the pot!" The "pot" was the word used for the bursar but also for the meals. "Having a pot" meant having a beer or a coffee in a bar; this expression has actually passed from the ENS to the general public! The laundry lady of the ENS, who worked for the pot, received the charming name of "hippo-pot-dame". This lady was in charge of having our sheets and handkerchiefs made up out of white material we brought from home: we didn't have uniforms. I remember that we had to bring 1-meter by 1-meter pieces of material for the handkerchiefs, and 90 cm by 6 m 50 for the sheets. Some students received a scholarship of 300 francs a month, but my family was too well-off for me to be eligible. Today, the students at the Ecole Polytechnique and the ENS are very well paid, independently of their family's income; they receive six or seven thousand francs a month. The toilets were just holes in the ground, which I always found highly disagreeable. (Well, as an adult travelling in third world countries, I encountered a lot worse.)

Groups of four to six students from a given class were given a study to work in, called a "thurne". The floors were not waxed; they were dusty and had a musty smell. I didn't like working in the same room with others – I didn't like working shut up in a room at all. It got worse and worse; I lost the desire to work. There was the school garden, where I sat to work in beautiful weather, but it wasn't sufficient to make my life at the school a success.

In the middle of the courtyard garden there's a little fountain with goldfish swimming in it, which we had called the Ernests. The fountain of the Ernests was often a most interesting place, thanks to Delange, from the class of 1932, who used to run round and round it at top speed without ever falling. To combat centrifugal force, he used to have to lean way inside as he ran. Delange also liked to take apart and put together clocks in his spare time. One day, he did this to the great clock of the ENS, but after his ministrations, it took to running backwards: the caretaker nearly went mad. The ENS pranks, which we called "canulars", amused us a lot; some of them have remained famous, but they weren't sufficient to fill our existence.

The most famous canulars had to do with the entrance exams. We would get false examiners, older students who would take the place of the real examiners, and invent a day for the tests. Naturally, the false examiner enjoyed tormenting the student who thought he was passing a real exam. The student would come out utterly depressed, but then we would tell him it was all a fake; usually he would end up laughing, which had a salutary effect. That happened with Chevalley. We also recruited fake students: we'd get young people from outside the school and send them in to the real examiners. They would answer the questions with utter nonsense and miserably fail their exams. The real examiners were used to canulars and didn't make too much fuss when they realized what was happening. Once we actually sent a fake student in to a fake examiner, each thinking the other was real. The fake examiner asked absurd questions; the fake student gave ridiculous answers. The best thing was when we confronted a real examiner with a real student, having informed each of them that the other was a fake. We certainly had plenty to talk about in our spare time!

In spite of these entertaining aspects, I really found life in the ENS difficult. Finally I asked to be allowed to live at home. They granted me permission to do this, and I didn't come to the ENS that often any more. The following year (1935–1936), my parents moved, because Daniel, who was two years younger than I was, was entering the Classes Préparatoire to prepare the entrance competition of the Ecole Polytechnique. He switched from the Lycée Janson-de-Sailly to the Lycée Saint-Louis and my parents decided it was necessary to live closer to there, which was perfect for me. We lived at number 70, rue Madame, and one of our windows gave onto the rue Assas, just over a bus stop. Daniel used to fill one of the rubber pears we used to push around the balls at Nicholas billiards with water and pour it out of the window onto the heads of people standing on the open platform of the bus, just as it was pulling away.

I was a bourgeois at the ENS, amidst students coming from backgrounds mostly very different from mine; I felt unstable and precarious. Politically, my parents were more or less on the right; they read *Le Temps*, a right-wing newspaper, and *Le Matin*, even more right-wing. I knew nothing about politics. When I was in my last year of high school, someone asked me if there was a lot of politics in school. I said no, just a few Communists selling *L'action française* at the door – only that happened to be the royalist newspaper! I was impregnated with my family's ideas, convinced for instance that the responsibility for starting World War I was exclusively Germany's, and that an entirely innocent France was struggling to uphold liberty in the face of barbarianism. It's true that the rising wave of Nazism in Germany supported this view, which I had always heard. I thought that France's colonial activities were spreading the word of civilization to the backward populations of darkest Africa.

And suddenly, I discovered a completely different reality. Not the story of the progress, particularly in medicine, which the great empires certainly did introduce into the colonies, but the story of the pitiless oppression of the natives, the arbitrary imprisonments, the executions, the massacres. A universe collapsed around me and I felt utterly confused. I began by separating myself from the friends who told me these things. Then I thought about it, and little by little, I returned to them; after a few months we were even closer than before. I had to revise all my prejudices about colonialism and the war. As soon as I was strong enough to do so, I evolved naturally towards the ideas of my fellow-students. The ENS recruited more democratically then than it does today. In our class there was the twenty-six year old son of a worker, Mathieu. The attitude of the public high schools at that time was to push elites to the fore, in the hope of raising the level of the whole class and discovering brilliant talents in the working class students. And many working class students, who would never be discovered now, profited from this merit system. Over the last thirty years, there's been a real improvement in education thanks to the fact that schooling has become mandatory till age sixteen; yet the renewal of the elite seems to me to be less effective than it used to be. The proportion of working class students in higher education has not increased as we hoped; in fact the proportion of students in the ENS whose parents are professors has increased. Research is becoming hereditary!

## In love

There was something else, just as fundamental, which contributed to diminishing my capacity for work. Marie-Hélène had repeated her "taupe" as she hadn't been accepted to the Ecole Normale after her first year there. The second time around, she succeeded, and thus found herself in the same class as me. One of my dearest desires was thus realized. At that time, there was an Ecole Normale for boys and another one reserved for girls, with separate entrance exams. Quite frequently, students from the one school married students from the other. But a few girls chose to pass the entrance exam to the boys' ENS and entered the rue d'Ulm. There were never more than half-a-dozen or so of these altogether. Marie-Hélène was one of them. To say that I let myself be submerged by emotion would be an understatement. I had adored Marie-Hélène in silence all through high school, and now she was not only in my class, but she had spontaneously chosen to be in my "thurne". The tutelary divinities had arranged things extremely well. My love was about to spread its wings. I saw myself assiduously courting her, modestly but deliciously, all the more light-heartedly as our families were old friends.

One day I couldn't keep it in any longer and confided to my mother my intention of declaring my love for Marie-Hélène. I wouldn't have dreamed of declaring anything without telling my mother beforehand. The two mothers were indispensable. I suppose this may cover me with ridicule in the eyes of my younger readers, but sixty years ago, it was quite normal, especially in the bourgeois society I came from. Marie-Hélène's father was the mathematician Paul Lévy, whom I have already mentioned. My mother went to see her mother, and informed her of my feelings and my desire to marry her daughter. When her mother solemnly announced to Marie-Hélène that she had received an offer of marriage, Marie-Hélène understood from her tone that it was serious. When she learned that the proposition emanated from me, she reacted with a pleased amazement which could hardly escape her surprised but sagacious mother. It turned out that no other than myself was the lucky object of her dreams, but she had considered this dream quite removed from reality. Certainly we were acquainted, but we had never had a real conversation. I was nineteen and a half, and probably looked completely absorbed in my studies, and very naive for my age. She was twenty-one and already quite mature. In any case, the day after learning of my proposal, she felt that I was certain to eventually become her husband – and yet that we were not yet ready for a full-fledged engagement.

She accepted to meet me, but alas, during the eagerly-awaited meeting we were as awkward as could be: I stammered out a short, embarrassed sentence, and she answered that she didn't feel ready to take such a decision, without revealing any of her feelings about me. I took this reply to be a definitive refusal and left utterly crushed and miserable. I had been dreaming about her for three years.

That day, our mothers did everything necessary. Mine had telephoned hers in the meantime; hers already knew that her daughter had given me a kind of no, a no which according to her meant yes. It was simply necessary to wait for the situation to ripen. She was not ready, said her mother, but it was nevertheless the happiest day of her life – and Marie-Hélène later confirmed this. The happiest day of her life was the one on which she refused me! I was just a boy and couldn't understand it. But between them, the two mothers succeeded somehow or other in giving me back my courage, and all things considered, I never regretted following their advice to be patient. From that time on, we saw each other all the time.

Often, we walked together to Marie-Hélène's house, rue Théophile-Gautier in Auteuil, carefully avoiding any mention of my love or our engagement. That is when we actually learned to know each other. We often stopped at the Louvre. Our classmates naturally thought we were already engaged. Only we were not aware of any such engagement. Several times I asked my friend Revuz: did he think Marie-Hélène was in love with me? "It leaps to the eye," he would answer. But I was perfectly blind. By and by, Marie-Hélène stopped seeing me as a youth, but came to think of me as her future husband. Now it was she who undertook to move things along. In March of 1935, her parents invited me to spend Easter vacation with them in a village in Normandy called Hennecqueville, and I understood that the long-awaited moment was approaching.

## The light of Venus

We became engaged in Hennecqueville in April 1935. Together, we took agreeable walks in the forest. During one of these walks, I remember suddenly surprising a fox. Her parents wondered when we were going to come to a decision. But we never actually needed to speak openly. One evening, we were sitting side by side, on a point jutting out over the sea, and we drew together, deeply moved. We had just become engaged, without speaking a word. The reader could laugh at this point, if it were not for an amazing phenomenon which occurred that evening, and which I have never seen again since. A long and luminous ivory streak clove the sea, striking towards us from a great distance. It was night – I searched for the moon. No moon. But the silvery path led straight to Venus. The light from Venus cannot possibly reach Earth; it isn't strong enough. But on the sea, its faint light is reflected on each and every wave, and it can leave this impalpable trace; thus we were engaged by the light of Venus. I never remember this small miracle without deep emotion. It doesn't take much perspicacity to note that we were thenceforth under the protection of the goddess of love.

My parents and my brothers came to spend a day with us. The engagement was sealed by the ritual family kisses. My uncle Jacques Debré warmly con-

Schwartz and Marie-Hélène

gratulated me on my return to Paris. "Wait," I told him with my usual candor, "it isn't official yet." He looked at me ironically and laughingly replied "Ah! I don't suppose you ever intend to make it official?"

Fifty years earlier, Jacques Hadamard had become engaged in quite similar circumstances. He was very much in love with Louise Trénel, known as Lou, who was the sister of my maternal grandmother. But he didn't dare express his feelings. Lou, for her part, was in love with him, but it was impossible for a young girl to speak first. Finally, Lou had enough of waiting and began to think of getting engaged to someone else. When Jacques heard about it, he was desperate. On his mother's advice, he raced to Lou's house, declared his love for her and asked for her hand in marriage. They were engaged immediately and married soon after. Uncle Jacques and Aunt Lou were a marvelous couple. But customs had not changed much between their generation and mine.

We did end up announcing our engagement to our classmates, who all thought that we had been engaged for ages already. I found myself together with Marie-Hélène for the physics lab classes. We were so hopeless that the young preceptor who helped us finally declared "Those two are the worst idiots I've ever met." I am not proud of this, but I can't help finding it funny. My colleagues who are physicists know that I am one of the mathematicians the most deeply interested by physics; they would be surprised at my ridiculous behavior in a physics lab. Once, we had to manipulate an electrical arc. A negative and a positive conductor have to be brought together and briefly touched, then separated; a big spark bursts out and then they become joined by an electrical arc which remains present even if you pull them quite far apart. I was bringing them together; they touched, and like an idiot I stood there holding them together. The result was a giant short-circuit which burnt out all the fuses over a large area (my schoolmates claimed the whole arrondissement, but I think they were exaggerating), and the electrical wires burst into flames which reached to the ceiling. Fortunately, Marie-Hélène had the presence of mind to leap over and shut off the electric switch.

We were very happy for some time, then the situation became difficult because our parents were insisting that we remain engaged for three years, until Easter of 1938, chastely it goes without saying. They were certain that if we were married, and perhaps had a child, we would not pass the agrégation. We couldn't imagine living separately, at our respective parents' houses, for all that time! We were to pass the agrégation in June and July of 1937, but right after it I would have to take up my military service, which would begin with six months of special training in Metz, to make me a sub-lieutenant in the reserves. We absolutely didn't want to wait until Easter of 1938 to get married. Relations between our parents and ourselves cooled. The tension began almost as soon as we became engaged. Engagement, to Marie-Hélène, meant marriage and children as soon as possible. She was mature and was not afraid of having a child before the agrégation. Of course, nowadays young people decide when to get married without even talking it over with their parents – or they live together without getting married, or sometimes they even get married and don't bother to tell anyone about it until later on.

After taking a vacation with Marie-Hélène's parents, in the Dolomites – it was the first time I had ever taken such a long train trip alone and I was twenty – we obtained their permission to marry in December 1935. Only two more months to wait! But suddenly, as we returned to the ENS in October, Marie-Hélène fell ill with pulmonary tuberculosis. At that time, everyone was terrified by this illness. I caught a primary infection from her, whose main symptom was a state of exhaustion which lasted two months. Most people go through this minor type of tuberculosis, which changes the skin-test reaction from negative to positive, at a certain age. Usually, it has no consequences and simply tires one out for a short period, but sometimes it leads to a case of full-blown tuberculosis. Shortly afterwards, Marie-Hélène's sister Denise contracted a primary infection. As for Marie-Hélène herself, she hadn't realized she was ill, and had overworked herself until the true nature of her illness was revealed by a burning fever. In those days, there was no cure for tuberculosis. Nowadays, thanks to antibiotics and chemical products such as Rimifon, it's not a big deal. But at that time, the doctors prescribed a pneumothorax, which consisted in the injection of air between two layers of the pleura, to suppress breathing in the affected part of the lung. Unfortunately, several years earlier, Marie-Hélène had contracted pleurisy, and doctors forbade her to have a pneumothorax. She left hastily for the sanatorium in Passy, on the plateau of Assy in Haute-Savoie. It was a disaster for us and for our families. Marriage in December was out of the question. Dr. Arnaud, the director of the sanatorium, was a remarkable man: discreet and understanding. He was executed by the Nazis for protecting his Jewish patients. As soon as Marie-Hélène arrived at the sanatorium, he explained to her that she would be there for a long time, to be measured not in months, but in years, and that in fact she should altogether give up wondering when she would be allowed to leave. He told her to make herself ready for a long reclusion. As much as possible, she was to mix with other patients in the sanatorium, as every semblance of social life helped them preserve a state of mental equilibrium. Actually, the reality was not quite as he presented it. Sexual relations were strictly prohibited, but the young men and women had them anyway. Marie-Hélène decided to avoid them at any cost.

Very soon, she realized that she had been told the truth. She met an old friend of hers, Baldenweg, who had lived for seven years in the sanatorium; he died the year she arrived there. Fernande Lévy-Alvarès, one of the friends she made when she got there, died soon after, suffering from tuberculosis of the throat. A young woman who was still superb when Marie-Hélène arrived died shortly after. In his memoirs *Are you laughing, Mr. Feynman?*, the physicist recalls his young wife, who contracted tuberculosis in 1941 and died in 1946, just before the appearance of the first antibiotics. In 1935, there was just no serious treatment for tuberculosis at all. Cures were frequent but not always definitive, and the illness was often fatal.

## Like a dragon

Marie-Hélène and I made resolutions: she would not participate in the social life of the sanatorium, she would use all her energy to care for herself, to be cured and stay in the sanatorium for as short a time as possible. And indeed, she mingled very little with the others, making only a few friends. She asked to take her meals in her room, and spent her days reading and resting all wrapped up on her balcony in the cold air. She took care to eat all her rations. We firmly believed that it was her time in the Classes Préparatoires and at the ENS which had tired her out and made her ill. Once the "pot" had said to me: "At least a third of the young girls who come here don't make it." Marie-Hélène understood from her weekly X-rays where Dr. Arnaud and another doctor examined her lung that her case was very serious. One entire lung was affected. There was an actual hole in it, and what she coughed up was loaded with Koch bacilla. For months, the hole kept growing. Dr. Arnaud informed Marie-Hélène's family, and they kept me up to date. I downplayed the situation a little bit, but did not completely hide it from Marie-Hélène herself. Anyway she was lucid enough to understand it on her own. The hole was as big as a tangerine. Seeing death lurking around the corner, Marie-Hélène took to caring for herself with astounding will-power; "like a dragon", said Dr. Arnaud. At first, her mother stayed in the sanatorium with her, but when she left, Marie-Hélène redoubled her efforts.

We wrote long letters to each other every single day, throughout the whole time she remained in the sanatorium. Our letters were sentimental and even sensual, but I also told her every single thing about my mathematical studies and my political activities, so that she could evolve along with me. In this way, she became passionately interested in the same grand political causes as me. Of course, we also read enormously, which also nourished our abundant correspondence. We should have kept those letters; they would have been quite interesting. Unfortunately, during the war they remained in her parents locked apartment, rue Théophile-Gautier, which was later looted by the Nazis. We were horrified at the idea that they might get our precious letters. With incredible temerity, a cousin of Marie-Hélène's called Annette Dufour crept onto the Lévy's balcony from the balcony of a neighboring apartment and grabbed the compromising suitcase of letters, which we prudently burned. The Nazis didn't get them – alas, neither did we.

The doctors told me that even if Marie-Hélène survived, she would spend many years in the sanatorium, that she would not be able to get married and even if she did, she would not be able to have children; she would live the life of an invalid. Pregnancy, which leads to decalcification, is a serious threat to the life of a tuberculotic woman. In fact, decalcification is the main risk of tuberculosis, because tuberculosis is cured when the Koch bacilla (KB) become enclosed in little pockets of calcium; there is a permanent danger of relapsing since all those bacilla want is to spurt out of there and overwhelm the antibodies, and fatigue and decalcification can lead to this. Thus, a former tuberculotic was always in danger.

Our whole future was collapsing. The situation soon led to a serious conflict between my parents and Marie-Hélène's. Her parents naturally tried to minimise her illness, so as not to envision the terrible consequences, except for times when her mother panicked and believed her daughter was about to die. My parents, like the doctors, advised me to stay friends with Marie-Hélène, but to slowly increase the distance between us and to renounce all idea of marriage. I was deeply revolted and told them it was out of the question. The bourgeois customs make a curious difference between what happens before and what happens after marriage. If his fiancée is ill, a man can decide not to marry her without breaking the unwritten rules of the bourgeois community. But from the moment they are married, on the contrary, it is his duty to devote himself to her care and her cure for the rest of his life, if necessary. Divorce, even because of a serious illness, remained a shameful stain.

I considered the situation realistically, and decided to persist. If Marie-Hélène had really remained ill for an indefinite length of time, without being able to marry nor have children, perhaps things would have changed eventually. But at that moment, I felt that I must persist and above all, help and support her. The idea of abandoning her horrified me, and I clearly saw how important my firm support and the strenth of my love were to her. My feelings, my moral sense and even my sense of reality all agreed on this. After several quarrels, my parents stopped mentioning Marie-Hélène's name and never spoke of her in the house. But once a week I had lunch with the Lévys, where I felt more at ease than with my own parents, although their attitude of simplifying everything also occasionally embarrassed me.

Our intimacy had grown very deep, and Marie-Hélène immediately understood that I would not abandon her. In fact, the idea never occurred to her, and my presence gave her strength to resist the illness. We supported and sustained each other mutually. It was almost impossible to telephone; you had to wait at least half-an-hour and then the connection was terrible; it quelled even the greatest desire for conversation. But our letters wove an indestructible thread between us.

For a long period, I did not see her. My primary infection was quite serious. For a long time, I was exhausted and could not work at all. I just stopped by the ENS from time to time – where my classmates fully approved of my attitude – and I read a lot. For a time, I thought I would have to give up preparing the second-year certificate (rational mechanics and probability). I was about to officially request the school to allow me to postpone them, when my close friend Revuz advised me to wait, and let the first part of the academic year 1935–1936 pass without mentioning my problem to anyone. And indeed, as I also took care of myself energetically, I felt much better and was cured by January 1936. But morally, the situation was very difficult. Marie-Hélène's sister Denise was also cured, and we called each other the co-tyrants, because we directed things a bit at the Lévys'.

Marie-Hélène's illness revealed to us how much we loved each other. In spite of our fear and anguish, we were very happy. We were stubborn, and the consequences soon made themselves felt, convincing us (if we needed it) of the fundamental role played by the psychology in illness. At the beginning of March 1936, Dr. Arnaud noticed that the hole in Marie-Hélène's lung was becoming smaller. Then a kind of miracle occurred: within a few weeks, the hole disappeared completely! There were still many traces of illness in the lung, but it was on the way to a cure. I remember that by the end of spring 1936 there were no traces left at all, and analyses showed no presence of KB – we had been saved by our tenacity! After all, Venus had spread her wings over us . . . in the manner of goddesses, much like Ulysses, protected by Athena, returned from Ilium to Ithaca – after a ten year voyage. I've always wondered about those verses of du Bellay "Happy he who, like Ulysses, has taken a long voyage." I feel certain that Ulysses, like myself, could very well have done without it!

Ours was really an exceptional case. Of course, the story was not completely finished; we knew that we mustn't ignore the fact that relapses were common. Marie-Hélène spent another eighteen months in the sanatorium. During the summer vacation of 1936 I went to live in a hotel in Saint-Gervais, together with her parents, and finally I could visit her every day. She did not leave the sanatorium until near the end of the spring of 1937, but then she was considered completely cured. She spent the rest of the spring in a pension in Fontainebleau where I used to come and join her. I also accompanied her and her family to Igls, in the Austrian mountains, over the summer.

In October 1937, I started my military service, and Marie-Hélène came to live in a pension in the mountains near the sanatorium, where she had weekly check-ups. At one of these, "some KB" were detected. This was not at all impossible – we were terrified. But there had been a mistake: we had received the results of another Miss Lévy's check-up. This information took a tremendous weight off our hearts. We were so relieved that I don't believe we gave a thought to the poor girl who really did have some KB. Such a complete cure after such a serious case of tuberculosis was quite rare. Marie-Hélène still has check-ups of her lungs every year or two, and when she shows doctors the X-rays of her lungs taken in 1936, they can't believe it: "You're not telling me that this is an old X-ray of your own lung!" And to this day, she has never had any relapses at all.

We chose ourselves not to get married immediately, because of my military service, the first part of which took place in Metz. After it was over, we were sent to various anti-air defense units throughout France. I ended up in Laon. There we got married, on May 2, 1938, seven months after I left the ENS, seven years after the birth of my love.

For my grandfather, the rabbi Simon Debré, I accepted a religious marriage. I had already had a bar-mitzvah, although I felt some repugnance at the idea of pronouncing words promising an eternal faith in the Lord. I felt that I was being pushed to lie, since I had no faith at all. But the marriage ceremony did not contain any such compromises. It did not take place in a synagogue but in my grandfather's house, and we were blessed by my uncle Mathieu Wolf, also a rabbi and the husband of my aunt Delphine (during the Occupation, Mathieu and Delphine remained in Paris and wore the yellow star, since he was a rabbi; they were both deported and never returned). Thus, our marriage was a private affair. During the ceremony, the bridegroom must smash a glass on the ground. I must have been impressed by everything; I tossed it too timidly, and it bounced up again, and broke only upon falling back down – I suppose it must have cracked the first time. We like to look back on that as a symbol of our own story: the marriage fixed for December 1935 hadn't taken place, but was bounced to the marriage of May 1938. In fact, we never took the date of May 2, 1938 very seriously – we considered we had been married for a long time, and left immediately for Laon,where finally we could live together! We paid no attention to May 2, 1963 – the date of our silver wedding anniversary, or even May 2, 1988, our golden one. Indeed, from the start of the whole story, as a protest against the lengthy engagement we were forced to undergo, I had decided to consider us as husband and wife; I addressed every single one of my letters to the sanatorium to Mrs. Laurent Schwartz. Marie-Hélène liked it, and the sanatorium considered that if anything, it was helping her get better.

I think I have spoken enough about my love life. We never stopped fearing a relapse with all its terrible consequences. Our political evolution towards the extreme left also liberated us from the strict laws of marriage followed in our bourgeois society. We knew perfectly well that even the most happily married couple can undergo extramarital affairs with strong feelings, and we accepted that, while always considering conjugal love as the highest ideal. We believe that what is really necessary is a sense of responsibility and a deep sincerity.

Marie-Hélène's illness did have consequences. Firstly, she was told not to have a child for at least four years. Then we had to wait four years before having a second child. After that, the doctors said an interval of two to three years would suffice, but alas, no other children came. So we had two children,

Marc-André and Claudine, though we would have liked to have at least three or four. Marc-André died in 1971; it was a terrible blow, which I will talk about later. Now we have only Claudine.

My mother had very strict principles. Marie-Hélène had married me; she was to be integrated into the family. A certain tension always remained between the two families, but my parents welcomed Marie-Hélène warmly. There was never any mention of the past. My conflict with my parents simply evaporated. I have always felt love and admiration for them.

## Paul Lévy

Marie-Hélène's father Paul Lévy was not a professor at the Sorbonne, but at the Ecole Polytechnique. He was a very great mathematician, one of the main founders, with Kolmogorov, of probability theory, and one of the greatest probabilists of his time, and even of all time. His name and his work are still cited in every seminar on the subject, throughout the world. I don't mean to give his biography here, but I wish to say a few words about him. He had always been an excellent mathematician. He was first at the ENS entrance competition and second at the Ecole Polytechnique; he chose to attend the second school, without really knowing why. He was ranked first on leaving. Very prestigious mathematicians sat on his thesis jury: Poincaré, Hadamard and Picard. One day, after the death of Henri Poincaré, he was asked to give a series of six lectures on probability for the polytechnicians. He knew nothing about probability at that time. He had three weeks to prepare the course. In a letter to his wife, he wrote "Three weeks is not enough to read all that's been written, but it's enough to rediscover it by myself." And that is what he did. Even more than me, he felt that reading something was almost a personal assault. He read almost nothing, and his colleagues kept him in touch with whatever was being done. He remained a probabilist for the rest of his life.

He never had a thesis student at the Ecole Polytechnique, because no one did research there. Apart from Henri Poincaré, he was the only research mathematician to come from the Ecole Polytechnique. The professors at the Sorbonne ostracized him tenaciously, partly because he was so much better than they were (with the possible exception of Denjoy; they were the two greatest French mathematicians of their generation). He had the intuition of a genius, but sometimes lacked rigor. Not that he made actual mistakes, but some of his proofs were incomplete, which did not prevent him from making great advances in probability theory, particularly in Brownian motion, a subject on which he was the universal expert. Gauss' law and Brownian motion are the central themes of his work.

He didn't have a very good memory. I remember asking him one day if he knew a simple proof of Lebesgue's theorem of density, which is a theorem of analysis, not probability. "I've seen several proofs, but I don't remember them any more," he told me, "but I'll try to discover one." After half-an-hour, he came out of his office and showed me a beautiful, elegant proof. Six months later, he happened to ask me: "Do you know a nice proof of Lebesgue's density theorem?" Of course I remembered what he had told me. But he himself had forgotten it completely. When I showed him his own proof, he said "Oh, that's a magnificent idea, very astute! I'd never have thought of that." I informed him that he had found it himself, but he couldn't believe it.

In spite of his occasional absence of rigor, Paul Lévy's incredible intuition got him out of some delicate situations. He never really defined the meaning of two independent variables; he used to say that they are independent if they are randomly and independently drawn. Then he would add: they're independent if knowing something about one of them gives no information at all about the other, which is almost perfect. Conditional probability remained vague, but he did consider the Brownian bridge. The foundations of probability remained uncertain from their beginnings in the seventeenth century until quite recently. At the International Congress of Mathematicians in 1900, in Paris, David Hilbert posed 23 famous problems: one of them was to find a correct foundational basis for probability theory. This problem was solved by the Soviet mathematician Andrei Kolmogorov in about 1930, thanks to his introduction of the set $\Omega$ of proofs, sigma-algebras over $\Omega$ and the systematic application of measure and integration theory. Paul Lévy never accepted Kolmogorov's set $\Omega$, nor sigma-algebras, but today they are taught everywhere, even in high school. However, Lévy succeeded in working without them, thanks to incredible mental gymnastics. He was very sure of himself and his opinions, but at the same time extremely sweet and gentle. I never heard him raise his voice.

During my last two years at the ENS, I had lunch at the Lévys' once a week. I had already spent a lot of time at their house in 1934, interested in both the daughter and the father. He was superior to all the professors who had taught seminars at the ENS. He gave me a first-class education in probability and also in analysis. He taught me about $L^p$ spaces and the best proofs on the orders of entire functions. I read in detail every one of his books published before the war. I learned about his work in probability from them, and I enrolled for the certificate in probability and was ranked first (the professor was Emile Borel, a very good probabilist who had done important work, particularly on the strong law of large numbers, one of the main theorems of probability, which has applications in every branch of science). Lévy's most important work from this period deals with sums of random variables: he found the necessary and sufficient condition for such a sum to be Gaussian.

In 1937, he was the world's foremost expert on everything to do with Gauss' law. He truly understood why so many error laws have Gaussian behavior: it is because the error is the sum of many small independent errors, of which the even largest is very small (or they are all uniformly small) with respect to the total error. To complete the proof of his great result, it remained only to prove a result which he was personally convinced of, namely that if the sum of two independent variables is Gaussian then each one of them is. He tried to prove this result for a long time without success. I was audacious enough to work on it at the same time as he did. Finally, the Swiss mathematician Cramer succeeded in proving the result, so that the final theorem is known as the theorem of Lévy-Cramer. Paul Lévy never really got over his failure to find the proof, which turned out to be so elementary that I could even have discovered it myself.

During the war, existence was too complicated to do mathematics. His most sensational discoveries on Brownian motion, and his book on the subject which is a real masterpiece, date from 1948. Brownian motion, discovered by the physicist Robert Brown in 1827, is the disordered motion of a microscopic solid particle in a liquid due to collisions with the molecules of the liquid. This theory has been much studied by mathematicians and has become fundamental to diverse branches of mathematics and other sciences.

It was thirty years before I turned again to the study of probability. After my two years of military service followed by a year of war, during which I could not do any mathematics at all, I met Bourbaki in 1940, and was projected in a new, radically different direction, from which sprung my discovery of distributions. But what we do when we are young always leaves its traces. In the 70's, I returned to probability and remained with it. Analysis and probability, which are closely related subjects, have been the guiding forces of my research life.

## The agrégation

All during these years of sentimental upheavals, I continued my studies. The third year of the ENS is usually devoted to preparing for the agrégation, which opens the door to a career in high school teaching. I don't think I actually turned in more than four or five of the agrégation assignments which punctuated the year. They were quite difficult; the competition consisted of several seven-hour exams. We also had to prepare oral "teaching lessons", corresponding to real classroom situations, for example: "introduction to trinomials of degree 2 for eleventh graders", or "first lesson on the ellipse in twelfth grade", or "determinants in the second year of Classes Préparatoires". The oral part of the competition consisted in two of these lessons, delivered to a jury, on subjects drawn at random; each lesson took a whole afternoon. We had to come at two o'clock to receive our subject, then we were sent for four hours to prepare the lesson, alone in a classroom with books and papers, and finally, the lesson itself was delivered from six to seven. During the third year of the ENS, each student was supposed to give several of these lessons before a general inspector or a university professor, and some other students, who would then criticize it. Like the others, I gave several of these lessons. In principle, we were supposed to know the subject already, and only the pedagogical approach was judged. It was an excellent preparation for a teaching job, and everyone profited from it. Personally, I've always found this kind of pedagogical job quite easy. The lessons did not take much of my time during the year, and I had a lot of free time to learn other things and to ponder my favorite subjects.

Finally, at the end of June 1937, the day of the agrégation arrived. I had spent two whole weeks before the written exam resting in Autouillet, and I did the same before the oral part; I always did this before competitions so as to arrive with a fresh spirit. As usual, I didn't do too well on the written exam. After it I was ranked twenty-eighth (I think there were about fifty places). But I wasn't at all discouraged – I thought I would move up a lot at the oral exam.

The day of my oral exam was as full of adventures as an epic poem. In the morning, I was still in Autouillet. I was supposed to drive back up to Paris in my father's car, with his chauffeur. At the last moment, the car broke down. At that time, people did not have country houses; there wasn't another car in the village. It wasn't easy to get to Paris from Autouillet! I was going to miss my examination for a stupid reason! At the last minute, a great friend of my mothers, Annette Lazard from La Queue-lez-Yvelines, was able to send us a little truck used for transporting animals, and my two brothers and I stood in the back and were transported to Paris in this manner. We carried a little cold meal with us. I had brought along one of my favorite treats, a raw egg to swallow whole. But suddenly – the truck bumped, and the egg smashed and flowed over my pants and jacket! I couldn't possibly show up for my oral exam in that condition. We barely had the time to rush to my grandmother's, whose cook hastily washed and ironed my clothes, before scrambling to the examination, where I arrived at two twenty. The lateness did not have any negative consequences; I just had twenty minutes less to prepare the lesson in. I drew an envelope containing the following subject: "the surface generated by revolving a curve around a fixed axis, for the second year of preparatory class". It was a subject I hadn't previously prepared, and quite difficult, but all right for a strong candidate. I knew what I should say, but I didn't have time to prepare it properly. There were several things I couldn't remember how to do at at first. But just on time, I was finally ready. I don't think I would blush today at the lesson I gave then, a mixture of pure geometry and analytical geometry with cartesian coordinates. We had been told to speak to the jury exactly as though they were school students, so I did: "as homework, you will prove the following property." The members of the jury were astounded. After I'd raced through my lesson in fifty minutes, they spoke together briefly, and then – unheard of event – the president publicly congratulated me. I was first at the oral exam, and they asked me to write up the text of my lesson, which was actually published in some journal!

## The Last Judgment

In the end, I came in second behind Choquet, who was one of my best friends – today we are neighbors at the Academy of Sciences. He was born on March 1, 1915, I was born on March 5. Our class (1934) was one of the most brilliant; we have three Academicians (Choquet, Blanc-Lapierre and myself) and one correspondent (Félici). But although I wasn't as competitive as I had been, that old desire to come out first hadn't disappeared altogether – it came out as soon as I tried something where I had a really good chance of being top. I had an amusing dream about this a couple of weeks after the competition. I was taking the competition of ... the Last Judgment – I suppose no one will dispute that it is more important than the agrégation. The president of the jury was Jehovah himself. I was asked two questions: what is the purpose of the stairs in the subway, and what is the relation between the revolution and the Jewish question? I can't think what the first question could mean. But the second was an odd mixture. At that time, I was a Trotskyist, the Jewish question was dramatically revived by the Nazis, and revolutionaries had their own answers to it. On the other hand, my oral exam had also concerned "revolutions". Dreams are sometimes ingenious constructions. Anyway, apparently I gave an excellent lecture, and Jehovah himself publicly congratulated me – an unheard-of event. Thus, all boasting aside, I came in first at the Last Judgment!

# Chapter III
# Trotskyist

## The chrysalis breaks

My years at the ENS radically determined my subsequent political evolution and indeed my whole life. Coming from a right-wing background, I evolved all the way to the extreme left. My ideas about the First World War changed completely. Many studies showed me that what I had hitherto considered to be a war for liberty against barbarism was something else entirely. Germany was certainly an empire, but it was a constitutional empire, where the most important decisions were taken by a House, the Reichstag, and where Social-Democrats and unions played an important role. Germany may not have been a republic, but it had a free press. The difference in lifestyle between France and Germany was no greater than the difference in their cultural levels. After all – Mozart, Beethoven, Heinrich Heine (many of whose poems I had memorized), Goethe, Kant, the mathematicians Riemann, Gauss, Hilbert – these were not representatives of barbarism! Naturally, there was a deeply rooted and reactionary aristocracy consisting of various German generals and landowners, which had much more weight than in France, particularly in the Reichstag where one third of the voices were attributed to it. Another third was attributed to representatives of the rich and the final third to the remainder of the population. It is true that Germany had violated the neutrality of Belgium and, in 1871, annexed Alsace and part of Lorraine (an act against which German Socialists such as the famous August Bebel had strongly protested). In spite of this, the war was a monstrous, useless and unjustifiable butchery. For four years, soldiers waded in the mud of the trenches in order to massacre each other, while the "rearguard" got high on chauvinist patriotism. And yet, each side was deeply convinced that it was defending its motherland. The German soldiers, who were occupying part of French territory, believed it just as much as the French and defended their motherland with incredible courage. Did the fighters ever stop to think that they could make peace instead of war, and that the territories were not threatened to the point of legitimizing such a massacre?

The war was murderous: nine million dead, millions more wounded; the battle of Verdun was apocalyptic. It was the first manifestation of the incredible barbarism of the twentieth century. If civilization should disappear from the face of the earth, the 1914–1918 war will have been its tomb. Chauvinism wore the same hideous mask on both sides. The medias of the time were notable for their diffusion of false information and shameless brainwashing (similar practices can be observed today except that disinformation methods have now become extremely sophisticated). Reading a wartime newspaper is quite an edifying activity. Propaganda attained unheard-of levels. In Germany, it was possible to speak of military success from the beginning, but this was hardly the case in France. Fortunately, the Russian "bulldozer" was advancing with giant steps towards Berlin. Germans invented gases and chemical weapons, but the French sent trench cleaners to finish off the wounded by slitting their throats, and introduced saw-toothed bayonets. For four years, France and Germany faced off to exhaustion, in an absurd war. All of my comrades evoked the war in these terms, although without denying the valiance and devotion of the fighters.

The discovery of this reality stupefied me at first, and contributed to my isolation. For me, it was a deep shock, a break with ideas which I had been taught were unquestionably obvious. The blatant contradiction between what I had been told and the bare facts was even worse. Chauvinism was expressed even more violently in France than in Germany. It is true that our country was partially occupied, and that many French people dreamed of reconquering Alsace and Lorraine. When the French mathematician and academician Darboux died in 1915, David Hilbert gave his obituary speech in the Academy of Sciences in Berlin. The contrary would have been unthinkable in France. French mathematicians went so far as to oppose the invitation of their German colleagues to the International Congress of Mathematicians in 1920, and again to that of 1924! Only in 1928 were mathematicians from other countries able to counteract French chauvinism.

French political leaders and unionists who spoke out against the war can be counted on the fingers of one hand. The most famous one, Alfred Rosmer, wrote a remarkable book about Zimmerwald, a place in Switzerland where pacifists from different countries met clandestinely in September 1915. The desolate atmosphere which impregnates this strange picture of the calm before the tempest made a strong impression on me when I read it. The Zimmerwald meeting was followed by one in Kienthal. Romain Rolland, who was a refugee in Switzerland, published *Au-dessus de la mêlée* (*Above the Fray*). My parents blamed him severely for the book, although they respected him for other things; I on the contrary was full of admiration for his book. These rare individuals were filled with despair and completely impotent; although they met regularly, they were never more than a handful. A revolutionary (essentially unionist) review, *La Vie Ouvrière* (*The Worker's Life*), was published by Pierre Monatte. It had been founded in 1909 and had two thousand subscribers in July 1914. They had to stop its publication, but the original group remained together: Monatte, the uncontested leader, Rosmer, Alphonse Merrheim, who had been in Zimmerwald, Marcel Martinet, Robert Louzon, Georges Dumoulin. *La Vie Ouvrière* appeared again only in April 1919. These militants observed the 1917 revolution with enthusiasm and all joined the new Communist Party except for Merrheim and Dumoulin. But they left the Communist Party in 1924, and since *La Vie Ouvrière* had become the newspaper of the Communist union, the CGTU (Confédération Générale des Travailleurs Unifiés)[13], they left it as well, and founded a new review in 1925, called *La Revolution Prolétarienne* (*The Proletarian Revolution*), which still exists. I read this review regularly in the period after the war, starting in 1936. I didn't know Monatte and Rosmer personally, but I knew the engineer Louzon. He had been a classmate and a friend of my uncle Robert Debré in his youth, and they remained close in spite of obvious differences of opinion. I learned of him through Uncle Robert. I don't believe we ever actually met, but we exchanged letters. Louzon was very pleased that a nephew of his old friend shared his political views. He was an anarchist, which pleased me. It is true that one day we had an argument about the famous mathematical problem of Achilles and the tortoise which he misunderstood completely, like many other of my left-wing friends.

To this handful of men, I must add the internationalist Mensheviks Trotsky and Martov, as long as they remained in France.

## Meeting with remarkable men

The situation in Germany was quite different. The Spartakusbund, made up of some energetic Social-Democrats such as Karl Liebknecht and Rosa Luxemburg as well as Clara Zetkin, had rejected the war from the very start. A whole wing of German Social-Democracy opposed the war. At the end of 1914, ninety deputies in the Reichstag, led by Karl Kautsky, refused to vote in favor of war credits. On the Russian side (and thus internationally since many Russian Social-Democrats lived in foreign countries), the Mensheviks adopted the chauvinist attitude of the European Socialist parties, with the notable exception of the internationalist Mensheviks I mentioned above. Only the Bolsheviks resolutely

13) CGTU: General Confederation of Unified Workers

opposed the war. Lenin was intransigent on this subject. His historical article "Against the current" defined the conflict as an "imperialist war" whose goal was to share out the world between the major capitalist countries. According to the doctrine of revolutionary defeatism, the task of the revolutionaries was to oppose their own imperialism and favorize their defeat in order to transform the "imperialist war" into a "civil war"and provoke a revolution. Trotsky, an internationalist Menshevik, agreed with Lenin's analysis of the war: "blood and mud of imperialist butchery". Lenin prophesied the future revolution which would be provoked by Russian military defeat, and a world revolution as the result of the war. "We, also, will live happy days" was the conclusion of his article. His article and his refusal to meet or shake hands with one of the social-chauvinists provoked my admiration. But although these attitudes were logical and comprehensible in the given situation, they played an important role in the total break between Socialists and Communists. In politics, as in life, it is necessary to conserve a certain tolerance, and to keep from digging uncrossable trenches.

My next step was to interrogate myself on the subject of colonialism. The book *Indochine SOS* by Andrée Viollis revealed to me the atrocious repression and contempt imposed by French colonists on the Indochinese people: she described the Poulo Condor prison. According to my parents, colonial conquests were made for the purpose of bringing democracy and civilization to backward peoples. Certainly, colonialism brought non-negligible industrial progress to non-European countries, and also gave rise to undeniable ameliorations in the medical infrastructures. Later, people of the ex-French and English empires remained in contact with their former capitals for cultural reasons. Yet, taken altogether, colonialism was accompanied by abominable practices of torture, massacre and constant repression.

When the party Front Populaire (Popular Front) was founded in France, a new newspaper appeared; it was called *Vendredi* (*Friday*) and was run by André Chamson, Jean Guéhenno and Andrée Viollis. It denounced colonialist repression, and was aimed at the whole of the intellectual left-wing. After a few months at the ENS, I became a faithful reader. I only met Jean Guéhenno much later, at the beginning of the Algerian war in 1955. From 1958 until the end of his life, he lived in the same apartment building as I did.

## The events of 1934 in a new light

On February 6, 1934, the French agitators led by Colonel de La Rocque attempted a coup, and on February 12, the groups of affiliated trade unions responded massively. At that time I was in taupe, and at first I understood nothing of these events. Their descriptions in the newspapers *Temps* (*Time*) and *Matin* (*Morning*) were twisted. The speeches on the wireless alluded to "moral forces", giving indirect support to La Rocque, and I also heard speeches of this type given by Mademoiselle Sainte-Claire Deville, daughter of the famous chemist whose name I had learned in high school. These speeches influenced me and retarded my evolution. It was a period of major scandals in France, such as the Stavisky scandal, and the assassination or suicide of the counselor Prince. A radical Minister like Chautemps could be openly presented as a rascal in *Matin*, and the same was even true of Léon Blum, about whom various anti-Semitic rumors circulated, such as that his name was really Garfunkelstein, that he possessed magnificent silver, etc. Back then I also believed that Léon Blum was corrupt. Marie-Hélène was wiser than I was and distrusted all such rumors. My comrades at the ENS were much more lucid about this famous morality we constantly had to hear about from the reactionaries, and which they considered as a morality of submissiveness to the army and to war. Gaston Doumergue, who was a right-wing man, was called to the presidency after the troubles of 1934. At that time he appeared to me as a neutral man who might be able to calm the conflicts .... February 1934 also saw the fascist coup d'état of chancellor Dollfuss in Austria, which overthrew the workers' institutions in Vienna, which were supported only by pacific although armed militias. In left-wing circles, these radical Social-Democrats were called Austro-Marxists. They had effected many social transformations, and played a non-negligible role in the rehabilitation of reformism in revolutionary circles. I heard all these things for the first time at the ENS.

Morality in politics has always been a major subject of preoccupation and reflection for me, and the attitude of part of the left-wing on this subject shocked some of my deeply-held convictions. On the pretext that the right-wing was defending old reactionary values (Pétain's future "work-family-fatherland"), the left-wing liked to display sovereign disdain for the word "morality". Even if the right-wing did often use the word to achieve retrograde political aims, there were still people for whom morality was a serious affair, and I always felt close to them. I don't want to establish general rules here, but to look reality in the face. All these facts contributed to perturb the politically ignorant young man that I was.

At the ENS, my understanding of the complex events of 1934 improved. At first, I had considered Nazism as the expression of a revengeful and anti-Semitic party. Then I became aware of the terrifying nature of the oppression which the regime exercised on the German people, and of course on the Jews.

People spoke of one million prisoners in Nazi concentration camps; in reality, the number at that time was probably closer to a hundred thousand, but all the democratic structures – left-wing parties, unions, public freedoms – had been eliminated, and Nazism was undoubtedly already public enemy number one. I was horrified by the brutality of June 30, 1934. Although it was recounted in the newspapers we read at home, I wasn't able to understand the meaning of the events until I entered the ENS, where conversations with my classmates gave me the necessary information and analysis to come to an understanding of the situation. It was more than a simple demonstration of strength; it was a precise plan aiming for the total elimination of former political opponents such as General Schleicher, the chancellor before Hitler, and also the most Socialist elements of the party, the ones who took the National-Socialist doctrine seriously, such as Gregor Strasser. On that day, National-*Socialism* died. June 30 was meant to show what Hitler was capable of to all of his internal and external opponents. Terror remained a constant feature of the Nazi regime. Even before taking power, Hitler approved the political assassinations committed by Nazi militants, which were a clear sign of violence to come. In a propaganda meeting in Nuremberg, he shot his revolver at the ceiling upon entering. "I want everyone to know what goes on in my concentration camps" he once said. It was only in a later phase of the war that he attempted to hide his crimes. This component of ostentatious terrorism has not been sufficiently emphasized: it was incredibly efficient. Islamic fundamentalists use it today. And I don't accept the generous excuses for fundamentalist violence explaining that it is due to the very real misery of the population, any more than the misery of the German population could be taken as an excuse for Nazi violence. Desperation may explain violence, but it doesn't justify it.

## The wind of revolution blows through France

The Front Populaire fought for "bread, peace, freedom". The cinema newsreels showed demonstrations of unemployed people, shouting the slogan "Work and bread!" A powerful current was flowing through the country. It was impossible not to ask oneself questions about the capitalist regime. Some texts by Marx in which he analyses the pre-eminent role of economic forces, and the book *The Accumulation of Capital* by Rosa Luxemburg, ended up shaping my ideas into authentic Socialism. Yet I approached the book by Rosa Luxemburg with a critical eye. In this book, she uses mathematical formulas to prove that capitalism could not survive without colonialism, which provides an indispensable external market. This is only partially true: colonialism has in many cases been replaced by a sort of neo-colonialism playing a similar economic role, but which is not the same as pure and simple colonialism, and Keynesian theories about the necessary growth of the internal market have saved capitalism from the type of periodic crises which used to occur.

These reflections awoke in me the desire to understand world economic geography, a subject which I had neglected in my senior year of high school. I borrowed numerous books, starting with the little book by Maurette on the great markets of raw materials. I learned about world productions of coal, steel, minerals, kilo-watt hours of electrical energy, textiles, agriculture, basically the whole subject of world commerce. For years, this subject remained one of my chief amusements. From time to time, to refresh my spirit, I would spend two or three weeks studying economic geography, reading books and acquiring a quantity of personal documents. I became quite competent in the domain. Unfortunately, the culture I obtained through this work was not only rather limited, but even soon became obsolete since world production obviously evolves, on top of which I have mislaid most of my precious documents. I firmly believed in the doctrine which stated that Socialism must *inevitably* succeed capitalism. The ever more numerous working class would rise against the bourgeoisie and establish a new power, first Socialism, then Communism. The "inevitability" of Socialism, which was practically engendered by capitalism independently of human will, lent itself to a language of formulas which was very comforting, and I believed in it. I was a witness to the spectacular rise of the Front Populaire in 1935, and I willingly let myself be pulled along with the flow.

I underwent a metamorphosis even more spectacular than the one my parents had undergone when they abjured their religious upbringing to become atheists (my mother was actually converted to atheism, together with her brothers and a whole group of youngsters, by Jacques Maritain who then himself became a militant catholic). My parents were not horrified by the changes which were occurring in my views, as my mother and her brothers had themselves been influenced in their youth by the Socialist movement led by Jean Jaurès and Charles Péguy. My parents subscribed to *Temps*, which after the war became *Le Monde* (*The World*), as well as to *Matin* which was a right-wing newspaper. Personally, I took to reading left-wing newpapers. *La Flèche* (*The Arrow*), edited by Gaston Bergery, represented a moderate Socialism, or populism, much appreciated by young people, and it impressed me more than the others. I was particularly struck by an issue devoted to the "two hundred families" said to run the country. It was, I suppose, a little facile to oppose two hundred families against the rest of the country. The simplicity of the theory made it attractive. Although Marxism is a more highly developed theory, it nevertheless ends with a similar theme: the final struggle in which a crushing majority will defeat a handful of exploiters. This issue of *La Flèche*, which denounced the war industry and its profits, in particular Schneider, the Wendel cousins (one of whom was French and the other German) and the machine gun maker Basil Zaharoff, influenced me deeply. At the ENS in those days, we were worlds away from imagining that after the war, all these industries would be nationalized and the war industry would flourish more than ever. Bergery's populism was actually rather unstable, and he ended up fascist. A monthly magazine called *Le Crapouillot* (*The Little Cannon*), edited by Jean Galtier-Boissière, also influenced me. It was about a hundred pages long and each issue was devoted to a particular subject. I particularly recall one issue devoted to the origins of the First World War, which insisted on the sharing of responsibility and concluded "Ten million men were to die." I am not sure about the precision of the facts mentioned there, nor about the historic rigor of the issue, but the tone and the position taken appeared justified to me at that time. I also read *Le Populaire* (*The Popular*) and *L'Humanité* (*Humanity*) as well as the anarchist newspapers sold in kiosks, for instance *La Patrie Humaine* (*The Human Homeland*) and *Le Libertaire* (*The Libertarian*). These papers were pacifist and called for total, unilateral and unconditional disarmament. I understood that this call was utopic given the rising Nazism in Germany, but the articles interested and troubled me nonetheless.

A book called *Notre jeunesse* (*Our Youth*) by Charles Péguy played an ambiguous role in my early leftwards evolution. The author was a Socialist before becoming an opponent of Jean Jaurès, and he makes a distinction between mysticism and politics, praising the first and condemning the second. This theory held some attraction for me, but it completely lacked pragmatism. As I was somwhat fearful of slipping irremediably to the left, this book was useful in providing a kind of alibi for the metamorphosis, helping me to a smooth transition.

When Marie-Hélène left for the sanatorium, I scrupulously let her know every detail of my political evolution, sending her numerous newspaper cuttings in my daily letters. She became deeply interested in social questions, which she had never previously really thought about. However, she was at less of a psychological distance from the idea of internationalism than I was. Born in 1913, she lived the first years of her life in the house of her maternal grandmother, with her cousins Marc and Françoise, whose father, Guy Stein, was a Jew from Prague (and thus an Austrian subject), who immigrated to France and became naturalized. He fought in the French Army, but his two brothers, of whom he was very fond, had remained in Prague and fought on the Russian front under the Austrian flag. Furthermore, a grandfather of his mother, Henri Weil, who was a deeply respected man both in the family and in the world of erudition, was of German origin. He was a celebrated Hellenist, who was born in Frankfurt and emigrated to France in 1848, where he became perpetual secretary of the Academy of Inscriptions and Literature. For all these reasons, Marie-Hélène's family was patriotic but not chauvinistic as mine was. When she heard the common remark that all Germans were scum, she naturally thought it was ridiculous. She was spontaneously ready to accept the idea of internationalism, which it took me a long period of mental progression to absorb.

I was overjoyed at the victory of the Front Populaire in the 1936 elections. I had become close to the Socialist Party, and of course voted Socialist in May 1936. In the 6th arrondissement of Paris, a right-wing candidate called Wiedemann-Goiran held an electoral meeting which confirmed my views. Conversely, the Socialist meeting was directed by an enthusiastic young woman called Mireille Osmin, whose anti-capitalist speech was so inflamed that I didn't feel the need to move any further leftwards.

I interpreted the crushing victory of the Front Populaire as the culmination of a secular struggle. In other countries, particularly Spain, analogous struggles were occurring. The Frente Popular, which *Le Matin* referred to as the "Frente Crapular"[14], had many similarities with the French movement. During the previous decades, Spain had lived through alternating periods of dictatorships (Primo de Rivera) and vaguely democratic regimes under the monarchy of Alphonse XIII. But the Frente Popular threatened to go much farther and just as in France, progress in the direction of Socialism seemed inevitable. But we know that such peaceful transitions are historically extremely rare. When General Franco came to Spain on July 18, 1936, I immediately took sides with the republicans, with a leaning towards the leftmost parties on the Spanish political spectrum. French anarchist journalists exalted heros such as Buenaventura Durruti and Francisco Ascaso, who died very early in the civil war. Although anarchists in France represented practically nothing, in Spain they had considerable political power and in particular a central union, the CNT (Confederación nacional del trabajo)[15], which was particularly strong in Catalonia, the bastion of the Spanish left.

The whole of the French left-wing was shaken by the tragedy of the Spanish civil war. Mussolini and Hitler immediately sent arms and logistical aid to Franco, whereas in France, Léon Blum decreed a policy of non-intervention, citing a European pact which had actually been signed by all European nations. I don't need to give an analysis of these well-known events here. Some of

14) "Frente Popular" means *Popular Front*; the wordplay is on "crapule" which means "scoundrel"
15) CNT: National Workers' Confederation

Léon Blum's arguments were comprehensible, and it is true that French military intervention would have necessitated considerable forces and could have caused a rupture with Great-Britain.

Although I felt close to the ideas of *La Patrie Humaine*, I could not share its pacifism. Almost all the students at the ENS had demonstrated against the mandatory two-year military service. The situation was complicated. One the one hand, the majority of the French population didn't care about Nazism. The two-year service was required in order to reinforce the French army, which to us was reactionary. Yet there was an uncontestable blindness in our struggle against military service right in the face of Nazism. The dilemma was a real one: having to vanquish Nazism with a reactionary army sympathetic to Franco! My father-in-law Paul Lévy favored Franco early on: of course he detested Hitler, but he feared that if France did not support Franco he would become an ally of Nazism.

In 1936, Nazism was still weak militarily. Hitler was rearming Germany at a prodigious rate, but he had only held power for three years and he needed at least another three to start a war. Non-intervention deeply disappointed me and appeared to me as a major political error. I absolutely disapproved of Léon Blum and from that point, my path led me away from the Socialist Party which I felt no longer represented my ideas.

This context, tragic in Spain, was a moment of incredible development of the workers' movement in France, and of the general popular conscience. We remember the strikes of June 1936 and the essential conquests made by the working class: salary increases, the forty-hour week, paid vacations. During the summer of 1936, the beaches of France were invaded by innumerable workers taking their first vacations and seeing the sea for the first time in their lives. In spite of the terrible threat which hung over the country because of Nazism and Franco, France was uplifted by a feeling of incredible enthusiasm. It was quite amusing to read the headlines of the pacifist newspapers. *Le Libertaire* and *La Patrie Humaine* both believed in total unilateral disarmament, and previously, they had loudly proclaimed that war itself was an evil greater than any of the evils it struggled against. Yet as soon as Franco arrived on the scene, the headlines of *La Patrie Humaine* screamed "Cannons and Airplanes for Spain!" Their pacifism was dead and its bankruptcy openly admitted.

The Spanish republican government assembled a wide coalition containing a party similar to our radical party, Socialists led at first by Largo Caballero (who observed a careful silence on the subject of the execution of non-Communist leftist militants by the Spanish Communist Party) and then by Negrin, the POUM

(Partido obrero de unificación marxista)[16] and even anarchists. The anarchists, who had always refused the compromise of belonging to a government, participated in this one, led by a woman, Federica Montseny. Day by day, we followed the defeat of the Spanish republicans and the crushing victory of the nationalists. I lived these events as a personal tragedy and felt deeply that they were the prelude to a world war. One of the long list of Pétain's ignominious misdeeds was the sending of Spanish republican refugees in France to concentration camps (created by Daladier) followed by their deportation to the Nazis after 1940. A remarkable description of the life of Spanish prisoners was given by Arthur Koestler in his book *La Lie de la Terre* (*The Dregs of the Earth*). Spanish republicans were not the only ones to be given up to Hitler; Pétain deported anti-fascist German refugees as well. The German Social-Democrat Rudolf Hilferding, famous for his book *Das Finanz-Kapital*, took refuge in France in 1938. After the French defeat, his friend the mathematician Lipman Bers, a Jewish refugee who left Lettonia in 1939 with incredible difficulty and who succeeded in obtaining an American visa in 1940, and who told me this story, tried desperately to persuade him to come to America as well. "No," Hilferding kept replying. "France would never deliver me to the Nazis." He was deported and executed.

## The Moscow trials

In my letters to Marie-Hélène, I told her every detail of the changes that were occurring in me. She had no access to direct information, because there were no left-wing newspapers in the sanatorium and we had no money to subscribe to them ourselves. I read as much as possible and then told her about them or sent them to her. We shared our every thought.

My work in mathematics slowed down at this time. My father was horrified to see that I was "hanging out doing nothing". In fact, I wrote my daily letter to Marie-Hélène and read the newspapers in the morning (except on days when there was a seminar at the ENS). In the afternoon, I went to the ENS until about five o'clock to talk about mathematics and politics. From five to eleven, except for dinner, I worked at mathematics, adapting my working methods to my rather weak capacity for hard work. I worked in my room, wearing only a shirt, with the window wide open even in winter. Everyone else put on a coat to enter my room. I did exactly what I now often reproach students at the Ecole Polytechnique with doing: I lived on my previously acquired knowledge.

---

16) POUM: Workers' Party of Marxist Unification

Perhaps this is a slight exaggeration since I obtained the best grades in the final examinations in differential and integral calculus and in probability.

During the summer of 1936 which I spent in Saint-Gervais near the sanatorium, with Marie-Hélène's family, I saw Marie-Hélène every day and we discussed politics from every angle. After endless reflection and discussion, we decided that on my return to Paris in October I would join the Communist Party. Of course, the language of the Communists lacked elegance. I learned plenty about their formal discourse and sectarianism, but I imagined that a grand-working-class-party-preparing-the-revolution simply had to make use of a little propaganda. Many French intellectuals joined the Communist Party around that time. The grand party of the revolutionary left seemed to have a bright future before it. At first it was small, but eventually it obtained seventy-three seats in the Front Populaire, elected in 1936.

In the spring, my uncle Jacques Debré and Auguste Detœuf, who were fashionable industrialists and close friends, were invited to the USSR. Jacques Debré was a remarkably intelligent man. He had been a student at the Ecole Polytechnique and was the CEO of the General Electric Company. He was familiar with music, literature, painting, mathematics, physics, chemistry, biology, medicine and natural sciences; like my mother, he adored animals and butterflies. One could discuss anything with him, and he loved to talk. He had a fantastic sense of humor and laughed loudly at his own jokes, with a warm and communicative laugh. One day when I offered him some chocolate he confided to me "I have no sympathy for people who don't like chocolate". Detœuf, a progressive, modern, reformist industrialist, was a man with wide views. The two of them were charmed by the USSR. Upon his return, Jacques Debré wrote a very positive little book about what he had seen there (he changed his mind rather radically later on). My uncle said that he had seen Bukharin in public meetings. Bukharin had wanted to meet him, but it hadn't been possible. We understood why shortly afterwards.

The incredible Moscow trial burst like a thunderclap (at least for me) on August 14–19, 1936. The principal accused was absent: it was Leon Trotsky. His book *My Life* had been innocently offered to Marie-Hélène by a rather right-wing family friend who thought it might interest me. He had no idea of the influence the book would have on me! At first I didn't pay particular attention to it; I attributed the quarrel between Trotsky and Stalin to a divergence of political views such as may occur in any revolution. But Trotsky was accused of being a traitor, a Nazi agent, of having betrayed the revolution from the beginning and even in 1905, when he was president of the Petrograd Soviet! The insane confessions of the other accused such as Zinoviev and Kamenev, striking their breasts and declaring that they had also been traitors from the beginning, rendered the scenario absolutely absurd. The twenty-eight of them were condemned to death and executed.

The chief prosecutor Vychinsky called them "viscous rats" and "lewd vipers" and demanded that "these mad dogs be shot down without exception". These expressions, which later acquired an almost humorous resonance, added humiliation to Machiavellism. The newspaper *L'Humanité* took to copying them. It was vomitable to read it during the months of August and September. The real mad dogs were the Communist newspapers, who were completely directed by the IIIrd International and the Communist Party of the Soviet Union. Stalin, whom until then I had considered as a political leader of some intelligence, revealed his true odious nature. I'm not mincing my words. The Communist intellectuals who observed the Moscow trials in the summer of 1936 and continued to read *L'Humanité* and to accept its criminal inventions displayed a cowardly and unacceptable blindness.

I saw two aspects of this tragic farce: some intelligent people seriously believed in the trials (even though their fakery leapt to the eye), and others, believing that the dice were loaded, considered that such "accidents" should not be held against the "realizations" of the Soviet Union. History showed that an event of this kind, an enormous trial based on inventions and fabrications from A to Z, could not in any way be considered as merely a minor episode. The amazing dishonesty of the accusers recalled the Dreyfus affair which my parents often spoke about. It was clear to me from the start that the trials were fakes. It's even interesting to notice that this was obvious to me, although I had no political experience whatever, whereas experienced Communists in France and elsewhere didn't notice anything; they were apparently blinded by their credo and their personal relations. Even the League for Human Rights, founded after the Dreyfus Affair, justified the trials, claiming it wasn't interested by this "quarrel between Bolsheviks". Françoise Basch mentions and deplores this in her biography of her father. From that point on, it was clear that no political group in France could hold any attraction for me. I felt an anguishing loneliness, just at a time (right after the Spanish war) when everything in the international situation demanded political involvement.

There were three big trials, in 1936, 1937 and 1938. Zinoviev and Kamenev were executed by firing squad after the first one, Piatakov and Radek after the second and Bukharin and Rykov after the third. The great Rakovsky, author of the admirable text *The Professional Dangers of Power*, was condemned to twenty years at the third trial, and assassinated in prison on the orders of the KVD in September 1941. At each of these trials, the victims of the following trial were announced. As the height of infamy, Iagoda, who had been the grand inquisitor of the previous repressions and had actually presided the firing squads of the first two trials was himself accused, along with Bukharin and Iekov, at the third one. He had just been replaced by Yejov. Associating the judgment of revolutionaries like Bukharin and Rakovsky to that of a murderer like Iagoda was disgusting. Unlike the others, Bukharin, one of the most courageous of the accused, refused to admit that he was a traitor: he restricted himself to stating that he had made mistakes. After the first Moscow trial, Lenin's widow Nadezhda Konstantinovna Krupskaya, sent congratulations to Stalin. Since she had a lot of international prestige, her gesture particularly disturbed us. It's clear that she never believed in the farce for a single instant, but she must have been forced to sign, or perhaps her opinion was not even consulted.

## Meeting with some remarkable Trotskyists

I knew nothing about the existence of a Trotskyist party nor of the IVth International. I learned about these things thanks to an interview of Fred Zeller, an important militant Trotskyist, in the popular newspaper *Le Petit Parisien* (*The Little Parisian*), which even mentioned the address of the POI (Parti Ouvrier Internationaliste)[17], in the French section of the IVth International: it was 4, passage Dubail, near the train station Gare de l'Est. I went there and met with Trotskyist militants whose political intelligence struck me.

I wouldn't be able to list the names of all the people I met there. My first contacts were with Paul Parisot, Albert Demazière and Marcel Beaufrère, who initiated me to Trotskyism. These men were sensible and generous; they guided my first steps as a militant. The first two played an essential role in the logistical organization of the post-war Trotskyist electoral campaigns. Marcel Beaufrère had a job with France-Presse. He became a close friend of mine and often helped me. Much later, he told me a touching story. Trotskyists detest the French anthem *La Marseillaise*, which represents to them the triumphant chauvinism of the First World War (and later, the betrayal of Pétain), and they refused to sing anything except the *International*. But when Beaufrère returned exhausted from deportation, he was placed in a camp for survivors, and just for him, they played the *Marseillaise*. After the beatings and horrors he had lived through, the anthem moved him to tears. He died just a few months ago; I regretted him deeply. An old past is disappearing.

17) POI: Internationalist Workers' Party

Marcel Hic was one of the major figures of the party. He was my age, but had been a Trotskyist since 1933, and he immediately made a profound impression on me. I will talk more later about his activities during the war: he died in deportation. I was deeply saddened by his death, and indeed it was a terrible loss for the party and for the future of the French left-wing. I also knew David Rousset, another important leader of the party; he was a Gaullist deputy in 1968, after rupturing with Trotskyism in 1947. After surviving concentration camp, he wrote some important books, *L'Univers Concentrationnaire* (*The Concentrationary Universe*) and *Les Jours de Notre Mort* (*The Days of our Death*) which won the Théophraste-Renaudot Prize. After his break with Trotskyism in 1947, he became a senator. He combated the Soviet and Chinese system of concentration camps ceaselessly, and he was also one of the best analysts of the evolution of the USSR, as can be seen from his book *La Société Eclatée* (*The Exploded Society*).

Pierre Naville was the main theoretician of the party, and he had a lot of influence on the younger members, particularly during the Spanish war. He was a prisoner during the Second World War, and I hardly saw him afterwards. He also left the Trotskyist Party, and later, together with Maurice Nadeau, Gilles Martinet and Charles Bettelheim, he founded *La Revue Internationale* (*The International Review*). He became a research director in the National Center for Scientific Research, in sociology. He died a few years ago.

I always remained very close to Roland and Yvonne Filiatre, who were among our very best friends. Both workers with a high level of political culture, they were among the first Trotskyists in France. They were deeply honest people, and Filiatre preserved this same straightforwardness in his relations with company directors. When he organized a strike, he informed the directors precisely what the strikers' requirements were, avoided every kind of demagogy, and stopped the strikes when his demands were satisfied. The couple evolved rather like I did: they fully realized the transformations of the proletariat and the political situation, and the divorce between Trotskyism and reality eventually became as clear to them as it did to me.

Filiatre had had a difficult youth. He was the second of a family of nine children. His father, a director at the Opera-Comique, died in 1920 when Roland was just 20, leaving seven children. The eldest girl was already married, but the last child was a little girl of two. After the death of his father, Roland became the head of a family of five small children. He stopped his studies and got a job as a qualified worker in a telephone and electric company.

His wife Yvonne was an orphan, and at twelve was placed in a family by her uncle to work as a maid and take care of a baby. At the age of eighteen, she asked her uncle to free her and worked as a stenographer in a telephone company, where she met Roland. He was taken prisoner in Dunkirk, and made to work in a factory in France, building war material for the Germans. As much as possible, he tried to sabotage the work. He also came in contact with Germans against Hitler, and worked with them to transmit information to the English. Roland and Yvonne were denounced and arrested. She was sent to Ravensbrück, and then to Bergen-Belsen. He was horribly tortured. What makes a man give way or not under torture is not necessarily related to the physical pain. He told me the following story. During a particularly horrible scene of torture, Roland felt that he was about to give way, although he had said nothing until then. But suddenly his torturer tossed out "We just arrested Craipeau." It was a terrible shock. Seeing that it was working, the torturer added "We arrested Francis as well." But Francis was Craipeau's alias. Roland felt sure that they had not understood this, and that nobody had been arrested, and he suddenly found in himself the strength to resist. He was deported to the Dora camp. Both Roland and Yvonne survived deportation; they struggled to survive by thinking constantly about their little girl, Rolande. We also knew Rolande well, and later, her husband. They are both teachers and worked with Bertrand on the new teaching programs in the Ville Neuve in Grenoble. Roland died in 1991.

I was and always remained very close to Ivan Craipeau, one of the main figures of the party. He was a remarkable theoretician, whose political divergence with Trotsky on the subject of the defense of the USSR was the subject of interminable discussion for years. Trotsky never ceased explaining that the USSR was a "degenerate workers' state" which had to be defended in case of war, whereas Craipeau considered it to be a form of "state capitalism" to which Lenin's revolutionary defeatism applied perfectly. I believed that one should defend the USSR, but I was more or less indifferent to these subtle efforts to define the nature of the Soviet Union precisely. Craipeau participated in almost all of the important discussions within the party, which he left around the same time as I did. He wrote an autobiography for his grandchildren, partly in order for them to understand how much of his existence he had devoted to fighting for an illusion. From what I know of certain episodes, his life must have been extremely exciting.

After the war, I became very close to a couple of Greeks, Michel Raptis, whom we called Pablo, and his wife Hellê. They were brilliant, deeply analytic thinkers, sympathetic and generous, but they remained pure Trotskyist, absolutely convinced that a world revolution was about to take place. Pablo directed the IVth International at one point, and a "pablist" group was actually formed.

In spite of his incorrigible optimism and many of his ideas which I really cannot share, we remained close friends. I also conserved excellent relations with his comrade Gilbert Marquis. Pablo was well-known in Greece where he worked with two major newspapers. He died in February 1995, mourned by numerous friends in France. His burial in Athens was accompanied by a dense crowd, including Papandreou and two government Ministers.

Gerard Bloch was in high school when I met him in 1936. His mother was a friend of Marie-Hélène's family and had been her piano teacher. During the summer of 1936, seeing me reading Trotsky, he figured that we must be on the same side, and we quickly became close. He was highly intelligent, and particularly good in mathematics. He won a first prize in the Concours General, and we all thought he would become a mathematician, but he sacrificed his scientific ambitions to politics. He suffered from a serious case of asthma which he treated with cortisone, but which eventually caused his early death. After the agrégation, he taught in high school where he displayed enormous pedagogical talent. During the Occupation, he was deported, but he survived. He remained Trotskyist until his death a few years ago. He never forgave me for breaking with Trotskyism. Looking me deep in the eyes, he would say "When *I* look in the mirror, *I* don't see the face of a traitor." This typical radical left-wing formula did not lessen my deep affection for him.

There were two Trotskyist parties in France in 1936, both of which belonged to the IVth International: the POI (to which I belonged) and the PCI (Parti Communiste Internationaliste)[18], many of whose members I came to know: Pierre Frank, my friend Roger Foirier, Grimblat called Privas, and the eternally young Jean-René Chauvin, among others. The PCI was even more sectarian than the POI, but the two groupuscules insulted each other incredibly. At the end of the war they merged, but only to resplit again later on. The Trotskyist organizations were divided into a whole archipelago of different tendencies, to the point where the situation became untenable, as the whole activity of the party consisted in the quarrels between factions. Since 1968, members of Trotskyist parties have no longer been exposed to a shower of amazing Stalinist insults which no one would accept any more; indeed, the nebulous Trotskyist of today has nothing to do with those of an earlier time. Although I know the Krivines well, I've never met Arlette Laguiller, although I have on occasion marched in demonstrations of her party, the Lutte Ouvrière (Workers' Struggle) which is the most worker-oriented, both in theory and in practice, of the Trotskyist organizations. The group "Socialism or Barbarism", founded by philosophers Claude Lefort

18) PCI: Internationalist Communist Party

and Cornélius Castoriadis, was an improvement on the other groups thanks to its greater concern with reality as well as a subtler political understanding. After I broke with Trotskyism, they continued to play an important role in the French left-wing, and I always appreciate their opinion on any question.

In October 1936, I found my own path by joining the IVth International. I remained Trotskyist for eleven years, until 1947. Every week I bought the Trotskyist newspaper *La Vérité* (*The Truth*) (which took the name of the pre-revolutionary Bolshevik paper *Pravda*, which is still published in Russia). After reading it, I sent it to Marie-Hélène. She says now that she was never as convinced as I was, and that she never really believed that a world revolution was about to occur. It may be true, but it might also be that with hindsight, the idea is so bizarre that no one can really believe that they ever actually believed in it. At any rate, there was no real divergence between us at the time.

## Trotskyism as knowledge of the world

Lies did not cease. I remember reading a little brochure edited by the Communist Party: *Moscow, Nuremberg, two conferences, two worlds*. Nuremberg represented total slavery and regression, and Moscow the future of humanity – a future which I totally rejected. By opposing these two poles, the propaganda intended to convince intellectuals that in reality, there were only two choices, Nazism or Stalinist Communism, and no other way.

I bought a copy of Trotsky's book *The Revolution Betrayed* in a little bookstore downstairs from my apartment, shortly after its appearance in September 1936. It is an excellent book, one of Trotsky's best. The criticisms of Stalin are more political than personal, particularly those of his doctrine of "Socialism in a single country". I read and reread it. Almost all of my political culture comes from reading; reading is always a source of personal reflection. Trotsky writes that Socialism is international by its very essence, and should triumph first in the most advanced countries. Germany should have been the cradle of Socialism, as Marx and Engels hoped. But no, it ended up being born in Russia, a backward country, and Trotsky maintained that it was impossible to establish Socialism in a single country, and all the less so if that country was undeveloped, with a small proletariat and an enormous peasantry. It was necessary to live, and since Russia had had a revolution all alone and was completely isolated in the capitalist world, it seemed that there was no other alternative to constructing Socialism in a single country. But Trotsky maintained that such a Socialism was bound to degenerate, slowly absorbing the Communist Parties of the IIIrd International and setting them, not to the service of their own proletariat, but to the service of the internal and external politics of the USSR. This phenomenon was already noticeable in 1936 and became flagrant in the following years.

This is why, when Trotsky was asked why he had not tried to seize power by using the Red Army, with which he was very popular since he had created it, he always answered that degeneration was inevitable and that he did not want to be its instrument. I'm not absolutely convinced of the truth of this answer. The USSR was attacked on every side in the years following the revolution. Fascist bands attacked from the interior, perpetrating massacres and pogroms: Koltchak, Denikine, Petlioura, Wrangel committed mass killings. Furthermore the Allies, after their victory in Germany, send troops to besiege the new Soviet Republic on every front. At one point, the republic consisted in hardly more than the ex-grand-duchy of Moscovia. Those years were terrible and ended only in 1921. "War Communism" and Bolshevik repression have their origins in these tragic episodes. It is certainly true that if Trotsky had taken power in the years following the death of Lenin (a practically impossible scenario, since Stalin was already master of the situation), he also would have had to have recourse to repression in the untenable effort to establish Socialism in a single country. But I argue with Trotsky's reasoning when he "refuses to transform his prestige into power", saying that he would have been unable not to behave "like Stalin", which he didn't want to do. The two men had fundamentally different characters: Stalin was astute and dishonest, Trotsky almost too honest, expressing his point of view so straightforwardly that he offended everyone. He was an intellectual, a theoretician and a great orator, but one cannot say that he was a statesman, even less a politician or an apparatchik. His activity during the revolutions of 1905 and 1907 owed a lot to Lenin's support: with Lenin dying or already dead, Trotsky had no power compared with Stalin, the ultimate apparatchik.

I believe that the trials which took place over several decades were specific to Stalin. The Stalinist trials, which lasted for a long time, bore no resemblance to any tribunal erected by other dictators. Trotsky would not have resorted to torture. Perhaps he would have founded some kind of regime of Terror, because he did have some resemblance to a new Robespierre (like him, he was "incorruptible"). But he would never have had recourse to individual or massive assassinations. Perhaps he would have been intransigent, but he wouldn't have been Stalin. Anything, even an authoritarian regime installed by Trotsky, would have been better than the Machiavellian and brutal dictatorship of Stalin. It's not really possible to speculate about these things; the fact is that Trotsky was absolutely unable to face up to the situation and take control in the struggle for Lenin's succession. He could only passively accept his destiny until his exile to Mexico. There, he remained the pugnacious leader of a faction defending his views against intellectuals, while showing himself to be a sociable and cultivated human being.

The falsification of history is another aspect specific to Stalinism. The past is annulled to make way for a purely fictitious past. Revising history was not enough for him; Stalin liked to create versions of photographs of the successive sessions of the central committee of the Communist Party, in which former members fallen from favor were eliminated and replaced by newly favored members supposed to have participated in the sessions from the first days of the revolution. These mystifications even reached the domain of science under Stalin's reign, and even more under Brezhnev. For instance, in the most anti-Semitic period of Brezhnev's reign, the names of Jewish authors were quietly eliminated from the bibliographies of scientific articles.

In addition to these modifications of history, Stalinism saw a fair number of political about-faces. Thus, just a few years after the defeat by the Nazis, a united front tactic was substituted for the "third period" of the Communist International, in which Socialists were called social-fascists. These lies were reinforced by an incredible "personality cult" around Stalin, the father of the peoples, the leader of genius, etc., etc. and accompanied by the usual formal discourse which every knows, and which made Communist newspapers impossible to read, and later affected even Communist popular democracies. Without intending to vex anyone in particular, I must say that this kind of formal discourse is typical (although less caricatural than among Communists) in Trotskyist and Socialist circles and even in unions, and even within the right-wing government. There are heaps of examples: "social", "exclusion", "change", "equal opportunity", and today's new expression: "social fracture", are phrases which everyone rushes to pronounce and which are accompanied by every possible political sauce, so that in spite of their absolutely legitimate original meanings, they end up as pure formulas. Whole talks are often based on this comfortable system of formal rhetorical phrases.

Stalin particularly emphasized the development of heavy industry in the USSR. I think Trotsky was right not to criticize this. Today, we consider it with a certain irony, but one should consider the reality of the situation. All the big western countries went through a period of powerful development in heavy industry until quite recently. Cars, railroads, airplanes, boats, electricity, agricultural machinery; the whole of daily life was based on steel and metals. They remained fundamental in the years following 1929, in spite of the economic crisis. It was really necessary for the USSR to reach the same level of industrialization. Lenin said "Socialism here consists in the Soviets[19] and the electrification of Russia." Only later did the country attain the leisure to develop light industry, quite late with respect to western countries.

Trotsky also considers the problem of the collectivization of land. I will talk more later about the disaster of the Russian effort to eliminate kulaks (rich peasants). Russian ideas about the ownership of land were rather confused. The slogan "Bread, peace, land for the peasants" had dominated the Russian revolution. The Russian mode of production was the so-called "Asian" system, where original "primitive" rural communities, transformed into groups called "mirs", formed the main framework for the collective farming of state property. Marx actually asked himself if this structure could serve as a basis for the Communist development of the countryside. Bolshevism had promised land to the peasants and had delivered on its promise; the forced collectivization of 1929–30 took it away from them again. It was a total failure. From the point of view of Communist ideology, giving land to the peasants was a bourgeois measure. With the exception of the Israeli kibbutzim, which were set up in a society which was particularly highly developed technologically, intellectually and culturally, collective farming in the form of sovkhozes and kolkhozes has been a disastrous failure in every country where it was attempted. The nationalization of industry and banking, which was once considered as a necessary step towards social progress, is now widely doubted. Industries which were once nationalized are being privatized. This is obviously part of the "class struggle": the left nationalizes, the right privatizes. I don't particularly approve of the rash of privatizations going on now in France, guided mainly by considerations of profit. I did hope for total nationalization in 1936, but I no longer consider it a panacea. Every case is different.

Although it contains many remarkable ideas, some of the assertions in *The Revolution Betrayed* are rather surprising. Stalin described the new constitution of the USSR in 1936 as "the most democratic in the world". This may have been the case but the constitution was never really applied. In reality, Soviet citizens voted with a pistol to the head for the candidates presented by the Party authorities, who invariably won with 99% of the vote. The constitution required that all votes be secret, which of course is eminently democratic. But Trotsky disapproves of this: he says that the only justification for a secret vote is that in a society which still contains different social classes, the bourgeoisie must accept that the working class protect itself against it, just as the bourgeoisie accepts

19) The Soviets were elected government councils on the local, regional and national levels, culminating with the Supreme Soviet (the Russian word "soviet" means "council".

other democratic principles erected against itself. Trotsky explains that in a society which really has no more differences in social class, between workers and exploiters, all votes can be open. But the differences between people do not stem uniquely from differences in social class! For instance, a vote to accept or reject an applicant for a job as a university professor will never be based on such reasons, and since there will obviously be differences of opinion, it's normal that the vote be secret. To consider that every difference in voting comes from a difference in social class is outrageously simple. It is true that Marx already affirmed that the entire history of society was the history of the class struggle. I never accepted such absurd reductions. Trotsky considered that the secret vote instituted by the constitution of 1936 proved that the Soviet Union was not Socialist. This concept reminds one of what biologists call mechanistic concepts, according to which all the phenomena of life are reduced to a small number of physical and chemical phenomena. Fortunately, we are understanding ever more deeply the real complexity of life.

I avidly read the other books by Trotsky, particularly the four volumes of his *History of the Russian Revolution* in the summer of 1937 and also *The Permanent Revolution*. This book contains Trotsky's personal doctrine, and it impressed me a lot. It was adapted to the period just preceding the Second World War, but afterwards, this doctrine became doubtful. It is true that the revolution which began in Russia never really stopped throughout the century. It is also true, as he claims, that in underdeveloped countries, the bourgeoisie was also underdeveloped and thus incapable of starting a bourgeois revolution. This bourgeois revolution was to be undertaken by the proletariat, which should then continue with a proletarian revolution. Trotsky then foresaw the difficulties which would arise: bureaucratic degeneration and a peasantry with practically no role. In fact, after the Russian revolution, there was no revolution of any kind until 1945. But after 1945, there were numerous peasant revolutions directed by the Communist Party or similar parties. There are still uprisings on the part of the peasantry, but the working class is absent from them. Apart from this, Trotsky's scenario of the bureaucratic dictatorship of a unique, so-called proletarian party remains absolutely valid.

I learned a great deal about Stalinism and Trotskyism from the books of the Russian writer Victor Serge. This author survived concentration camp only because of a strange circumstance. His friends regularly sent him books which he never received. The functioning of the complex bureaucratic game allowed him to be paid damages for each package he did not receive, so that he was able to survive. Thanks to an international petition of writers, he obtained permission to leave the country, but he was the last person to obtain this favor. His book *If it is Midnight in the Century* is captivating; it appeared just before the war and gives an accurate picture of the tragic and complex political situation in which Nazism and the Stalinist degeneration of the Russian revolution was plunging the entire world. Around the same time, I read the book *Stalin* by Souvarin, which I found extremely instructive although Trotskyists criticized it severely. As for the work of Arthur Koestler, I was fascinated by every bit of it. After the war, I enthusiastically read *Le Zéro et l'Infini* (*Zero and Infinity*), the analysis of a Stalinist trial, and *La Tour d'Ezra* (*Ezra's Tower*), a book about Israel.

## Trotsky and Nazism

Trotsky's famous article about Nazism, written shortly before Hitler took power, remains a monument to his clairvoyance. The IIIrd International had inaugurated what was known as the "third period": politics of class against class. According to this theory, which was established by Stalin himself, two classes were essentially confronted: the proletariat and the bourgeoisie, the latter containing the whole of German Social-Democracy. Socialists were called social-fascists, the Social-Democratic Party was called the "moderate wing of fascism", Social-Democracy and fascism were twin sister and brother, representing the principal enemy of the proletariat. I still possess a French electoral poster from the third period, an absolute monument to sectarianism and imbecility. The German Communist Party refused any collaboration with the "social-fascists", causing the inextricable confusion of the last presidential elections in Germany, in 1932. Apart from Hitler, the candidacy of Marshal von Hindenburg made the extreme left (Communists and Trotskyists) say, with some justification, that a vote for von Hindenburg was a vote for Hitler. Then there was the Communist candidate, Ernst Thälmann. The Socialists called for von Hindenburg, in the hopes of blocking Nazism. The German Trotskyist Party, which didn't have much influence but was made of brilliant militants, voted for Thälmann who obtained four million votes. A union of Socialists and Communists would have generated new dynamics, but the Social-Democrats were not ready to accept a united front any more than the Communists. They were so hostile that at the referendum on the Social-Democratic government in the Land of Prussia, on August 8, 1931, the Communist Party joined the Nazis in voting against. Of course, one can't be sure that a different strategy on the part of the Communists would have avoided Nazism, but the example of the Front Populaire in France suggests it. There were, however, other factors. Major sources of German capital (Krupp, Thyssen) massively supported Hitler before he came to power; Daniel Guérin gives a masterly proof of this in his fundamental book *Fascisme et Grand Cap-*

*ital* (*Fascism and Grand Capital*), one of the books which most influenced me at the time (together with his brochure *La Peste Brune est Passée Par Là* (*The Brown Plague Passed Here*). I knew Daniel Guérin very well until his death. He was an anarchist with Trotskyist tendencies, and later joined the PSOP (Parti Socialiste des Ouvriers Paysans)[20] of Marcel Pivert. He was kind enough to welcome a meeting of the Bourbaki group on his property near La Ciotat after the war. For two weeks, he shared our life and our meals in the friendliest way. I even went butterfly-hunting one night and caught a death's-head hawkmoth (*Acherontia atropos*).

At any rate, this calamitous situation progressively brought Nazism to power, while the German left slowly decomposed. It's instructive to recall the number of votes obtained by the Nazis in each of the successive elections: 2.6% in 1926, 18.3% in 1930, 37.2% in 1932 (although with a loss of seats which made Léon Blum say, curiously enough, that Hitler had lost not only power but any prospect of power). With a total of less than 40% of the vote, and faced with insurmountable difficulties, president von Hindenburg was obliged to name Hitler chancellor of the Reich. I learned of these events only at the ENS and not in the years 1933–34, when I was still an unpolitical high schooler. As soon as he took the position (on January 30, 1933), large-scale political repression became the order of the day, without any objection on the part of von Hindenburg, and without the workers or the democrats being able to lift a finger against it. Thälmann sent out a desperate last-minute appeal, speaking for the united front of workers' organizations, exactly on January 30, 1933. He was arrested, sent to concentration camp and executed by firing squad in 1944. Nazi troops marched that day in the streets of the towns. Hindenburg, whose ideas were completely unclear at that point, said "I didn't know we had taken so many prisoners". He died soon afterwards, leaving Hitler with full powers. The destiny of the world was signed and sealed. After this historic defeat Trotsky considered that the IIIrd International had failed, and that it was absolutely necessary to found a IVth International.

It is remarkable how lucid Trotsky was about the necessity for a united front in Germany, before Hitler's arrival to power, and that the Trotskyists were always so lamentably incapable of forming a united front with anyone at all. It is true that it was not so easy for them, since anti-Trotskyist calumny was so powerful that Trotskyists would have been unable to ally themselves with Communists even had they wanted to. And it was even less possible for them to ally themselves with the Socialists, but without the Communists. Still, they

20) PSOP: Socialist Peasant Workers' Party

should have worked towards these goals. They never even envisaged an alliance with Marcel Pivert's leftist group in France, or with the POUM in Spain. Even the Trotskyist theories on the subject were wrong. I became conscious little by little of my own hesitations on the subject. Trotsky's formula "March separately, strike together" meant that even if common actions were decided on, free expression of criticism was to be allowed. But a united front of this kind cannot work. In the rare cases in history where Communists and Socialists presented a united front, criticisms were hushed and even scandalously eliminated (I'm thinking about the Algerian war). It is necessary to establish a contradictory dialogue, but the Trotskyist Party thought that they could go on with their polemics during a period of united front, just as though the united front didn't exist. I remember one discussion during the war on the necessity of proposing to the Communists in the Resistance the formation of a united front. Once again, this was impossible, because Communists didn't hesitate to assassinate Trotskyists. Once I asked a comrade on what a united front could possibly be based. He answered in stentorian tones "On proletarian internationalism". On such a basis, an alliance with the Communists who like to recommend "taking out the Boches"[21)] appeared purely fictional. A text from the newspaper *La Vérité*, in the middle of the war, proposed to the Communists to form a united front, in a message beginning "Stalinist comrades . . . ."

After Hitler came to power, most of the German Communists who had not been arrested fled to the USSR. One of the most remarkable of them, Heinz Neumann, was the victim of one of the numerous liquidations which touched almost every foreign former Communists after the Moscow trials. His wife, Margarete Buber-Neumann, imprisoned in a Soviet concentration camp under Stalin, was delivered to the Gestapo after the German-Soviet pace and sent to Ravensbrück, where she survived. In a book written in 1947, she compares the Soviet camp to the Nazi camp. She soon realized that the essential difference was, that in the Nazi camps, even those which were not death camps, the prisoners died massively from starvation, exhaustion and ill-treatment, whereas in the Soviet camps they died somewhat less (except in 1938, when all the Trotskyists were executed in a single blow). It's a miracle that she was able to survive. I gave her book to everyone around me who refused to believe in Stalin's infamy. In Ravensbrück, Margarete Buber-Neumann became close to another prisoner, Milena, who had been Kafka's beloved in the 1920's. In the biography she wrote about Milena, she describes her incredible courage. Milena died in the camp. Apart from the fascination and the emotion of this friendship,

21) "Boche": pejorative term for a German

born in unimaginable circumstances, between these two exceptional women, one is also struck by their resistance and by Margarete's penetrating inspection of her torturers in two different concentration camps. Margarete Buber-Neumann's book was widely read and certainly drove many people to leave the Communist Party.

## Trotsky and Spain

The role of the Communists in the Spanish war has been insufficiently discussed in France. The Spanish Communist Party, which was relatively weak at the beginning, became little by little all-powerful within the republican government, essentially thanks to the presence of international brigades dominated by Communists, and to the growing influence of the USSR in the support of Spanish republicans. The international brigades played a considerable military role in the combats. The volunteers in these brigades were political militants with unlimited courage and devotion. Even in the darkest night of this period, they obstinately believed in the future. But these Brigades were directed and efficiently exploited by direct agents of Moscow, who organized numerous political assassinations of anarchists and members of the POUM. These horrors, which we knew of already in 1936, have been amply confirmed by the Soviet archives opened in the last few years. Many revolutionary militants of the POUM were liquidated. Communists arrested and tortured its main leader, Andrès Nin, who had been Minister of Justice of the Generality of Catalonia, hoping to obtain from him various confessions in the style of the Moscow trials. But he wouldn't give in, so they assassinated him. His disappearance provoked international protests, then the affair was forgotten. But there were hundreds of other similar disappearances, such as that of Erwin Wolf, a secretary and bodyguard of Trotsky in Norway, of Czech origin, arrested in Spain in 1937 and never seen again. Communist movements in the Stalinist period were covered in silence. If, in a group which met daily, some member suddenly disappeared, prudence recommended under the Stalinist regime that nobody remark on it nor even notice anything. Obviously, such methods ended up by demoralizing the republican army and reducing the politics of the republican government to nothing. Communist repression also broke the Catalan resistance. It may be too much to assert that these internal massacres caused the defeat of the Spanish republicans, but they certainly contributed to it. Certain historians of the Spanish war have not explored this aspect sufficiently. Elena de la Souchère even succeeded in the exploit of writing an entire history of the Spanish war in which the Communist assassinations are not even mentioned.

In an article which is still a standard reference, Trotsky analyzed the politics of the republicans and showed why their choices led to defeat. Unfortunately, the politics of Trotsky and the IVth International with regard to the POUM were also filled with errors in my opinion. The POUM was a revolutionary, Marxist but not Stalinist party, quite close to Trotskyism. Maybe it did not push revolutionary ideas as far as Trotskyism, but that's still a grotesquely far cry from claiming that the POUM was a "centrist" party destined to be annihilated. The POUM resisted powerfully, both openly and clandestinely, throughout the duration of the Spanish war, and was only eliminated along with all the other republican parties, by Franco's victory. Instead of offering its support to this strong group, the IVth International supported the "Spanish Trotskyist Party", a phantom party which apparently never contained more than two members, Munis and Carlini, who paled beside the numerous leaders of the POUM (it is true that Carlini spoke of a dozen Spanish Trotskyists ...) This is just one example of the inglorious sectarianism of Trotskyism. An excellent and deeply moving film, *Land and Freedom*, is based on these tragic episodes of the Spanish war and the fate of the POUM.

In 1938, Marcel Pivert was excluded from the Socialist Party and founded the PSOP, the Socialist Peasant Workers' Party, a party decidedly to the left of the Socialists. I met Marcel Pivert and he struck me as a person with no real political force. But that was no reason to despise him and his party. Even Trotsky advised entering the PSOP, but the French Trotskyist Party refused.

## Sizeable Reservations

I deeply admired Trotsky, whom we in the Party referred to affectionately as the "old fellow". Lenin and Trotsky, probably Lenin even more than Trotsky, were certainly the main artisans of the October Revolution. Lenin was mainly a man of action, Trotsky a theoretician. He was exceptionally intelligent and deeply rigorous; I find his reasoning quite mathematical, and indeed, before becoming a professional revolutionary, he had considered becoming a mathematician. His arguments appeared irrefutable. In general, Trotskyist arguments were well-constructed logically. Unfortunately, while Trotskyist analysis of world events were remarkable, they were not very talented for action and their efficiency always remained close to zero.

After a short visit to France (Barbizon), Trotsky left for Norway and then for Mexico. During a trip to Mexico in 1953, we tried to meet his widow, Natalia. A Mexican Trotskyist who knew her invited us to lunch (the lunch was served by his wife, who did not have a place at the table and ate in the kitchen in spite of our vigorous protestations). We did meet her later, in Paris, through a common acquaintance, Marguerite Bonnet, a brilliant professor of literature who wrote her thesis on *Les Chants de Maldoror* by Lautréamont. We went to see *The Choeophores* by Aeschylus together. Several of my comrades from the French Trotskyist Party, in particular Ivan Craipeau who was always deeply involved in a discussion about the "defense of the USSR" knew Trotsky personally and corresponded with him.

Even when I joined the Trotskyist Party, I conserved my intellectual independence, and often thought differently from the party and even from Trotsky himself. Quite soon, I began to agree with several objections expressed by various people, particular Victor Serge. He was a great admirer of Trotsky, but he never hesitated to criticize him, especially on the subject of the revolt of the Cronstadt sailors. It was this revolt, in 1917, which brought the Bolsheviks to power in Petrograd. When Russia was on the edge of famine in 1921, they revolted again on March 2, but this time protesting against Soviet power. Lenin and Trotsky organized the repression of the revolt on March 18, and it was bloody and pitiless. Trotsky claimed that many of the original sailors of the 1917 revolt, who were advanced Communists, had died during the civil war, and the sailors of the 1921 revolt formed in his opinion a retrograde political force. This was only partly true and in any case certainly did not justify the repression. The revolution was passing through a difficult moment, but this event stained it with blood.

Trotsky's position on unions in 1920 was also erroneous. He thought that in the context of proletarian power, the role of a union was not to defend workers against the State, but to represent the State to the workers. There was a lot of opposition to this idea, because the so-called Workers' State was not what it claimed to be. And even if it was, every power should have counterpowers with which to contend. Trotsky soon understood this and changed his mind. In reality, all of the world's Communist parties in the following decades used unions as "transmission belts", and Trotsky very naturally opposed this. Even worse however, was his defense of the use of terrorism in his pamphlet *A Defense of Terrorism*, which horrified me. Of course, Trotsky was referring to the Terror of the French Revolution and not to the individual terrorism of our time. According to him, it was absolutely necessary to terrorize the adversary in order for the revolutionary party to conquer. Such ideas obviously have extensive theoretical and practical consequences. His conception of the militarization of work, which certainly had its origin in the role he played as creator of the Red Army, is absolutely unacceptable and condemnable. Along with perfectly pertinent arguments, his well-known brochure *Their Morality and Ours* contains some remarks which froze my blood as soon as I read them: "We clearly say to the bourgeoisie that when power will be ours, in the name of our principles we will refuse them the liberty that we now demand of them in the name of their principles." These germs of tyranny did not escape me, and they hurt, but I still conserved a positive approach; I considered them as errors left over from the past. I did however fully understand how real was the risk that the revolution could degenerate into ferocious dictatorship. The Trotskyist Party had its own system of formal discourse; I admit that I even used it myself for some time, even while thinking independently.

Rosa Luxemburg, whom I deeply admired and who was assassinated with Karl Liebknecht on the orders of the Social-Democratic government in 1919, passionately believed in the Russian revolution. Her principal criticism of Lenin and Trotsky concerned the dissolving of the Assembly. After taking power, the Bolshevik government had organized elections for this Assembly throughout Russia. The elections were apparently democratic, but the party which won a majority of seats was the Socialist Revolutionary Party, a democratic, non-Marxist peasants' party. Shortly afterwards, the Assembly was dissolved by the Bolshevik government. Rosa Luxemburg thought that it would have been better not to organize general elections so soon, since the victory of the Socialist-Revolutionaries was perfectly predictable. But once the elections had taken place, since they were democratic and the Socialist-Revolutionary Party had won a majority, they should have been obliged to share power with it. She criticized many other undemocratic and violent activities of the Bolshevik leaders. These leaders respected her deeply for having maintained a position of extreme left-wing internationalist Social-Democracy throughout the war. Victor Serge in particular agreed with her criticisms, and I agreed with him.

The politics of the French Trotskyist Party also caused me some feelings of doubt. During the strikes in June 1936, Trotsky wrote "The Socialist revolution in France is beginning". But the party, which was willing to join a Socialist-Communist united front, refused to join the Popular Front because of the presence of a bourgeois party, the Radical-Socialists. Because of this, the Trotskyist Party remained isolated and had no influence on the government of the Popular Front. This was, as usual, a consequence of the party's sectarian attitude. The Trotskyist Party predicted the defeat of the Popular Front, in which they were correct, but they attributed it to the presence of Radical-Socialist Ministers, whose activity negated all of Léon Blum's attempts at advanced politics. Even if some of their criticisms were justified and their predictions were certainly correct, they would still not have been dishonored by joining the government of the Popular Front, and they could always have left it if the situation had become untenable. Instead of this, Trotskyists heeded Trotsky's call for something called "entrism" into the Socialist Party, which they admitted as a political tendency. But quite soon they were accused of "plucking the fowl" – an expression invented in the 20's by the Communist Treint in regard to the Socialists – and excluded.

Trotsky was first exiled to Alma-Ata (in Kazakhstan) from 1928 to 1929, and then to Prinkipo (the island of Princes) in Turkey, where he remained until 1933. His prestige in the USSR was so high that when he arrived in Turkey he was received with honor by the USSR embassy there. He left for France, and then in October 1935 for Norway, from which he was ignominiously expelled by the Social-Democratic government shortly after the Moscow trials, in December 1936. He finally took refuge in Mexico. The Mexican president Lázaro Cárdenas, a great statesman, had undertaken to oppose American imperialism on the question of Mexican petrol. He had also sent thousands of rifles to the Spanish republicans. Cárdenas had sympathy for Trotsky and installed him in a little town near Mexico City called Coyoacán, where he lived in a house lent by the famous Mexican painter Diego Rivera, which was protected like a fortress by Mexican police and a force of Trotskyist volunteers. There were several efforts to assassinate him, for example the one organized by the Mexican Communist and bandit (and painter) Siqueiros, who was arrested, judged and condemned to two years of prison. Diego Rivera, who had been seduced by Trotsky's ideas, left frescos and murals immortalizing Trotsky. Marie-Hélène and I saw them during our trip to Mexico in 1953, when we were no longer Trotskyists. It's the only place I ever saw murals representing Trotsky. Unfortunately, Trotsky and Rivera eventually broke off relations for some futile reason. Some claim that it was because Trotsky insisted on considering himself a connoisseur in painting, whereas Rivera felt himself to be a connoisseur in politics ....

Trotsky led a reclusive but very active political life, meeting with intellectuals, artists and statesmen. We now know all the details of his assassination on August 20, 1940, by the Stalinist agent Mercador, with a blow from a ice axe. The event caused a shock in Mexico where Trotsky was very popular. He was buried in the Mexican Pantheon and three hundred thousand people attended his funeral.

I learned of his death from the French newspapers in the summer of 1940. There was no commentary, just the facts. For a long time, I knew nothing of his latest writings, particularly the ones following the French military defeat. I was still mobilized in August 1940, and had no contact whatsoever with the Trotskyists. I had met an engineer who was a member of the French Communist Party, who interrupted a quite unpleasant discussion with the remark "Anyway there's no point talking so much about it; we're well rid of the dear man". We separated on rather poor terms. After the war, this man played quite an important role in the unions, representing a non-Communist left, and became quite close to the PSU (Parti Socialiste Unifié)[22]. And as in many other instances, I ignored our past differences.

One of our comrades, Jean Van Heijenoort, followed Trotsky into exile. He was a young Frenchman of modest origins, who had lost his father very young. He was a first-rate student and meant to enter the ENS to study mathematics. But then he met Ivan Craipeau who converted him to Trotskyism. He decided that a world revolution was more important than his studies and joined Trotsky in Prinkipo in 1932, then in France, and then in Mexico at Coyocoán. He remained there until 1939 as his bodyguard and secretary. In the French party, we often spoke of comrade Van Heijenoort. A budding mathematician turned professional revolutionary, just like Trotsky himself, he ended up returning to his first love. First he proved a beautiful result on convex surfaces, then turned to logic. I don't think he proved original results in logic, but he became internationally known for a remarkable book about the history of logic and published a lot on the subject. He was an extraordinarily original man. I met him in 1950; he had left Trotskyism in 1948. Our relations were mostly through correspondence. He was very sentimental, had a complicated love-life and ended by being stabbed in his sleep by his wife who then killed herself. Anita Burdman Feferman has just written an excellent book about him called *Politics, Love and Logic*.

## Stalin's Crimes

Executions in the USSR did not only take place in 1936, but continued throughout the whole history of the USSR, from 1929 until Stalin's death in 1953. On December 9, 1935, Kirov was assassinated. He was the secretary general of the Communist Party in Leningrad, and was an intelligent man who had a lot of prestige in spite of the crimes committed in the name of Stalinism. Stalin became afraid that Kirov would supplant him or even assassinate him, and it could have happened. For a long time, it was believed that Stalin arranged to have him assassinated. Recent revelations have shown that in fact, he was assassinated by a certain Nikolaiev for little-understood personal reasons. Naturally, although Stalin was apparently not responsible for it, the assassination must have pleased him. However, like the consummate opportunist that he was, he immediately claimed that Kirov was the victim of a vast conspiracy and had a

22) PSU: Unified Socialist Party, see chapter 9

large number of people arrested of whom sixty-six were executed. I learned all this at the ENS where the students, even the Communist sympathizers, vigorously protested these massive repressions. At that time I knew nothing about this purge which preceded the Moscow trials.

The main purges took place in 1937. The Soviet mathematician Gohberg, now living in Israel, told me about that fateful year. His father was a member of the Communist Party and he himself belonged to the Communist Youth. One evening, his father did not return home. His wife believed it must be an error and that he would soon come home. But they never saw him again, nor did they ever receive the slightest explanation. Such events occurred in hundreds of thousands of families. It must have been a basic terrorism tactic. Millions of Soviets occupying important positions were made to "confess", which meant admitting to crimes of high treason, after which they were deported or executed. This selective repression was aimed at people in important positions who might criticize the regime. But with Kafkaesque absurdity, quotas of victims were fixed in each town and the victims then drawn by lot, so that terror should penetrate absolutely everywhere. This sinister atmosphere is described in *If it is Midnight in the Century* by Victor Serge and *La Société éclatée* by David Rousset, written much later. Among the victims of 1937, Marshal Tukhachevsky, a major figure of the civil war which followed the revolution and one of the founders of the Red Army, earned the right to a (non-public) trial, after which he was condemned and executed. He was rehabilitated in 1961 under Khrushchev. Nearly fifteen thousand officers and two hundred generals were executed. I remember a friend of mine, Emile Kahn, general secretary of the League for Human Rights, telling me after the French defeat in 1940: "Our generals betrayed us. The Soviets were smarter: they executed the ones who were liable to betray them." This was a deep error: the generals executed by the Soviets were actually the best ones, Stalin's potential rivals. This partly explains the weakness of the Red Army during the Nazi attack in June 1941.

In the Soviet concentration camps, of whose existence we were aware, a great many prisoners had belonged to Trotsky's former left-wing Opposition. In 1938, the year of Munich, all these Trotskyists were liquidated; we suspected it, but it was only confirmed much later, in the sixties, when Soviet archives were opened.

Revolutionaries were not only assassinated in Spain. Rudolf Klement, a Trotskyist of German origin who was a refugee in France and had been one of Trotsky's bodyguards, was assassinated in July 1938, probably with the aid of the leaders of the French Communist Party. His body was found in the Seine, cut into separately packaged pieces. Although this was an isolated incident, the threat hung heavily over France's two hundred odd Trotskyists. Even without being assassinated, we were always welcome to the flood of insults which Communists in every country poured on Trotskyists, the worst one being "Hitlerites" or "Hitler-Trotskyists". After the Moscow trials, they took to accusing the IVth International of being a Gestapo agency and later, an agency of fascist Italy and the Japanese mikado. I remember a headline in *L'Humanité* about the "Hitlerite Schwartz" or the "Hitler-Trotskyist Schwartz". This myth about the Trotskyist Party being an agency of the Gestapo lasted for a long time.

Frédéric Joliot-Curie told me about the following amusing episode. A delegation of French scientists, containing Jacques Hadamard, Elie Cartan and so on, was visiting the Museum of the Revolution of the Soviet Union just after the war. Jacques Hadamard observed that there was no portrait of Trotsky and asked the guide why. Hadamard was a Communist sympathizer and the question appeared quite natural to him. "He was a counter-revolutionary," was the guide's equally natural reply. A bit naively, Jacques continued "That may be, I don't care, he played a role in the revolution and he ought to be represented in this museum." The people around him stared at him in disbelief. Maybe he was naive to think he could convince the guide, but I wonder what would have happened if his group and the people around him had supported him. Too many Western intellectuals supported Stalin. Fascist dictators don't care about foreign intellectuals, but leftist dictators need them. Once again, I was able to witness how easily intellectuals are able to accommodate the truth to what they want to believe. Frédéric Joliot ended his tale with the following question: "Of course, I know that you yourself are perfectly honest, but seriously, don't you think that the majority of the members of the Trotskyist Party are agents of the Gestapo?" I asserted the contrary but I don't think he believed me.

The Communist Ignace Reiss was assassinated in Switzerland in 1938. The Trotskyist press wrote lengthy commentaries about the event, because Ignace Reiss had been the head of the information service of the Red Army and had chosen to join the IVth International. Later, I discussed the whole event with his widow, Elsa Reiss, whom I visited frequently towards the end of her life, when she lived in our street. She was very politically involved. During the difficult period of her youth in the USSR, when she remarked that some of the measures taken by the government were too harsh, her husband used to lecture her "You're reasoning like a member of the petite-bourgeoisie. The terrible situation in the country obliges us to take these measures." She admired her husband immensely all her life, played an important role during the investigation of the circumstances of his death and died a few years ago.

We must also mention the massive assassinations of the republican veterans from the Spanish war. The members of the International Brigades and the Communist sympathizers (some of whom had themselves carried out the Stalinist injunctions to assassinate non-Communist republicans) took refuge in countries such as France, where they were then given up to the Nazis. The destiny of those who took refuge in the USSR was hardly better. In order to eliminate embarrassing witnesses, Stalin had practically all the Communist refugees in the USSR killed. I was obsessed by these millions of assassinations, at a time when most people didn't care at all and even the newspapers only expressed vague and indifferent judgments. From the beginning of the war I was aware that the combat against Nazism was for me an issue of life or death, but I also knew that if the Soviets were to invade Europe, I would be one of the first to be deported and executed. We were in danger from every direction. It was midnight in the century ...

## Former Trotskyist for life

Several of my comrades at the ENS considered my Trotskyist tendencies with sympathy. There were also Communists among the students, particularly Leopold Vigneron. Before I became Trotskyist in 1935, he gave me a book by an English Communist, Palm Dutt, called *Fascism or Socialism*, which was partly the cause of my abandoning the SFIO (Section Française de l'Internationale Ouvrière)[23]. Later, after 1936, I understood that this book was in the style of the third period (with social-fascists all over the place) but at the time I didn't understand the significance of this. Thus, step by step, I completed my political education. After I became Trotskyist, I had a lot of lively but tolerant discussions with Vigneron; each of us considered the other as a poor lost soul. Naturally, I think I was right and he was wrong. He tranquilly accepted all the flip-flops of the Communist Party; he would say "Well, we were wrong yesterday, but today we are right." Later he actually joined an anti-Communist left-wing group and worked hard at various reasonable causes, such as the defense of Soviet dissidents. In 1936, he invited Maurice Thorez to give a talk at the ENS. Hearing that there was a Trotskyist in the audience, Thorez expected to be heckled. I did heckle him, on a subject which today I find a little ridiculous, but at that time was one of the war-horses of Trotskyism against Communism: Trotsky called Stalin "the Thermidorean", the Communist Party had long ago ceased to be a revolutionary party. This was of course perfectly true, but with hindsight the whole question appears much more complex. Yet

23) SFIO: French Section of the Workers' International

it was striking in the period of the Popular Front; the Communist Party never tried to support a revolution in France. (Would it have succeeded, even if it had tried?) Similarly, the Communist Party did not defend Algerian independence, nor Viet-Nam against American aggression. It should have expressed powerful opposition to these things, but instead its voice remained timorous. Thorez listened to me patiently and then replied that I merely had revolutionary whims, the Communist Party had revolutionary will. I think we both came out of it well.

However, he thundered quite violently against Doriot. The case of Jacques Doriot is quite special. After the victory of Nazism in Germany, Jacques Doriot, a member of the central committee of the French Communist Party and the general secretary of Communist Youth, declared at a meeting of the IIIrd International that the politics of the third period, the politics of social-fascism, had been disastrous and contributed to bringing Hitler to power. He advised a united front with French Socialists. He was excluded from the Communist Party and became a kind of isolated Communist, constantly and shamefully calumniated by the party. I remember talking about it to my friend Jean de Neyman, who came to visit us in Autouillet when I was already a Trotskyist and he was a Communist. "Doriot is a nut," he answered, "how can someone with so little experience oppose revolutionaries like Stalin?" (I remember that Neyman rode the forty kilometers separating Autouillet from Paris on a bicycle arranged to meet every eventuality, like professor Cosine's anemelectrobackwardpedaliwindbreakumbrellaparanailcycle; he looked great riding along on it.) But then the roles changed. The French Communist Party and Stalin adopted Doriot's views on the united front, while continuing to calumniate him, and he himself evolved more and more towards fascism, which is exactly what the Communists had been accusing him of. Under the Occupation, he was a firm supporter of Nazism, and together with Marcel Déat founded the Legion of French Volunteers against Bolshevism, which fought on the Russian front.

The comrade who was politically closest to me was Paul Ruff; we continued to communicate with each other for many years. Paul Ruff often agreed with my ideas. In my eyes, morality is one of the conditions of politics. It is unthinkable that a party in power, capable of setting up a phenomenon like the Moscow trials, can undertake any activity which is really salutary for society. At the time when I understood their meaning and felt the blow, I knew of no explicit crime of Stalin's, but I immediately became convinced that he had committed crimes and would continue to commit crimes. The future, unfortunately, proved me only too right. In the years 1929–30, millions died in the the "dekulakization" (liquidation of the rich peasants) of the countryside; in 1936, thousands of people were killed following the assassination of Kirov on December 9, 1935; the immense purge of 1937 was one of the most tragic events of Soviet history. The period after the war only brought more of these facts. Stalin is uncontestably one of the greatest criminals of history, and the Moscow trials of 1936 alone revealed this.

It appears incredible that the Communist intellectuals of the period considered the trials as negligible. Certainly, some of them said, the trials are fake, but this doesn't affect the justice of the political line adopted. This is obviously an erroneous and extremely dangerous attitude. My uncle Jacques Debré, who had been quite favorable to the USSR, illustrated it by a sort of parable: "Suppose I admire a certain universally respected person, but that suddenly I see him, believing himself to be quite alone, piss in the soup tureen. Immediately I believe him to be an impostor capable of anything." Others said "The end justifies the means." Paul Ruff used to answer with the inverse formula: "Sick means can only lead to sick ends." Ruff and I were excellent friends and his vision of political events played a considerable role in my political evolution. In more elegant language, one might state his proposition as "Not only the end does not justify the means, but the means form an intrinsic part of the end, which they ineluctably influence."

After the victory of the Popular Front, there was a workers' demonstration in Clichy against the Socialist government, organized by the Communists and supported by a large section of the left. The Trotskyists decided to participate. The leader of my cell recommended us to go there, if possible, with a revolver. I disapproved of this grotesque suggestion so much that I decided not to go at all.

Of course, I had contacts at the ENS with people in other classes besides mine. But since I did not live there, I only spent part of each day there and maybe took less advantage of the possibilities of meeting people than others. I was close to Jean-Toussaint Desanti, a student in literature of the class of 1935 and today a well-known philosopher, whose life, together with his wife Dominique Tania, took many strange turns. At the ENS, Desanti had become a Trotskyist shortly before me. During the war, he and his wife sheltered us for a day when we were fleeing. After the war, they became convinced Communists, and believed absolutely in the veracity of all the trials taking place in the various popular democracies; Tania wrote most disagreeable articles in the newspapers about them. At that time I didn't feel like seeing them any more. Then they distanced themselves from Communism. Tania wrote an self-criticism. Later, when we met unexpectedly in the Balearic Islands, she gave me a detailed explanation of her political evolution, and we became reconciled.

The biologist and butterfly collector Raymond Croland was a little older than I was, in a higher class at the ENS. He had a beautiful collection of French butterflies (I met him through butterflies), from every family, all remarkably fresh and perfectly set up, because he had a huge caterpillar nursery. He was an extreme leftist, near the anarchists (although not very active), and we used to have passionate discussions. He was one of the earliest resistance fighters, and together with a chemist from our class, Pierre Piganiol, he organised a resistance group called the Velite network, whose ramifications stretched throughout the country. Out of prudence for themselves and for the school, they decided never to meet there. There were, however, a few exceptions. He was arrested in his office in the biology laboratory on February 14, 1944, and deported. We know that he survived until April 1945 and was kicked to death with several other prisoners just before the arrival of the Americans. This horrible end was a blow for me and also, I must say, the loss of an extraordinary intellect for science. I went to visit his parents in Fontenay-aux-Roses, in the street which now carries his name, rue Raymond-Croland-et-d'Estienne-d'Orves, in the house he lived in before the war. Piganiol organized a commemorative ceremony at the ENS on November 18, 1994.

What was the meaning of my joining the Trotskyist Party? At the beginning of the summer of 1936, after a lengthy political evolution, I intended to join the Communist Party. The Moscow trials of August 1936 pulled me up short and I joined the Trotskyist Party in the autumn. Logically, by becoming a revolutionary, I believed in Marxism, Leninism and the Russian revolution of 1917. I wore the guise of an internationalist Communist, member of the IVth International and partner of Soviets everywhere and of the Socialist world revolution. We believed in the individual Trotsky, not the leader or the general secretary, but the intellectual creative theoretician of genius yet fallible, the prestigious revolutionary of 1905 and 1917. In order to avoid a seeming personality cult, we did not call ourselves Trotskyists but Bolshevik-Leninists. We could have used the hammer and sickle as our symbol, but it would have created inextricable confusion, so instead we used the "spark" (the one used to indicate dangerous electrical appliances), which recalled the name of the oldest Bolshevik newspaper, the *Iskra*. We were totally against every kind of imperialist war, and believed in civil war for the transfer of power. According to Marx' analyses, we wanted to establish the dictatorship of the proletariat, the immense majority of the nation, against the bourgeoisie, the exploiting minority, in order to reach the ideal of ending the exploitation of man by man. To accomplish this, we adopted the ideas expressed by Lenin in *What Should We Do?*, the necessity of power belonging to a unique party, representing a single social class, the proletariat (this was a serious simplification of existing social reality). It was impossible for such a dictatorship not to degenerate, especially with a unique party. It is true that Marx and Lenin said that the State should die away (according to the same principle as before, since Socialism would suppress class differences, the State should begin to die away as soon as the proletariat took power. A sorrowfully naive idea!) We were radical internationalists, believing in a world government, and anti-colonialists supporting the emancipation of what later came to be called third world peoples. I felt myself to be individually "responsible for the whole world"; my only country was humanity itself. This program was a kind of charter which united us. Theoretically, it was also the charter of the Communist Party. It was frequently mentioned but its reality was entirely altered, and our criticisms of Stalinism were fundamental. We may have forgotten this, but it was also more or less the charter of the IInd International of Socialist Parties. At the beginning of the XXth century, Eduard Bernstein had tried to replace the revolution, the official goal of the International, by reformism. Reformism was condemned as inefficient by the IInd International. With hindsight, this is rather striking. Today, practically everyone supports reformism.

I am deeply non-violent, I hate violence. But the world which I discovered during my years at the ENS was filled with terrible violence: capitalist exploitation, colonialism, fascism, Stalinism, the 1914–1918 war and the war which was to come. Revolution was the in the air of the time, inevitable and certain, and the prestige of the Russian revolution was immense. It was still recent, and wore a halo of romantic heroism which is beautifully rendered in John Reed's book *Ten Days that Shook the World*. The 25th of October and the 7th of November 1917 are historic dates, like the 14th of July 1789.

I devoted a large part of my life to politics, throwing myself into the "career" of a politically involved intellectual. But still, of course, mathematics remained my primordial interest. I always wanted to "change the world," to change life. I am still a reformer who is bothered by every defective or degenerate structure. I broke with Trotskyism in 1947, after eleven years. But my political culture comes almost entirely from that Trotskyist period, with all its lighter and darker moments. Pierre Vidal-Naquet likes to say that I'm still a "former Trotskyist". I will be one all my life and don't regret it.

# Chapter IV
# A Researcher in the War

On leaving the ENS, in July 1937, I decided to do my military service at once. In October, I was inducted into the army at Ban-Saint-Martin, near Metz, in an anti-aircraft defense unit. Technically, our weapons were rather good. It's not easy to shoot at a rapidly moving aircraft. The whole thing functioned automatically and was quite interesting scientifically. It's true that its efficiency was limited, it was much less efficient than the radar which allowed England to win the battle of London, but it was sufficient to shoot down several airplanes and frighten all of them.

## A bad student

I had no problems with this part of my new job. Things did not go so well with the installation of country telephones, or the handling of cannons, machine-guns and rifles, which in particular we were supposed to know how to take apart for cleaning. I was entirely incapable of doing this. I could, if necessary, take a machine-gun apart, but I was just unable to put it back together again. My comrades, who were all quite practical, had no problems at all. But for me, seeing nothing at all in space, as usual, I just couldn't visualize the positions of all the different parts. There were plenty of other similar things which I couldn't do. And those weren't the worst of my mistakes. Once when we had to spend a day doing exercises in the countryside, all the instruments had been transported to a large field, from which we were supposed to measure fictive shots at airplanes. I was supposed to take care of moving the telemeter, a very large apparatus. It stood on four legs, and to move it, it had to be lifted up and have wheels placed on it and screwed in, after which it was lowered and towed by a car. I put the wheels in place, but forgot to screw them in. The driver of the car which towed it away didn't notice anything; the wheels came off, and the telemeter rode all the way back to base camp on its axles, which wasn't particularly good for it. We had to ride all the way back to the field, looking for the wheels; we had no idea where to find them or even if they had all come off together. I felt rather meek. For the first time in my life, I found myself in the position of the bad student. I tried really hard – yet ended up among the very worst at our final exam. Perhaps after all, it wasn't so bad, psychologically speaking, to feel what it's like to be a poor student, dragged along forcibly by circumstances.

Because of this, on leaving Ban-Saint-Martin, I was sent to one of the least desirable spots, Laon. Even though I wasn't separated from my friends from the ENS, the whole semester was far from enjoyable. It was the last semester before our marriage, and celibacy became extremely difficult for me. Marie-Hélène was finishing her cure in a little hotel near the sanatorium, with weekly checkups.

We moved to Laon together in 1938, and at first we lived very comfortably. Marie-Hélène was absolutely obliged to rest during the day, and Mme. Ruby's little hotel was a perfect place; it had a ravishing garden where we liked to stay when it was warm. We lived in a lovely room for which we paid one thousand eight hundred francs a month, including meals. We only had lunch together on Sundays since the other days, I had lunch at the officers' mess, but we always had dinner together. Officers had the right to go home in the evening, whereas privates had to sleep in the camp dormitories. For us, it was a period of incredible happiness. We were finally together, after years of waiting. We enjoyed each and every evening together; we talked, we read, we ate together, we took walks, we led a normal love life. The captain once told me slily that I had been seen in town after dinner kissing my wife in the street, which was not proper behavior for an officer. I don't think I took his scolding too seriously.

My captain was severe with the soldiers, but he rather liked me because I was a scientist. He himself was preparing for the Ecole de Guerre, which, very democratically, allowed officers from the ranks to rise quickly through all the echelons. The intellectual level of the lieutenants around me was rather low. The sub-officers were in general very competent people, who often knew the practical functioning of the weapons better than the officers.

My friends from the ENS had been dispersed to various places, and I didn't know anyone. I don't even have very precise memories of this period spent in Laon. I only remember the horrible boredom of the entire day spent in the camp. Boredom, boredom, boredom. I tried everything. It was impossible to do mathematics in the camp; I wasn't allowed. In the evening, I was too tired. There was no one with whom I could talk about the things that interested me: science, politics, society, the books I was reading. I had absolutely no work to do in the camp; nobody gave me any. I could sit dreaming in my office or walk around the camp, both of which represented lost time in my eyes. I was a Trotskyist, and the camp was full of career officers. Quite rapidly, I realized that we did not speak the same political language, and probably not the same language at all. They weren't interested in politics, only in the daily routine. I rather think most of them didn't think about anything whatsoever.

Eventually I learned about the workings of the DCA (Défense Contre Avions)[24] battery, which plugged some of the drastic holes in the knowledge I was supposed to have gained at Bar-Saint-Martin (although I still didn't know how to take apart and put together a machine-gun). I followed the courses given by sub-officers to soldiers; they were usually quite good, but occasionally contained some curious explanations. A sergeant taught his soldiers that the tube of the cannon was graven in a spiral in order to impart a rapid rotational motion to the shell coming out of it. The goal of this was to turn the shell into a gyroscope which retained a high level of stability during its flight, remaining more or less tangent to its trajectory throughout. But it was impossible for him to explain this properly to the soldiers. So he told them that the shell was supposed to screw its way into the air, and for this they started it off with a screwing motion in the clockwise direction. This is obviously silly – it would work just as well in the opposite direction! I couldn't resist asking the soldiers what they thought would happen if the shell was fired turning the other way. They answered unanimously – that it would fly out backwards, from the other end! The marshal himself didn't see anything wrong with their answer. I wasn't trying to humiliate him; he was really a nice, interesting fellow. And I didn't even succeed in explaining the thing properly to him, although I tried to compare it to a spinning top.

The fact that I didn't know how to drive did not stop them from naming me director of the parking lot. I knew absolutely nothing about cars. It didn't even occur to me to take advantage of the occasion to learn to drive, the whole thing was so foreign to me. It's true that my total absence of spatial orientation probably renders me inapt to drive anyway. So, I went every morning to the parking lot where I met the adjutant, an agreeable and competent gentleman who respected me and didn't require a stroke of work from me. He would toss out "Everything's fine", I would leave satisfied and that was the full extent of my inspection of the parking lot. Fortunately, nothing out of the way ever happened during my tenure.

Apart from our wonderful evenings in our cocoon, we were totally isolated. We both read enormously. We borrowed books from the library of Laon. I spent the entire war reading and learning about all kinds of different areas: physics, chemistry, biology, natural sciences, geography, economic geography, economy,

---

24) DCA: Anti-Aircraft Unit

history, literature (I remember reading *Les Thibaut* by Roger Martin du Gard in Laon). I had only just learned enough geography in high school to avoid total dishonor at the exams, but I really didn't know much. I bought the high school classics by Gallouédec and Maurette and studied geography from them. I studied the world markets of raw material, the major industrial zones and world commerce, borrowing all the necessary books. It gave me a lot of pleasure, and I spent a lot of time on it. I learned a great deal. It was much easier for me, in my situation, to read books of this kind than to do mathematics.

## Munich

I immediately understood what Munich[25)] represented and violently refused the idea of capitulation. The Spanish war was turning into a debacle. The ignominy of Czechoslovakia only postponed the inevitable outcome. Mobilization sent us to a place which I hardly remember at all.

The army was in a lamentable state of disorganization; I saw it well enough in Laon and even more later on. For example, for a whole week I taught soldiers how to work our anti-aircraft defense weapons. But on the day we were supposed to go out and exercise, I saw that my group had been replaced by a group of new recruits. I protested, but I was told that my usual group had been sent to peel potatoes. On the other hand, I did successfully teach trigonometry to a dozen soldiers. On paper, we measured angles with a protractor; in nature we used our hand. I taught them sines and cosines and tangents, and they computed sides and angles of right triangles. I showed them that $\sin^2 + \cos^2 = 1$ and that if you have an angle of $x$ degrees with $x$ less than 45, then the sine of the angle is roughly equal to $x/60$ (indeed, the radian is equal to about $57^o$ and $\sin y/y$ is near 1 when $y$ is small measured in radians). For $x = 30^o$, the sine is exactly equal to $1/2$! When $x = 45^o$, we have $45/60 = 3/4 = 0,75$, whereas the actual sine of $45^o$ is equal to $\sqrt{2}/2$, roughly 0.71. Anyway, they could always use trigonometric tables. My "assistant" was the only soldier who had passed his baccalauréat. Everyone had fun and so did I.

The Front Populaire had failed. Its Chamber remained unchanged, but they had to accept a particularly reactionary Daladier government. The former Trotskyist predictions about the role of the Radical-Socialists were coming true. In November 1938, the CGT (Confédération Générale du Travail)[26)] (which was still unified) started a strike: Daladier fired every single government worker

25) The Munich Pact gave the Sudetenland to Nazi Germany on condition that it respect the remainer of Czechoslovakia, a condition it later violated.

26) CGT: General Confederation of Work, a Communist union formerly called CGTU, cf. chapter 3

who participated in the strike, so that a huge number of people suddenly found themselves jobless. There were even pro-Nazi French people such as Georges Bonnet, who circulated around the ministries. Without knowing it, France was getting ready for Vichy, and the war against Nazism was conducted in deplorable circumstances. When I shared my fears about losing the war to my parents and my uncle Robert Debré, they couldn't see why I was so worried. Uncle Robert said to me "Don't worry, it was like this at the beginning of the first World War as well, but then we had the Marne battle and the situation straightened out. The French army is never ready at the beginning of a war, but it always catches up." There was no battle of the Marne this time. The atmosphere became unbreathable to every democrat.

The material conditions of our life in Laon and later were quite precarious. My salary was 1,200 Francs per month – equivalent to about 3,300 Francs today. (It's not that I did any better in my career at the university: a high school teacher earned 2,400 F per month, equivalent to about 6,400 F today, and as an assistant professor with a Ph.D. and a family in 1947, I earned 7,500 F per month, equivalent to about 1,800 F today). My father-in-law helped us with about 750 F per month. The whole thing barely covered our monthly payment to Mme. Ruby. If I needed to eat in a restaurant, I always went to the same one and paid 5F for a mackerel and a fruit (and a "fresh", i.e. a bottle of cool tap water – there were no refrigerators). Everything was inexpensive, but the quantity and quality of goods was much lower than it is today.

In the autumn of 1938, after Munich, I was transferred to the garrison of Mont Valérien, near Paris. The strong winds of Laon were not good for Marie-Hélène's health, so shortly after Munich, we moved to Châtillon-sur-Bagneux. From the autumn of 1938 until the declaration of war in September 1939, we led the same kind of life as in Laon, in a little house surrounded by flowering roses in spring.

I didn't sympathize with the officers in Mont Valérien any more than I had with those of Laon, but I don't really remember anything about it. The spring and summer weather was incomparably beautiful, and we felt well. I had abandoned my butterfly collection at the beginning of my military service, but I started raising caterpillars again that spring. I found a lot of caterpillars of the butterfly *Lasiocampa quercus*, the oak bombyx, whose name is a mistake since the caterpillars live on lilacs, privets and brooms. The name comes from the fact that the cocoon resembles an acorn. I raised as many as two hundred and fifty caterpillars in our living room. In the camp, I had had a large box constructed; it measured about one cubic meter, and had holes punched into the bottom. I stood my box on jam pots filled with water and a bit of earth. I gathered lilacs and stood them up vertically in the box, their stems passing through the holes and standing in the jam pots. They remained fresh for several days, but still I had to renew them every week or so. I used to go into the town and get an enormous armful of lilacs which I carried home. I had to be discreet, because officers were not allowed to carry anything, even a package or a bag. They had an aide (or their wife) to do such things, but this idea embarrassed us extremely!

I continued to read intensively. As in Laon, we didn't know anyone around us. But I did get back into contact with Paris and with my former Trotskyist friends, and with my family and Marie-Hélène's family. Steve, one of my Trotskyist friends, came to see me once a month, because he needed money to put out the monthly bulletin *La Vérité*. My comrades from the ENS had been dispersed throughout France, and there were almost none in Mont Valérien. I didn't manage to make any new friends. In winter, everybody took notice of me because I wore mittens, and worse, kept them on while saluting senior officers. I also went to a session of exercises in the countryside dressed in town pants and a kepi instead of full army regalia, with leggings and a helmet. The captain was furious and as the general was going to come and inspect us, he told me to keep hidden behind a truck and to keep moving around it so that the general never set eyes on me. I must say that on the whole, they were relatively tolerant of my absent-mindedness. I'm not really all that absent-minded, but it was caused by boredom.

The announcement of the German-Soviet pact, called the *NichtsAngriffsPakt zwischen Deutschland und Soviet Russland*, didn't surprise us. We knew Stalinism. But the circumstances were depressing. Molotov exchanged his famous handshake with von Ribbentrop, and the Soviet newspapers spoke of nothing but the agreement between Nazi Germany and Soviet Russia, two proletarian countries united against the capitalist states of France and England. With the division of Poland, war became inevitable. Naturally, we could *imagine* that Stalin wanted to take advantage of a certain period of peace in order to let his country recover from the purges of the preceding year, and improve his armaments in view of a future war with Germany. Unfortunately, it was clear that the leaders of the USSR believed in the German-Soviet pact and in the sincerity of Germany. For them, an alliance between National-Socialism and Soviet Socialism against capitalism was in the nature of things. I didn't meet any Communists at that time, but I can imagine what most of them must have felt.

It was becoming more and more difficult to breathe in the political climate in France. The general tendency was towards capitulation to fascism. We had to endure the invasion of Czechoslovakia in May 1939. Immediately after the German-Soviet pact, anti-Communist repression began in France – although the

Chamber of the Front Populaire was still in place! The Spanish war ended with the victory of Franco; Marie-Hélène and I felt this as an immense personal tragedy. Fascism settled into Spain for decades. Europe was entirely open to fascism; Germany, Italy, Spain, Portugal. I felt like a minuscule pawn on the gigantic chessboard of the world, a drop of water in the middle of an ocean of events which totally escaped me. These two years of military service appeared to me like a total waste of time.

## Ballancourt on a volcano

September 3, 1939 was a magnificent, sunny day. I had been sent to the Bouchet powder factory in Ballancourt a few days before, to defend its arsenal. It was a dangerous position; a massive bombardment would have provoked immense explosions which we would have had no chance of surviving. The DCA was not at all sufficient to prevent such an attack; it might have shot down a few airplanes, but not all. (The powder factory did actually explode accidentally in 1952, causing a fair amount of destruction.) The USSR then entered the war; Poland was divided and the Jews who found themselves in the zone occupied by the Soviets must certainly have breathed a sigh of relief. There was plenty of anti-Semitism in the USSR, but at that point it had not yet taken the form of violent and systematic repression (starting in 1948, Jewish writers and intellectuals were arrested, thrown in prison and assassinated, and this went on until 1952), and was certainly not comparable to Nazi anti-Semitism. There were massacres, in particular the Katyn massacre which was discovered only after Hitler's attack on the USSR. Without actually being able to guess the dimensions of the horrors the war was to bring us, I already foresaw a dreadful cortege of events. The civilian population, in particular, was to suffer far more than it had in the 1914–1918 war. How could this be imaginable on such a beautiful day?

I was a sub-lieutenant under the orders of a captain and a lieutenant who were both quite nice; both of them were mobilized civilians, an industrialist and a banker (I myself was no longer active, since my military service had ended in September 1939, but I remained on duty as a reserve officer.) But there were plenty of other officers in our unit. The place was agreeable and the weather magnificent. I had to spend the entire day in camp, eating at the mess and even sleeping in the camp, like everyone else, all dressed and covered only with a simple woollen blanket, ready for any alarm, day or night.

Marie-Hélène had followed me and was living in a little hotel in Ballancourt, where I had the right to visit her every day from five to seven, accompanied by a car and a chauffeur. She hadn't thought about mathematics for a single minute during three entire years, from October 1935 to October 1938. During the academic year 1938–1939, spent in Laon and Châtillon, she set to work again, and passed the certificate of rational mechanics (she couldn't take the courses, but she worked partly with me), and thus obtained her Licence. To pass the agrégation, she would have had to wait till the end of the war, so she gave up that idea altogether. In Ballancourt, she completed her studies which had been interrupted by her illness, and it gave her tremendous pleasure. For the first time in years, she felt that she was completely cured and that she could really devote herself to mathematics. In spite of the situation, I think she was really happy in Ballancourt. I don't know if all human beings are like this, but we in any case always succeeded in being happy and good-humored in the midst of the most awful circumstances. It was our way of enduring things. I was the only officer who had his wife there. I also continued to read a lot. The situation was very different from Laon and Mont Valérien; our country was at war, we were obliged to remain at the ready, but there was nothing to do, and nobody forbade us to read. As for mathematics, it was really impossible to work on them. Never, except at Laon, did I feel so absolutely isolated; fortunately I had some interesting conversations with my captain and my lieutenant, and was able to chat occasionally with other soldiers of the contingent.

Disorganization reigned within the army. I was supposed to organize the shooting and the training of the contingent. I never fired anything except a single cannon, once. One night, we had the impression that an unidentified airplane was flying nearby. I wasn't really sure and we could hardly make it out. But because it hadn't identified itself, we presumed it to be an enemy aircraft. So I got the company into firing position and shouted "Fire!" Four cannons, four shells. They didn't hit anything. Honestly, I'm not sure there was any airplane out there. I gave an account of the event to the authorities, and gained great prestige in the eyes of the contingent . . .. We were so bored during the "drôle de guerre"[27] that we were deeply grateful for any action at all. Another time, we saw an airplane passing about ten kilometers away, near an advanced observation post. A soldier from the post gave us its position in the following laconic terms: "It's to the right." It wasn't very helpful since we had no idea where the observer was standing, but I couldn't get anything more precise out of him. And the airplane continued on its way, leaving no trace.

A curious coincidence occurred in Ballancourt. Our friend Suzanne Augonet, the wife of the Trotskyist Marcel Hic, came to see us for a day. She wanted to

27) the "funny war"; the French used this term to describe the months during which French soldiers were mobilized but completely inactive, preceding the French defeat.

work in a factory, to gain some first-hand experience of being a worker, and she was looking for an employer with ideas broad enough to accept her knowing who she was, naturally under the condition that she would not agitate. We thought of Auguste Detœuf, whose son Pierre had been a classmate of Daniel in high school. He was politically very broad-minded and believed in industrial reform; he had accompanied Jacques Debré to Moscow in 1936, and seemed like the perfect person. But how could we reach him? In the political situation of the time, we certainly couldn't write or phone him explaining the situation directly. We climbed up a steep hill together, thinking and talking about what to do, when who should we see coming towards us but . . .. Pierre Detœuf! He had been mobilized nearby. We introduced Suzanne to him and told him what she wanted to do. He promised to speak to his father about it, and his father immediately accepted her proposition. Providence has strange ways.

Life moved on monotonously during the "drôle de guerre". A colonel from Paris arrived unexpectedly one day: "How is it possible that you didn't fire? An airplane flew almost directly over you!" We assured him that it couldn't be possible, our instruments would have detected such an airplane. He was so sure of it that we finally asked him how he could know, since he came from Paris. "I know it by divination. I hung a pendulum over a map and it stopped here." He took advantage of his visit to inspect my instruction of the military personnel. He liked what he saw, and this had its consequences. A few weeks later, I was transferred to Biscarrosse, in the Landes, as a professor for mobilized reserve officers.

## Biscarrosse. Monsieur le Professeur

It was very unlikely that Biscarrosse would be bombed by German planes, even though thousands of officers were trained there. Marie-Hélène settled into a little bungalow in Biscarrosse-Plage. Some of the officers of the camp also lived there with their families. Every officer had to be in his office or his classes from eight in the morning to seven in the evening. However, a little later, officers obtained permission to go spend the evening and the night in Biscarrosse-Bourg, sometimes to meet their wives in agreeable places; it was quite nice for us.

The period we spent in Biscarrosse-Plage was also nice for us because there were many students from my class and nearby classes of the ENS among the mobilized officers. In spite of the dramatic situation, I breathed more freely amongst my friends, particularly Paul Ruff. We had dinner every evening with him and his wife and quite naturally picked up our political discussions where we had left off. Progressively, I came to agree with many of his arguments.

He was convinced that the German army was infinitely more powerful than ours, and that it would simply sweep our army away, even if we succeeded in organizing it better; in this, he was perfectly lucid. At that time, I was reading the book by Hermann Rauschning, a former friend of Hitler, who was pessimistic about the immediate future of Nazism, but even though we wished to agree with him, Ruff refuted all his arguments, and convinced me that he was right. Unfortunately, the Ruffs remained only a few weeks in Biscarrosse.

I had to teach firing at observed airplanes, which I was perfectly competent to do, and my pedagogy was good; my teaching was well-liked. However, there were some officers of the reserve who were scandalized to see me come from Paris to teach, whereas they had been there much longer than I had, and they complained to the colonel in charge of the camp. He told them that I had been sent to him as a remarkable teacher (this was due to the divining colonel) and that he was simply following orders. The tables turned completely when I gave an "inaugural lesson". The officers who had complained returned to the colonel to tell him that they had been wrong and that my nomination was fully justified. They told me the whole story themselves. Both the permanent officers and the mobilized ones showed me great respect. Among the higher ranked ones, some of them who were themselves experienced military teachers called me not only "lieutenant" but also "Mr. Professor". I had an office where I could work. My duties did not absorb all my strength, and I was able to get back to mathematics. I remember working on continued fractions, a subject which had intrigued me at the ENS, but I've completely forgotten what I did with them. Naturally, I didn't have any mathematics books. I should have been more enterprising; the ENS still existed, with its library, and if I had asked them to lend me some books by mail, I'm sure they would have. But I didn't have a lot of free time.

In Biscarrosse-Bourg, we lived in a charming, tiny old house; I spent every evening with my wife. She ate lunch, and we sometimes had dinner together, in a little hotel kept by a Mr. Septima Bidouze whose culinary talents were unforgettable (like his foie gras). The only disadvantage of this nest was that every morning at six o'clock, a small train passed right under our windows, and emitted a long and fantastic whistle. I didn't work at mathematics very much, but Marie-Hélène worked a lot, and actually published two notes in the mathematical journal *Comptes-Rendus de l'Académie des Sciences*, presented by Paul Montel. The intuition she developed during that period later led her to solve a problem which had been open for years in spite of its very simple statement: "Is a deficient value (in the sense of Ahlfors) of a meromorphic function on **C** necessarily asymptotic?" She found that the answer was negative, and her proof was fine and elegant. Dieudonné was very impressed by it. The proof of the result had a curious destiny: a German mathematician, Teichmüller, who later died fighting on the German front, had proved the same result just a little earlier, though we didn't know it. He was a convinced Nazi (which was rare among German mathematicians) and actually went so far as to consider his result as a success of Nazi ideology! Later, a Chinese mathematician again proved the same result, and considered it a proof of the success of Mao Ze Dong's thought. We did not attribute Marie-Hélène's proof to Abraham, Isaac and Jacob, nor to Trotsky.

I was called "lieutenant La Rafale" because that was the name of my bicycle and because of my legendary slowness. Morning and evening, I biked the ten kilometers separating Biscarrosse-Bourg from Biscarrosse-Plage. I remember the delicious odors floating in the air, emanating from plants I didn't recognize; I was bathed in an intense feeling of liberty. The only shadow lying over my daily bike ride was that the road was scattered with corpses of large toads from the many ponds of the region, killed by passing cars. Even at the beginning of spring, it was very warm. I used to ride wearing only a straw hat, shorts and espadrilles on my feet. Arriving at the camp at seven thirty, I would hastily rush to my office, which fortunately was rather out of the way, throw on my officer's uniform, and arrive on time for my classes. I was never caught, but it was against the rules.

One day, posters appeared in the camp announcing an inspection at eight o'clock sharp the following morning. Unfortunately, I didn't see them. Each person was told exactly where he should stand to salute the general, and at seven, everyone was at his post. The soldiers must have gotten up at five, because in the army nothing can be gotten ready in less than three hours. I arrived innocently at my usual hour, in the (lack of) clothing I mentioned above. The general's car was supposed to pass in front of the officers directly across from my office. I immediately realized the extent of the catastrophe. Fortunately, the general had not arrived yet, but there was no way for me to avoid bicycling in my shorts right in front of the whole row of officers, just as the general was to pass in front of them twenty minutes later, to get to the door of my office. I changed my clothes and took my assigned position, under the furious eyes of my captain, who condemned me to fifteen days of detention. The review went without a hitch, then I presented a firing sequence to the general (there were no airplanes). My demonstration must have been convincing because the general left us saying that he had understood everything. "You really exaggerate," huffed the captain after the general had left, "but since your teaching performance was so good, I guess I can suppress your detention. Don't do that again – read the posters we put up for your benefit!" In reality, I was quite well-treated.

Life in Biscarrosse was as calm as life in a university. Sunday was a day off, we didn't go to the camp on that day. An attack was really unlikely. So I went fishing in the fishy ponds of Biscarrosse. I fished innumerable pikes and perches. In the ponds, I was limited to catching pikes weighing only two or three hundred grams. But if I got up at six o'clock, I managed to get quite a number of these, and we were very happy to eat them. They're actually even better than older pikes. One Sunday, several of us went fishing together, in a boat on a bigger pond. There, I fished a pike weighing a kilo and a half, which was really delicious. It was my biggest catch ever, even if it was not so extraordinary – one officer from the camp caught a twelve-kilo pike.

I didn't have anyone to whom I could really talk in the little world of active and reserve officers, apart from Joyal, an ex-student of the ENS, professor of physics in taupe. The group of mostly reserve officers essentially prefigured what Vichy was later to be. Some of them were even royalists! Their conversations at table were positively odious. It wasn't rare to hear that we would do better to become allies of Hitler against England. The obscenities they proffered at the ends of meals had nothing to envy today's most pornographic films. I tried hypocritically to hide my disgust. I possessed an inexhaustible store of jokes which I used to tell; they would probably fill hundreds of pages. Most of them were just funny, some of them a little risqué. I don't remember ever hearing a single conversation in the mess about the war and what it really meant, or about democracy and Nazism, or about any interesting subject at all.

The "drôle de guerre" was not funny at all. France and England did not want to declare war and remained passive until the German offensive took place. We could and should have attacked Germany on its own territory, in spite of the Siegfried Line, which was not all that strong in September 1939, and indeed only became so impregnable because the "drôle de guerre" gave it time to reinforce itself. Such an attack would have slowed down the blitzkrieg in Poland. Instead, France contented itself with constantly playing Chopin on the radio. The whole winter passed without a movement, demoralizing the French and above all the soldiers. Daladier and above all Chamberlain, the men of Munich, led the war. Fortunately they were replaced by Paul Reynaud and Winston Churchill, men of an entirely different spirit. But by the time Paul Reynaud appointed Colonel de Gaulle to the War Ministry, it was too late to change the face of things; we could only begin the massive construction of tanks. The Allies mined the Norwegian coasts. The Germans received iron from the Kiruna mines of Northern Sweden, exported via the septentrional port of Norway, Narvik. I can still remember hearing Paul Reynaud speaking on the radio, during the officers' mess in Biscarrosse: "We have cut off German access to iron. We will win because we are stronger." He reminded me of a beginner at chess, who all proud of his offensive tactics, doesn't anticipate the response of the adversary. I'm hardly exaggerating. But it must be admitted that when the Nazis invaded Norway, the French sent an expeditionary corps and the English a naval expedition with artillery to Narvik. The Germans tried in vain to reach Narvik over land. The Allies only evacuated Narvik after the French defeat; German access to iron really had been cut off for some months. A little later, the Nazis invaded Belgium, Holland and France. France didn't have enough tanks. We had as many airplanes as Germany, but they were dispersed throughout the country, non-operational, and our generals didn't want to fight, nor did the majority of the officers, nor the majority of the French people. Defeat was unavoidable.

## The defeat. Aire-sur-l'Adour

After the Nazi victory, the installation of the Vichy regime and the speech by Philippe Pétain, the appeal of June 18 came like a light in the night[28]. Memory is a fragile thing: I'm certain that I heard it on the radio, together with Marie-Hélène, and she is equally certain that we were told about it. One piece of evidence in my favor is that the next morning, a captain came to get me and led me to the camp; I told him enthusiastically about the call of the night before, expecting him to be grateful – my illusions on the subject of humanity are boundless, and even though I have evolved, I'm still hopelessly optimistic. Instead, he furiously shouted "What is this guy talking about?"

Shortly before German troops arrived in Biscarrosse, we were evacuated to Aire-sur-l'Adour. That is where I was demobilized in August 1940. Marie-Hélène spent several more weeks in the occupied zone.

The decision to evacuate Biscarrosse was a last-minute one. The colonel phoned Paris incessantly and never took any decision by himself. Just a few of us officers decided together to flee and hide in the forest, even if we had to go naked, to avoid being taken prisoner, if the Germans came as far as Biscarrosse, and to return only when an armistice was signed. But finally we were allowed to withdraw. After the armistice, and right through the middle of the month of August, there were no more officers to instruct; only the group of active and reserve officers who were permanent instructors of the camp remained. We lodged in the homes of people of the town. I lived in the house of a furiously anti-English Alsatian: "If the English come, I'll take my rifle." Life in our little

28) In his famous appeal over the radio from London, General de Gaulle told the French that "France has lost a battle, but not the war."

group of anti-aircraft shooting teachers was as calm and monotonous as before. I didn't even have any books left to read. For days together, we played bridge as a distraction. The group of officers around me were not exactly Nazis but they displayed strong Vichyist tendencies. One of them who was very young and quite nice, confided to me that he was afraid for his wife. He adored her, but she was English, "from the wrong side". I told him that quite soon, he would probably decide that after all, she was on the right side. I don't know what he ended up doing. The climate varied according to the different groups. Fortunately, I came into contact with a group of Gaullist officers. Some of the active officers from Biscarrosse eventually left for England.

The negative predictions of my friend Paul Ruff, which had convinced me, turned out to be quite right: there was no battle of the Marne. I think it's useless to express here my feelings about our defeat; everyone can imagine them.

## In the free zone. Toulouse

I was demobilized on August 15, and rushed to Toulouse where my parents had moved. Marie-Hélène joined us there. I finally was able to return to the civilian world, after three years of military service! Three years of unbearable boredom, social life devoid of interest, constant hypocrisy, deprivation of mathematics and of every spirituality, finally came to an end. I don't know what military service is like today, but I certainly hope it has changed. I still have the impression that intellectuals do not get much out of it ....

Marie-Hélène and my mother looked out a reasonable little house for us, containing three furnished rooms, on the banks of the Garonne, rue du Pont-de-Tounis. It was a little bit dark but still quite agreeable. My parents lived there throughout the war and during a large part of the Occupation, with difficulty because Toulouse suffered from severe food shortages. My mother was really courageous about confronting the endless lines to obtain food. My father, a doctor-colonel in the reserve, was sixty-eight years old. He entered the Purpan Hospital as a voluntary surgeon. At first he took orders from Prof. Dubost, who was responsible for the surgical division of the hospital. They were both doctor-colonels, one civilian, the other military. After looking down at my father from a great height, Dubost, who was really a talented surgeon and a man of great moral value, realized that my father was a very great surgeon, and changed his attitude. The two men became close and stayed excellent friends even after the war. The Dubost family visited us in Autouillet several times.

At that time, Toulouse was a dirty town, invaded by fleas which it was impossible not to catch in the public transportation. I woke up sometimes absolutely devoured by itching, whereupon I would plunge bodily into the bathtub, drowning as many as eight fleas simultaneously. The fleas provided us with an inexhaustible subject for conversation or letter-writing. They only bothered me and never Marie-Hélène, which seems very unjust to me, as though I get twice as many bites. It's the same with mosquitos. Young people who are thinking of getting married, inquire first whether your fiancée is attractive to insects ....

In 1939, my brother Daniel had graduated from the Ecole Polytechnique with the title of Tobacco Engineer. He was immediately mobilized to a fighting unit. Maxime, the son of Daniel and Yvonne, is today a well-known biologist, director of the Pasteur Institute and a corresponding member of the Academy of Sciences; he was born on the inconvenient date of June 1, 1940. Daniel soon came to Toulouse and went to see my parents. As a Jew after 1940, it wouldn't have been easy for him to take up the job which he would normally have been entitled to, and which would have led to a directorship of a factory. Instead, he was sent to a research center in Bergerac in Dordogne, which was exactly what he wanted anyway. He and his wife spent the war there. He began research on a viral illness of the tobacco plant, called the tobacco mosaic; he discovered in particular that the illness was transmitted by the person who planted the tobacco, which explained why whole rows of tobacco were ill and others not. Through him, I learned quite a lot about this virus, which was very little known (we spoke respectfully of "filtering viruses", without knowing anything further). Only later, after having studied the dangers which people pose to tobacco, did Daniel become interested in the dangers which tobacco poses to people. For years he has worked, using rigorous statistical tests, on proving that tobacco is harmful. These statistical methods were inaugurated with the study of tobacco, and are now taught to every doctor in France. Daniel's role in the development of these statistical techniques has no equivalent. He was given the post of professor of epidemiology at the Medical Faculty even though he was not a doctor – he founded the science of epidemiology in France. A former director of the laboratory of the INSERM, he has just been appointed corresponding member of the Academy of Sciences. Thus, the discrimination he suffered as a Jew caused Daniel to end up a successful researcher and professor instead of the director of a tobacco factory.

It's an interesting and curious fact that a single substance, tobacco, can possess so many features which are bad for mankind, encouraging all kinds of serious maladies, particularly lung cancer and heart problems. Sometimes things happen the other way; nowadays people are finding more and more benefits in aspirin, originally a simple medication. With a little attention to the dangers caused by its hemorragic properties (aspirin is an anti-coagulant), it can be taken in a great many cases. It has the triple property of being analgesic, anti-pyretic and anti-inflammatory. For these reasons, it can be very useful to help rheumatism, for example. Furthermore, it helps prevent heart attacks. The discovery of aspirin is almost a joke. A researcher working on rheumatism observed that willows grow in water and yet "never suffer from rheumatism". The idea of a willow having rheumatism is obviously bizarre – yet he deduced that the willow must have a substance warding off rheumatism! And aspirin was fabricated from a willow extract called acetylsalicylic acid; salicylic means coming from the willow (salix in Latin). Grotesque idea and beautiful discovery.

Bertrand was supposed to enter the Ecole Polytechnique in 1939, but he was mobilized in the North and could not start immediately. From Dunkerque, he was sent to England together with his comrades, and they remained there almost until the armistice. Returning to France, he was taken prisoner with his entire unit, but in such absurd conditions that they all easily escaped. Immediately after the armistice, he set to searching for his family. He located Marie-Hélène's address in Biscarrosse-Bourg, and from there he was able to locate us and let us know he was alive. Soon he came to visit my parents in Toulouse. I remember one political discussion we had at the end of September 1940, after the British victory against the German bombardments (the famous battle of London). We both believed that whatever was to happen in the immediate future, Hitler would eventually lose the war. None of us, not even our parents, ever doubted this. What optimism! The grand speech of Churchill to the British nation filled me with immense admiration.

Bertrand entered the Ecole Polytechnique in October 1940 and remained there until 1942. The school had been transferred to Lyons so as not to be a hostage of the Nazis in Paris. It was extremely difficult for Bertrand to live amongst a group of students and professors with strongly Pétainist leanings. Some of the professors were even collaborators. One of them tried his best to have the Ecole Polytechnique remain in occupied Paris; the same professor tried unsuccessfully to have Hadamard excluded from the Academy of Sciences after he left for the US, on the grounds that he no longer participated in their meetings. After leaving the Ecole Polytechnique, Bertrand came to Toulouse and spent some months in my parents' house, then he went to Spain in 1942, in order to reach North Africa, and then England. He spent six months in a Spanish prison before being able to realize this goal. We had no further news of him until the liberation of Paris, in which he participated as a member of the Leclerc division.

Toulouse contained other guests. My grandmother lived with my parents until the end of 1942. The three-room house in the rue de Pont-de-Tounis contained my parents, my grandmother and Bertrand for several months. The Hadamards spent at least two years in Toulouse after the armistice. Jacques Hadamard was faithful to himself and his legendary absent-mindedness. Once, he went with Bertrand to buy eggs in a store which had just received some. Displaying an admirable civic spirit, the grocer refused to sell more than half-a-dozen eggs to any one home. Bertrand and Uncle Jacques came from two different homes, so to make sure the grocer didn't doubt this, they decided that they "didn't know each other" while in the store. Everything went fine for a while, but then Uncle Jacques forgot all about it and began to chat to Bertrand about everything and everyone in a loud and clear voice. Suddenly, when his turn came to buy eggs, he remembered, and after having paid, he turned to Bertrand and said ceremoniously "Sir, it was a pleasure to make your acquaintance." Once out of the store, he turned back and called cheerfully "See you later at home!" In 1942, it became clear that the Hadamards were going to have trouble living in hiding; Hadamard was already 79 years old. We managed to obtain a temporary invitation for them to go to the USA; it took a very long time to obtain the visa. They went to Spain to take a boat. But in Madrid, they suddenly realized that Uncle Jacques had forgotten the precious visa in Toulouse! Interminable visits to embassies were necessary before this situation could be rectified. Even this wasn't the end of his tribulations. In the US, in spite of his age, he had to earn his living. He looked for a job in a small university; it was quite a usual thing for retired professors to do this, in order to complement their pension. At the first university he visited, he was received by the chairman of the mathematics department, who unfortunately had never even heard his name, and explained to him that it would be difficult to find a job for him. Hadamard pointed to the wall and said to him "Look at that row of portraits of famous mathematicians. That one over there is me." The chairman recognized him and eight days later, he asked him back, only to tell him that unfortunately, they were really unable to give him a job. In the meantime, his picture had been removed from the wall . . ..

At first, the Lévys lived near Lyons, with their daughter Denise Piron, her (non-Jewish) husband Robert Piron and their daughter Hélène, born in August 1940, another awkward time to be born! Later, Hélène married Uriel Frisch. They are both astrophysicists, and Uriel is a corresponding member of the Academy of Sciences. It seems that being born at such drastic moments was not too harmful for Maxime Schwartz and Hélène Piron. They are the oldest of our seventeen adored nieces and nephews. In March 1943, the Lévys and the Pirons were able to reach the Italian zone, together with another Piron child, Olivier, in Montbonnot, near Grenoble.

We meant to spend the whole Occupation in Toulouse. It seemed unthinkable to return to Paris. I had begun to work alone, borrowing two or three books from the University of Toulouse, in particular one by Emile Borel on divergent series. I was very attached to these series and obtained a few original results on them, not enough however to obtain a doctorate. In fact, I began by going to visit the rector of the university, Deltheil. He had been a member of the jury for my agrégation, and so he knew me. "If you want to teach," he said, "I can take you on immediately in hypotaupe."

But I preferred to apply to the Caisse National des Sciences (the future CNRS)[29]. I don't remember my actual title, but I was accepted in the month of November. I retained my job with no problems until the end of 1942. The members of the Caisse National were not in government service, and the statutes concerning Jews simply forgot about them. At least, until the end of 1942 – then I suddenly lost my job. Fortunately, the ARS, Aide à la Recherche Scientifique, an organization founded by Michelin, was not concerned by the status of Jews and paid me a salary equivalent to my salary from the Caisse from the beginning of 1943 until the end of the war. After the war, I wanted to express my gratitude to the ARS by working for it for a few years, and I remained on its board. The administration of the Caisse National was so disorganized that although I was given my job in October 1940, I hadn't received a penny of my salary in March 1941. I had to travel to Vichy in order to meet the director, who apparently understood my situation – he even asked me if I needed money urgently – but seemed unable to pay salaries. I had been living for months by borrowing and using my tiny savings. Eventually, however, I was paid; I believe that the fortuitous presence of the rector Deltheil had something to do with it. This custom of not paying new recruits for months and months continued in France for several more decades. When a foreigner was invited to France, he often was not paid until after his return to his country, which was extremely difficult particularly for people invited from the East. These heavy administrative methods disappeared for a while, only to return, including in the US. Today, in France, it's become even worse in some cases. For instance, the fixed payment for a one-hour conference is one hundred sixty seven francs and thirty-five centimes, whereas a room in a hotel costs at least three hundred francs. Result: the president of the university is obliged to multiply the true number of talks by two or three, organizing a national system of high-level cheating.

---

29) The CNRS, Centre National de Recherche Scientifique, is a still-existing French institute providing research positions with no specific location.

## Clermont-Ferrand: searching for a university

I worked on divergent series for several weeks. We knew that Henri Cartan and Jean Delsarte were in Toulouse organizing entrance examinations for the ENS. Marie-Hélène, who needed advice on her own work, took the initiative to go and meet them. We knew them by reputation and I had taken courses by Elie Cartan, but we had never met them. They had also heard of us. They welcomed Marie-Hélène warmly. Cartan gave her the advice she needed and asked what we were doing in Toulouse, whose university was equivalent to a scientific desert, in which we were unlikely to find an interesting guide. After one unsuccessful try, I hadn't even attempted to contact anyone in the mathematics department of the university, and I was working alone. Information was difficult to obtain during the Occupation, and I don't know what I would have done without this miraculous meeting. Cartan energetically advised us to go to Clermont-Ferrand, where the scientific university deserved its name, on top of which the whole of the faculty of the University of Strasbourg, the best university in France at that time, had taken refuge there. The union of the two faculties made Clermont-Ferrand the main mathematical center of France, better than Paris. The main mathematicians working there, apart from Cartain, who was a professor at the University of Strasbourg (although he only remained there for a few months and was then appointed professor in Paris) was Georges Cerf (he had been president of the League for Human Rights in Strasbourg, until he was fired for being Jewish), Claude Chabauty (arithmetic), Jean Dieudonné (algebra and functional analysis; he was a professor in Nancy but staying in Clermont), Charles Ehresmann (algebraic topology), André Lichnerowicz (differential geometry), Szolem Mandelbrojt (quasi-analytic functions) and René de Possel (analysis). The dean Danjon, who was an astronomer from the University of Strasbourg, was a person of considerable importance, who constantly protected his professors. I remember having a conversation with him about mathematics useful for astronomy. A French policeman entered and started to interrogate him about a student at the university. Danjon replied firmly "I'm very sorry, but I can't give you any information whatever, it is not my role. If you need information from me, then the order should be transmitted to me by the rector of the university, who should himself be solicited directly by the police headquarters, upon which he can take upon himself the decision of whether or not to transmit a written order to me." The policeman left empty-handed and did not return.

Danjon told me a story about mathematicians which is worth retelling here. The great mathematician Boussinesq lost his wife. The day of the funeral started out quite beautiful, but ended under a rainstorm. Everyone was soaked. Boussinesq married again, but again became a widower. The same meteorological phenomenon occurred during the second burial. When his third wife also died, the funeral took place under a beautiful blue sky, but everyone there brought an umbrella. Emile Borel, the grand guru of probability theory at the Sorbonne, turned to Polya, a foreign mathematician who happened to be visiting France just then, and said "Look, Polya, isn't this ridiculous? We're all university professors, I'm a probabilist, I know perfectly well that there is no relation at all between the weather and the funeral of Madame Boussinesq. Yet, I brought my umbrella." "Well," replied Polya, "we're scientists, we work on observed facts. And it's a scientifically observed fact that it often rains at the funeral of Madame Boussinesq."

There were only three graduate students for the whole illustrious faculty of Clermont-Strasbourg: Jacques Feldbau, a student of Ehresmann, Gorny, a Polish refugee in France and a student of Mandelbrojt, both brilliant, and myself. We could hardly imagine a better pedagogical situation. Feldbau had been an auditor at the ENS – he had not succeeded in entering the ENS because, as a practicing Jew, he had refused to pass one of the exams on a Saturday). He had passed his agrégation in 1938 and like me, he had arrived in Clermont in 1940. He taught me a lot of algebraic topology and we became very friendly. I had studied algebraic topology in a brochure on homology written by Ehresmann, and Feldbau taught me cohomology, invented by Kolmogorov. Unfortunately, he was arrested in the round-up of Strasbourgers in the dormitories, on November 26, 1943, and deported to Auschwitz. In the camp, he taught mathematics to Raymond Berr, the father of my sister-in-law Yvonne, and to Jean Samuel. He survived in camp thanks to the help of Professor Weitz, who had him admitted to the infirmary. He somehow found the strength to write a fifty page treatise on topology for his listeners. But he was among the prisoners transferred by the Nazis before the liberation of the camp by the Red Army on January 27, 1945, and he died during the transfer, in April, just a few weeks before the end of the war.

Gorny was a Polish Jew who had taken refuge in the free zone. He worked on classes of quasi-analytic functions. We helped him as well as we could; his existence was highly precarious. We learned one day, from a letter written hastily before his departure (and which ended with the words "Hopefully I'll come out of this alive") that he had been deported to Drancy. The packages we sent him took months to arrive and when they came, of course they were in a miserable state, particularly the bread, which was moldy. To our astonishment, our last letter was returned to us with the words "Left Drancy without leaving an address". The cynicism of the Post Office – after all, deportation to Germany was no secret – disgusted us. We never saw him again. The crimes of the Nazis and the Vichy regime were becoming palpable realities to us, and we were fully aware that the fate of foreign Jews in France would soon become the fate of all French Jews, and all European Jews. Of the three of us, two died in deportation; I myself only barely escaped it.

## Boisséjour et Ceyrat

Our first home near Clermont was in Boisséjour; a tram took us to Clermont in one hour. We had a poor little room, with a water closet in the courtyard. A little restaurant whose proprietor was usually drunk prepared our lunch. For a modest sum, we dined daily at a neighbor's house. We were desperately short of money. Dieudonné said that our room reminded him of the study rooms at the ENS. Beaufrère, an old Trotskyist friend who came to see us when passing through Clermont, was so horrified by our miserable room that he organized a subscription within the party for us, believing that we were dying of hunger. Soon, however, we found a better place in Ceyrat, the terminus of the tram. It was a charming little village with a nice climate. The landlady was slightly mad, but very generous. Our room was quite cute and had a bathroom and a kitchen. The walls were covered with stencil paintings of parrots and flowers, which attracted butterflies, who entered the window to gather pollen; this actually furnished me with an incontrovertible proof that sight plays a role just as important as smell in guiding butterflies to flowers. I've also seen butterflies try to gather pollen from the flowers on ladies' hats. On our wall, we watched them try three, four, five flowers before finally understanding that they'd been had. We were mediocrely heated by a large heater which we affectionally called by its brand name, Nesto 12, but we didn't have the right kind of coal for it.

This period was tranquil enough for Marie-Hélène to be able to work on mathematics. It wasn't a big problem to get to Clermont-Ferrand, but it took time. I communicated a lot with the Clermont and Strasbourg mathematicians. André Weil spent some weeks there and I encountered the collective mathematician Nicolas Bourbaki, whom I'll discuss in more detail later on. All in all, our move to Clermont was extraordinarily beneficial.

The inhabitants of Ceyrat were nice on the whole, and we made a few friends among them, particularly with people from Strasbourg. From our house, we often heard troops of young people passing by, singing *Marshal, here we are*. For our part, we emphasized our disgust with the Vichy regime by calling our toilet paper "marshal paper". We hid our political books, and I also carefully

Schwartz in the garden

buried a revolver under a tree near the house. When we left Ceyrat, I wanted to take it with me, but I never found the tree or the revolver again . . ..

## The meeting with Bourbaki. A life outside of the war

After three years of military service and total isolation, Clermont-Ferrand represented a real liberation and a renaissance to mathematics, and I found my old enthusiasm once again.

I had a solid memory, and I had learned mathematics without ever taking a single note, organizing my knowledge in my head. Only in 1972, following

Grothendieck's advice, did I begin to write up everything systematically, whether my own ideas or those of others. And it is true that my memory did not always remain as good as it was when I was young. On the left-hand side of my office, there is a pile of three hundred and eighty manuscripts, synthesizing the results of my reading and publishable or unpublishable results; they are my written memory. Now that in my old age my memory sometimes fails me, these pages are very precious to me. If I forget some result or other, I can go fish it out. My manuscripts, numbered 1 to 380, are all listed on my computer. To find a result, I just have to click on "search" and give a word from the title of the manuscript. If the document is easy to find, I read it over, and until three or four years ago, it was usually enough for me to glance over it for a few minutes to understand and remember the results. Unfortunately, now, it sometimes takes me hours or even days to recover a theorem proven many years ago; 1972 for instance is already 25 years ago .... My mathematical life has become more laborious than ever before. I lost three years from 1937 to 1940 and I was very worried that my memory might have been affected. But in just one month in Toulouse, I was able to recall everything I had ever learned, and I was reassured. My stock of acquired knowledge had not suffered, but it was still three lost years, which is a lot for a mathematician. Alas, I lost many more over the course of my existence.

The topological vector spaces I was interested in were absolutely new to me since I had never considered any vector spaces beyond the plane and "space", i.e. the three-dimensional space we live in; I'd never dared consider 4-dimensional space. But now, I encountered topological vector spaces in complete generality; it's one of the first things I learned from Bourbaki. Once again, I was enthusiastic. Dieudonné was teaching a course on topological vector spaces at the university; Marie-Hélène went to it for both of us, traveling up to Clermont for the purpose once or twice a week. On her return, she would tell me everything she'd learned, so I could learn it too.

Two years later, I used what I had learned in Dieudonné's course in my thesis. The great theorems of the theory: the Hahn-Banach theorem, the Banach-Steinhaus theorem, Baire's theorem, bounded sets and neighborhoods, duals, biduals and reflexivity, not to speak of some more delicate theorems, were major tools for me throughout my work on distributions. *A posteriori*, I realize that the whole theory is quite easy, and has even become banal nowadays, but at the time it was all new to me. It was not difficult for me to assimilate it; I helped myself by borrowing books from the university, particularly *Theorie und Anwendung der Laplace Transformation* by Dœtsch and *Orthogonale Reihen* by Kakzmark-Steinhaus. I learned a lot of results in functional analysis, which I had already seen, in truncated form, at the ENS. Now the theory took on its full value. I may have trouble visualizing in three dimensions, but I'm perfectly at ease in infinite-dimensional spaces, since there's really no need to visualize inside them.

Bourbaki was a revelation for me. In 1935, André Weil and Henri Cartan, both professors at the University of Strasbourg, decided to put some rigorous order into the mass of mathematical notions, so as to improve teaching at the secondary and university levels. At the university level, they wanted to modernize Goursat. At first, their main objective was didactic, but later this changed. They got together an initial group of "founding members": Henri Cartan, Claude Chevalley, Jean Delsarte, Jean Dieudonné, Charles Ehresmann, René de Possel and André Weil; a few other people, for instance Jean Coulomb, Paul Dubreil, Jean Leray and Szolem Mandelbrojt made brief, temporary appearances. The nine official founders of the first conference, in Besse-en-Chandesse, were those of the above list except for Dubreil and Leray. They started mathematics again from zero, and set them solidly on an axiomatic basis. To define an object, one had to define the axioms it must verify rather than the object itself. Groups, rings, fields, vector spaces, topological spaces, metric spaces, uniform spaces were all redefined in this way. Then came topological groups, topological vector spaces, topological varieties, differential varieties, fiber spaces .... Whatever problem one considers, one should always know in exactly what axiomatic class it is situated. Here is a simple example. In taupe, I had learned two theorems by Cauchy on real continuous functions of the line. First of all, the restriction of any continuous function of the line to a bounded set takes a maximum and a minimum. Secondly, a real continuous function of the line cannot pass from one value to another without passing through all the intermediary values. The whole structure of the line is deeply contained within these two theorems, each of which corresponds to a different partial aspect of the structure of the line. Firstly, every continuous function on a compact topological space admits a maximum and a minimum. But a bounded subset of the line is a compact topological space. Secondly, a real continuous function on a connected space cannot pass from one value to another without passing through all the intermediate values. The line is a connected topological space. Each theorem is thus replaced in the general framework corresponding to the structure it is describing. The distinction of different structures is one of the great innovations of Bourbaki. When a set is equipped with several different structures, they have to satisfy certain conditions with respect to each other. For instance, a topological vector space has the structure of a vector space and the structure of a topological space, but the two structures are not unrelated: the operations of the vector space must be continuous for the topology.

After meeting Bourbaki, I took the habit of determining the structures I was dealing with before attacking any problem. As I showed above with the example of Cauchy's theorem, there's no reason to state a theorem about the real line if the theorem is really only referring to one part of the structure of the real line. All mathematicians have now essentially adopted this mode of reasoning. The books by Bourbaki correspond to the different structures which can be put on mathematical objects. For instance, there's a book about topological vector spaces, and a book about commutative fields. Each structure is entirely characterized by its axioms. This makes it possible to classify all mathematical objects with respect to each other. Furthermore, one is rid of the absurd constraints which existed before. I found this intensely satisfying.

In this way, the language of mathematics also becomes extremely pure. The type of book found in the Borel collection, reminiscent of novels, disappeared completely. Writing styles became more like what the Germans already used. Writing styles have varied a lot with the centuries and have gone through various transformations. In Greece, the mathematician (or group of mathematicians) Euclid completely reorganized the study of geometry, thanks to a more or less axiomatic theory and a very precise style. The Greek texts are written with infinitely more precision than the texts of the Borel collection or Lebesgue's writings, or the texts of the seventeenth and eighteenth centuries. The writings of Euclid or of Archimedes, a Greek from Sicily, are as limpid as pure water. The lemmas, theorems and corollaries follow each other in impeccable order and refer to each other with no possible ambiguity.

A similar reform took place in the nineteenth century. At the beginning of that century, mathematics had become completely unrigorous. Series were freely manipulated without attention to their possible divergence (see Fourier, for example!) and even if they converged, no one checked to see if the convergence was uniform. Continuous functions were used although the notion of continuity was not fully understood or even defined at that time. Infinitesimals were defined and yet no one understood them. Cauchy himself, in his lectures at the Ecole Polytechnique in 1821, "proves" that the sum of a (simply) convergent series of continuous functions is continuous. In spite of the fact that Abel published a counterexample to this in 1826, Cauchy continued repeating it even in 1833. Most of the great mathematicians of the nineteenth century reflected on the fundamental notions of analysis, but universal rigor was not acquired before the last quarter of the century. This rigor was generalized mainly by Cauchy in France and Weierstrass in Germany. Cauchy imposed a rigorous form of language and expression on the other mathematicians of his time, sometimes by rapping them on the knuckles. When he attended lectures by well-known mathematicians, they used to tremble for fear of the numerous observations he made on their style, expression, rigor and even their use of concepts. The theory of infinitesimals was one of the greatest stumbling blocks of the eighteenth and the beginning of the nineteenth centuries. Cauchy rendered it all rigorous, and in spite of a few errors, his lectures in analysis of 1821 represent the high point of the theory. The necessity to add considerations of teaching to those of research was the motivation for this reform. Today, we almost have to fight to add research to teaching.

The famous mathematician Lagrange had written a treatise on "derivatives without infinitesimals", in which he made truly incredible errors (the advent of a Cauchy was really necessary after that). He defines the derivative of a function $f(x)$ at the point $x$ as follows: consider $f(x+h)$, take its expansion (what does this mean?); it gives a series in integral powers of $h$ (sic); the coefficient of $h$ in this series is $f'(x)$. Such an expansion exists if $f$ is a polynomial, since then the series has only a finite number of non-zero terms, or even if $f$ is an analytic function, a concept unknown in the time of Lagrange, and in those cases, it is true that his definition of the derivative works. Lagrange added with amazing innocence: "One might object that fractional powers of $h$ could occur in the expansion. But this is impossible, since the root of a real number always has several complex values." Apparently, he forgot that he was working on real functions of a real variable, and that in general, such functions have no expansions at all. It is true that Robinson's recent theory of non-standard analysis introduced infinitesimals again, which are in some sense analogous to the intuitions of the analysts who came before Cauchy. Cauchy's reform was fundamental, but it was not as deep as those of Euclid or Bourbaki.

Bourbaki's was the third major reform: mathematics began to be written as they never had been before. Gone were the vague references to "what we said above"; references were only to a precise theorem or corollary. This contributes to establish a certain unity within mathematics. One reasons the same way in all branches of mathematics. Previously, mathematics was divided into different chapters each of which obeyed different rules, having nothing to do with the others. Now, a quantity of mathematical theories can be built on a single notion, for instance the notion of topological vector spaces. Only at a later stage do theories differ from each other.

Bourbaki's classification of mathematics can be compared to the immense revolution introduced by Linnæus in his *Systema naturae* in 1758, in which he classified all animal and vegetable living organisms. Branches, classes, orders, families, genera, species, subspecies, which make it possible to situate a given organism at every degree, replaced the chaos which reigned amongst animals and vegetables before. The subkingdom of vertebrates contains five classes: mammals, birds, reptiles, batrachians and fish. Previously, whales were thought to be fish because they lived in water. But its internal structure makes the whale into a mammal, not a fish. The study of whales is based on this fact. Whales do not have fur, like other mammals, but fish have scales. The skeletons of mammals have thirty-three vertebrae, unlike those of fish. Whales have lungs and breathe air, fish have gills and breathe water. Fish lay eggs and do not care for their young – indeed, they often devour it – female whales carry their young and nurse them. The classification of Linnæus situates whales in the subkingdom of vertebrates, in the class of mammals, in the order of cetaceans. Cetaceans are divided into several families; whales correspond to a particular genus, this genus contains several species of whales. *Systema naturae* first appeared in 1758 – by just the tenth edition, whales are found classified as mammals, whereas in 1753, Daubenton still called them fish. Similarly, a bat is not a bird, but an insect-eating mammal. Classification becomes quite easy. For instance the lion belongs to the subkingdom of vertebrates, the class of mammals, the order of carnivores, the family of felines, the genus *Panthera*, the species *Panthera leo*. Tigers have the same classification right down to the genus, but their species is called *Panthera tigris*. Cats are *Felis felis*, pumas are *Felis puma*. Dogs share the classification as vertebrates, mammals and carnivores, but they belong to the canine family, their genus is *Canis* and their species is *Canis domesticus*, whereas wolves belong to the species *Canis lupus*. Bourbaki is the Linnæus of mathematics.

Mathematical theories proceed more by crossings than by vertical lines. Topological vector spaces are at the intersection between the category of vector spaces and the category of topological spaces. The two things are related as we said above. Certain structures are quite poor, for instance the structure of topological spaces, which allows one only to study continuous maps and open and closed sets. But the structure of a Lie group is a very rich structure. A Lie group is an object which possesses a group structure on the one hand, and on the other hand the structure of a differential variety, which is itself a very rich structure; the group operations must be differentiable for the differentiable structure. The richest structures are usually the most interesting ones to study. However, this is not absolutely true, since sometimes it is just a single object which holds the most interesting structure. For instance, the orthogonal group (the group of rotations) of $\mathbf{R}^n$ is unique. It is called $O(n;\mathbf{R})$ and has all the properties of less rich categories, and yet, or because of this, is a deeply interesting object to study.

My mathematical nature is deeply attracted by classification, and intensely appreciates this organization which I have always deeply respected. Distributions are studied first on $\mathbf{R}^n$ and then on arbitrary finite-dimensional affine spaces, and finally, as currents, on arbitrary infinitely differentiable varieties.

Bourbaki also introduced a very enriching vocabulary. The role of vocabulary in mathematics is really interesting. Some objects, before earning the right to a name, have to be expressed by stating some very long property they are supposed to satisfy. For example, if $f$ is a function on the set $X$ with values in the set $Y$, we often use the following object: if $B$ is a subset of $Y$, then $A$ is the set of $x \in X$ such that $f(x)$ lies in $B$. This phrase was repeated very often. At the ENS, I had discovered, without ever reading it explicitly in books, that if $X$ and $Y$ are topological spaces (in my mind they were 1, 2 or 3-dimensional spaces), then $f$ is continuous if and only if whenever $B$ is an open subset of $Y$ then the set $A$ defined above is also an open set. Every time, the definition of $A$ had to be repeated starting from $B$. At some point, it became indispensable to introduce the notions of the image and the preimage of sets under a map $f$. If $B$ is a subset of $Y$, then the set $A$ of points $x \in X$ such that $f(x)$ lies in $B$ is called the preimage of $B$ under $f$, and it is written $A = f^{-1}(B)$. The vocabulary simplifies everything incredibly, and now we can say that $f$ is continuous if and only if the preimage under $f$ of every open set of $Y$ is an open set of $X$. The preimage has interesting properties with respect to unions, intersections and complements. The inevitable and obvious notions of images and preimages were not introduced until 1932, by the German mathematician Dedekind! Of course, Bourbaki adopted them at once. I still remember the difficulties I had at the ENS to state a certain number of propositions, in which I had to give all the definitions over and over again. From the day that I saw the Bourbaki vocabulary, I progressed much faster. The notions introduced by Bourbaki were already familiar to me, but from a different point of view. I quickly assimilated set theory and the theory of topological spaces, and the notion of a group; once I saw these notions, I learned them in two or three weeks. The same would hold for any mathematician suddenly brought into contact with the books by Bourbaki. After the first two books I read, *Set theory* and *Topological Spaces*, I real *Linear Algebra or the Theory of Vector Spaces* and *Multilinear Algebra or the Theory of Tensor Products and Tensors*. Tensors, in particular, were previously defined only by coordinate changes. The tensor product has an intrinsic definition. Some mathematicians are easily able to understand the tensor product of two vector spaces, but others have a lot of difficulty, and some quite reputable mathematicians have never succeeded in really using it. For myself, though I have various other problems, tensor products have always been very helpful.

It's the same with manifolds. Many applied mathematicians study open sets of $\mathbf{R}^n$ whose boundary is a regular surface, but Bourbaki studied differential manifolds with boundary. The notion of manifold is a difficult one because its definition is long. But once the definition is really understood, the use of the word manifold simplifies everything. Bourbaki used manifolds very early on, but never published a full account of them, and this is because a complete account with every definition and detail would be monotonous and pedantic, not to say incomprehensible.

Every question about a manifold comes down to a question about charts, and one has to use expressions like $\phi_2\phi_1^{-1}$ to reduce the question to a question about open sets of $\mathbf{R}^n$. Beginners are obliged to do this step by step in order to understand. Then, quite soon, one doesn't need to do it any more, because one has understood how it works, and it is enough to write "The proof is obvious via the charts". Mathematicians who haven't really mastered manifolds cannot write this, precisely because to them, it is not obvious. Bourbaki never wrote a book on manifolds, because in the Bourbaki style, it would be impossible to write "obvious by the charts"; you would have to write all the $\phi_2\phi_1^{-1}$ and then you obtain a formulation which is too sophisticated, even typographically, for the idea it is supposed to represent. Bourbaki always tried, generally successfully, to give intrinsic definitions to all objects associated with manifolds (tangent vectors, or the coboundary or exterior differential of a differential form), without using charts, even if a chart (but not two) is sometimes used to prove their properties (dimension of the tangent space, etc.). This is one of the main reasons for their success.

At what point does a mathematician introduce a new term or a new definition into his work? I generally try not to do it. When I develop a theory, I state certain properties explicitly, even if they are a bit long. But when I see that the same property comes back again and again, and that it's becoming annoying to repeat it continually, I go back and give a name to the property. For example, we have the notion of a hypocontinuous bilinear map on a product of topological vector spaces. At first, I refused to use the word hypocontinuous and kept on repeating the definition. I only introduced it when it became really inevitable. Naturally, every mathematician must be ready to use a fair number of new words. One part of his work actually consists in conceiving suitable new names and notations whenever necessary. For example, when studying a partial derivative of a function on $\mathbf{R}^n$, one used to write: $(\partial/\partial x_1)^{p_1}(\partial/\partial x_2)^{p_2}\cdots(\partial/\partial x_n)^{p_n}$. Now, one writes $\partial$ for the $n$-tuple of partial derivatives $(\partial/\partial x_1), (\partial/\partial x_2), \ldots, (\partial/\partial x_n)$, and $p$ for the $n$-tuple $(p_1, \ldots, p_n)$ of integers greater than or equal to 0, and the partial differentiation given above can be written $\partial^p$. A differential operator on $\mathbf{R}^n$ of order less than or equal to $m$ can then be written as though the dimension

$n$ were equal to 1: $\sum_{|p|\leq m} a_p(x)\partial^p$. If we agree that for $x = (x_1, \ldots, x_n)$, $x^p$ denotes the scalar $(x_1)^{p_1}(x_2)^{p_2}\cdots(x_n)^{p_n}$, and that $p!$ means $p_1!p_2!\cdots p_n!$, then the Maclaurin series in $n$ variables can be written just like a series in one variable: $f(x) = \sum(x^p/p!)(\partial^p f)(0)$.

This very elementary simplification is in fact an idea of genius. It seriously contributed to the grand development of the theory of partial differential equations. I used it often in my book about distributions. It's even often been attributed to me, but this is wrong; it was really invented by Hassler Whitney, who used it systematically in his beautiful work on differentiable functions. An even more striking example is the invention, by the Chinese and the Arabs, of the zero; not just the number 0, as one often hears, but of the digit 0 within the writing of larger numbers. It considerably simplifies all operations with these numbers.

Without these continual simplifications, mathematics would become very confused, even impossible. However, such syntheses represent an augmentation in the level of abstraction. Mathematics is the most abstract of all the sciences, and this is what makes it difficult. From the moment one introduces distributions, one can speak of the space $\mathcal{D}'$ of distributions, then of continuous linear maps from the space of distributions into itself (for example derivation), then the space $\mathcal{L}(D;;D')$ of these maps. The degree of abstraction involved here is already quite high, although familiar to any mathematician. This abstraction is an important feature of mathematics; one gets used to it quickly. It's difficult at the beginning, but then becomes easier. The major portion of mathematics is abstract. If for example we write the equality

$$\int_{-\infty}^{+\infty} e^{-x^2/2}\,\mathrm{d}x = \sqrt{2\pi}$$

(the Gaussian integral, fundamental in the study of probability), it contains a large number of "abstract" ideas: the number $\pi$, the number $e$ (once I remember that in order to explain the work of a mathematician, I proposed a text containing the number $e$ to the newspaper *Le Monde*; the journalist didn't understand it and went to see the director, who didn't understand it either but said "I guess Schwartz knows what he's talking about ..."); the exponential function, which was unknown until the eighteenth century, the integral, discovered by Riemann in the nineteenth century, except that here the integral is improper but absolutely convergent. Today, not only mathematicians, but also physicists, chemists, engineers, biologists and doctors, know more or less what this formula means, and they also know that the result of the integration really is $\sqrt{2\pi}$. This number is banal (even for journalists of *Le Monde*, doubtless), but mathematicians of the olden days had difficulties dealing with square roots and the number $\pi$. If we begin from zero, the Gaussian integral is a really difficult concept. Abstraction rises to dizzy heights; it is possible for us only because each of the objects has come to seem familiar and concrete to us from the habit of using it frequently. It's a question of time and energy, but it always works eventually. This is what makes it so difficult to explain mathematics to the layman. On this subject, a professor once defined an abstract object to be an object which one can neither see nor touch, and a concrete object as an object which can be seen and touched. Yes, agreed a student; my underpants are concrete because I can see and touch them, but yours are completely abstract, since I can't. More seriously, a concrete object is an abstract object which one has gotten completely used to. Any physicist who manipulates atoms, protons, electrons, quarks, gluons, quantum mechanics or relativity lives in a world of extreme abstraction. The same goes for a chemist who writes down a long formula, or a biologist who understands the double helix, the laws of immunity or the HLA system. But he forgets this and considers these things as concrete objects.

Bourbaki progressed slowly. In the beginning, there were successive write-ups and regular meetings in a restaurant on the boulevard Saint-Michel. Then it became a real "secret society" which ended up containing a good fifteen members. More members would probably have slowed down the dynamics of the movement. To recruit a new member, the association would first decide if he had the necessary qualities, then accept the new member as a guinea pig, and if he was interested and the other members were interested in him, he became a member. Bourbaki has existed since 1935. Its members are not supposed to reveal that they belong; the reason for this was supposedly so that members cannot use Bourbaki for purposes of self-aggrandizement.

Where did the name of Bourbaki come from? The group needed to find some kind of signature. André Weil tossed out "Why not Bourbaki?" Bourbaki was a general of Napoleon III who took refuge in Switzerland during a battle in order to avoid being taken prisoner with his troops. He was the son of Sôter ($\Sigma\omega\tau\eta\rho$) Bourbaki, who probably came from the island of Souli ($\Sigma o\upsilon\lambda\eta$), and who clandestinely led Bonaparte from Egypt to France. As a recompense, Bonaparte promised him to look after his children. All this had nothing to do with mathematics, but that was not in the least bit important.

We sent a note by Bourbaki to Elie Cartan, to be presented to the journal *Comptes Rendus de l'Académie des Sciences*. A note of this kind must be signed with a first and a last name. The founders of Bourbaki thought that Bourbaki sounded Russian, so they called him Nicolas, which sounded simultaneously

Russian and Greek. By pure chance, a real person called Nicolas Bourbaki existed, and when he arrived in Paris from Greece, he was amazed to see books published in his name. When Elie Cartan asked about the author of the note he had presented, he was told that it was a young man full of promise, and he never thought to inquire further. The whole thing was an important aspect of Bourbaki; they would never have accepted a member too austere or too rigid to play the game.

The Bourbaki books were written in the following manner. First, one member was given the task of making a "zero-th version", the first stage in the organization of what was yet an ill-defined mass of information. The manuscript was then read collectively by the group; this was very difficult for me, because I really couldn't follow the reading fast enough, while others such as Dieudonné had no problems. Cartan was more like me. I remember one reading where he gave up, saying: "This is impossible, I have go away and redo it all by myself." Half-an-hour later, he came out with a complete write-up which turned out to be practically identical to the one which we had just been discussing. But he had needed to figure out the details of it himself. Generally, the zero-th version was completely demolished by the group, who criticized it severely. To belong to Bourbaki, one had to accept that one's work was universally rejected; it took a very special type of pride. After the general criticism, the group noted down several suggestions for the next version, and some other member of the group was chosen to write it up. This procedure would continue through a certain number of meetings. Around the seventh or eighth version, everyone would be completely tired of it, and more or less in agreement, so the decision would be taken to publish. Bourbaki published nineteen volumes of different lengths. After each conference, each person was sent a bulletin called *La Tribu* (*The Tribe*) containing minutes of the conference and the decisions taken there, and the list of newly chosen writers.

Bourbaki's success in the mathematical world was slow but considerable. It's quite easy to judge by the organization and the style if a given article was written before or after Bourbaki.

The flexibility and lack of rigor in texts like Goursat or Lebesgue is just as bad as those of the Borel collection. The algebraic geometry of the Italian school is a striking example of what happens when requirements of rigor are relaxed. They did plenty of excellent and striking work, but the lack of rigor is characteristic – André Weil used to say that for them, "a counterexample constituted an interesting addition to a theorem." This generally meant that the specific hypotheses of a given theorem were not properly determined. And indeed, one finds the following kind of remark in their articles, following the statement of a theorem: "For this theorem to be true, it is necessary that certain conditions be satisfied" (these conditions are not specified); "here is an example where the theorem is false." Or else: "The notion of a variety is difficult to define. Let $V$ be a variety."

The great mathematician Hermann Weyl wrote the following sentence somewhere: "the group $G$ is open or closed, in the sense that the plane is open and the sphere is closed." Open, here, means locally compact but not compact, and closed means compact. After Bourbaki, all these deficiencies disappeared radically. Young people everywhere in the world flung themselves on the first volumes of Bourbaki, to learn the new mathematical methods. Older mathematicians were much more sceptical, sometimes even hostile and contemptuous.

It's important to remember that Bourbaki almost never actually proved new theorems. Each member of the group published his own research. The collective books are textbooks, not discoveries. Bourbaki simply wrote mathematics in a radically new way, always going from the general to the particular, which can actually make it rather difficult to read sometimes, if one doesn't know what particular objects one is aiming at. Examples usually follow the definitions quickly, but there are not very many of them. Bourbaki wanted to clean out the stables of Augias.

There were not many Bourbaki conferences when I was a guinea pig, from 1940 to 1942. At the first conference I went to, the members were invited by a telegram sent by Delsarte, the oldest member. In 1942, the police, believing it must be a political meeting, asked Delsarte for an explanation. After 1944, Bourbaki became more active, and the conferences took place in Nancy. We worked in the University of Nancy, rue de la Craffe, and we used to meet for meals in the restaurant Panarioux. To relax, we used to say so many idiocies that the other clients of the restaurant were shocked to learn how stupid honorable university professors could be. My brother Bertrand lived in Nancy for a while and once happened to eat at a table near ours. He did not spare me his opinion.

Bourbaki had an "adjutant"; it was Dieudonné, the real organizer. He used to check if write-ups obeyed the injunctions their author had been given, and he wrote *La Tribu*, which he termed a "ecumenical, aperiodical and bourbachic publication"; he included several poems by Pierre Samuel in it, along with the mathematics. Once, he even published an announcement of the marriage of Betti Bourbaki. Only a mathematician can really understand the wit of the thing, but nevertheless I can't resist reproducing it here. Betti numbers are used in algebraic topology; they are named after their inventor.

*Left-hand page*:

Monsieur Nicolas Bourbaki, a canonical member of the Royal Academy of Poldavia, Grand Master of the Order of Compacts, Conserver of Uniforms, Lord Protector of Filters, and Madame née One-to-One, have the honor of announcing the marriage of their daughter Betti with Monsieur Hector Pétard, Delegate Administrator of the Society of Induced Structures, Member of the Institute of Classfield Archaeologists, Secretary of the Work of Sou du Lion.

*Right-hand page*:
Monsieur Ersatz Stanislas Pondiczery, retired First Class Covering Complex, President of the Reeducation Home for Weak Convergents, Chevalier of the Four U's, Grand Operator of the Hyperbolic Group, Knight of the Total Order of the Golden Mean, L.U.B., C.C., H.L.C., and Madame née Compact-in-itself, have the honor of announcing the marriage of their ward Hector Pétard with Mademoiselle Betti Bourbaki, a former student of the Well-Ordereds of Besse.

The trivial isomorphism will be given to them by P. Adic, of the Diophantine Order, at the Principal Cohomology of the Universal Variety, the 3 Cartember, year VI, at the usual hour. The organ will be played by Monsieur Modulo, Assistant Simplex of the Grassmannian (Lemmas will be sung by the Scholia Cartanorum). The result of the collection will be given to the House of Retirement for Poor Abstracts. Convergence is assured.

*Lower left*:
After the congruence, Monsieur and Madame Bourbaki will receive guests in their fundamental domain; there will be dancing with music by the Fanfare of the VIIth Quotient Field. Canonical Tuxedos (ideals left of the buttonhole). QED.

Bourbaki's intentions were not directly didactic. It would be catastrophic to try to use it in high school. Certain too zealous mathematicians, apparently inspired by Bourbaki, tried to introduce something called "modern mathematics" into high schools. This had nothing to do with Bourbaki itself. Many abstract terms were introduced instead of the usual concrete words familiar to young people. Pierre Samuel called these people "hyperaxiomatizers desperately seeking generalization." The result of the introduction of "modern mathematics" into high schools was an international large-scale catastrophe, which took its most serious toll in France. A whole generation of young French people was basically sacrificed from the point of view of learning mathematics, and mathematics itself lost a lot of credit with the public. The whole of the mathematical community, including myself, was guilty of ignoring what was going on in high school and even elementary school education. Today, mathematicians in universities are much more concerned with what happens in elementary schools, junior high schools and high schools.

Each volume of the works of Bourbaki is followed by a sketch of the history of the subject. Parts of the text contain the warning sign for "dangerous bend". The parts considered easy are stated and proved simply and straightforwardly, without short cuts. We called these parts "the trotting donkeys". We wanted to indicate them with a little sign of a donkey, but the printer had scruples. The most difficult exercises are marked by a ¶ in the margin. Dieudonné checked every single exercise. Unfortunately, he never took a single note, and towards the end of his life, he frequently found himself in the embarrassing situation of having someone write to him to ask him the solution to a problem and not having any idea of the answer. Now that he has passed away, solutions of the exercises are lost forever. Some good soul should probably write them all up.

Dieudonné had a powerful voice, and he dominated the discussions at conferences. He regularly resigned from the group in a thundering voice. It didn't mean much and usually he went right on as before. However, one time he actually stormed off to his room and went to bed. There were subjects on which he resigned automatically. For example, he wanted topological vector spaces to come before integration since they are needed in the theory of integration. If anyone dared suggest that they come after, Dieudonné immediately tossed out his now legendary resignation. Sonia Godement, the wife of the mathematician Roger Godement, didn't believe our stories of the automatic resignations, and asked to sit in on one. We told her to enter our next meeting at ten o'clock precisely, at which time we would provoke the usual scene. At three minutes to ten, someone let fall "We absolutely have to put integration before topological vector spaces." Dieudonné immediately and violently resigned, just as Sonia was entering the door.

The group also organized "Bourbaki seminars" three times a year. These had nothing to do with the internal conferences. Each seminar consisted of six talks at the highest level, and mathematicians from all over the world came to them.

We laughed a lot and had tremendous fun in all the Bourbaki conferences. There was no hierarchy at all, and irony was strongly encouraged. At one conference, in Pelvoux-Le Poët in the Alps at the end of June, we were the only people in the hotel, a little hotel kept by a charming lady called Madame Roland. It would probably have been rather disagreeable for anyone else, we shouted and laughed so much. Once a violent quarrel broke out about some mathematical point. André Weil actually hit Henri Cartan over the head with the notebooks containing the write-ups. The situation seemed irreversibly catastrophic. Poor Madame Roland whispered in my ear: "I suppose it's all over, and you'll leave

Dieudonné expostulating

now." I asked her why. She said "You're quarreling so much, I suppose you won't stay on after this!" I reassured her, and indeed, ten minutes later everyone was laughing again. She didn't worry at all about our quarrels after that.

Several of the members brought their families to Pelvoux. I brought Marie-Hélène and our children. Claudine, who was barely seven, was very timid, and was actually frightened of the Bourbaki meetings. Once, Dieudonné proposed to take her around the mountainside to get plants for her plant garden. She followed him, secretly terrified of the huge man with the enormous laugh and the violent temper. She really believed he was taking her, like Tom Thumb, to

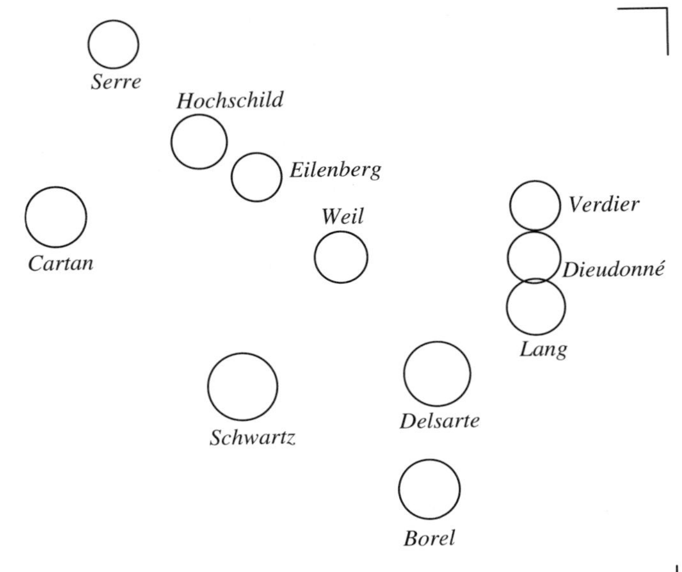


Group photo from Pelvoux

The Bourbaki children

lose her in the forest, but she was afraid to refuse. Nevertheless, she memorized the path they followed with care. On their return, Dieudonné helped her classify her plants using the book of Flore Bonnet. After that, Claudine was not afraid of him any more.

The influence of Bourbaki was considerable and lasted for a long time. Bourbaki still exists; it has about fifteen members, all new since members were required to leave at the age of fifty (by fifty, Dieudonné explained, one has become an idiot) and were replaced by new ones. Some universities, and in particular the Academy of Sciences from which one is not obliged to retire, do seem to resemble old-age reunions. The Bourbaki team was supposed to renew itself and remain young. I don't even know the list of its members now. But I think that nowadays, the work of Bourbaki is not so useful. Its members do meet and write successive versions, but they have not published anything in the last ten years (although a new volume is apparently about to appear). The whole world has adopted Bourbaki's general ideas about definitions, concepts, reasoning, written proofs and so on, and the work of Bourbaki is thus in some sense over. It is read and reread by young people, but there is no more real reason to go on with it, in the sense that a book by Bourbaki is no longer really different from a book written by any other good mathematician.

Whole areas of mathematics were not examined by Bourbaki. For instance, one notebook of results was written on the theory of manifolds, but it never went further, I already explained why. Bourbaki never really considered partial differential equations either (except for some recent notebooks on pseudo-differential operators, which are very good, but have not been published). It seems that the subjects avoided by Bourbaki are those where it is necessary to admit a large quantity of prerequisites; even though they may be well-known to everyone, Bourbaki will not use them if he has not written them himself. Furthermore, Bourbaki's tendencies were always more algebraic than analytic. The most important thing I gained from Bourbaki was to become "algebraized". I'm an analyst by nature, and all my work is on analysis and probability. These are not algebraic theories. But I use algebra and algebraic methods as much as possible. I prefer intrinsic formulations to methods using local coordinates, for instance when discussing manifolds or connections. This is typical Bourbaki style, whereas most analysts write formulas with indices as Riemann and Elie Cartan did. I am one of the most algebraic of analysts, and it comes from Bourbaki.

## Errors of Bourbaki

In the evaluation of probabilities, Bourbaki made some real errors. Before Bourbaki, one had abstract or Borel measures on sets equipped with a sigma-algebra. Under the influence of André Weil and his remarkable book *Integration in Topological Groups and its Applications*, which appeared in 1940 and which I studied at length, Bourbaki introduced Radon measures on locally compact spaces. Thus, there are two measure theories, equally noble but strangers to each other: abstract measure and Radon measure. (The term "abstract" is a bit ridiculous here. Fréchet's "abstract spaces" are just "general" spaces, generalizing all the "usual" spaces such as the line, the plane, and the three-dimensional space "we live in".) Bourbaki adopted Radon measures and rejected the others. But it is the others which are necessary in probability theory.

I adopted Radon measure theory on locally compact spaces enthusiastically in 1940, all the more easily because the probability theory I had studied with Paul Lévy before the war did not really mention general measures, and certainly not sigma-algebras. Natural, practical circumstances had driven me away from probability theory: I had stopped studying it when I left the ENS in September 1937, and abandoned it completely during my three years of military service, and then I met Bourbaki, and then I was absorbed by the discovery of distributions. Bourbaki rejected probability theory, considering it as non-rigorous, and by its considerable influence, it led the young generation away from probability theory. Bourbaki bears a heavy responsibility, which I share, for the slow development of probability theory in France, at least for everything concerning processes and other modern developments. I still remember a lecture by the famous American probabilist Joe Doob, in Paris, in which he discussed probabilities on the non-locally compact space of continuous functions; he was talking about Wiener's probability and Brownian motion. The members of Bourbaki present at the conference interrupted him constantly, loudly and unpleasantly, on the pretext that his space was not locally compact so what he was saying "didn't make sense". I was really bothered by their attitude and indignant because of their rudeness. I already mentioned how Paul Lévy was ostracized. Bourbaki should have recognized the legitimate equality of the two categories of measures and thus neutralized his ostracism, especially as its members individually respected and esteemed Paul Lévy and considered him one of the best mathematicians of his generation. The biased influence of Bourbaki and the ostracism of Paul Lévy, together, caused a historical slowdown in French probability theory, which fortunately has recovered today.

Bourbaki was also not interested in applied mathematics. In earlier periods (the nineteenth century, Poincaré's generation, Hadamard's generation), applied mathematics had flourished, but later it slowed down and by the time I was a student, it had almost disappeared. Bourbaki always ignored it deliberately, from the beginning until the present day. We know the reason: it simply do not lend itself to the Bourbaki style. The first work in applied mathematics dating from after the war was rather low-brow, and its authors were not very brilliant, whereas splended work was being done in the US and in Russia. Again, Bourbaki's influence prevented applied mathematics from picking up speed in France. I did not work much in applied mathematics myself, but I did introduce some new ideas which have been useful to applied mathematicians, and I was very interested in applied math, unlike the other members of Bourbaki. However, my research concerned fundamental mathematics, and I always write in the Bourbaki style, except maybe more pedagogically and with clearer statements of my intentions. I don't actually do research with the aim of being useful in such and such a domain, but I do feel particularly pleased if my work has wide-ranging applications, and I address physicists, mechanics and engineers in their own language. For nine years, I taught "Mathematical methods in physics" in Paris, and my course was taken by young mathematicians and physicists alike. Already when I taught the Peccot course in the Collège de France (1945–1946), half of my twenty-five auditors were electrical engineers. Jacques-Louis Lions did his thesis with me in 1954 on limit problems in partial differential equations. He became the principal founder in France of modern applied mathematics, on an international level, and yet he is not a member of Bourbaki. His style, however, is impregnated with the Bourbaki influence, even if he doesn't like to admit it. There are frequent conflicts between pure and applied mathematicians, but each side knows that they should know the mathematics of the other. I still feel this conflict is too palpable in France, but on a world scale it is slowly diminishing.

One final criticism: Bourbaki causes people, especially young people, to think that mathematics is an absolutely rigid, finished product with no vague regions, and that everything is deduced from other things, via a unique royal road (i.e. that there exists a unique good proof for each property), and that every problem is well-posed. But this is completely untrue, and indeed, badly posed or badly solved problems are the richest sources for new discoveries. A similar reproach can be addressed to every publication nowadays. I would like to add that Bourbaki certainly caused an explosion in the number of new discoveries, and that the novelistic books of earlier periods were an expression of the dormant state of the mathematics of their time, whose awakening is partly due to Bourbaki.

## A thesis in extremis

Bourbaki inspired me to a great leap forward (which became a little more moderate later on, as I began to observe the aspects which I have criticized above), and in spite of the horrible international circumstances which struck me like a whip, mathematics made me really happy from 1940 to 1942. I began to make interesting discoveries in 1941, on the sums of real or imaginary exponentials. Dieudonné was enthusiastic about them and encouraged me to use them for my thesis. After two years of work, around the end of 1942, he pushed me to finish it quickly; he knew that as a Jew, I was going to run inevitable and terrible risks, especially after the German invasion of the free zone consequent to the Allied landing in Northern Africa. So I finished my thesis in two years. Marie-Hélène and I had now taken shelter in Vernet-la-Varenne, thirty kilometers from Clermont, and I wrote my thesis there. It was entitled "Etudes des sommes d'exponentielles réelles", and I also published a complementary article in the *Annals* of Toulouse, called "Etudes des sommes d'exponentielles imaginaires". The latter is one of the most complicated papers I ever published. My thesis was published by the Librairie Hermann. At that time, it was the custom for the thesis candidate to pay for the publication of his thesis; it cost twenty-three thousand francs. For me, with my salary of two thousand francs a month, this was obviously out of the question. But my father was so happy to see me a Doctor of Mathematics that he paid for it himself. It took courage for Freymann, the director of the Librairie Hermann, to publish a thesis by a Jew, and display it in his shop window in 1943! He also displayed books by Einstein, whose publication was outright forbidden. Now and then, the Gestapo visited his store in the rue de la Sorbonne and tried to exercise discreet pressure. But he would simply answer: "There are four Mexican Jews in France and ten thousand Germans in Mexico, so leave me alone."

I defended my thesis in January 1943, officially at the University of Strasbourg, but actually in Clermont since the university faculty had taken refuge there. The jury consisted of Georges Valiron, one of the readers (he had been our professor for functions of one complex variable at the ENS), invited down from Paris, Charles Ehresmann and André Roussel. Dieudonné had left Clermont to return to Nancy in December 1942. I defended my minor thesis orally, with Ehresmann, on the subject of algebraic topology, on the duality theorems of Poincaré-Pontryagin and Alexander-Pontryagin. In spite of the role Poincaré played in them, I believe these theorems were only superficially understood in France. My minor thesis was my introduction to algebraic topology. My parents came expressly from Toulouse; my thesis defense was followed by a party in a restaurant in Ceyrat. At that time, customs were different from nowadays: Valiron was a professor at the Sorbonne, and it would have been considered insulting to reimburse him for his trip to Clermont! My major thesis was a pertinent application of function analysis and topological vector spaces to a classical problem in analysis. Let me describe the subject precisely.

Let $(\lambda_n)$ be a sequence of real numbers $\geq 0$, and consider finite linear combinations $\sum_n a_n x^{\lambda_n}$. If the $\lambda_n$ are sufficiently "frequent", then the linear combinations are dense in the space of continuous functions on $[0, 1]$ (if $\lambda_n = n$, we obtain the polynomials). If on the contrary they are "rare", then the linear combinations are no longer dense. The precise condition had been given by

Charles Müntz: it is necessary and sufficient for the series $\sum 1/\lambda_n$ to be divergent. If now this series is actually convergent, then how can one determine the set of continuous functions which are approximated by the linear combinations? I showed that the only continuous functions on $[0, 1]$ which are approximated are analytic functions (multiform, with a critical point at 0 if the $\lambda_n$ are not integers) in the open disk of center 0 and radius 1, which can be expanded as Taylor series according to the powers $x^{\lambda_n}$, convergent by groups of terms. I called this thesis "Sums of real exponentials" in order to avoid multiform analytic functions; one replaces $x$ by $e^{-t}$, which replaces the powers $x^{\lambda_n}$ by real positive exponentials $e^{-\lambda_n t}$, and the open disk of radius 1 punctured at 0 becomes the half-plane $\mathrm{Re}\, t > 0$ of the complex plane. I made essential use of the Hahn-Banach theorem, and Fourier's transformation theorem, and the Laplace transform, all of which became extremely familiar to me. My thesis only considered real positive $\lambda_n$, but the other article, published in the *Annals* of Toulouse, considered arbitrary complex $\lambda_n$, actually angularly very close to the real line; then convergence only works by groups of terms and exponential Abel factors. The proofs of both theorems were very sophisticated, and the article is one of the hardest things I ever did.

The German Occupation rendered our situation more dangerous every day. As soon as I finished my thesis, we felt obliged to disappear, melting into our surroundings with false names and false identity cards. My possibilities for mathematical research were over. Just protecting myself as a Jew and a political militant took constant precautions. As the reader must understand, when living amongst the kind of dangers which threatened us, constant vigilance was a question of life or death. Mathematical research does not really make people become as absent-minded as legend would have it, but it does demand a high degree of concentration. My survival and that of my wife and our new baby required my full attention. I continued to read all kinds of books, including scientific books, but the period from March 1943 to September 1944 represented a fourth year of mathematical inactivity for me.

The richness of my mathematical thought during the period of my thesis made it comparable to my experiences in twelfth grade, hypotaupe and taupe, which my years in the ENS had not really matched. I lived the whole time within the warm atmosphere of stimulating university professors. By passing my thesis, I acquired a new situation; once the war was over, I could apply for a job as assistant professor, and then as professor, in a university. But before then, a period of clandestinity and danger yawned before me. During the period in which I was working on my thesis, Marie-Hélène was working on generalizing Ahlfors' theorem – a theorem she had studied in Biscarrosse – and this work eventually led to her thesis. She did a remarkable piece of work on tilings of the complex plane by fundamental domains for a meromorphic function. Unfortunately, it turned out that the result had already been discovered by a Japanese mathematician, Shimizu.

Politically, we were not as isolated as before. The faculty of the university in Clermont and even most of the inhabitants of Ceyrat hoped that the Nazis would be defeated. Thus, the period we spent in Clermont held many contradictory features. The war continued with all its varied horrors; the round-up of the Vel d'Hiv organized by Vichy and the French police on July 16, 1942, before the American landing in North Africa, the invasion of the USSR in June 1941, the Japanese attack in December 1941. Barbarism ran wild. And yet in spite of all, in the somewhat confined atmosphere of university life in Clermont, my life flourished. That is how the world goes. I learned that Japan had entered the war in a rather curious way. In a tram, one woman called to another: "Look, the king of Belgium has gotten married again!" I lifted my eyes to her newspaper and saw the front-page headline: "Japan declares war on England and the United States." Everyone's horizon is different ....

## Supply expert

I had become a real expert in the art of obtaining supplies. It was essential for Marie-Hélène to rest a great deal. I don't recall for exactly how many years she needed to live a slow and restful life, after her recovery from tuberculosis, but I am convinced that it was partly because she took such care of herself that she never suffered a relapse. She took care of the small shopping which could be done in Ceyrat, and I did the shopping which had to be done in Clermont. I would bring along a large shopping bag and our ration cards, and take the tram to the town. I didn't worry about bread, butter, milk and sugar, all of which could be obtained in Boisséjour or in Ceyrat, I attacked the main job: cold cuts, cheese, sweet things and sometimes even meat. I learned to recognize good cuts right away; I became quite competent at it. Once, I was delighted to bring home an enormous greasy sausage which, I really can't think why, the butcher sold without a ticket because he found it too full of fat. But fat was precisely what our diet was desperately short of. We had the right to three hundred and fifty grams of bread per day, but very little butter, sugar and meat, if I remember rightly. I returned home triumphantly after two hours of shopping, proud of my acquisitions. We had to put up with a lot of ersatz. There was almost no coffee; we had a barley substitute. It wasn't exactly bad, but it had a special taste. Instead of sugar, we used saccharine or grape sugar, which was not great.

I sometimes bought cheese so lacking in fat that it was inedible. Naturally, we sacrificed as frequently as everyone else to the god of rutabagas, which is used as animal food in normal times, but which is actually perfectly edible, slightly resembling turnips. When we had no lentils, we used vetches, which have a similar taste but are so hard you have to cook them forever. Like almost everyone else, we used the "Norwegian pot" for anything which needed to be cooked for a long time; it was just a big cardboard box lined with newspaper. As soon as the contents came to a boil, one isolated the whole thing by wrapping it well and covering it with paper, and it stayed hot for a very long time. It was an astute little invention for saving energy.

After a while, it became impossible to live on what we could procure by shopping locally. One day, I learned from my hairdresser in Ceyrat, Mademoiselle Paulet, that she obtained food from an aunt of hers who lived in Saint-Germain-l'Herm. I asked her if she would feed us as well if I went down to see her, and she said there wouldn't be a problem. I soon made the trip, taking the train to Issoire and then a bus to Saint-Germain, and went to meet Madame Paulet and her obliging neighbor, Madame Choisy. Since Mademoiselle Paulet had sent me, I was warmly welcomed. It was there that I discovered the extraordinary riches hidden in the countryside. First, I asked for a little butter. "Sure, I have butter," she said. "How much do you want? I have a kilo six hundred grams here." One kilo and six hundred grams of butter! Our ration cards allowed us only one hundred twenty-five grams per person and per week. I bought the kilo six hundred grams, as well as several cheeses, in particular an entire Saint-Nectaire. Then I was introduced to an old, toothless gentleman who sold goat cheese. It smelled very strong, but it was real cheese! I bought at least half-a-dozen. Then I had to return to Saint-Germain-l'Herm; it was at least a three-mile walk. I put everything into the lining of my old military coat, which had been made into civilian clothing and was extremely warm; it was very cold in Saint-German-l'Herm (at least minus thirty degrees). The difficulty was to return to the hotel without passing in front of the police station next to it. The police were particularly vigilant of anything smelling of the black market. But I wasn't carrying anything very visible. As soon as I had finished my shopping, I packed everything into a suitcase and returned to Ceyrat. But in the train, the goat cheeses began to emit an odor so powerful that no policeman could possibly have avoided noticing it; certainly everyone else in my compartment did. We ate the butter freely for two or three weeks (we had no refrigerator), and melted the rest, which was not so good on bread, but very good for cooking. The cheese was delicious; Madame Choisy's cheeses became legendary. I never see a Saint-Nectaire without remembering her with emotion.

From time to time, I returned to Saint-German-l'Herm, always with equal success. Thanks to those visits, we were sure of being well-nourished. City dwellers everywhere were trying to find people in the countryside who could provide them with similar services. It was not easy for poor people, unless they had family. One day, Ehresmann actually obtained a small pig. He carried it to town himself and pushed it around in a small caddy, covered with a tarpaulin. We shared it between us, half for him and half for us. We ate part of it immediately, but most of it we salted away for long-term keeping. Alas, in our inexperience, we used too much salt, and in the springtime only bones and briny liquid remained. We were bitterly disappointed.

We remained faithful to Saint-Germain-l'Herm for a long time; the place was full of good memories for us. After the war, we used to return there with Marc-André, and later with him and Claudine, for winter vacation. It was freezing cold; we used to visit all our old acquaintances. Bertrand also came several times. In 1946, while visiting some farmers, I remember a lady thinking Marc-André was very sweet, and saying to him: "Oh, what a nice little boy." Marc-André drew himself up and said firmly: "I'm so not small I'm already three years old!" On a winding road, he would say: "Oh, a turner." And a little farther down: "Oh – another turner. The other way." Towards the end of the fifties, we still spent vacations there. After dinner, we sometimes played guessing games. We've always called this kind of evening "Saint-Germain-l'Herm evenings", even when they took place elsewhere. During the war, when I went to purchase supplies from farmers, I sometimes had to use persuasion or even ruse, but usually we soon established simple and excellent relations, giving rise to friendships which outlasted the war.

## Revolutionary defeatism

The doctrine of revolutionary defeatism was applicable, perhaps, in 1914, in every country, since there nearly was a revolution in Germany in 1919, when soldiers returning from the war revolted, tore off the epaulettes of reactionary officers and even formed some committees. But "order" was soon reestablished, Karl Liebknecht and Rosa Luxemburg were assassinated on the orders of the Socialiste Noske. After 1918, Communists in every country fought courageously against the imperialist, colonialist and anti-Soviet undertakings of their own governments. But this attitude was no longer applicable to the new war. It was not an imperialist war of the same type, because Nazism was a very particular and radically new phenomenon. Certainly, Nazi Germany and Japan intended to conquer the world, and the Allies opposed these intentions and attempted to conserve their own dominion. From this point of view, the situation had some parallels with the war of 1914–1918. But the essential point of the new war was elsewhere: for the Allies, the whole survival of democracy and of humanity itself was in danger. The capitulation of the Socialists in 1914 never ceased to haunt Trotskyists. Yet, the situation was not at all comparable. In 1914–1918, Socialists of all countries began by being strongly pacifist and fomenting the idea of a general insurrectional strike in case their country declared war, but they ended up taking the side of their bourgeoisie and adopting the chauvinist attitude of support for a total, purely imperialist war. In 1939 and for quite a while afterwards, the bourgeoisie of the Allied countries did not direct their countries towards war. It was even to some extent defeatist, and at any rate not particularly decided, unready to undertake any action against Germany even after the invasion of Poland, and it probably would have remained indefinitely inactive if Nazi Germany had not attacked. These facts formed the core of my many discussions with Ruff; Ruff was one of the first group of officers I taught in the Biscarrosse camp. He explained to me that faced with Hitler's power, the role of the proletariat and in particular of Trotskyists was to take over the direction of the war and the resistance, against the will of the pacifist and future collaborationist bourgeoisie. I was quite impressed by his analysis. In that period of forced inactivity, I spend many mornings in Ballancourt and in Biscarrosse, battling with myself over this question of revolutionary defeatism, cut off from the world.

Already soon after 1933, Trotsky had modified Lenin's abrupt formulation, realizing that it was inapplicable to the new situation. He was accused by the Communists of betraying Lenin, but he could not radically separate himself from the old dogma: even if a Nazi victory came to seem probable, he said, it should still be the main effort of the working class to fight against the bourgeoisie, because in the long run, "revolutionary contagion" would be more dangerous for Hitler than a military offensive. There was some truth in what he said, but it was still not altogether realistic or convincing. Just after our defeat in 1940, a text by Marcel Hic called "Letter to English workers" came to my attention. Marcel Hic was one of the most intelligent members of the POI, and I always felt great admiration and friendship for him. He wrote: "Don't submit to your bourgeoisie to the point of ceasing all your demands, but take the responsibility for the war on yourselves; work ten or twelve hours a day, if necessary, to make arms and munitions so that Great-Britain, allied to De Gaulle's France, can win the war against Nazism." This letter was approved by a large part of the POI (particularly by Craipeau, Parisot, Demazière, all members of the central committee), but it was rejected by the PCI, which had a majority in the unified party, and so it ended up being buried. I fundamentally agreed with the contents of the Letter, but I didn't dare make any clear statements about it, I was so full of inner contradictions. In some weak sense of the term, I was almost schizophrenic. On the one hand, the ten commandments of Trotskyist, and Lenin's statements concerning the previous war, clearly stipulated the necessity of precipitating the defeat of imperialism within one's own country, with the purpose of transforming the imperialist war into a civil war. On the other hand, such a transformation was quite obviously unthinkable in the present situation. The combat against Nazism seemed to be an imperious necessity, beginning with an adaptation of our language, which had still remained within the terminology of one war ago. The result of this inner battle was to make the actions of many Trotskyists sterile and inefficient.

In fact, Trotsky himself, just before his assassination in August 1940, had radically changed his position, and adopted a rather Churchill-style language to counsel a universal defense against Nazism: "We must defend our towns, our villages, our churches, the lives of our intellectuals, our workers and our peasants, against Nazi barbarism." He even wrote an article asking American workers to push Roosevelt towards declaring war on Nazism. But in France, none of us were aware of these late writings of Trotsky, and in any case, they came too late. None of us was mentally prepared. I was surrounded by pacifist, fearful semi-Nazi officers, and in theory, I was supposed to hope for defeat in order to provoke a workers' revolution!

In spite of my belief in the ideas expressed by Marcel Hic, I must admit that in practice I was too pliant and too used to the usual Trotskyist language of formulas. And furthermore, although we realized the horrors of Nazism, we also realized the extent of French, British and American imperialism in the world and in particular in the colonies, and even worse, the nature of Stalinism. Thanks to excellent articles by Trotsky, we could foresee the future, dominant and aggressive role of American imperialism. Naturally, Nazism was still our enemy number one, but how could one fight against Nazism by submitting to the orders of its capitalist adversaries? Naturally, we couldn't express our ideas in public, not only in France under Daladier or Paul Reynaud, but even within the Resistance, because Communists would not have hesitated to liquidate us.

There were three major groups within the POI. Firstly, there were those who thought the primordial necessity was to organize fraternization with German soldiers, who would be persuaded to become anti-fascist through our teachings, and to help us in the struggle against Nazism. This heroic enterprise traced a path into the future, but it was certain to fail because there were far too few of us. Thirty or so Trotskyists, among which were Marcel Hic, Roland Filiatre and his wife Yvonne, David Rousset, Marcel Beaufrère, Maurice and Renée Laval and others, devoted themselves to this fraternization, and published a bulletin in Bretagne, together with a certain number of German soldiers, called *Arbeiter und Soldat* (*Worker and Soldier*). This bulletin expressed strongly anti-Nazi views, more or less close to Trotskyism. Several dozen German soldiers joined the group, which really constituted a remarkable success. Calvès tells how he formed a group of twenty-seven German soldiers, with a newspaper called *Arbeiter im Westen* (*Worker in the West*). But these initiatives were obliged to remained on the small scale of Trotskyism itself. Moreover, these groups of soldiers always fatally contained an informer, who revealed the identity of the other soldiers in the group. Those of the group formed by Calvès were shot, one of them was horribly tortured, and many of the members of the POI involved in the story were tortured and deported. Rousset, the Filiatres, Beaufrère and the Lavals returned from deportation.

After the war, Rousset continued militating as I have already described. However, Marcel Hic fell ill and died very soon, which saddened me extremely. For some time, I tried to send him packages, but it was never possible. I was in contact with his father who lived in Saint-Hilaire-du-Touvet. The death of Marcel Hic, after his clairvoyant activities and our deep friendship, left a deep trace within me. The final outcome of his actions was terrible. How can one judge such an activity, which furthermore remained almost completely unknown? If you count how many lives it cost, it was absolutely ruinous. But as a revolutionary activity, it was exactly the idea which Vercors expressed in *Le Silence de la Mer* (*The Silence of the Sea*), the idea that all Germans were perhaps not irremediably bad.

On the level of principles, the action of this group was grandiose. It's a good thing that it was attempted, and that we saw that it was possible. If more people had devoted themselves to the final goal, it would have had a chance of success, but with so few militants, it was impossible. The Trotskyist Party was bled white by the war. Our propaganda was essentially anti-Nazi and never anti-German. We were totally internationalist, as we had always been and remained afterwards. I myself was internationalist when I entered the Trotskyist Party and remained so always. It was so true, that in speaking of France I never used the words fatherland, patriot, patriotism. My fatherland was humanity. Later, I became deeply involved in the anti-colonial struggles for the independence of Indochina and Algeria. From this point of view, we were well in advance of our time. We can also say this: during the battle of Stalingrad, the Soviets never tried to suscitate resistance amongst German soldiers to their own regime. If they had tried it, on the scale of their combat, maybe some change would have been possible. They only did it after Stalingrad, with the help of Paulus once he was taken prisoner.

It is also necessary to remember that highly-ranked officers of the German army were opposed to Hitler, if not to Nazism. Clandestine newspapers circulated constantly between most of the generals and many officers. Their program was certainly not democratic, but we know today that the group directed by Colonel Count von Stauffenberg, which almost assassinated Hitler in July 1944, was opposed to the Final Solution of the Jewish question and repression of the Jews, restricting themselves to demanding their expulsion from Germany. Their objective was to obtain a peace treaty with the Allies under which Germany would conserve its borders intact and all of its power, without accepting democracy. The power of the German generals was to replace Hitler's. It was a kind of opposition, which could have translated into a democratic offensive on the part of the soldiers.

Almost all the generals were involved in the 1944 conspiracy against Hitler: von Manstein, Paulus before he was taken prisoner, von Rundstedt, Rommel. After the failure of the putsch, most of them retained their positions because the Gestapo didn't know about their involvement. But two officers went to see Rommel, who was highly admired in Germany, and handed him a flask of cyanide in order for him to end his days, which he duly did. General von Stülpnagel, who commanded the German army in France and was cordially detested by all the French, was deeply involved. On the day of the attempt, he had the whole of the Gestapo in Paris arrested and imprisoned! He was absolutely ready for action. Although he did not enter into direct contact with the Allies, his intention was to open the front to them, as was Rommel's. If he had tried, he would not have been well-received; no one in the Resistance was aware that the Paris Gestapo had been dismantled. When it became clear that the attempt had failed, the generals never tried to keep up their contacts, to revolt or to seize power, even though they held the power of all the armies of the Reich. They were blocked by the force of personality of the Führer. If he had been eliminated, the conspiracy would have worked; as long as he remained alive, he constituted an insurmountable obstacle, although this appears amazing. The personal power exercised by a tyrant is frequently observable throughout history. Even if the generals and marshals had attempted something, the Gestapo remained blindly faithful to Hitler. Von Stülpnagel freed the Paris Gestapo, and reinforcements were sent to it. He was not treated as well as Rommel; he was arrested, tortured and killed. The commander of the Paris Gestapo imprisoned by von Stülpnagel was the same one who organised the last deportations of Parisian Jews, before the liberation of Paris. Hitler reigned over Germany as powerfully as before. In fact, the conspiracy of the generals had been going on since the beginning of Nazism. At first, the Wehrmacht held considerable power in Germany, and Hitler needed to rely on landowners and Prussian generals who were not always necessarily Nazis. They had decided that if the remilitarization of the Rhineland failed, and if France and England attacked Germany at that point, which would have been possible during the period of rearmament, they would depose Hitler by force. They were also ready to do so if the Anschluss with Austria in 1938 failed, or if the invasion of Czechoslovakia in May 1939 failed. It would have still been possible if the war in Poland had not taken the form of a *Blitzkrieg* and if Germany had been attacked by English and French forces on the west while still fighting in Poland. After all this, it was too late. After the war with the USSR had begun, it was impossible to act against Hitler, until very near the end. Yet the whole of the army staff continued to circulate clandestine newspapers. These facts were recounted by people perfectly informed about the situation in Germany. One mustn't believe that solidarity with German soldiers was inconceivable. The German soldiers were not all fanatics, and they were not all ready for a war of extermination. Many had had enough and longed for peace, already in the middle of the war. At any rate, those whom we saw leaving France in 1944 were completely demoralized. If the Communist Party had remained internationalist, the results of a movement of fraternization could have been extraordinary. But it took the opposite line, recommending the arbitrary killing of Germans. I still remember a public meeting of the Communist Party in Grenoble, after the end of the war, in which one of the leaders spoke out saying: "All our militants have adopted the simple principle: take out the Boche – as much as possible, take out the Boche." Their choice was the opposite of ours. I think this should be recalled as much as possible. Ours was the way to the future. At any rate, after the war and even more nowadays, the French people showed itself ready to renew relations with the Germans.

The Trotskyist workers who formed unions within industries formed a second group. These people were able to mobilize workers in the combat against collaborationist employers, and they could be quite efficient, especially if they were well-known within their unions. Several of them were arrested and deported. Alfred Bardin of Lyons, who was a union member working in Berliet, acquired so much influence that after the Liberation he was able to lead a union. This group of people also contained members who spread propaganda among young people in Youth Hostels; I myself participated in this activity several times.

A third group, to which I also belonged, was simply made up of little, isolated knots of Trotskyism. The Clermont group which I joined had only eleven members. Among these were Suzanne Augonet, the wife of Marcel Hic (they were separated), and Georges Glaeser, a great friend of mine, a mathematician and the son of Léo Glaeser, who was executed in Rillieux-la-Pape; he played an important role in the recent trial of Paul Touvier. We were very determined, but what could eleven people do? Derailing trains and murdering German soldiers was out of the question; in any case, our internationalist views prevented it. We had weekly meetings and spent most of them discussing, believing that little by little, we would conquer the masses who would open their arms to us. We really had no sense of reality at all. In some sense, we were actually more resistant than the rest of the country; we had already begun writing tracts and putting out our newspaper, *La Vérité*, at the beginning of July 1940. Our newspaper probably echoed a certain popular feeling, in a time when Socialists no longer existed and the Communist Party was mixed up with the German-Soviet pact, and was sliding towards collaboration with the occupying forces: Jacques Duclos actually asked their permission of the Germans to publish *L'Humanité*. Our newspaper *La Vérité* never ceased appearing during the whole of the Occupation. But our propaganda was necessarily inefficient. It just wasn't very credible to denounce Nazism and its ally Vichy as public enemy number one, while simultaneously condemning Anglo-Saxon and Gaullist imperialism.

We were always going around with tracts, and this was very risky. Marie-Hélène took a large package of tracts right across Clermont-Ferrand; they were sticking out of her pocket and the whole packet fell out in the place de Jaude. Several helpful people stooped down to help her pick them up, without apparently even looking at them. I remember one distribution of tracts in mailboxes in the workers' residence of Michelin; it was just before curfew, and we kept kissing in order that anyone noticing us should not suspect us, just like the young couple in *Guichets du Louvre*. But nobody bothered us. We didn't do anything much besides that, and yet we were already risking the maximum penalty. I often had to pass German checkpoints, and sometimes I even had tracts, in a box where they were carefully covered by inoffensive papers. After 1943, I took to always carrying a German mathematics book. A soldier looking through my bag said: "Ah, Techniker" and closed it respectfully. In this way, we handed out many copies of *La Vérité*. When much later, more people joined the Resistance, I could have entered the underground movement, hiding my ideas, or maybe even expressing them in veiled and reasonable form. But resistance groups were usually closely linked to the English, received material and arms from the English; in a word, they were dependent on the English, and we weren't ready to accept that. To put ourselves militarily at the service of the British army was violently in contradiction with our principles. Earlier, in peacetime, we had strongly criticized the Communist Party for remaining inactive whilst we were preparing the revolution. But during the war, it was the Communists and other moderate groups who did what had to be done, while we were paralyzed. My own judgment of my actions and those of all the Trotskyists at that time is extremely severe. Every time the Communists thought they had discovered a Trotskyist, they covered him with somber accusations of collusion with the enemy, which of course were pure calumny, but made it very difficult for us to communicate with each other. During the first weeks and months of the war, while we were in Ballancourt, I struggled with myself, in the midst of inextricable psychological complication. In any case, my general weakness due to polio would have rendered me inapt to fight in the underground resistance.

The Trotskyists' resistance remained completely inefficient in spite of their indeniable courage. There were less than two hundred of us! Naturally, we could have joined the underground in Vercors or in Aiguillères. We could have done so without mentioning our political views, and maybe even tried to spread some "healthy" ideas amongst the others, without necessarily brandishing any specific label, because there was a grand tolerance for variations in personal opinion. But Communists assassinated Trotskyists, and they wouldn't have hesitated to assassinate us. Several times I risked being deported as a Trotskyist – and I wasn't even doing anything at all useful! During my years at the ENS, Trotskyism had furnished me with a remarkable political education, much more advanced and sophisticated than most of the young people of my age. But its extremism, sectarianism and addiction to formulas completely neutralized me during the war.

I chose not to fight for another reason as well: I had a wife and a baby. There was not much serious fighting until the Germans invaded the free zone in 1943. Marc-André was born in March 1943. It seemed to me out of the question to abandon Marie-Hélène and the baby in the tiny village of Saint-Pierre-de-Paladru, where we were living, to go and fight. The inhabitants of the village would have immediately understood where I had gone, and they might have denounced Marie-Hélène. Maybe I would have accomplished useful things, although I rather doubt it, given my physical weakness, but at what price? And history showed only too clearly that the greatest risks we ran, throughout the war, stemmed from our being Jewish.

Starting with the German Occupation in 1942, I almost never set foot in a cinema since round-ups were frequent. But I did go once and never regretted it. The news was so late, with respect to events, that it was a real pleasure to watch it. It was in January 1943, just before my thesis. According to the newsreel shown in the cinema, the Nazi armies were flying from victory to victory in Egypt, whereas in reality, the American landing and the El-Alamein victory had already taken place. Rommel's troops were fleeing from Egypt to Tunisia, thence to Germany. The possession of the French colonies in Africa were supposed to give Germany a base for military intervention in all of Africa. But Morocco, Algeria and Tunisia were all occupied by the Americans! Finally, we were gratified with a speech by Hitler announcing that the fall of Stalingrad was only a matter of hours. The battle of Stalingrad was already over (although officially, it ended on February 2, 1943), and Marshal Paulus had surrended with his entire staff and four hundred thousand prisoners. In spite of the danger, the audience couldn't help laughing. The newsreel must have dated back to the autumn of 1942, before the Americans invaded North Africa on the 8th of November, but it was still being shown. I left the cinema feeling deeply comforted, but not understanding how the Nazis could allow such a thing.

# Chapter V
# The War Against the Jews

## A narrow escape

In 1942, we still didn't have false identities. That year, we decided to have a baby. Some people might think it was pure insanity. But we thought about it carefully. We had three reasons. One of them was that we thought that if Marie-Hélène and I were deported, we could give the child to someone from the family, and it would survive us. Actually, children had already been deported from the Drancy camp and the Vel d'Hiv, but we were not fully aware of the situation. Secondly, Marie-Hélène had suffered deeply on account of the doctor's orders to wait four years before having a baby. She desired one ardently, and couldn't bear the idea of waiting until the end of what appeared to be an endless war. My desire for a child was not as pressing as hers, but I understood her. Finally, because of the "Work, Family, Fatherland" atmosphere enforced by Marshal Pétain, pregnant woman and young mothers enjoyed extraordinary prestige. People opened doors for them, they were allowed to be first on buses, their alimentary rations were increased. In a word, we thought that in the free zone (and we didn't expect the free zone to be occupied any time soon, since we expected the Americans to land on the western coast of France and not in Africa), a baby could afford us extra protection. As for the baby itself, we didn't think its life would be in any real danger. Maybe we were crazy and irresponsible, but that's how it was. In July 1942, Marie-Hélène became pregnant.

In June, something had happened which had almost cost us a heavy price. I had gone to Paris for three days to meet various members of our family who had remained there, and to contact some of my Trotskyist friends. I went to an extended central committee meeting of the party, in a well-protected little room, where everyone was wearing the francisque[30]. During my three-day absence, the police came to our little apartment in Ceyrat, at seven in the morning. Marie-Hélène received them in her dressing gown.

---

30) The francisque, a sort of medieval hatchet, was the symbol of the Vichy government.

Two French policement informed her that she knew very well why they were there. She showed them into the main room of the apartment – our combined office and bedroom – and told them that I was in Paris, that I had crossed the separation line legally (I had a fragile throat, and a doctor had given me a permit to travel to Paris to consult a doctor there, whom naturally I never consulted). She asked them if I had been arrested: that was not it. They had a search warrant and started hunting through the (mathematics!) books and digging through the desk drawers, which fortunately had been put in order quite recently. In particular, they came upon several eulogistic mathematical texts and a letter from the surgeon G., who had been a student and then a collaborator of my father, and had great respect for him. He was Pétain's Minister of Health. Seeing these things, they became somewhat politer. Things improved still further when they found that we were both from the ENS, like ... Pucheu, another Pétain Minister, who was later captured in Algiers by the Americans, condemned to death at the end of 1942, and executed in spite of a letter written by General Giraud. They continued their search in the kitchen, where they discovered the ladder leading to a trapdoor into the attic. Marie-Hélène had the impudence to invite them up, albeit with a pounding heart. A few days before, that attic had been strewed with banned books, in particular Marx's *Das Kapital* (which we had hardly read) and other books by Lenin and Trotsky. We had hidden everything under a loose floorboard. If they found it, or if we had accidentally left a book lying around .... However, it turned out that we had hidden those books so well that we never found them again ourselves. So the police saw nothing but some banal food stores, and left, telling Marie-Hélène to come to the Clermont police station later that morning. She went, guessing that she must be followed. I was also told to let the police know as soon as I returned home.

In front of the police chief, Marie-Hélène pretended to be a perfectly normal Jewish bourgeois girl whose husband was visiting for three days in the occupied zone. But she guessed that it must have something to do with our relations with some PCI (Trotskyist) friends from Clermont-Ferrand. Shortly before, we had thought that we ought to come up with some non-political reasons for knowing each other, but we hadn't gotten around to doing it! But Marie-Hélène recovered her aplomb when the police chief asked her a question about Gerard Bloch, who had just been arrested in Lyons. "Oh, it's Gerard!" she remembers saying. And she explained the simple truth, that Gerard was a young and brilliant mathematician whom she had known since childhood and whose mother had given her piano lessons. He had received a prize in the Concours Général in mathematics, and was preparing his Licence. At that time, he lived in Lyons. She asked if there was anything we could do for him, praised him (apolitically) and made no difficulties about admitting that he had come to see us for mathematical reasons, with an enormous backpack. She considered it normal that with his backpack, he had preferred not to go see his father-in-law, a professor of geography at the University of Clermont. Both this gentlemen and Gerard's mother belonged to the traditional bourgeoisie, and both had children who were not at all bourgeois. After that, the atmosphere relaxed. "You Alsatians," said the police chief, "you're like us Corsicans, we're patriots." And he accompanied her to the door of the police station and even went so far as to warn her that she might be followed.

On the way home, she first went to Gerard's father-in-law, who already knew about the situation. We had been invited to lunch the day of his daughter Lucienne's marriage to Gerard. The couple had three children after the war. Later, Marie-Hélène visited our mathematical colleagues and then went home, feeling that she had narrowly escaped some threat, and hoping to avoid the errors one makes so easily in these situations. But she was very worried about Gerard.

As requested by the police, I went to see them as soon as I returned. They interrogated me politely. My answers were simple, as Marie-Hélène's had been. Gerard Bloch belonged to the Trotskyist group in Lyons, and his visit to our house, with his enormous backpack stuffed with political documents, was not of a mathematical nature. It was insanely imprudent: his backpack singled him out from afar and he hadn't even realized he was being followed. "Do you know Ripot?" the police chief asked me? The Clermont and Lyons groups communicated by secret ink. It was easy to make it; when poured on starch, a few drops of water colored with iodine turns bright blue. Bread is essentially made of starch, so we would boil a few pieces of white bread in a little water, dip a pen in the mixture and write with it; after drying, the letters were completely invisible. In order to read them, we wiped the paper with a cotton ball dipped in water with a little iodine in it, and out came the bright blue letters. I used to write little messages in secret ink within my mathematical texts, which I could later find or communicate at the right moment. That was how I corresponded with Gerard Bloch, with secret messages between the lines of perfectly inoffensive letters. There was quite a regular correspondence going on between Paris, Clermont-Ferrand and Lyons. Now, Ripot was the signature I myself used on my letters. The police chief's question meant that when they arrested Gerard, they had found Ripot's letters. I doubted that they had discovered the secret ink, though. They just wondered who this Ripot was who corresponded regularly with Gerard Bloch, whom they had been following for a long time. After a flash of hesitation, I understood that if I said that I knew Ripot, I'd have to go into long explanations. So I just said "No really, I don't know him at all." "But," the police chief said, "he's a philosophy student in Clermont-Ferrand, where you yourself are on the faculty." I laughed and bluffed: "Look, how should I, a scientific researcher refugee from Strasbourg, know a student in the philosophy department of the University of Clermont-Ferrand?" There wasn't any good answer, and the interrogation ceased, but I felt threats in the air. I wrote with my usual pen and didn't disguise my handwriting. The Lyons police had Ripot's letters, the Clermont police chief had my address book and intended to keep it (it only contained inoffensive addresses). The two handwritings could easily be shown to belong to the same person.

Later, I learned that Gerard Bloch had been deported to Dachau. Gerard met Georges Charpak there – he spoke to me about it recently – and taught him a little set theory. After returning from deportation, he took up his Trotskyist activities again. At the time of my interrogation, I only knew that he had been arrested. One thing was obvious to me: they would surely compare those handwritings. I was on the edge of a catastrophe. Only the unbelievable lack of organization and the underdevelopment of the police can explain how I was not deported. The German police was incredibly violent but had no subtlety. The French police was less brutal, as can be seen from my story, but equally lacking in finesse and logistics. It eventually became clear that not a single photo of a letter by Ripot was ever sent from Lyons to Clermont, where they were trying to identify him. It's true that today's modern photocopying machines did not exist then, and taking a picture was a more complicated and less frequent operation. Still, it's incredible. They never did it, and I guess they never did establish that identification, since they left me in peace for months. The incident was over. Like Marie-Hélène, I was warned that I might be followed from time to time in the next months, and that I should remain available to the police at all times and let them know if I changed my address, and that I would be summoned again if the situation required it.

However, we didn't move from Clermont-Ferrand, and kept our real identities. Marie-Hélène learned only shortly afterwards that she was pregnant. Our plans for the future did not yet include any changes in our lifestyle; the whole thing was incredibly imprudent.

Gerard Bloch was not the only victim. Almost the whole Trotskyist group in Lyons was arrested after being followed for a long time. Furthermore, as Gerard had visited Marseille, a good part of the Marseille group was also arrested. I was almost arrested myself. If the police had been smarter, they would have followed me for a few weeks before interrogating me, and they could have arrested the whole Clermont group. But I also think we took many more precautions than the other two groups, and we often met in the university where it was impossible to follow us. When the Germans arrived in the free zone after the Allied invasion of northern Africa, the various prisoners were dispersed to different camps prior to their deportation. Gerard was deported very quickly. Several members of the Marseille group (including Albert Demazière, the Trotsky leader of the free zone) were imprisoned in the prison of Puy-en-Velay, together with dozens of independently arrested Communists; Abraham Sadek from Lyons also was kept there. All the prisoners in Puy-en-Velay, five Trotskyists and dozens of Communists, escaped with the help of the guard, Chapelle, and reached the Wodli maquis[31]

During their imprisonment, the five Trotskyists were constantly disturbed and threatened by the Communist prisoners, particularly by their leader, named Ollier. In the Wodli maquis, the Trotskyists were imprisoned in a little prison guarded by armed Communists, then assassinated in October 1943 by certain members of the FTP (Francs Tireurs et Partisans)[32], on the direct initiative of the maquis or on orders from Paris or from Italy, we don't know. Only Demazière escaped death, because he had left Wodli immediately on an expedition to find food and gotten lost. All trace of the four others was lost after their execution, and full details of the affair came out only long afterwards. One of the four assassinated Trotskyists were Pierre Tresso, known as Blasco, a member of the Italian Trotskyist Party, and the former founder together with Gramsci and Bordiga of the Italian Communist Party, which made him a highly important political personage. The others were Maurice Ségal, known as Salini, Jean Reboul and Abraham Sadek, a Jewish grocer from Lyons. I got to know Sadek well during a committee meeting in the Southern zone in Lyons, at the beginning of 1942. I only learned that he had been assassinated after the end of the war. But we had already witnessed the assassination of Rudolf Klement in Paris in 1938, and of anarchists and members of the POUM in Spain, and we knew the danger which threatened every Trotskyist. We were in danger from every side: from the Nazis, from the Vichy police and from the Communists. Relations between Trotskyists and Communists are completely different today, and the LCR (Ligue Communiste Révolutionnaire)[33] has officially requested Robert Hue to clarify the affair of the assassination: he has agreed.

31) The term "maquis" literally means "shrubs"; it is used to refer to French resistance groups during the second World War, which were generally located in the countrysides.

32) FTP: Irregulars and Partisans

33) LCR: Revolutionary Communist League

## An entomological interlude

I was particularly interested in a certain family of moths, the Attacidae or Saturniidae, because of their unusual habits. They don't possess a proboscis or a digestive tube; they don't eat. They only live for a few days, for purposes of lovemaking, i.e. to reproduce, and then they die of exhaustion. The female is different from the male from birth; she's much bigger and fatter, remains immobile and emits pheromenes, a kind of perfumed substance made up of minuscule particles, which attract the males. These particles spread out through the air even in the complete absence of wind. The male is sensitive even to a single molecule of the substance. As soon as he emerges, he seeks the female. The sense of smell plays a fundamental role throughout the animal kingdom in seeking for a partner of the opposite sex, but in this case, the male can follow the pheromenes for surprisingly large distances, hundreds of meters or more. The impregnation takes place with no amorous parade; the female remains immobile and accepts the male upon his arrival, ceasing immediately to emit pheromenes. Any other male who shows up too late in his discovery of the rare pearl is obliged to depart empty-handed and seek another job. Then the female waits a few hours, and goes to lay her eggs on a leaf which is edible for the caterpillar. Her abdomen stuffed with eggs (no digestive apparatus) is heavy. The sense of smell of these moths, like that of many other animals, is in their antennae. The Attacidae males have long antennae with bilateral ramifications, like a double-sided comb, which they use to seek for the female. The antennae of the females are sometimes ramified, but less than the male antennae, and sometimes just straight. They use them to seek out suitable plants to lay their eggs on. Generally, the male dies shortly after impregnation of the female.

The large night peacock, *Saturnia pyri*, the largest French Saturniidae moth (fifteen centimeters wide) played an important role in one lovely experience of mine. It's less beautiful than other moths, dark brown with a round transparent patch on each of its four wings, but prides itself on a majestic flight style. It's quite common even around Paris; you can find – or used to be able to find – their cocoons in Sceaux, Robinson, Châtenay-Malabry, Enghien etc. In October 1941, near Clermont-Ferrand, I had obtained a cocoon which, to judge by its size, had to be that of a female. The caterpillar is a striking apple green color, covered with sparse hairs carrying little blue bumps; it usually lives on fruit trees (*pyri* means of pears) and becomes enormous. After having lived for some time, around the end of July or the beginning of August, it weaves a beautiful silk cocoon for itself, much like the silkworm's, and becomes a chrysalid. The moth appears a long time later, in the month of May or June of the following year. I kept my cocoon in a large hat box, which I kept closed (but not airtight, of course), on my mantelpiece. I expected my night peacock female to emerge in May or June 1942. One night, towards two in the morning, we heard a noise in the room, as though something was brushing against the furniture. We were a bit scared – it was June and I was just back from my trip to Paris during which the police had interrogated Marie-Hélène and searched the apartment), and finally we turned on the light. Three night peacock males were bumping into the walls (the window was open). I immediately said to Marie-Hélène "Oh, my female must be born!" I ran to the box, where my female, all developed, was enthroned next to her cocoon. The three magi had come to announce the birth of a beautiful, nubile young woman, captive in a cardboard box on the second floor of a house . . . .

## The invasion of the free zone

Since we managed to eat well thanks to our acquaintances in Saint-Germain-l'Herm, we imagined that Marie-Hélène's pregnancy would take place in the most favorable conditions. But everything changed with the invasion of the free zone. We were on the police lists, and there was no way we could remain in Clermont-Ferrand. We left our little apartment in Ceyrat and moved to Vernet-la-Varenne, between Saint-Germain-l'Herm and Issoire.

We moved into a little room, and took our meals in the unique restaurant of Vernet-la-Varenne. The chef was a very agreeable lady and an excellent cook, so that Marie-Hélène was still able to eat properly. But the village postman started telling everyone that we were surely Jewish. As soon as we heard about these rumors, we decided we had better put a stop to them, by energetic counter-propaganda. Most Schwartzes in France are Jewish, but some are Protestant, so we told everyone we belonged to Protestant high society.

We insisted on keeping our own name of Schwartz, since we wanted our baby to have this name at birth. The family name is on the family documents of any couple, and also registered in the respective town halls of the parents, so any changes would look suspicious and probably cause dangerous investigations. We thought that the police would become interested in us any day, but that they wouldn't track us down in Vernet-la-Varenne. We had already had an example of their lack of organization. So we spent several perfectly peaceful months.

We were in Vernet-la-Varenne on February 2, 1943, when Paulus surrendered to the Soviets. In our enthusiasm, we decided to eat "Stalingrad chicken", a special recipe of our chef in the restaurant, who shared our delight. In the newspaper *La Vérité*, we read "This is not the beginning of the end, but the end of the beginning!" The decisive victory at Stalingrad owed a lot to the people, to the soldiers, to the generals, but (should one say this?) it was engineered by Stalin. We can't deny him that. After the crushing victories of Hitler on Russian territory, which had plunged us into gloom, Stalingrad organized a formidable turnaround. Unfortunately, the spectacular advances of the Japanese army around the same time, which we carefully followed on our maps, continued to be alarming. After having swallowed up all the Anglo-American possessions in the Far East, they advanced into China where they massacred the population. But China was too big for them. They never took the route to Burma or to Southern China. The Allied armies succeeded in arming Chiang Kai-Shek's army, and indirectly, the Chinese Communist Party. After Singapore, the Japanese, at the door of India, tried to provoke a liberation movement there with Subhar Chandra Bose at its head. They didn't really expect to conquer India, but they wanted to find a fifth column in this movement. The Indians, led by Mahatma Gandhi, put up a pacifist struggle for independence; the elite was far too cultivated to take the crude Japanese propaganda seriously, and they never accepted Chandra Bose.

## Marc-André

Marie-Hélène was due to give birth in early April. Three weeks before, on March 16, we decided to return to Clermont-Ferrand, where she was registered in the Michelin Clinic, in Aubière. But the descent from Vernet-la-Varenne and Issoire was incredibly bumpy, and the ride provoked the birth; we spent just one night in the hotel in Clermont before Marie-Hélène went to the clinic.

Marie-Hélène's maiden name, which they asked for at the town hall, had the effect of making the nurse in the clinic become disagreeable. Marie-Hélène stayed in the clinic eight days. Since she was a former tuberculosis patient, she did not take care of the baby herself, and she certainly could not nurse it. Once or twice a day, she was allowed to see it, but without coming too near. All these precautions seemed excessive, but were protections against the possibility that she might have a relapse of tuberculosis and infect the baby, which would certainly die. For those eight days, I stayed in the hotel, and Marie-Hélène joined me there when she came out of the clinic. Marc-André remained in the care of the nurses for a while longer. Suzanne Augonet (divorced from Hic) and her friend Charles Boudinot (really called Schechter) then invited us to their house for ten days or so. They were very close to us. Marie-Hélène went to the clinic regularly to visit the baby. But the Clermont police were still looking for us, and had noticed that we had left Ceyrat. They realized we were there by the registration of the baby at the Clermont town hall and went to the clinic in Aubière. Marie-Hélène had just left after visiting the baby. We had given our address to the clinic in case of anything; it was the address of the hotel, even though we were staying at Suzanne and Charles' place. The police went to the hotel. Now, this hotel had been recommended to us by Madame Martinet, the mother of Gilles Martinet; the Martinets were close friends of ours who lived in Clermont. Gilles Martinet, a former Communist, had become a "Communist of the left", quite close to Trotskyism, and during this period we had a warm friendship with many fruitful discussions.

The hotel, which hadn't seen us after the first eight days we spent there, gave the police the name of Martinet. The police went to see the Martinets, but Gilles sent them packing, saying he had no idea where we were, while his wife came secretly to warn us at Suzanne's. The threat was approaching dangerously. Two or three days later, Marie-Hélène's mother fetched the baby from the clinic; together with the whole Lévy and Piron families, she had just moved from Saint-Bonnet near Lyons to Montbonnot near Grenoble, in the Italian zone. Marc-André had not been well-fed in the clinic, and was covered with a rash. A few days of healthy food restored his good looks, after which, cuddled and spoiled by his grandparents and their extended family, he became a laughing and cheerful baby. A friend of ours from the ENS, Colette Rothschild, traveled all the way to Montbonnot with baby clothes for him. Marie-Hélène was told to stay away from the baby for three months, to avoid any possibility of contagion. It goes without saying that she obeyed without pleasure. She knew he was in good hands, but it was a difficult privation for a new mother, and the baby had already been ill in the clinic.

Now we needed to find a new house. We tried to leave Suzanne's house by bicycle, but Marie-Hélène hadn't been on a bicycle for ages, she was weak from all that had happened and barely managed to stay on her wheels. After some telephoning, we went by car to stay with other friends, Dominique and Annie Desanti, who also lived in Clermont-Ferrand. The mathematician Lichnerowicz was also staying with them. They put us up for two nights, and then sent us to the Pluvinages. Pluvinage was a very nice young student from the ENS, and they kept us for another two or three nights, and during that period, Madame Pluvinage forged new identity cards for us. We wanted to leave the Clermont region as soon as possible, and we left early in the morning, by bus, for Riom, and then by train for Roanne, where we chose a nice hotel at random. We registered under our new names and unpacked our suitcases. Then I strolled out of the hotel and the first thing I heard was "Schwartz! Hello! How are you? I haven't seen you for a long time! You were my instructor at the DCA camp in

Biscarrosse." Catastrophe! I hadn't gone to all the trouble of obtaining a new identity just for this to happen. To make it worse, he added "I see you're staying in this hotel. I know the owners, I'll tell them to take special care of you. I can't do it right now because I'm shopping, but I'll stop in on my way back, in half an hour or so." I rushed back up to our room and announced to Marie-Hélène that we were leaving at once. I don't even remember where we ended up that day.

It was in Roanne that I first noticed certain posters on the walls. At first we didn't know what they were about, but they looked sinister to us. In the tiny village of Saint-Priest-la-Prugne, we asked if there were rooms to let, and the answer was "Oh dear yes, there are a great many, because we used to have a lot of Jews living here, and they've all been deported." We did not move to Saint-Priest-la-Prugne.

It became obvious that we would have to pass into the zone occupied by the Italians, around Grenoble. So we left Roanne for les Adrets, a few dozen kilometers from Grenoble. We remained there for a whole week, studying the map and wondering what to do. The mayor of les Adrets was a schoolteacher, and it didn't take me long to figure out that I had better not confide in him. During that week in les Adrets, I encountered six or seven *Vanessa Polychloros* caterpillars on an elm, and took them home in a box. No matter what the situation, I never lost my sense of priorities! I no longer collected butterflies, and I brought up those caterpillars for the sheer pleasure of it. When the butterflies came out, I let them go.

I decided to explore the Paladru region north of Grenoble. Two villages attracted my attention, Paladru and Montferrat. A tiny hamlet called Saint-Pierre-de-Paladru lay between them, and was inhabited by just a few dozen peasants. It looked as though it would not be difficult to find a place to live there. I don't know whether it was because of the nice little restaurant kept by Madame Marthe Carus, another excellent cook, but I decided that that was the place where we should live. And I soon managed to rent a little house with two floors and an attic, and went to fetch Marie-Hélène from les Adrets.

In Saint-Pierre-de-Paladru, the entire chore of finding food fell to me. There was no grocery store there, no store at all. You had to cycle to Montferrat or Paladru. I also explored the countryside; I rode an incalculable number of kilometers over country roads. I offered cigarettes and chatted lengthily with new acquaintances, and little by little, I learned about their problems and difficulties. One very intelligent lady, whose husband was a prisoner, had to handle the farming with the tractor and the heavy instruments all by herself. The one big tractor was owned by the village and shared out among the peasants. I tried to understand the difficulties they encountered, and to be useful. Sometimes I'd do some shopping for them. I remember even fetching a toothbrush once. Marie-Hélène used to tease me about it; someday I'd be buying a bra for some lady, she said. I always managed to procure some butter, some cheese and some lard for us. The peasants I met were mostly hostile to the Germans – they accused them of stealing all the produce, which was largely true.

In the middle of winter, the temperature of our bedroom on the second floor, next to the old-fashioned heater, barely reached 7° C. The kitchen was on the first floor, and in the attic we stored nuts. The house was full of mice. When I went down to the kitchen at night (driven by hunger) I always found at least half-a-dozen dancing around happily. Oddly enough, they didn't come to our bedroom, but up in the attic they made a terrific racket running over the nuts!

In Saint-Pierre-de-Paladru lived a count and a countess, whose name I don't remember. They kindly gave us wood to burn and some potatoes. One day, the count invited me in for a chat. In a little toweret, he kept a remarkable library which he cared for jealously, marking each and every book. It was admirably kept in order. After having talked about this and that, he told me "You know, I think the present situation is absolutely abominable. Charles Maurras used to be my professor, so I was an anti-Semite, but a moderate one. I'm horrified by what the occupying force is doing to the Jews, it's a dreadful crime. What do you think about it?" I agreed with him, and he continued "You are from the ENS, so I suppose you're not an anti-Semite." I confirmed with a smile that most students from the ENS were not anti-Semites, but didn't add anything further. I felt all right with him, given the sentiments he had expressed; I knew it was all right to say I wasn't an anti-Semite. I visited him occasionally the whole time we stayed in Saint-Pierre-de-Paladru.

Our families were dispersed, searching for safe places to live, just as we were. Daniel briefly joined a maquis in Dordogne. He and his wife were separated from their three-year-old boy Maxime. He was too little to understand the meaning of a change of identity, so he was placed in an institution where he simply wouldn't have to say his name. Maxime was a smart child and probably understood more than his parents realized. One day, though, when a pig was being slaughtered nearby, he asked what the noise was, and when they told him, he remarked "They're killing a pig, or maybe a Jew." Fortunately, he didn't understand what he was saying.

As for Bertrand, he left for England and we had no news of him until he returned to Paris with the Leclerc Division.

Yvonne's family suffered a tragic destiny. Raymond Berr had decided, as a witness, to stay in Paris and wear the yellow star, together with his wife and his daughter Hélène. All three were deported and never returned. Yvonne mourned terribly. My parents left Toulouse and moved into the house of a surgeon, a former student of my father who had a large apartment. He belonged to a militia. They lived there, hidden and apparently completely ignored by the rest of the world. One day, the surgeon had difficulties with an operation and sent the nurse to fetch "the chief". My father came down and finished the operation. On the way, he realized that every single person in the medical service knew who he was. He lost no time searching for another place to hide. My grandmother entered a convent whose superior was a deeply charitable woman, who protected her until the end of the war. She told the other nuns that my grandmother was an old lady who was pious in her own way, who prayed in her room, and who was not to be asked to participate in mass.

Marie-Hélène's brother Jean-Claude hid in Saint-Marcellin, in the Italian zone.

## Laurent-Marie Sélimartin

Our friends did not know where we were hiding. We had the false identity cards forged by Madame Pluvinage. Three hundred thousand French people were using false identities. Marie-Hélène fabricated false cards for my parents and her parents. It wasn't that difficult. Occasionally, it was even possible to simply modify the old identity card, but usually, it was best to buy a new card and write in the desired identity, then get some friend to sign for the police (Marcel Brelot signed ours). The difficult point was to get the police stamp. What we did was to spit on a piece of toilet paper and then apply it to a real stamp; the stamp was absorbed into the humid paper and could then be reapplied onto the new card. Madame Pluvinage and later Marie-Hélène became quite expert at these maneuvers. The new stamps looked really authentic. Sometimes Lichnerowicz would enter a police station on some pretext or other and take advantage of a momentary absence of the policeman at the desk to hastily snatch his stamp and stamp one of our cards. The situation of the Jews demanded prodigious inventiveness at all times.

The name we chose for ourselves was Sélimartin; it sounds much more distinguished than Schwartz. We only learned later that there was even a count Sélimartin in the middle ages. The transformation from Schwartz to Sélimartin is quite subtle: you replace c with é, h with li, w with m, leave art and replace z with in. It's the kind of change that can even be made on old papers. I managed to alter my demobilization papers this way. The best thing is how similar my old and new signatures were. I just threw an accent onto the new e and a dot on part of the h to make it into an i; I left my w which already looked like an m, and wrote in just like my usual long-ended z. Even if I'd forgotten and written Schwartz by mistake, it would have looked all right. These details were important since certain papers could only be modified directly; not only the demobilization papers, but also our ration cards, for example. When we left Clermont, we were the Sélimartins.

I took advantage of the change to become five years older. It was useful because at age thirty-two, I avoided the mandatory Work Service for men between twenty and thirty years old. For a while, I was afraid that I would be detected, and pretended to be old. I did such a good job that one day, when Paul Lévy came to visit Saint-Pierre-de-Paladru, a peasant said to him "Your brother-in-law is in poor shape." When Paul Lévy explained that I was his son-in-law, the peasant was amazed. Maybe after the end of the war, I said to myself, when we'll be legal again, I can become five years younger and act like a young man again. My first name, Laurent Moïse, didn't necessarily mean that I was Jewish – Protestants often use Biblical names, and I was supposed to belong to Protestant high society. But I thought it was better to replace Moïse by Marie. Laurent-Marie Sélimartin: admire, if you please, the perfection of this name – except for the lack of a particle, it sounds nearly noble! Marie-Hélène's maiden name, from Lévy, became Lengé; her parents also used this name for their false papers. My parents took the name of Simonet. It was easy for us to live illegally, we had no scruples about it. But it was hard for my parents, and even worse for my parents-in-law. Paul Lévy couldn't change his very legalistic habits. For a while, he went around carrying both his true and his false identity cards. With all due respect, this was really stupid. One day a German officer in a train asked for his identity card and he accidentally handed him the one in the name of Lévy. By a stroke of unheard-of luck, the German paid no attention and handed it back. And Paul Lévy immediately got rid of it.

For Jews, life had become a constant series of dangers. You had to keep your eyes open and stay lucid. For this reason, I decided to stop mathematical research during this period, even though I was perfectly free. I did some reading, even scientific reading, but I kept myself from the distraction of research which might have led me to relax my vigilance. We spent over three hours a week discussing strategies to avoid the traps which arose around us constantly. For instance, we had to renew our ration cards every month, and when we moved, we had to reregister. And we never knew if the information on our cards wouldn't be communicated to some control center where they would immediately notice that Sélimartin didn't exist. I was always terrified when the time to renew the ration cards came around. Nothing ever happened, but we always had to be incredibly prudent, you never knew.

Except for the one time I've mentioned, I never set foot in a cinema. When I was obliged to go to a hotel, I always wondered if there wouldn't be a round-up the next morning, or a visit from the German police. When I traveled around, I kept looking to see if I wasn't being followed. The best method was the following: I stuck to the walls whenever I had to turn right or left, and stopped right after the turn, pretending to think for a minute or two: anyone following me would then suddenly appear beside me. He could of course arrest me, but generally people were followed for reasons of long-term observation which failed if the follower found himself face to face with the followee. If we had to go into a house, we first walked past it, and then suddenly turned back and went in. If I took a taxi, I always gave it an address some thirty or forty houses away from the one I was going to. If I took the tram, I never waited in the line to get in, but slipped on furtively at the last possible second, secretly observing the people around me. Once, leaving the place Jaude where I had had a talk with Suzanne Augonet, I saw that I was being followed. That time, I shook off my pursuer without too much trouble. We also had the habit of never, ever greeting a Trotskyist comrade if we saw one in the street.

Our ration cards ought to have borne the word "Jew", even in the free zone, where we were not made to wear the yellow star. But naturally, I hadn't said that I was Jewish, so it wasn't written on my card. I think I can safely say that no non-Jew can understand what this existence of constant peril was like. I saw that my colleagues in Grenoble, whom I almost never saw and who knew my new name but not my address, led relatively tranquil lives. I guess that it's equally difficult for French Jews to imagine what foreign Jews had to undergo. Those French Jews who were deported and who returned also had the impression that it was impossible to communicate their experiences. And it was impossible for a Frenchman to guess at the life of a Pole or a Russian during the war.

The fact that I was circumcised made the situation even more dangerous (naturally, we didn't have Marc-André circumcised). This factor was independent of my will; I never attached the slightest importance to it, but it put me in danger of death. It made me particularly furious to think that I could be deported for that one reason. My parents were atheists, I was an atheist, I never really felt Jewish. Obviously, there are plenty of traditionally Jewish characteristics which can be found in my personality, but I didn't feel, and still don't feel "Jewish". I never paid any attention to the question when choosing my friends, for instance I only learned after many years that one of my closest friends, Pierre Vidal-Naquet, was Jewish. Some of my closest friends like Henri Cartan and Michel Broué are not Jewish. The whole question leaves me indifferent. My daughter married a non-Jew, Raoul Robert. If possible, she feels even less Jewish than I do. However, after the Shoah, we did all feel somewhat more Jewish than before, in the sense of knowing that we belonged to a community which was suffering. If I met a Jew during the war, I knew that he must be thinking the same secret thoughts as I was. I am very attached to Israel, but not in a truly fundamental way. The two big countries in whose defense I've fought have been Algeria and Viet-Nam, during their struggles for independence. The fact that I could have been deported because I was circumcised – and it almost happened several times – appeared simply absurd to me. The Nazis didn't actually hunt for circumcised men, they wanted to exterminate Jews. But for men, it represented a determining criterion for deportation.

In the early part of our stay in Saint-Pierre-de-Paladru, I was very nearly caught. I was supposed to send a trunk containing all our possessions by train and bus from Grenoble to Saint-Pierre. I went to the Grenoble train station with my trunk and waited my turn to register it. I looked around mechanically, because every crowded place was ideal for round-ups. It was a permanent problem. I suddenly noticed three Germans, rifles on their shoulders, hanging around in front of the newspaper stand without buying anything. Another glance around the station revealed three more, standing around in front of the restaurant. There was going to be a round-up any moment. I had no choice: I abandoned my trunk and left the station calmly, passing right in front of those Germans, and took the first tram passing in front of the station to the other end of Grenoble. I returned to the station four hours later, and learned that there had been a round-up. The station had been locked and every identity card examined and counterchecked with a phone call to the relevant town hall. Every man was undressed and those who were circumcised were deported. One minute of negligence or distraction and I would have been with them. My trunk had remained in the station, nobody had wondered about it, and I registered it and sent it off. My name, Sélimartin, was written on it, and my address in Saint-Pierre-de-Paladru. I doubted that the Germans had taken note of that. In those days, survival for a Jew was based on constant lucidity and quick reactions, along with a fair dose of good luck.

## Saint-Pierre-de-Paladru, Monestier-de-Clermont: a multiple identity

Now I needed to find some convincing justification for moving to Saint-Pierre-de-Paladru. Any city family who suddenly arrived and moved into a country village was considered to be certainly Jewish. A denunciation resulted in immediate deportation. I wanted to avoid this risk at any price. So I organized a multiple life for myself. I was a professor at the Scientific University of Grenoble, I was thirty-two years old, which seemed normal. Every week, we decided that I would leave Saint-Pierre-de-Paladru for three days, Monday to Wednesday, "to give my lectures", except during vacations. We always obeyed that rule. We said that we had moved there because the fresh air and healthy food were necessary to my wife, a former tuberculosis patient. The fable worked perfectly. At that time and later, there were massive numbers of denunciations. Another family which had taken refuge in Montferrat, but hadn't been able to take similar precautions, was denounced and deported. One day, I purposely dropped our identity cards into some oil and went to the town hall to ask for new ones, which were made immediately. Thus, even our identity cards were genuine, unless one went so far as to check with our original town halls. After each of my three-day absences, Marc-André, who was just a few months old, didn't recognize me until I'd spent a good quarter of an hour reminding him of all my typical manners and gestures.

During my three-day absences, I usually devoted one day to political activity with the Trotskyist group of Grenoble. This activity was much like what I had done before; sometimes I worked with the Lyons group. Furthermore, with the help of a colleague from the faculty of the university, Marcel Brelot, I found a reasonable job.

Several families from Monestier-de-Clermont, south of Grenoble, had sent their children to boarding schools in Grenoble, but had taken them back home because they were insufficiently nourished. They needed someone who could teach their children roughly the equivalent of the high school program. I introduced myself as a researcher from the CNRS, writing a doctoral thesis, and gave an address in Moirans. No one (apart from our nearest relations) was to know our real address. The half-a-dozen families looking for tutors adopted me and a young philosopher – who was quite probably as Jewish as I was.

I had to be careful not to get mixed up. In Saint-Pierre-de-Paladru, I was a professor at the university, in Monestier-de-Clermont, I was a CNRS researcher. I lived in Saint-Pierre and Saint-Pierre was written on my identity card, but in Monestier I said that I lived in Moirans. I gave lessons to a dozen children of different ages. There were three little girls in the eighth grade, in whom I was to inculcate the rudiments of geometry and who giggled constantly; they were adorable and understood nothing. On the other hand, one tenth grader was quite good in mathematics, whereas a certain little sixth grader was positively surprising. For example, I asked him this question: Suppose I take the number 2517 to the power 513, what is the final digit of the result? and he would answer quite quickly. The problem is not that difficult, but you have to understand it. All that matters about 2517 is the last digit, 7. If you look at the successive powers of 7, you see that $7^0 = 1$, $7^1 = 7$, $7^2 = 49$, $7^3 = 343, \ldots$ In fact, all the successive powers of 7 end with these digits: 1,7,9,3,1,7,9,3, etc. It's a periodic sequence of period 4. So all you have to know is the remainder of 513 after division by 4. You have $513 = 4 \times 128 + 1$, so the remainder is 1, so the last digit of 2517 to the power 513 is the same as that of $7^1 = 7$, namely 7. That boy understood this and a lot more, but he was no good in Latin.

One of my students, named Lachaud, was in eleventh grade and preparing his baccalaureate. I guess I taught him well, because he obtained it. I believe he became a doctor. He found me under the name of Schwartz, after the war, thanks to a picture of me in the newspaper, and came to visit me in my office, where I was temporarily a scientific researcher at the University of Grenoble. We remembered those days in Monestier with a lot of sympathy. However, I never went back there. It's too bad, because I would have had a lot of pleasure in meeting those families again.

Since I was supposed to be living in Moirans, I had to register as a resident of Moirans in the hotel where I slept in Monestier, and I had to have an identity card describing me as a resident of Moirans. In Saint-Pierre-de-Paladru, I had to renew our ration cards in the town hall. So I had two false identity cards in the name of Sélimartin, one from Moirans and one from Saint-Pierre-de-Paladru. Any mistake would be disastrous, so when I went to Monestier, I carried only the Moirans card. If anything were to happen to me, at least no one would be able to track down Marie-Hélène and the baby.

One day in January 1944, I arrived at the hotel in Monestier only to learn that the anti-Jewish brigade was at work in the village. This German brigade was aided by French militiamen. There were two hotels in Monestier-de-Clermont, one heated, the other not. The latter one was cheaper and I always went there. The rooms were freezing. The three or four members of the anti-Jewish brigade were of course lodged in the other hotel. They examined the papers of every passers-by, and then ordered the population to go home. They seized the few Jews in the village who had been denounced and brought them in a truck to the neighborhood of their hotel. Then they entered the hotel, ordered every man there to undress and took one seventy-year old who happened to be circumcised. Among the people they deported, I remember seeing one child whose fate shattered me; I can't forget it even today. It was a little twelve-year old girl suffering from jaundice, who was dragged violently to the truck. I had had jaundice myself a few months before and I knew how exhausted she must have felt. And they dragged away that sick child like cattle, to send her to the gas chambers. For the parents it was unbelievably horrible, for they were Jewish and had learned just the day before that their son had been arrested in Grenoble. That day, they had gone to Grenoble to try to find out something about his disappearance, and when they returned the following day, the anti-Jewish brigade was gone, together with their daughter. I know that there were thousands of similar stories, but this one, in which I saw lives vanishing in front of my very eyes, terrified me and still haunts me. How I would rejoice if I could learn that that boy and that girl were not gassed, and returned alive to France, but it's so unlikely. Thank God, the anti-Jewish brigade was unable to capture a five-year old boy who had taken refuge in a village which could be reached only by a road deeply covered in snow. They sent a car there but it was forced to turn back without the boy. It was a slight consolation.

By some unheard of luck, the anti-Jewish brigade did not visit my hotel to check identity cards and circumcision. They didn't do their job completely, and that little flaw saved me, once again, from deportation. But I was horribly frightened. When I learned about what was happening, I went up to my room directly without eating. I lay down under the bed, wrapped in my warm winter coat, and laid my suitcase on my stomach. Then I got up and checked that my hiding place was invisible from the door of the room, and opened the window wide. It was freezing cold. What I thought was, if the Germans came, the proprietors of the hotel would probably not want to be particularly helpful – they didn't attract a lot of sympathy – and send them upstairs to check the rooms by themselves. Opening my door and turning on the light, they would see that the room was freezing and empty. I don't know if my plan would have worked since the Nazis didn't even visit the hotel. But the reasoning was probably all right. I remained petrified in my hiding place until two in the morning. Then weakness took over and I got up and lay down on the bed to rest. It was imprudent since they might very well have come at dawn. But as luck would have it, they left during the night with their cargo. They continued their work in France for several more days, and then moved on to Hungary with the goal of capturing the eight hundred thousand Hungarian Jews. About half of these did fall into their clutches. I pitied the Hungarians with all my heart, but right at that moment, it was a relief that the Nazis were over there and not in France. In the dining room, where I went at about seven o'clock, I breakfasted normally, knowing that they had left. Everyone in the hotel was talking about them, angrily. I mingled in the general indignation; I suffered intensely from the destiny of those poor people. The fact that I was myself a potential victim didn't affect my feelings of pity and compassion in any way. The others were also horrified by the sight of the old man and the little girl. The owner, a generous lady, had given them all hot soup, although she had not gone so far as to offer them a real meal or some sweaters – which would have been taken from them in the camps, anyway. I gave my lessons as usual, and in fact my students had not been informed of the events. I only learned something about the fate of the Hungarian Jews much later, on a trip to Hungary in 1959, from the Jewish mathematician Renyi. He himself only survived the final days of the Occupation because he found a dead horse in a field. Together with a friend, they buried the horse under a heap of snow, and they were able to eat its meat for several days. Just before the Soviets entered Budapest, in February 1945, half the town was occupied by the Red Army, which was doing a house-to-house search for Germans, while the other half was occupied by Nazis and the Gestapo was doing a house-to-house search for Jews.

It goes without saying that I was stabbed by a single thought until my return to Saint-Pierre-le-Paladru: had the anti-Jewish brigade gone to do its sinister work in the village? Had they taken Marie-Hélène and Marc-André? The second I arrived, I learned that they had been to Paladru and Monferrat, but not to the tiny hamlet of Saint-Pierre. Marie-Hélène asked me all about what had happened in Monestier. Our strategy for avoiding denunciation appeared to have worked.

One silly incident could have had incalculable consequences. Returning from Grenoble to Saint-Pierre, I saw the mathematician Elie Cartan in the bus. The Cartan family lived in Dolomieu over the summer, and he went from there to Grenoble. He saw me from afar and we vaguely gestured at each other, but didn't move from our seats. However, I had to pass him getting off the bus at Saint-Pierre. This man knew the situation of the Jews perfectly. His entire family was strongly anti-Nazi and anti-Vichy. One of his sons was actively involved in the resistance movement, and was later captured by the Gestapo and decapitated. "Why, hello, Schwartz!" he said to me. "How are you doing?" Now, during this period, it was an unwritten rule that you didn't call Jews aloud by their names. However, he continued imperturbably, "Are you all right? Are you having any problems?" I said "No no, everything is fine." But worse and worse, he went on, "And how is that other Jew doing, that stateless fellow, Krasner?" I answered "I don't know!" and the conversation ended there; I leaped off the bus and looked nervously around me. People were all moving around, getting off, and I desperately hoped no one had heard. "How is that other Jewish fellow?" It was insanely naive! I wasn't angry with him, because obviously he thought he was being kind and considerate. But he could have gotten us all deported! If someone had been listening to us, we would have been obliged, like the wandering Jew, to be on our way once again, like in Roanne.

Another rather funny incident involved Marie-Hélène, her sister and her mother. Once, coming back from Monestier, I stopped as usual in Grenoble to meet with my Trotskyist comrades. The group had more or less decided to dissolve itself. Several of its members wanted to join the maquis, in the Alps, I believe. I agreed with their project entirely although I myself was not able to imitate them. During the meeting, I felt very ill and feverish. One of my comrades proposed that I spend the night at his place. I had a high fever and decided to stay in Grenoble for a couple more days, so as not to bring home the flu. The comrade who invited me over was to leave for the maquis within a few days, but I lay ill for five days in his house. When I took aspirin, my fever went down, but soon sprang back up to 40° C. On the third day, I finally understood that I had measles. From the start, I had asked a friend to send a telegram and a letter to Marie-Hélène. But he forgot! Seeing that I didn't return, Marie-Hélène began to panic. I was suffering from measles by pure chance, because at thirteen we had been vaccinated against it with a vaccination which had just been discovered by the group of Robert Debré. The new vaccination had been tested on about fifteen youngsters. Now, unlike modern vaccinations, this one remained valid for about fifteen years only. All fifteen of those first guinea pigs came down with measles at the same time (my two brothers among them). But childhood diseases contracted by adults are extremely exhausting. When my host left for the maquis and locked up his apartment, I took a taxi to the hospital of La Tronche. My friend's wife also left. Terrified, Marie-Hélène wrote to my friend, and received her letter back with the stamp "no longer at this address". She was convinced that I had been arrested and telephoned her family, who had taken refuge in Albenc. Her sister Denise left for Grenoble to try and find out something about me. She decided to call her mother back and let her know what had happened to me, using the following code: if I had been arrested by the Gestapo, she would say I had scarlet fever; if I had been arrested by the French police, she would say I had measles. First, Denise went to see Madame Conésil, the secretary of the physicist Félix Esclangon, who received their mail, and Madame Conésil gave her a letter I had written to them explaining my situation and asking if they could bring me some books. Overjoyed, Denise phoned her mother and announced to her that I had measles, completely forgetting about the arranged code! Fortunately, she then went into detail about my letter, so that my mother in law realized the situation and called Marie-Hélène to tell her.

I remained in the hospital from January 22 until February 9; I was afraid of contaminating the baby. In principle, there was no need for me to worry. Well, almost no need; I was in the ward for contagious diseases which was full of sick children, and the children wandered all over the place contracting all the childhood diseases at once! Some of those diseases were ones which I had never had, for instance scarlet fever. One room was even reserved for Malta fever. Apart from that, my nurses obviously had noticed that I was circumcised. Fortunately, hospital personnel everywhere was admirable during the war, and never let the Gestapo enter any hospital. After those fifteen days, I returned to Saint-Pierre-de-Paladru and rested there for several more weeks. Later, I took to cycling up to Grenoble, carrying shopping back on my small luggage-rack. I remember carrying back a big bag of flour which kept overflowing. In Grenoble, I regularly met with my friend Marchal. One day, we had agreed to meet in Voiron, a little town lying between Paladru and Grenoble, where we spent two perfectly surreal days. Marchal had found a hotel room for himself, but they had no room for me. I tried another hotel and rang and rang at the door until a passer-by addressed me: "Monsieur, don't you see that this hotel has been destroyed and that you're ringing at an empty façade?" It was true. There wasn't any other hotel, so I stayed in the house of Irène Pérard, who was a friend of Madame Esclangon, the wife of the physicist Esclangon and herself a friend of Denise. I went to her house rather late, but she put me up anyway. Already, my arrival in Voiron had been anything but calm. Just a few hours earlier, there had been an attack against the Germans and they had taken a bloody revenge, destroying the hotel and hanging twelve people from lampposts. They had chosen these people as follows: posted just at the entrance to the town I had come in by, they selected every single person arriving on a bicycle. If I had arrived two hours earlier, I would have been hung. When I did come, everything was over and the victims had been taken down and buried.

In the restaurant, Marchal and I told each other wild stories to take our mind off things. We did this a lot during the war, out of a kind of black and vengeful humor, all bent towards stigmatising Nazis. I told him that Hitler once went to an insane asylum, desiring to meet the patients who thought they were Hitler. He was carefully guided into a room where several people resembled him, and they all flung themselves on him shouting "Ich bin Hitler!" "I am Hitler!" He was left with them for some time and there was a row; when they opened the door to fetch Hitler, all the patients rushed forwards shouting "Ich bin Hitler!

Ich bin Hitler!" It was impossible to tell which one of them was the real Hitler. So they took one at random, and he is at this moment chancellor of the Reich.

After a few more stories of this kind, I thought of another, rather different one. A man driving in a car has an accident. He falls into the ditch and is decapitated. He picks up his head and rushes to the nearest pharmacy, carrying it in his hands. "Sorry, we're closed on Sundays," says the pharmacist. It wasn't very funny, but we laughed anyway. "Monsieur," said an elderly gentlemen sitting at the table next to ours, "your story is delicious, but it is not very realistic. In a case as serious as the one you describe, where a man actually comes to a pharmacy carrying his head, the pharmacist would open even on a Sunday." I guess a grain of surrealism had penetrated everyone's psyche by then.

Marchal spent the last months of the war in Grenoble, and worked hard in the Communist unions. His real name was Lucien Schmitt. He died of a heart attack in 1962, aged only fifty.

## Madame Carus

I used the Minitel recently to contact Madame Marthe Carus. She came to visit us in Paris with her whole family. Her granddaughter, Christine Reynaud, lives in Paris with her companion Emmanuel Camille-Bernard. They opened an antique bookstore which is doing very well and which they adore. The daughter of Madame Carus, Madeleine Reynaud, became a music teacher, apparently partly because of me! She reminded me that when she was ten years old and I was living in Saint-Pierre-de-Paladru, she was studying piano, and I gave her some music lessons. She remembers that she had the music of a Chopin waltz, and that I played Beethoven's Moonlight Sonata for her. The husband of Madame Carus is dead. She is now eighty-eight years old, six years older than I am, and she's very alert although one gets tired easily at that age. I would have recognized her anywhere. We left Saint-Pierre in 1945, fifty years ago, and I saw her again briefly after the Liberation. This visit was a tremendous pleasure to both of us. She told us quite a few things about Saint-Pierre-de-Paladru which we didn't know.

We remembered a certain Samson who often came to dinner in her restaurant. We didn't know, back then, that he was a member of the maquis who worked in the neighbourhood; this maquis was mostly made up of rebels from the STO (Service du Travail Obligatoire)[34]. The members of this maquis lived by

34) STO: Mandatory Work Service. The Germans imposed mandatory military service in Germany on French 20 year olds; many of them refused and worked for the underground resistance

requisitioning food from the village. She also reminded us of M. Pommier, a peasant from the village whom we knew well. In the village, there was a little church in which we never set foot, and which didn't even have a priest. Contrarily to what we had always thought, Madame Carus told us that she had supposed we were Jewish right from the start. All the families who came to take refuge in villages were, she said. It's true; that's why we had taken so many precautions to give good reasons for being there. She says I actually gave her a math book once – she didn't understand anything in it, but I suppose I must have wanted to convince her that I was really a mathematician. I'm not completely sure that she really guessed at first. Probably she just asked herself the question, but there was no way to find out for sure, what with our convincing story. On the other hand, she says that when she saw the state Marie-Hélène was in during the period when I had the measles, she became absolutely certain. We'll never know for sure what she really thought, but what is certain is that she was perfectly discreet. She claims that our precautions were helpful but not necessarily strong enough. There were other villages, for instance Monestier-le-Clermont, where informers worked for the Gestapo; the proof is the anti-Jewish raid I witnessed where they came to grab a little twelve-year old girl and a five-year old boy. They must have known all about the presence of Jews in the community. Informers like this were present in a lot of villages. But in Saint-Pierre-de-Paladru, there can't have been any, because according to Madame Carus, its fifty inhabitants lived in peace with each other, with only the most minor of conflicts. Any informer would have been found out and rejected immediately. We hadn't realized at the time that a very tiny village was actually less risky for this reason. Maybe there were even informers in Paladru and Monferrat – there were people deported from there. And if there had been any informer, it's clear that he would have figured out that we were Jewish; that would have been his job. He would have investigated us, followed me and ended by denouncing me to the Gestapo, and I would have been deported. So after all, I guess I better rectify my earlier opinion and consider that it was our luck in choosing Saint-Pierre-de-Paladru, as well as our precautions which saved our lives. We talked about the count a bit; the whole village liked him. Madame Carus reminded me of another Jewish family, the Lévys, who lived for a long time in Saint-Pierre-de-Paladru under the name of Lefèvre. Their situation was obvious to me at the time – I gave math and physics lessons to their daughter.

Madame Carus and we both preserved an old photo showing her daughter Madeleine together with Marc-André. We've promised to get together at Saint-Pierre-de-Paladru, the next time we go down to visit Claudine in Grenoble.

## The living and the dead

Grenoble was liberated on August 22, 1944; Paris on August 24. I was in Monestier-le-Clermont when the Americans came parading through with their tanks. The village was wild with joy. The Americans rapidly occupied the whole region; soldiers came to camp in Saint-Pierre-de-Paladru. A huge black soldier scooped up Marc-André and set him on his shoulder, to his utter delight. Madame Carus and Marie-Hélène both remember it.

Once we were liberated, we decided to live as though we didn't know the war was still going on. We felt so light-hearted that we only thought of picking up the course of our interrupted life. Yet, the war was still going on, and it did weigh on us. First, we had no news of Bertrand, who had gone to England in 1942. Returning to Paris with the Leclerc division, he tried to find out what had become of us. He got his first information from Philippe Monod-Broca, who was still living in Paris, rue de l'Université. His wife Claude is my first cousin. We're very friendly with them, he's our "family surgeon". But we had barely got back in contact with Bertrand, when he left for Germany. At the Liberation, he entered the Corps des Mines, thanks to his military activity. In Nancy, he began a campaign for educational reform, which he has never stopped since.

We were extremely worried about those who had been deported. They still had months of war to endure, including a tragic winter. After May 8, 1945, we waited for their return with deep emotion, but we were also seriously shaken by the atom bomb attacks on Japan. We had decided to return to Paris around September 1944. Cartan had arrived there before us and procured us an apartment into which we moved immediately. First, I went to see Montel, provost of the Scientific University of Paris, who played the role of grand master of the French Universities, a little like the Director of Higher Education does today. He told me that I had just about no chance at all of being appointed to the Scientific University of Grenoble, and guaranteed that it wouldn't happen that year. So we moved to Paris, to a little apartment with no heating (we wore our coats indoors), at 2, rue de Monticelli, not far from Cartan who lived at 95, boulevard Jourdan.

Shortly afterwards, we received the following piece of news from Paul Montel: I had been given a position at the Scientific University of Grenoble! I was to replace Küntzmann, who was a prisoner in Germany and would not return for another year. After having organized the whole move from Grenoble to Paris, we felt that moving back the other way was nearly beyond our strength! Furthermore, it was going to be complicated getting back to Grenoble: train seats were rare and only given out to people with serious business. I had a position, so I had the right to a ticket, but the police checked that only those with tickets got on the trains. So I couldn't take Marie-Hélène and Marc-André. Fortunately, I was used to getting around the rules in every way, so I simply got my ticket and put Marie-Hélène on the train with it. She sat in the seat and installed the baby in a little hammock hanging between the two parts of the compartment. He babbled and murmured through a large part of the night, until some wise traveler said to him kindly "Be quiet now, little rabbi!" Marie-Hélène arrived safely in Grenoble with Marc-André and went to join her family. As for me, I simply took a train to Montargis. I spent one night, without sleeping, in the station, and the next morning I leaped into the Paris-Grenoble train which stopped in Montargis, but where there were no policemen to prevent people without tickets from getting on. One day after Marie-Hélène's arrival, we were all reunited in Meylan where her family was installed. Soon we found ourselves a charming little house with a garden in Meylan, a pretty little suburb. Our daughter Claudine, who is a professor at the University of Grenoble, lives in Meylan now, and we often visit her there. But we never succeeded in finding the actual house where we lived for a year again.

We had been late in starting our clandestine lives, and we were late in getting back out of them. During the months we spent in Paris, we had never needed our identity cards since we lived in an apartment belonging to friends of Cartan, but we had ration cards. We went on using our false Sélimartin cards for at least two months! We didn't dare go back to being legal! After all, the Nazis were not vanquished yet (the von Rundstedt initiative was still to come). Only later did we go and openly explain our situation in the town hall of the fourteenth arrondissement, and ask for identity cards and ration cards in our true name of Schwartz. Nobody in the town hall asked us silly questions about what we had done during the two months we had already been in Paris. The employees were amazingly kind and helpful. And so, progressively, we went back to a normal lifestyle.

During this end part of the war, we seemed to spend all our time moving, and taking trains, and waiting in stations. It was that way for many Jews. Hélène, the oldest daughter of Denise Piron, who later married Uriel Frisch, was a ravishing little girl with whom we liked to play a great deal when we were in Montbonnot. After we left, she was sometimes asked where Laurent and Marie-Hélène had gone, and she always answered simply: "In the train." Another time, when Denise was moving, she asked "When will we stop living in train stations?" Marc-André was still too small to notice these things. But our life was extremely agitated. It must have been for most Jews, but I happened to be Trotskyist as well as Jewish, which made everything much more complicated. I had two false identity cards, two lives, two jobs. And it was all necessary just to survive.

If I consider the fate of our family during the war, I find that we were spared: we survived, as did our brothers and our parents. It is amazing, because everything, everything was organized in order for us to die. And we did have near relations who disappeared in the horrible conditions of the Occupation. Yvonne's family, her parents and her sister Hélène all died in deportation. My father's sister, Delphine, who had brought him food when he locked himself in the attic at the age of fourteen, had married Mathieu Wolf. He was a rabbi, and had worn the yellow star in solidarity with other Jews; it was he who blessed our marriage in my grandfather's apartment. Neither Delphine nor Mathieu returned from deportation. Three of my friends from taupe and from the ENS, Neyman, Martinelli and Croland – all remarkably intelligent – all dead. In the Trotskyist Party, my friend Marcel Hic and many others died. A cousin of Marie-Hélène, the son of her father's eldest brother, the grandson of Onésime Reclus, who had been brought up essentially by his mother Hélène Reclus (the big Protestant Reclus family was more or less anarchist) had entered the ENS in literature in 1934, the same year as Marie-Hélène and myself. He had converted to a very mystical form of Catholicism, hadn't tried to protect himself during the war, and wore a yellow star. Although he was pacifist and non-violent, they sent several armed members of the Gestapo to arrest him, on February 12, 1944. We learned that he remained serene and gentle throughout his captivity in Auschwitz. He never returned.

A cousin of my mother, Guy Iliovici, a professor of mathematics in the lycée Saint-Louis, also remained in Paris and wore a yellow star. Because he was Jewish, he lost his job, but his colleague and close friend Saint-Laguë gave him half of his salary. He was deported with his wife Suzanne and his daughter Janine; they never returned. Towards the end, he had become terribly somber and anxious, feeling what was to happen. The painter Davids, who had painted the magnificent portraits of my parents in Autouillet, had a lovely twenty-year old daughter. She was deported and never returned; her parents never got over it. A close childhood friend of my mother, Annette Lazard, whom we used to call Aunt Annette, was married to Christian Lazard, of the family of the directors of the Lazard bank. They had a beautiful property called La Couharde, in La Queue-lez-Yvelines, at the edge of a dream forest where we loved to go. I fished there and hunted butterflies. We used to play a lot with their children Miquette and Janine, who were also friends of Vidal-Naquet. Christian Lazard died in deportation. I mention all this because many Manicheaen revolutionaries of the time were only interested in the deported proletarians. Walter and Trudel

Goldstein, Jews of German origin, lived near Clermont-Ferrand when we were there. They must have been denounced; at any rate gendarmes came one day and took their eighteen-year old son. The father followed them, and saw the gendarme walking slowly, casually swinging his rifle, obviously with no intention of firing. It almost seemed that he wanted to allow the boy to flee, but he didn't. He was deported and never came home. We invited the parents over to Ceyrat shortly afterwards; they were desperate. The too numerous absences among our old friends cast a deep pall over our days. Our memories of these friends remain intact and live on in us.

## The extermination of the Jews: goal of the war

I have always thought that the Allies did not struggle hard enough against deportation. By threatening to take revenge on German prisoners, de Gaulle had obtained that the prisoners from the Free French Army were treatd like prisoners of war by the Wehrmacht. The Russian prisoners of the Wehrmacht were horribly mistreated until the Russians had taken a certain number of German prisoners. The German practice of deportation was in direct contradiction with the Geneva conventions, and the Allies knew of it. It would have been possible to threaten retaliation against German prisoners of war; there were more and more of them as the war continued. Perhaps then the Wehrmacht would have ceased the deportations. I realize that this double-edged system would not have been without its risks, neither would it have been in accord with the Geneva conventions. Even though he was not extremely well-informed, Churchill knew a lot. Gerstein, a member of the Gestapo, had told the Pope Pius XII about the use of Zyklon B and cyanide. People have said that the necessities of the war itself took precedence: the Germans were defending themselves desperately after D-day, and the Allied success was not absolutely assured at that point. The clear priority of the Allied armies was to conquer territory without "losing" time on humanitary tasks. But England was savagely bombing Dresden, where three hundred thousand people died (more than those killed by the atomic bombs!) in order to demoralize the German people. Dresden was not a military objective, and the bombings had no decisive effect on the German people other than indignation. The Dresden bombing was a war crime; it should have been judged in a bilateral Nuremberg. The Allies could have and should have taken earlier action with respect to the concentration camps. I don't mean bombing them; it would have been inefficient and dangerous for the prisoners. But surely it would have been possible to destroy the railroad system and the logistics leading to them, and the industries (such as Siemens) which depended on prisoner labor. Of course, the Nazis could have transferred prisoners from one place to another, as they did with horrific cruelty at the approach of the Americans. But they thought of the extermination of the Jews as a well-oiled, industrialized, prioritary system, and its destruction would have caused them considerable disorder and even humiliation in front of the Wehrmacht, who considered that deportation played too much of a role compared with purely military objectives. We could also have bombed the gas chambers. Anyway, none of these things were attempted. For the Allied governments, the extermination of the Jews, of whose true extent they were not completely aware, was not a cause of primordial importance. In fact, a most surprising discretion and lack of curiosity on the subject reigned for several years following the war; far-reaching studies on the subject have been quite recent. As Jews, we avoided investigating the subject too deeply, not wanting to appear attention-seeking. The Communist Party glorified itself muchly on account of its seventy-five thousand executed members (it is true that admiration is due to its courageous resistance). But the Jews of France never glorified themselves on account of their seventy-five thousand who died in deportation. Even I never really talked about it. And neither did the Russians or the Chinese exalt their dead. Today, we know the extent of that genocide, one of the most abominable phenomena in the history of mankind. It is essential to know the full number of victims, in order to formulate a political and moral judgment about the war and the crimes perpetrated during it.

At the beginning of 1944, I went to Paris for a second time, to attend an extended union of the central committee. We wanted to discuss the unification of the PCI and the POI. The discussion was extremely lively, and the unification only took place much later because of the wide divergences of views. The PCI (directed by Pierre Frank) subscribed to a completely Manichean view of the class struggle and defended the view that there would be no D-day because, for reasons of imperialism, the Anglo-Saxons would prefer to let the Russians and Germans mutually exhaust each other, or even to let the Nazis exterminate the Soviets. Nobody mentioned Jews at all. It is also true that, if it were not for one very noticeable speech by David Rousset, the maquis wouldn't have been mentioned either. These questions were simply ignored by the Trotskyist organizations, whose eyes were glued to the workers' and the housewives' committees, but never even cast a thought to Jewish specificity or the possible efficiency of the maquis. Tales of massive deportations were reported in *La Vérité*, but the IVth International, which produced a mass of texts containing discussions between the different parts of the world, used the word "Jew" in a document exactly one time in six years. It ignored the genocide in the most total way. This was not just an error. It was a conscious and inexcusable choice of position. And I repeat, in saying this, that I myself never "felt" Jewish, either before, during, or after the war. However, I did feel driven inexorably towards a predicted and announced death, crushed by a bulldozer like all other Jews.

I also always thought it would have been useful to create a Jewish army, partly recruited in Palestine, and extra-well armed. It would certainly have obtained victories in battle, and German soldiers, who despised Jews as a submissive and inferior race, would have been terrified to see Jews defending themselves, attacking, winning and taking prisoners. The psychological situation could have been reversed. But there were many geopolitical reasons against forming such an army. In fact, there was a Jewish Legion in the British army, made up of European Jews, but it was not allowed to play an important role.

During the entire war, I never knew of the existence of gas chambers. Georges Glaeser had given us the information in his possession: the deported Jews were sent to work in salt mines in Poland, where they were made to work so hard that they were decimated. We believed that for a while. But when they starting deporting children, we understood that people were being sent to their deaths, but we didn't know what kind of death. The horrors which were revealed at the end of the war were simply inconceivable. At any rate, what we understood very soon was that to be taken or not taken was a clear question of life or death.

The goals of the two sides in the war were not, initially, based on the Jewish question. In spite of *Mein Kampf* and the immediate anti-Jewish repression exercised by Hitler, his primary goal was the conquest of "Lebensraum" via the destruction of Bolshevism, the conquest of Russia and the domination of the world "for the next thousand years". For the Allies, the primary goal was the protection of their empire and the destruction of the military power of the Reich, and this goal justified the alliance with Stalin's USSR. The dogma of the fundamental importance of economic structures is a Marxist idea which is too Manichean to really work: the Allies also intended to defend the whole of Occidental civilization, threatened by a fundamentally destructive adversary. That was not the view which the IVth International chose to take. I thought for a long time, during the war, that when Nazism attempted to conquer a central European country, it began by associating itself to the local population in the goal of exterminating the Jews, as an easy way of beginning the conquest. Once this was obtained, it was not hard to force the country into submission. However, information which I learned only after the war showed that this judgment was incorrect. Several dictatorial countries of central Europe, for instance Hungary and Rumania, fought the USSR along with the Reich, and yet, they also protected their Jews. It's only towards the end, when the necessities of the war pushed the Reich to occupy these countries completely, that the extermination of the

Jews was begun. I mentioned Hungary a little earlier; the extermination of its four hundred thousand Jews was begun only in the spring of 1944. The most revealing case is that of the German defeat in the USSR: even though rail transport of fundamental necessities was necessary for the survival of the Reich, priority was given to deportation of Jews above the protests of the Wehrmacht who judged other objectives much more urgent. Between the beginning and the end of the war, Hitler changed his goals: the battle of London had no real connection with the Jewish question. The initial paranoia in *Mein Kampf* slowly came to dominate, as the imperial aims faded away. The extermination of the Jews, the "Final Solution", was only decided in its final form in the Wannsee summit meeting in 1942, but a start had already been made in June/July 1941, just before the USSR attack. That is why I called this chapter "The war against the Jews". I wouldn't have called it that during the war.

I was not in any more danger than the 150,000 other French Jews, 26,000 of whom were deported (50,000 of the 150,000 foreign Jews in France were deported). Maybe a little more, on account of being a Trotskyist. Gerard Bloch, who was arrested by the French police who then proceeded to deliver him to the Germans, was deported as a Trotskyist and not as a Jew. Danger also threatened from the Stalinist machine. Jacques Duclos was one of the most cynical and dangerous leaders of the party during the Spanish civil war as well as during the world war. It wouldn't be dishonest to say that during these terrible years, I thought more about the sufferings of the world and the struggle against Hitler than about our personal problems, at least insofar as a prey being pursued by a wild beast has time to think. But we never became absolutely obsessed. We led an almost normal life, joking and amusing ourselves, laughing, even laughing wildly on occasion, and playing with our baby. In spite of the circumstances, we preserved an incredible aptitude for happiness. Without this quality, how would humanity survive?

# Chapter VI
# The Invention of Distributions

The invention of distributions occurred in Paris, in early November 1944, while my identity papers and ration card were in the illegal name of Sélimartin. The discovery was quite sudden, taking place in a single night. This is a rather frequent phenomenon, which happened to me several times in my life, and to many other mathematicians as well. Poincaré said that he discovered the essential point of the theory of Fuchsian functions while stepping onto the running board of a tram. In five minutes, while preparing for bed, I discovered a theorem which I had been hunting for weeks; it was while we were on vacation in Pelvoux. It's obvious that this could not happen if the brain had not stocked up all the ideas from the previous weeks of reflection. But a sudden click, brought on by random circumstances, suddenly brings the rough sketches together into a single whole. This is one of the essential phenomena of discovery. Without antecedents, work done by previous mathematicians finding partial results, there would be no scientific discovery. In this chapter, I would like to retrace the history of the precursors of distributions.

Independently from each other, Newton and Leibniz discovered differential and integral calculus at the beginning of the eighteenth century. It is obviously impossible to imagine that these discoveries simply sprang from nothing. They had been prepared by numerous predecessors. Greeks already knew the construction of the tangent to an ellipse and a hyperbola; the tangent to a point on an ellipse is the exterior bissector of the angle defined by the rays joining this point to the two foci, and the tangent to a point on the hyperbola is the interior bissector. Tangent led to derivative. In the seventeenth century, Pascal had constructed the tangent to the cycloid, again providing a link to differentiation. Moreover, the concept of the *speed* of a vehicle moving at constant speed had been understood for some time. But instaneous speed is a derivative and it was not known; Galileo discovered the law of falling bodies, but in a primitive form containing neither **a** nor $g$: namely, the space traveled is proportional to the square of the time. Newton and Leibniz unified many previously understood concepts in a single global theory. Integral calculus essentially came from the computation of surfaces and volumes. When Archimedes computed the area of a segment of the parabola, he used essentially the same methods as those which later became integral calculus. The grand theory of integration is the Lebesgue theory, discovered in the early twentieth century, but in the nineteenth century, Riemann had discovered the simpler and less general theory of Riemann integration, which was used for decades before being replaced by Lebesgue's general theory.

## Percolation

I would like to compare the phenomenon of sudden scientific discovery to the percolation of coffee. If water is poured over a mass of packed coffee grounds, at first it doesn't penetrate. Then some little trickles of water are born, but each one soon comes to a stop because the physical conditions don't allow it to continue. More and more little trickles begin, and they pierce farther and farther into the packed mass of grounds; they are not all born from the first trickle, but from collateral conditions arising from the existence of the first few trickles. And suddenly, one trickle makes a path right through the whole mass of packed grounds, and an artery is formed through which liqued passes quite quickly from the top to the bottom of the mass. The phenomenon of electric sparks resembles the same thing. When you have two conductors one of which is positively charged and the other negatively, and they are far from each other, there is no electric shock, because the electric current does not pass through air. But it does pass through ionized air. If you push the two conductors towards each other, the potential created by their two charges ionizes the air around each conductor. Small, almost imperceptible sparks are born. Then suddenly, after several abortive attempts, a spark flashes between the two conductors. It doesn't follow a straight path, but an ionized one. And the electric shock occurs, and it is total and absolutely sudden. As soon as a single spark forms an ionized path, the entire electric charge of one of the conductors contacts the charge of the other. In fact, only electrons can behave in this way, since positive ions are too heavy. Lightning behaves similarly. The two charges are contained one in the earth, the other in a cloud, and as one commonly hears, "there's electricity in the air". Numerous microlightnings appear before the true lightning flash; the microlightnings are imperceptible, producing only very weak electric shocks, whereas the true lightning flash produces the equivalent of a powerful electric shock of several million volts.

Cerebral percolation is accompanied by an enormous quantity of subconscious work. Between the long period of reflection giving no result and the sudden discovery, the subconscious has done its work. This can be observed in many everyday events. Nowadays, very often, I can't remember somebody's name. I search and search, generally in vain; I don't expect to find it, I'm just forcing my subconscious to work. Then I give up. And several hours later, sometimes upon waking after a good night's sleep, the name suddenly comes to me, without my having given it any conscious thought in the meantime. This also happens with mathematics.

It happens that the preliminary work preceding an important discovery does not give any publishable results, and no one notices them. If the person working on them doesn't keep them in mind, nothing will come of them. But the human spirit actually collects and hoards them. The author of the preliminary work may sometimes be the one who pierces through to the main discovery. Other times, it takes several mathematicians. Each mathematician interrupts his work of discovery at a certain point, according to the possibilities of his time and of his brain. Some periods of time were more propitious than others for the production of complete discoveries. When I discovered distributions in a single blow, I was not only prepared for the discovery by all my previous reflections, but also by the previous work of many mathematicians. If I had not discovered them, it seems clear to me that somebody else would have within the next few years. I can't say how many; maybe four or five, or even less; certainly less than ten. I can't say if they would have been introduced by the same methods. Maybe other methods would have been used. Indeed, certain mathematicians did try to short-circuit my methods by looking for simpler ones, but none of them actually passed the test of time; they all really turned out to be less simple. I was in the same position myself, since my first definition of distributions, which I gave in November 1944, was not quite right and I only found the definition which we still use today in February 1945, in Grenoble.

## Differentiation

Distributions have many very different properties. Firstly, and essentially, they are a generalization of the notion of function, and their purpose is to solve problems of differentiation. By a generalization, I mean that the set of distributions is a larger set than the set of functions; every function is a particular distribution, but there are distributions which are not functions. Why does one need such a generalization? Gratuitous generalizations are not interesting. Differentiation poses serious problems: there are certainly differentiable functions, but Weierstrass was the first to give an example of a function which is nowhere differentiable. This example is extreme, but one frequently meets with functions which are differentiable at most points, but which are not differentiable at certain exceptional points (for instance, the function $|x|$). Differentiation creates a host of computational difficulties. Distributions solve the problems of differentiation in the sense that every distribution is differentiable and even infinitely differentiable, and the derivatives are also distributions. If a continuous function is not differentiable, then, considered as a distribution, it always admits a derivative, but the derivative is a distribution which is not necessarily a function. This is why distributions are used everywhere in calculus; they are always differentiable, even when derivatives in the usual sense don't exist. When looking for the solution of a differential problem, one looks for it within the theory of distributions, but when you do obtain the final result, it is often (though not always) interesting to check whether it is really a function.

Analogous phenomena occurred previously in mathematics. For instance, rational numbers were insufficient to provide the answer to the problem of square roots; integers such as 2 or 3 do not have rational square roots. But if the rational numbers are generalized to the real numbers, then every positive rational number has a square root which is real but not necessarily rational, and in one blow you get the field of real numbers, in which every positive number has two square roots of opposite signs. The problem is still not completely solved since negative numbers do not have square roots. For this, you need another generalization: the complex numbers. In the field of complex numbers, every real number has two complex, opposite square roots, and in fact every complex number has two complex square roots, and you even obtain d'Alembert's famous theorem: every polynomial with complex coefficients of degree $m$ has $m$ complex roots. This is analogous to the relation between functions and distributions described above: not every function has a derivative which is a function, but all functions have derivatives which are distributions, and every distribution has a derivative which is a distribution. The mathematician Peano wrote in 1912 on the difficulties of differentiation: "I am sure that something must be found. There must exist a notion of generalized functions which are to functions what the real numbers are to the rationals." This was a marvelous intuition, and it arose long before 1944. But the mathematical knowledge of the time did not make it possible for Peano to find the generalized functions, or even to conceive of them; at that time it would have been a superhuman task.

## Heaviside's symbolic calculus

One of the most important precursors of distributions was the electrical engineer Heaviside. He wanted to solve ordinary differential equations, and particularly those with constant coefficients. There are many formulas giving solutions to such equations. Heaviside introduced his symbolic calculus in 1893–94, well before Peano. There existed a "composition product" for functions which were zero outside of the semi-axis $[0, +\infty[$, denoted by an asterisk $*$, introduced by Mercer at the time of the integral equations of Volterra-Fredholm, at the end of the last century. For $t \geq 0$, the composition product is given by

$$(f * g)(t) = \int_0^t f(t-s)g(s)\,ds = \int_0^t f(s)g(t-s)\,ds;$$

this product is associative and commutative, and thus defines an algebra. Composition plays a fundamental role in "transitory phenomena" in electricity, i.e. those which depend on time and start at time 0, so that they are represented by functions which are 0 for $t \leq 0$. André Weil studied and used composition extensively in his 1940 book on integration on topological groups. The name of composition was too vague, and it was eventually abandoned and replaced by "convolution", from the English; in German, it is called "Faltung". It has been studied in dozens of books, and occurs everywhere in analysis, in the study of functions which are zero on the negative real axis; it also occurs in the study of general locally compact topological groups, the integral from 0 to $t$ being replaced by an integral over the entire group. Convolution played a central role in my research from the moment I read Weil's book in 1941. Since it was too long to say "take the convolution product of $f$ and $g$" when previously people had been able to say simply "compose $f$ and $g$", I introduced the (perhaps somewhat barbaric) neologism "convole $f$ and $g$". Heaviside introduced a symbolic calculus of convolution, where a function $f$ of $t$ is represented by its symbol, a function $F$ of another variable $p$, and he wrote convolution multiplicatively: $(f * g)(t) = F(p)G(p)$. He introduced the function which he called "the unit-step" and which is now commonly called Heaviside's function and denoted $Y$. The Heaviside function $Y(t)$ is a function of a real variable, and satisfies $Y(t) = 0$ for $t < 0$ and $Y(t) = +1$ for $t \geq 0$. This function is not differentiable at the origin. Heaviside called its derivative $Y'(t)$ the "unit-impulse"; it is equal to 0 everywhere, except at the origin where it is equal to $+\infty$, but its integral must be equal to 1, i.e.

$$\int_{-\epsilon}^{+\epsilon} Y'(t)\,dt = Y(\epsilon) - Y(-\epsilon) = +1$$

for all $\epsilon > 0$, and also

$$\int_{-\infty}^{+\infty} Y'(t)\,dt = +1.$$

This function was reintroduced by Dirac in 1926 for use in quantum physics, and it is now often called the Dirac measure and denoted by $\delta$, but Heaviside defined it much earlier. There is no function having the properties of $Y'(t)$; it is almost everywhere zero, so its integral ought to be zero. And furthermore, the double of the function is equal to the function, so that the integral of the double should be twice the integral of the function at the same time as being equal to it! The relation between the impulse and the step function is thus also given by

$$Y(t) = \int_{-\infty}^{t} \delta(s)\,\mathrm{d}s = \int_{-\epsilon}^{t} \delta(s)\,\mathrm{d}s.$$

In electricity, one could say that an electromotor force equal to $\delta$ is more or less a powerful electromotor force, equal to $1/\epsilon$, applied over a very short time interval $[0, \epsilon]$; if the circuit contains self-inductor and capacitors, the current will last until $+\infty$, and Heaviside computed it. An intensity of the current equal to $\delta$ is even simpler. Consider the passage of a single unit charge, animated by a uniform motion in a straight wire, at time 0 at the origin; the quantity of electricity $q(t)$ which passed through the origin at the instant $t$ is equal to 0 at $t < 0$ and $+1$ at times $t > 0$; it is the unit step $Y$. The old formula $i(t) = \mathrm{d}q/\mathrm{d}t$ gives $i(t) = \delta(t)$, the unit impulse. Heaviside had already introduced the successive derivatives of $\delta$ in his symbolic calculus, which went beyond all limits of decency! Applying the convolution formula, we have

$$\begin{aligned}(\delta * f)(s) &= \int_{-\epsilon}^{t} \delta(s) f(t-s)\,\mathrm{d}s \quad (\text{since } \delta(s) = 0 \text{ for } s \neq 0)\\ &= \int_{-\epsilon}^{t} \delta(s) f(t)\,\mathrm{d}s = f(t) \int_{-\epsilon}^{t} \delta(s)\,\mathrm{d}s = f(t),\end{aligned}$$

so that $\delta * f = f$, and $\delta$ is the unit element of the convolution algebra. Its associated symbolic function is thus 1, and in the symbolic calculus of convolution, $\delta * f = f$ is written $1F = F$. If $f$ is differentiable on the whole real axis, with value 0 at negative times, we have

$$\begin{aligned}(\delta' * f)(t) &= \int_{-\epsilon}^{t} \delta'(s) f(t-s)\,\mathrm{d}s \quad (\text{now integrate by parts})\\ &= \delta(t) f(0) - \delta(-\epsilon) f(t+\epsilon) + \int_{-\epsilon}^{t} \delta(s) f'(t-s)\,\mathrm{d}s;\end{aligned}$$

the two first terms disappear because $f(0) = \delta(-\epsilon) = 0$, and we obtain $\delta' * f = f'$, so that convolution with $\delta'$ is differentiation. Heaviside had written $p$ for

$\delta'$, so the derivative of the function whose symbol was $F(p)$ was $pF(p)$. If $f$ has "jumps", places where it is not differentiable, then $\delta' * f$ cannot be $f'$; it is necessary to add terms in $\delta$ at the jump points, and Heaviside saw this. The formula $DY = \delta$ indicates that the symbol of $Y$ is $1/p$ or $p^{-1}$. The symbol for the $n$-th derivative $\delta^{(n)}$ is $p^n$. Consider now the function $Y(t)e^{at}$, and let us compute its derivative:

$$D\big(Y(t)e^{at}\big) = \delta(t)e^{at} + Y(t)ae^{at} = \delta(t) + Y(t)ae^{at},$$

so that

$$(D - a)\big(Y(t)e^{at}\big) = \delta.$$

Replacing $D - a$ by $p - a$ (or $p - a1$), he wrote

$$\frac{1}{p - a} = (p - a)^{-1}$$

for the function $Y(t)e^{at}$. Thus, he introduced more and more functions of $p$, each one more extravagant than the one before. He used them to introduce formulas which were extraordinary, but perfectly correct! For example, suppose that $g$ is a continuous function of a real variable, with $g(t) = 0$ for $t < 0$, and consider the solution $f$ of the differential equation $(D^2 - 5D + 6)f = g$. The function $f$ should also be zero for $t < 0$. Mathematicians of the nineteenth century knew how to compute the solution to this equation perfectly well, because $f$ is necessarily a $C^2$ function, so that it and its first derivative must be zero at $t = 0$, i.e. the initial Cauchy conditions must be zero. But Heaviside had another, instantaneous and magic way of finding the solution, via his symbolic calculus. He simply wrote

$$(p^2 - 5p + 6)F = G$$

so

$$F = \frac{G}{p^2 - 5p + 6}.$$

He decomposed the rational function $1/(p^2 - 5p + 6)$ into partial fractions

$$\frac{1}{p^2 - 5p + 6} = \frac{1}{(p-2)(p-3)} = \frac{1}{(p-3)} - \frac{1}{(p-2)}$$

(how can you use partial fractions to decompose a rational function of $p = \delta'$ which doesn't exist?!), and wrote the solution as

$$F = \frac{G}{p-3} - \frac{G}{p-2} = g * Y(t)(e^{3t} - e^{2t}),$$

where

$$f(t) = Y(t) \int_0^t (e^{3s} - e^{2s}) g(t-s)\, ds,$$

which is precisely the correct solution. Heaviside did plenty of other computations and obtained difficult results, with methods which were not justifiable in his time. In itself, the symbolic calculus is perfectly justified; in an algebra with unit and without zero divisors, one can certainly decompose a rational function into partial fractions. A good mathematician today knows that. But it was inimaginable for the analysts of Heaviside's time. As for talking of $\delta$ and its successive derivatives, it was not justifiable.

Heaviside's computations had an extraordinary effect in his time. The majority of mathematicians were adepts of rigor, and they refused to accept his system. He defended himself like the devil, but was vanquished. His system was completely rejected, even though it gave correct results. It is curious that a scientific community such as the community of mathematicians should completely reject a method just because it is not justified (even not justified at all, and it must be said, practically diabolical), while realizing perfectly that it was giving rise to correct results. Human nature is made this way. Heaviside was so demoralized by the failure of his theory that he died mentally unstable. Some professions generally considered not to be dangerous can unexpectedly become dangerous .... In the Middle Ages, he would probably have been burned at the stake.

Heaviside's symbolic calculus was justified later via the Laplace transform, by Norbert Wiener and Carson in 1926 and by Vanderpol in 1932, both long after Heaviside's death. Wiener also used convolutions of $C^\infty$ functions with compact support. The symbol of a function became its image under the Laplace transform. But this still did not justify $\delta$ and its derivatives. Anyway, distributions provide a much better justification of the whole theory. There is a distribution $\delta$, infinitely differentiable like every distribution, and the formulas $DY = \delta$ and $\delta * T = T$ and $\delta' * T = T'$ simply work in the sense of distributions.

## Dirac: new discoveries

Dirac also introduced the famous function which carries his name, in an article in 1926–27. Nobody remembered Heaviside in those days. But Dirac used his function for other applications. He gave an approximation of his function: for a very small number $\epsilon$, he considered a function equal to $1/2\epsilon$ in the interval $[-\epsilon, \epsilon]$. And he said that as $\epsilon$ tended to 0, the function $1/2\epsilon$ tended towards the Dirac function $\delta$. Indeed, in the theory of distributions, there is a topology, implying a notion of convergence, for which it is true that Dirac's $1/2\epsilon$ functions converge towards the Dirac distribution $\delta$. Dirac himself reasoned in the space $\mathbb{R}^N$. He also gave another approximation of the Dirac function, namely the bell curve $(1/\sqrt{2\pi\epsilon})^N e^{-|x|^2/2\epsilon}$. As $\epsilon$ tends to 0, this function converges towards the Dirac function $\delta$. And again, it turned out that in the sense of convergence for the topology on distributions, he was right.

So there were several possible definitions of the Dirac function, and physicists displayed plenty of imagination in finding them. All of this took a while to become absolutely correct; it was actually unjustifiable in its initial formulation, because the notion of limit used wasn't really defined.

Moreover, even if several functions depending on a parameter $\epsilon$ converged towards the Dirac function as $\epsilon$ tended to 0, it was still important to see why, and to define exactly what was meant by convergence, and this was not done. The Dirac distribution was actually justified well before the invention of distributions, as the Dirac *measure*. But like Heaviside, Dirac introduced the derivative of his function, and it is not possible to speak of the derivative of a measure. He simply took the Dirac function to be the limit of the Gauss functions given above, and defined its "derivative" to be the limit of their derivatives! It's not easy to represent the derivative defined in this way. It is true that it has value 0 every except at the origin, but it should also be 0 at the origin since it is an odd function. Here, again, the passage to the limit was later justified by the theory of distributions. But Dirac and all the other physicists who worked on it went much farther, and that is why the Dirac function was such a success. They made variable changes on their generalized functions.

For instance, they considered the function $\delta(u)$, where $u$ was the square of the hyperbolic Lorentz distance, $t^2 - x^2 - y^2 - z^2$; this has no meaning at all in function theory. But in relativistic quantum mechanics, not only these functions but several much more complicated ones were used in the theory of elementary particles and in relativistic quantum field theory. These formulas were all eventually justified by the theory of distributions, and they were reproved with precise computations on these singular functions by Méthée, a Swiss mathematician, who published his work a few years after my publication of distributions (1954). Thus, physicists lived in a fantastic universe which they knew how to manipulate admirably, and almost faultlessly, without ever being able to justify anything. In this way they made great advances in theoretical physics. They deserved the same reproaches as Heaviside: their computations were insane by standards of mathematical rigor, but they gave absolutely correct results, so one could think that a proper mathematical justification must exist. Almost no one looked for one, though.

I believe I heard of the Dirac function for the first time in my second year at the ENS. I remember taking a course, together with my friend Marrot, which absolutely disgusted us, but it is true that those formulas were so crazy from the mathematical point of view that there was simply no question of accepting them. It didn't even seem possible to conceive of a justification. These reflections date back to 1935, and in 1944, nine years later, I discovered distributions. The original reflections remained with me, and became part of the accumulated material I was talking about earlier, which remained in a corner of my mind, only to explode suddenly the night I discovered distributions, in November 1944. This at least can be deduced from the whole story: it's a good thing that theoretical physicists do not wait for mathematical justification before going ahead with their theories!

## Vibrating strings, harmonic functions

I had made certain remarks in hypotaupe and taupe, in 1933–34. The equation of vibrating strings is given by

$$\left(\frac{1}{v^2}\frac{\partial^2}{\partial t^2} - \frac{\partial^2}{\partial x^2}\right)u = 0,$$

where $v$ is the speed of propagation of waves. We had learned that the general solution of this equation is of the form $u(t,x) = f(x+vt) + g(x-vt)$, where $f$ and $g$ are functions of one variable. Naturally, this presupposes that $f$ and $g$ are $C^2$, so as to be able to differentiate them. What should one think of a function $u$ which would be analogous except that $f$ and $g$ would be merely $C^1$ (continuous), or not even continuous? Is it a wave or not? I was obsessed by this question for some time, then I stopped thinking about it and relegated the question to a corner of my mind for future reflection. I thought about the equation for vibrating strings, but not about the wave equation in several variables

$$\left(\frac{1}{v^2}\frac{\partial^2}{\partial t^2} - \Delta\right)u = 0.$$

Things were considerably more complicated in the case of several variables, but I knew nothing about it then.

The theory of harmonic and holomorphic functions is another necessary prerequisite for the theory of distributions, and which clearly foreshadowed them.

Since the nineteenth century, for instance since Gauss, it was known that a harmonic function can be defined by the fact that it is continuous in space, and that its average on every sphere is equal to its value at the center. Thus, the harmonicity of the function, which translates differentially into the fact that its Laplacian vanishes, can be defined by a property which does not involve the existence of its derivatives. And furthermore, one can deduce from this definition that it does have successive, continuous derivatives of every order. The same holds for holomorphic functions: the theorems of Cauchy and Morera say that a continuous function of the complex plane is holomorphic if and only if its path integral along every closed rectifiable path is zero. Holomorphy means that the function is differentiable with respect to the complex variable, and it can be characterized without reference to differentiation. Here also, a holomorphic function possesses successive derivatives of every order. In the theory of distributions, a continuous function $u$ is called harmonic or holomorphic if $\Delta u$ or $\left(\frac{\partial}{\partial x}+\mathrm{i}\frac{\partial}{\partial y}\right)u$ vanishes in the sense of distributions (Cauchy-Riemann relation). Also this does not necessitate the existence of derivatives in the usual sense, and as above, it is a consequence of this property that a distribution satisfying these equations is infinitely differentiable in the usual sense. In 1933–34, in taupe, I had studied harmonic and holomorphic functions, and they had attracted me in the same way as vibrating strings. But there is an essential difference between the two cases: the Laplacian is elliptic, so the solution will be $C^\infty$ in the usual sense, whereas the equation of vibrating strings is hyperbolic, and I saw that the solution I was looking for was not necessarily a differentiable function.

## Bochner's formal functions, Bochner's generalized solutions and Leray's weak solutions

Several other mathematicians introduced theories which were completely justified, and which came to constitute part of the theory of distributions. I am referring particularly to Bochner, Carleman, Sobolev, Leray and Wiener. In the last chapter of his famous 1932 book on Fourier integrals, Bochner introduced formal functions, which are finite sums of "formal" derivatives of products of square integrable functions with polynomials. Formal derivatives means that these functions are not necessarily differentiable in the usual sense.

Bochner first defined when two formal derivatives define the same generalized function (i.e. he established an equivalence relation). He then defined the computations one can perform on such generalized functions, and in particular multiplication and convolution, neither of which is always defined (he didn't attempt to clarify exactly when they were defined, except in some simple cases).

When these simple conditions are satisfied, the Fourier transform exchanges multiplication and convolution. These generalized functions are exactly what would later be called tempered distributions; tempered distributions actually already existed in 1932! Bochner himself did not attach undue importance to his chapter, which he put at the end of his book, without any applications, and never used again. I don't think anyone ever really noticed it. And a lot of things needed to be added before it could become really usable.

Firstly, the definition was only given in dimension 1. Dirac's generalized function was not mentioned, even though one can write it as the formal second derivative of the function $x \to (1/2)|x|$, which is the product of the square integrable function $|x|/2(1+x^2)$ by the polynomial $1+x^2$. The precise conditions of existence of the multiplication and the convolution were not given. No topology is put on the space of generalized functions. Among the applications of the theory, there are cases in which one actually wants to represent tempered distributions in this form, but it's fairly rare as it's difficult to use. Another time, Bochner was just an inch from introducing distributions. He needed to define a function $f$ on $\mathbb{R}^N$, which would be a generalized solution to a partial differential equation with constant coefficients of order $m$: $P(D)f = \sum_p a_p D^p f = 0$. He says that a function is such a solution if it is a uniform limit on every compact set of sufficiently differentiable functions $f_n$ which are ordinary solutions of the equation. So $\sum_{|p|\leq m} a_p D^p f_n$ is zero, and we can write symbolically that the sum $\sum_{|p|\leq m} a_p D^p f$ is zero, but one can't give a meaning to an individual term $a_p D^p f$.

These remarks were made in a little article isolated from his other discoveries, and published in 1946. He doesn't establish the relation between his generalized solutions and his generalized functions for the Fourier integral. That is a typical consequence of two isolated trickles of percolation, each of which pierces quite far, which actually give publishable (and published) results, but which stop before creating something really generally useful in analysis. Just like the trickle of water sometimes comes to a stop in the middle of the coffee grounds. The result is considered more as a mathematical curiosity than as a new theory. This is particularly obvious from the fact that the author introduces these new notions (at two independent times) without emphasizing them, without relating them, without using them anywhere else. Bochner was a rather touchy person and he was not particularly overjoyed by the distributions I introduced, and he even wrote a review of them somewhere in which he described them as merely "not bad". He also told everyone around him that I had found some obvious and some difficult things, but that he himself had already found all of the difficult ones. I wasn't upset by this, and although we rarely met, we remained on good terms. When I discovered distributions, I didn't know about Bochner's two publications (the second one actually appeared in 1946, after the discovery of distributions in 1944). I only noticed them afterwards, and of course cited them in my book. However, they did not help me solve any of my problems.

Jean Leray had done fundamental work on the equation of viscous liquid. He also introduced the notion of a weak solution of a partial differential equation of order 2. It is obtained simply by integrating by parts. This is also how it is obtained in the theory of distributions: a function $u$ is a weak solution of the equation $P(x, D)u = 0$ if for every $C^2$ function $\phi$ with compact support, we have the following symbolic formula for integration by parts:

$$\int\int\int (P(x,d)u)(x)\phi(x)\,\mathrm{d}x = \int\int\int u(x)\left({}^tP(x,D)\phi\right)(x)\,\mathrm{d}x = 0,$$

where ${}^tP$ is the transpose of $P$. I had taken Leray's Peccot course at the Collège de France in 1934–35, and I had been quite enthusiastic about it because it was very modern compared to everything we learned at the ENS. He solved my problem of vibrating strings, which satisfied me very much, but it had the same defect as Bochner's method: the symbolic result of the differential operator on the function $u$ is zero, but the partial derivatives of $u$ do not appear definable, giving rise to a new torment inside me which lasted until I discovered distributions in 1944.

As for Norbert Wiener, one must remark that in a 1926 article, in order to approximate a continuous function by a $C^\infty$ function, he "regularized" it by taking its convolution with a $C^\infty$ function with compact support. It was not an introduction of new mathematical objects, but more an audacious new utilization of $C^\infty$ functions with compact support, for purposes of regularization, which was completely forgotten until 1944 when I introduced something similar, knowing nothing about what Wiener had done.

I should add that others, for instance Hilbert-Courant, Friedrichs, Krylov and probably a few others, introduced generalized derivatives of a type similar to Leray's.

## Sobolev's functionals

One of the precursors who came closest to distributions was the Soviet mathematician Sobolev. In an article which appeared in 1936, he defined functionals, which are finite order distributions. Instead of using the space $\mathfrak{D}$ of $C^\infty$ functions with compact support, he used the space $\mathfrak{D}^m$ of $C^m$ functions with compact support, equipped with the usual pseudo-topology. He considered the dual $\mathcal{D}'^m$ of $\mathcal{D}^m$, which he called the space of functionals of order $m$. He did not succeed in eliminating the index $m$, which complicates his article considerably. He never considers the intersection $\mathcal{D}$ of all the $\mathcal{D}^m$, nor the union $\mathcal{D}'_F$ of the $\mathcal{D}'^m$, which is the space of finite order distributions. He defines the topology on $\mathcal{D}'^m$ as the $*$-weak dual topology of $\mathcal{D}^m$, so that he cannot say that functionals converge without giving the order $m$ precisely. It's all the more surprising that he succeeded in showing that $\mathcal{D}^n$ is dense in $\mathcal{D}^m$ for all $n \geq m$, as I later did myself with $\mathcal{D}$ instead of $\mathcal{D}^n$, by regularizing the "canonical" sequence of functions with compact support, which are precisely $C^\infty$! This is also what Wiener did in 1926. Then, by transposition, he is able to multiply a functional of order $m$ by a $C^m$ function and to differentiate every functional of order $m$ to obtain a functional of order $m + 1$, and he gives Leibniz' rule for differentiating a product. He doesn't define $\delta$, and doesn't establish any relations with physics. Convolution of functionals is not defined, and he didn't write anything on the Fourier transform. There are no topological properties and no finiteness theorems; he always remained in the case of order $m$. Moreover, he did not develop his theory in view of general applications, but with a precise goal: he wanted to define the generalized solution of a partial differential equation with a second term and initial conditions. He includes the initial conditions in the second term in the form of functionals on the boundary and obtains in this way a remarkable theorem on hyperbolic second order partial differential equations. Even today this remains one of the most beautiful applications of the theory of distributions, and he found it in a rigorous manner. The astounding thing is that he stopped at this point. His 1936 article, written in French, is entitled "Nouvelle méthode à résoudre le problème de Cauchy pour les équations linéaires hyperboliques normales." After this article, he did nothing further in this fertile direction. In other words, Sobolev himself did not fully understand the importance of his discovery. He used it to solve a particular and very important problem which interested him, but he made no attempt to propel his invention in much more general directions, and to turn it into a powerful tool. He himself did not realize the extent of the possible applications. His article appeared just before the war, at a difficult time, but he could have continued after the war. The theory of generalized functions was published by Gel'fand and Shilov only in 1955, four years after the appearance of my book on distributions. Sobolev had taken a few steps on the right road, and then had been forgotten, like one of those percolation trickles which seems to start out so well. Naturally, I knew absolutely nothing about his work, since after I left the ENS I left for my military service and for the war. We became good friends after the war. I remember that he telephoned me once, on his way through Paris, asking me unexpectedly "Guess who this is?" I immediately answered "Sobolev", since one can often recognize people by their voice. We had many discussions about all sorts of things, for instance life in the Soviet Union and the disasters caused by Lyssenko under Stalin. He never made any claim to the full theory of distributions. He must have regretted not pushing his theory further than the initial discovery, but he did not do it even after the war. He did, however, introduce $H^m$ spaces, for integers $m \geq 0$, which are defined by differentiation in the sense of functionals. He studied them deeply and they became essential tools in the study of partial differential equations. He never discussed the case $m < 0$. The mathematicians of the Courant Institute of New York attribute the introduction of this case to me. I find this difficult to believe, but it really does appear that Sobolev simply never mentioned them, and perhaps he was not sufficiently equipped to deal with them. Indeed, he must have considered that the dual of the Hilbertian space $H^m$ was $H^m$ itself, and he wasn't ready to consider the dual $H^{-m}$ of $H^m$ inside $\mathcal{D}'^m$, which exists because the injection of $\mathcal{D}^m$ into $H^m$ is continuous and dense; he also was probably not ready to describe the structure of the elements of $H^{-m}$. I studied the duals of subspaces of $\mathcal{D}'$ embedded in $\mathcal{D}'$ extensively, for example the spaces $\mathcal{S}$, $\mathcal{S}'$, $\mathcal{O}_M$ and $\mathcal{O}'_C$.

## Carleman's generalized functions

Another important method was introduced by Carleman, in an article which appeared in Uppsala in 1944, entitled "The Fourier integral and questions attached to it". This article is exactly contemporary with distributions, and I had no idea of its contents. There are functions $f$ on $\mathbb{R}$ which are differences of two holomorphic functions $F^+$ and $F^-$, respectively on the open upper half-plane of $\mathbb{C}$ (i.e. the set of $z = x+iy$ with $y > 0$) and the open lower half-plane of $\mathbb{C}$ (i.e. the set of $z = x+iy$, $y < 0$). For example, if $f(x) = x$, one can set $F^+(z) = z$ and $F^-(z) = 0$, and for $z \in \mathbb{R}$, we certainly have $f(z) = F^+(z) - F^-(z) = z$. If $F^+(z)$ and $F^-(z)$ do not have limits as $y$ tends to 0, we can say that $F^+ - F^-$ formally represents a generalized function $f$. In the modern theory of hyperfunctions, an important generalization of distributions worked out by Mikio Sato (1959), one can take arbitrary functions $F^+$ and $F^-$. In 1944, the year of the invention of distributions, Torsten Carleman restricted himself prudently to the case where $F^+$ and $F^-$ are slowly increasing in a neighborhood of every point of the real axis; in this way, he obtained distributions. For example, if $F^+$ and $F^-$ are both equal to $1/z$, we find that

$$\lim_{y\to 0, y>0}\frac{1}{z} \quad - \quad \lim_{y\to 0, y<0}\frac{1}{z} = -2\pi i\delta.$$

Mathematicians who work on hyperfunctions (and specialists in partial differential equations work on them more and more) handle such formulas very well, but considering the function $\delta$ as the pair $\left(\frac{-1}{2\pi iz}, \frac{-1}{2\pi iz}\right)$ does not come naturally to everybody. Furthermore, as in Bochner's work, one needs an equivalence relation; if you add a function which is holomorphic on all of $\mathbb{C}$ to both $F^+$ and $F^-$, the difference obviously remains the same. Multiplication (only by an analytic function), convolution, topology and the Fourier transform are all rather delicate to express in this formulation.

I knew nothing of all the work I have just described, except for harmonic functions, Leray's course, Dirac's $\delta$ function and the equation of vibrating strings. But the enumeration of my many precursors is impressive, and the dates of their inventions are closer and closer to 1944. The discovery was bound to come, and I was merely an "instrument of destiny" (one must obviously have suitable ability to become such an instrument!). Further on, I will discuss the question of why destiny fixed itself upon me.

## Finite parts of divergent integrals

One of the main influences on the creation of distributions comes from Jacques Hadamard's theory of finite parts of divergent integrals. I said that while I was at the ENS, I discovered these finite parts for myself, in dimension 1, and that I learned from Hadamard that he had studied them completely in all dimensions. My own interest was in analytic continuations, but Hadamard was interested in second order hyperbolic partial differential equations. Hadamard's theory in several dimensions went much farther than I had, and I studied it passionately. It was a tool which was all ready for use, although I was far from understanding how it joined up with those I have previously described. Like everyone who studies partial differential equations, Hadamard introduced the elementary or fundamental solution of such an equation: $E$ is a fundamental solution if its transform by the operator vanishes, but only in an open set of the space, leaving aside an exceptional set which does not necessarily consist only in the origin. Elementary solutions are very difficult to find, but even more difficult to define. Hadamard did not define them precisely since they are often distributions and not functions, but he gave a complete solution of equations whose second term and initial conditions involve integrals or finite parts of divergent integrals. The general concept of an elementary solution could not, however, be deduced from this work. It was much more explicit in the work of Marcel Riesz, whom I met in Lund in 1947, or in the work of other Swedish mathematicians such as Zeilon (already in 1911), and Belgian mathematicians such as Florent Bureau. Although the different authors of theories of partial differential equations could always understand each other, it must be said that the true concept of an elementary solution was drowned in an inextricable mess. I didn't know anything about this, but I did know about finite parts.

## Georges de Rham's currents

I was also influenced by my meeting with Georges de Rham, in the autumn of 1942. I was in Clermont-Ferrand, in a period of mathematical expansion, learning about differential forms and their integrals from de Possel, when de Rham came to talk about harmonic differential forms. It was a very beautiful and coherent theory. He then introduced currents, a generalization of differential forms: he defined a current $T$ of degree $p$ on a manifold of dimension $N$ to be a sum $\sum_i \eta_i \Gamma_i + \omega$, where the $\eta_i$ are continuous differential forms of degree $k_i$, the $\Gamma_i$ are $C^1$ closed oriented submanifolds with boundaries, of dimension $N-p+k_i$, and $\omega$ is a continuous differential form of degree $p$. The currents of degree $p$ obviously form a vector space. We can multiply a current of degree $p$ on the left by a continuous differential form $\alpha$ of degree $q$, which gives the new current $\sum_i (\alpha \wedge \eta_i)\Gamma_i + \alpha \wedge \omega$, of degree $p+q$. We can also multiply on the right by $\alpha$, but then one obtains alternating signs $\pm 1$. Note that $\Gamma$, a submanifold as above of dimension $N-m$, is a current of degree $m$; thus every current $T$ of degree $m$ is of dimension $N-m$. One can extend the submanifolds by singular chains. Moreover, de Rham defined a coboundary operator $d$; for the current $T$ above, if the $\eta_i$ and $\omega$ are $C^1$, he set

$$dT = \sum_i d\eta_i \Gamma_i + (-1)^{p-1} \sum_i \eta_i b\Gamma_i + d\omega,$$

where $d\omega$ is the exterior differential of $\omega$ and $b\Gamma_i$ is the oriented boundary of $\Gamma_i$. Thus for $\Gamma$ of dimension $N-m$, we have $d\Gamma = (-1)^{m-1} b\Gamma$. For every current $T$ of degree $m$, he defined his boundary by $bT = (-1)^{m-1} dT$. The goal of all this was to establish a cohomological theory generalizing the one he had introduced in his thesis in 1936: the coboundary establishes real cohomological vector spaces; the cohomology of degree $m$ of currents for $d$ is the same as the cohomology of differential forms of degree $m$ for $d$, and so is the homology of chains of dimension $N-m$ for $d = (-1)^{m-1} b$. I was very excited by de Rham's talks. I knew nothing about harmonic forms but they later became an important part of my work; on the other hand, I was quite familiar with the homology of singular chains and the cohomology of differential forms (since my stay in Boisséjour, in 1941).

After the talks, we went to the Marquise de Sévigné, a beautiful and traditional tea-room in Royat, near Clermont-Ferrand, which specialized in chocolate. Thanks to very high prices, the tea-room had preserved its atmosphere of cottony excellence. In that period of restrictions, a delicious cup of hot chocolate was extremely favorable to ethereal discussions. I still remember its voluptuous taste. Cartan was in Clermont for a Bourbaki meeting, and he came with us. De Rham told us that he was hoping for a more general theory, because his present theory of currents was visibly defective. His space of currents had no natural topology, whereas all the vector spaces useful in analysis come equipped with a topology, in which they are usually complete, and also with a useful weak topology. He covetously observed the space of Radon measures on $\mathbb{R}$; it has a norm convergence and a vague, or $*$-weak convergence which serves all kinds of purposes and which, like him, I knew well and worked with often. A measure admits a unique decomposition into the sum of a measure with Lebesgue density (analogous to the term $\omega$ in a current), and a singular measure, supported on a Lebesgue-negligible set, but much more general than the $\Gamma_i$ of a current. He had not managed to obtain any results of this kind. And yet, considering his deep knowledge of differential forms, it was not complicated. That is the secret of discoveries. As long as they have not been made, they appear inaccessible. We thought about it together. He made this marvelous prediction: "It's not for us, it will be found by the next generation." It was in 1942 – the "next generation" came in 1944! This conversation with de Rham left me with a beautiful and stimulating memory. I kept the ideas we discussed inside me, they connected with Dirac's $\delta$ function and its derivative, and later inspired me. I should mention that a 1943 article by Gillis contained similar ideas.

## Duality in topological vector spaces

The latest of the many steps which acted as precursors to my discovery was a piece of work of my own, in 1943, on duality in locally convex topological vector spaces. I said that I stopped working on mathematics during the war, as soon as we left Clermont in 1943. And it is true, except for one short period, lasting just two or three weeks, in Saint-Pierre-de-Paladru, during the period when I did not need to leave every day to pretend to go to the Scientific University of Grenoble, and before I had my teaching job in Monestier-de-Clermont. I expressly limited this period of research to a couple of weeks, for the reasons of caution I explained earlier. Locally convex topological vector spaces had been recently studied with great precision by G.W. Mackey in the US, but naturally, I knew nothing of his results. I had no books, so I had to work out for myself a theory which did not actually end up penetrating as deeply as his. I studied the dual of the vector space and put on it a $*$-strong topology and a $*$-weak topology, introducing the notion of a bounded set in a topological vector space, and the Hahn-Banach theorem which I had learned from Dieudonné's course which Marie-Hélène had taken for us in Clermont; I also introduced the Banach-Steinhaus theorem. I considered what remained of these theorems in general topological vector spaces, and examined the bidual and reflexivity. I found some interesting theorems, and published some of the results in a book called *Théorie des espaces $\mathscr{F}$ et $\mathscr{LF}$*, written together with Dieudonné in 1949, when we found ourselves together in Nancy after the war; of course we referred to Mackey. As an example of duality in 1943, I had taken the topological vector space $E$ given by the space $C^\infty([0,1])$ of $C^\infty$ functions on the interval $[0,1]$. It is not far from the space $\mathscr{D} = C^\infty_{\text{comp}}$ of $C^\infty$ functions with compact support on $\mathbb{R}$, but it is of course much simpler because it is a Fréchet space, with no inductive limit. I examined its dual $E'$ which, contrarily, is much more complicated than the space $\mathscr{D}'$ of distributions on $\mathbb{R}$, since it is the space of distributions on the closed segment $[0,1]$, or of distributions on $\mathbb{R}$ with support inside $[0,1]$. I did not give any other name than $E'$ to this space. I observed that the space and its dual were both reflexive and satisfied divers other properties which I considered in a general manner. I was not particularly inspired by this dual space $E'$. And yet, it was a future space of distributions. The idea which escaped me at that time was to transpose to $E'$ the differentiations which already existed for the space $E$. This would have given the future differentiation of distributions. But I had no inspiration of the kind. I even remember thinking "This space $E'$ will probably never be useful for anything". All of that was to change less than a year later. My topological studies continued to work within me, and thus when I did discover distributions, I was immediately able to find all of their important topological properties, without which they would certainly not have been able to play such an important role.

Thus, I was in possession of a certain number of mathematical tools all ready to be used. And yet, I was infinitely far from realizing that they all formed parts of a single, coherent theory. Let me recapitulate them: the Dirac function and its derivative, generalized solutions to partial differential equations, finite parts of divergent integrals, de Rham currents, and finally, duality theorems in topological vector spaces. I must add certain other very positive elements: the definition of a Radon measure $\mu$ on a locally compact space via a continuous linear form $\phi \to \mu(\phi)$ on the space of continuous functions with compact support, and convolution, notions which I juggled with constantly after having read Weil's book. But I knew nothing of my predecessors Heaviside, Bochner,

Carleman and Sobolev, who had no influence on me. I believe that Dirac's $\delta$ function, generalized solutions and currents led in a natural way to the final result. Finite parts and topological vector spaces were extras, which led by sheer coincidence to a considerable amelioration of the result, and allowed me to take it much farther than my predecessors.

## The final catalyst: an article by Choquet-Deny

I began to do mathematics again only after the liberation, in October 1944 in Paris. In a very short time, I was able to relearn everything I had known before. Finally, I came upon a 23-page article by Choquet and Deny called "Sur quelques propriétés des moyennes caractéristiques des fonctions harmoniques et polyharmoniques", dating from October 1944. One of the goals of this article concerned continuous functions on $\mathbb{R}^N$, which have the property that the set of linear combinations of their transforms by all Euclidean transformations of the space $\mathbb{R}^N$ is not dense in the full space of continuous functions on $\mathbb{R}^N$. They found that these functions are exactly those which are polyharmonic, i.e. which vanish after sufficient iterations of a suitable Laplacian. As these functions are not a priori differentiable, they introduced a definition of polyharmonic functions by means of iterated averages, as one can define harmonic functions by spherical averages, without supposing differentiability. After recent work of Brelot, such definitions were in the air. I immediately thought how to generalize their article: instead of taking similitudes, I could take only translations and homotheties. But then the iterated Laplacian had to be replaced by an arbitrary differential operator with constant coefficients. The functions $f$ having the property that all of their transformations by translation and homothety failed to generate the entire space are exactly those functions which are generalized solutions to a certain equation with constant coefficients, $P(D)f = \sum_p a_p D^p f = 0$. But which were the generalized solutions here? Unlike the polyharmonic case, they did not turn out to be usual solutions. The solutions which necessity led me to were more Bochner's solutions than Leray's. If the equation is of order $m$ (the case $\sum_p$ for $p \leq m$), a generalized solution is a continuous function $f$ such that for every $C^m$ function $\phi$ with compact support, $f * \phi$, which is also $C^m$, is a real solution of the equation. This was Bochner's former solution, which Choquet and Deny knew of no more than I did. I perfected the definition by replacing a sufficiently differentiable $\phi$ by an infinitely differentiable $\phi$ with compact support. I had some scruples about publishing this result, which was a really interesting generalization of the work of Choquet and Deny, because Deny was very young and had not yet completed his thesis, and I confided to Cartan that

I hesitated to work on the same line of thought as him. Cartan was convinced that there was no problem, and that he could continue to reflect for himself on theorems which used the Laplacian in a fundamental way, and that is what he did. My article was only four pages long, and was written very quickly. Cartan objected. "Listen," he told me, "you mustn't try to work with these $C^\infty$ functions with compact support, they're monstrous." And indeed, at the extremities of the support, the $C^\infty$ function with compact support is certainly not analytic since it is not identically zero but all its successive derivatives are zero at these points. "Maybe," I said, "but it's just the right way to talk about generalized solutions of partial differential equations." He admitted this, but recommended prudence, and he was right. I discovered later that Norbert Wiener had himself used a regularization by convoluting with $C^\infty$ functions with compact support, in an article dating from 1926, as had Sobolev in his 1936 article.

I was vexed, as I have said, by the fact that in the generalized solutions of partial differential equations, for example in the vibrating string equation, one could define $\sum_p a_p D^p f = 0$ without giving any meaning to each of the terms $D^p f$. I now felt the same sense of frustration, but it ceased to be a formal frustration: I felt the need to solve this shocking problem. I still don't really understand how the sudden click took place – it was a matter of minutes or even seconds – because it was necessary to realize that $D^p f$ was not itself a function, but a more general being. It was necessary to do what Peano had said to do in 1912, but had not done himself (and which I had never heard of). To generalize these functions, it was necessary to overcome a powerful inhibition. But like Peano, I knew by heart the generalization of the rationals to the reals! Perhaps the true question is – why did I not discover distributions earlier?

## The click or the successful percolation: operators

In order to understand what follows, we need to introduce two new mathematical beings. A distribution will be a linear form $T$ on the space $\mathscr{D} = C^\infty_{\text{comp}}$ of $C^\infty$ functions with compact support on $\mathbb{R}^N$, i.e. a linear map from $\mathscr{D}$ into the scalar field $\mathbb{R}$ or $\mathbb{C}$, denoted $\phi \mapsto T(\phi)$ or $\langle T, \phi \rangle$. An operator $\mathbf{T}$ will be a linear map from $\mathscr{D}$ into the space $\mathscr{E}$ of $C^\infty$ functions with arbitrary support, denoted $\phi \mapsto \mathbf{T} \cdot \phi$, commuting with regularizations, i.e. with convolutions with $C^\infty$ functions with compact support. If $\theta \in \mathscr{D}$, we must have $(\mathbf{T} \cdot \phi) * \theta = T \cdot (\phi * \theta)$. We also require that $T$ and $\mathbf{T}$ be continuous, in the sense of topology on $\mathscr{D}$ and $\mathscr{E}$. Actually, I was unable to put a topology on $\mathscr{D}$, but only what I called a pseudo-topology, i.e. a sequence $(\phi_n)$ converges to 0 in $\mathscr{D}$ if the $\phi_n$ and all their derivatives converge uniformly to 0, keeping all their supports in a fixed compact set. I only found an adequate topology much later, in Nancy in 1946. But it doesn't matter for the main properties. If we call $\mathcal{D}_K$ the subspace of $\mathcal{D}$ formed by the functions having support in a fixed compact subset $K$ of $\mathbb{R}^N$, it comes down to putting the topology of a Fréchet space on $\mathcal{D}_K$, as I had on $\mathbb{C}[0,1]$ in my reflections at Saint-Pierre-de-Paladru in 1943: uniform convergence of the functions and all their derivatives. One says that functions of $\mathcal{D}$ converge to 0 if their supports lie in a fixed compact set $K$ and they converge to 0 in $\mathcal{D}_K$. In contrast, the topology on $\mathcal{E}$ is very easy: it is that of uniform convergence of functions and all of their derivatives on every compact subset of $\mathbb{R}^N$. The continuity we require of $T$ and $\mathbf{T}$ is the following: if $\phi_n$ converges to 0 in $\mathcal{D}$ for the pseudo-topology, the $\langle T, \phi_n \rangle$ converge to 0 in $\mathbb{R}$ or $\mathbb{C}$, and the $\mathbf{T} \cdot \phi_n$ converge to 0 in $\mathcal{E}$.

The spark shot forth one night in early November 1944 – I no longer remember exactly which. To find generalized solutions of partial differential equations, it was necessary to generalize the notion of function! And I immediately found how to generalize it; the very notion which Peano had vainly searched for in 1912. Let $f$ be a function, say a continuous function. Its regularizations are always $C^\infty$ functions. And the map $\phi \mapsto f * \phi$ is a continuous linear map from $\mathcal{D}$ to $\mathcal{E}$, which commutes with regularizations, so it defines an operator $f$ in the sense above, with $f \cdot \phi = f * \phi$. So I had found a space generalizing the space of functions: namely, the space of operators. In fact, operators generalize not only continuous functions but Lebesgue classes of locally integrable functions. I quickly understood that I had come upon everything I had been searching for for more than ten years. The so-called Dirac function was really the Dirac operator $\phi \mapsto \delta \cdot \phi = \phi$, i.e. it was simply the identity operator (coming from the fact that $\mathcal{D}$ is embedded in $\mathcal{E}$; this is exactly what Heaviside had discovered). But moreover, thanks to properties of convolution which had already been used for decades and even noticed by Heaviside, it was well-known that the derivative of a convolution product of two $C^1$ functions is obtained by differentiating either one of the two functions. Consequently, one can immediately define the derivative of an arbitrary operator $\mathbf{T}$: in dimension 1, the derivative $\mathbf{T}'$ is defined by $\mathbf{T}' \cdot \phi = (\mathbf{T} \cdot \phi)' = \mathbf{T} \cdot \phi'$, and in $\mathbb{R}^N$, we have $D^p\mathbf{T} \cdot \phi = D^p(\mathbf{T} \cdot \phi) = \mathbf{T} \cdot D^p\phi$. The derivatives of the Dirac operator are now obvious: we have $D^p\delta \cdot \phi = D^p\phi$, so $D^p\delta$ is simply the differential operator $D^p$ of $\mathcal{D}$ in $\mathcal{E}$. Every operator thus became infinitely differentiable, and it was possible to invert the order of differentiation (the Schwarz theorem). Functions were operators, but there were many operators which were not functions, such as $\delta$ and $\delta'$. The mystery of the Dirac $\delta$ function and its derivatives was solved.

If I considered the partial differential equation with constant coefficients

$P(D)f = \sum_p a_p D^p f = 0$, and if a function $f$ was a generalized solution of this equation, then it could be considered as an operator, and the sum of these operators was the zero operator. The difficulty of non-differentiable solutions to the wave equation was solved once and for all: for a generalized solution of the form $u(t,x) = f(x+vt)+g(x-vt)$, with $f$ and $g$ merely continuous, the second order derivatives of the two functions $(x,t) \mapsto f(x+vt)$ and $(x,t) \mapsto g(x-vt)$ did not exist in the usual sense but they did exist in the sense of operators. Everything was solved.

Later, finite parts of divergent integrals defined new operators, and thus gave a much greater value to what I had discovered while at the ENS, and which Hadamard had discovered previously. If $f$ is a function of the form $|x-a|^\alpha g(x)$, for a regular function $g$, one could define a "finite part" operator FP $f$ by (FP $f$)$\cdot \phi = \text{FP} \int f(x-y)\phi(y)\,dy$. This made it possible to introduce finite parts of divergent integrals exactly where Hadamard had introduced them, in the theory of partial differential equations. Moreover, since I was dealing with topological vector spaces, it was immediately possible to apply to them what I had worked out earlier in Saint-Pierre-de-Paladru, in the summer of 1943. I did not have a topology on $\mathcal{D}$, but what I called a pseudo-topology, and I had a very natural topology on $\mathcal{E}$; I could speak without difficulty of a bounded subset of $\mathcal{D}$ (bounded in a $\mathcal{D}_K$) and a bounded subset of $\mathcal{E}$. This made it possible for me to say that a set $B$ is bounded in the space $\mathcal{L}(\mathcal{D};\mathcal{E})$ of operators if it takes every bounded subset of $\mathcal{D}$ to a bounded subset of $\mathcal{E}$. This corresponds to the notion of strong boundedness, boundedness in $\mathcal{L}_b(\mathcal{D};\mathcal{E})$, whereas the simple notion of boundedness in $\mathcal{L}_s(\mathcal{D};\mathcal{E})$ signifies that the image of every element of $\mathcal{D}$ is bounded in $\mathcal{E}$. I had studied precisely these notions of simple and strong boundedness in Saint-Pierre-de-Paladru. $\mathcal{D}$ was more or less one of the spaces I had studied deeply during that short period, always with the slight difficulty of the pseudo-topology, which nevertheless did not stop me. I was well-equipped to study the topological properties of operators.

I then flung myself into the problem of defining the support of an operator. A continuous function has a support, the smallest closed set outside of which it is zero. So an operator must have a support. Shortly before, Cartan had introduced the "closed kernel of masses" for a Radon measure; I followed him. At that time, I used the term "kernel" for what is now called the support. But I could not simply copy Cartan's idea, because a Radon measure is a linear form, and an operator is not. This is not too important, yet I might as well prove it here since it is not actually proved anywhere in this form. Let an operator **T** be said to vanish on an open set $U$ of $\mathbb{R}^N$ if for every compact set $K$ contained in $U$, there exists a neighborhood $V$ of the origin in $\mathbb{R}^N$ such that for every function

$\phi$ with support in $V$, the function $\mathbf{T} \cdot \phi$ vanishes on $K$. We must prove that if $\mathbf{T}$ vanishes on a family of open sets $(\mathcal{U}_i)_{i \in I}$ of $\mathbb{R}^N$, then it vanishes on the union $\mathcal{U} = \cup_i \mathcal{U}_i$; then there must exist a largest open set on which $\mathbf{T}$ vanishes, and its complement is by definition the support of $\mathbf{T}$. But this is easy to see. Indeed, let $K$ be a compact subset of $\mathcal{U}$. There exists a finite subset $J$ of $I$ such that $\mathcal{U}' = \cup_{i \in J} \mathcal{U}_i$ covers $K$. Since a compact set $K$ is normal, there exists an open cover of $K$ finer than the covering given by the $\mathcal{U}_i \cap K$: let us call it $(W_i)_{i \in J}$. We have $\cup_{i \in J} W_i = K$ and $\overline{W}_i \subset \mathcal{U}_i \cap K$. But $\mathbf{T}$ vanishes in $\mathcal{U}_i$, and $\overline{W}_i$ is a compact subset of $\mathcal{U}_i$, so there exists a neighborhood $\mathcal{V}_i$ of 0 in $\mathbb{R}^N$ such that for all $\phi$ with support contained in $\mathcal{V}_i$, $\mathbf{T} \cdot \phi$ vanishes on $\overline{W}_i$. Thus, if the support of $\phi$ lies in the intersection of the $\mathcal{V}_i$ for $i \in I$, we have $\mathbf{T} \cdot \phi = 0$ on $K$, and this proves the result. This proof is simpler than the proof of the analogous result for distributions, for which it is necessary to use a $C^\infty$ partition of unity. But for most of the other properties of the support, distributions are simpler than operators. One can give a good property of supports: if the support of $\mathbf{T}$ is contained in a closed set $F$, and if $B_\epsilon$ is the ball centered at the origin of radius $\epsilon$, and if the support of $\phi$ is in $B_\epsilon$, then $\mathbf{T} \cdot \phi$ vanishes on the complement of $F + B_\epsilon$. One deduces immediately from this that the space of operators with support contained in $F$ is closed in the space of operators, which is itself closed in $\mathcal{L}_s(\mathcal{D}; \mathcal{E})$, so in $\mathcal{L}_b(\mathcal{D}; \mathcal{E})$.

The definition of the convolution product of operators is particularly simple because the operators are convolution operators. If $\mathbf{S}$ and $\mathbf{T}$ are operators, and $\mathbf{T}$ has compact support, we define the convolution product $\mathbf{S} * \mathbf{T}$ by $(\mathbf{S} * \mathbf{T}) \cdot \phi = \mathbf{S} \cdot (\mathbf{T} \cdot \phi)$, which is possible because $\mathbf{T} \cdot \phi$ is a $C^\infty$ function with compact support, so it belongs to $\mathcal{D}$. All the usual properties of convolution can be easily deduced from this, as can those of regularization; the operators really are convolution operators, satisfying $\mathbf{T} * \phi = \mathbf{T} \cdot \phi$.

## The most beautiful night of my life

I always called the night of my discovery a marvelous night, the most beautiful night of my life. In my youth, I used to have insomnias lasting several hours, and never took sleeping pills. I remained in my bed, the light off, and without writing anything, I did mathematics. My inventive energy was redoubled in those hours, and I advanced rapidly without tiring. I felt entirely free, without any of the brakes imposed by daily realities and writing. After some hours, I would end up getting tired after all, especially if an unexpected difficulty came up. Then I would stop and sleep until morning. I would be tired but happy for the whole of the following day; sometimes it took days for me to get back into shape. On this particular night, I felt sure of myself and filled with a sense of exaltation. In this kind of circumstance, I lost no time in rushing to explain everything in detail to Cartan, who as I mentioned earlier, lived next door. He was enthusiastic. "There you are, you've just resolved all the difficulties of differentiation! Now, we'll never again have functions without derivatives," he told me. If a function has no (Weierstrass) derivative, then this simply means that its derivatives are operators, but not functions.

Distributions, and thus operators, possess one truly essential property: on every relatively compact open set, every operator is a finite sum of derivatives (in the sense of operators, of course) of continuous functions. This finiteness theorem resembles many others in the theory. I gave several proofs of it in my book on distributions. I did not, however, succeed in proving finiteness theorems of this type, whose existence indeed I did not even suspect, until several months later, in Grenoble.

In order to generalize the operation of multiplication of two functions, it was necessary to multiply an operator by an infinitely differentiable function. Because generalized functions are defined by convolution, the operation of multiplication becomes rather difficult, since multiplication and convolution do not commute. I did find a definition, but it was quite involved. Here it is. Let $f$ be a continuous function defining the associated operator $\mathbf{f}$, with $(\mathbf{f}\cdot\phi) = \int f(x-y)\phi(y)\,dy$. If $u$ is a $C^\infty$ function, the function $uf$ defines a new operator $\mathbf{g}$, by $(\mathbf{g}\cdot\phi) = \int u(x-y)f(x-y)\phi(y)\,dy$. If $u$ is a function on $\mathbb{R}^N$, let $\check{u}$ denote its transformation by the symmetry whose center is the origin, i.e. $\check{u}(x) = u(-x)$, and let $\tau_a u$ denote the transformation of $u$ by the translation $a$ of $\mathbb{R}^N$, i.e. $\tau_a u(x) = u(x-a)$. Then

$$\begin{aligned}(\mathbf{g}\cdot\phi)(x) &= \int \check{u}(y-x)f(x-y)\phi(y)\,dy \\ &= \int f(x-y)(\tau_x u\check{u})(y)\phi(y)\,dy \\ &= \big(\mathbf{f}\cdot((\tau_x\check{u})\phi)\big)(x).\end{aligned}$$

If $\mathbf{S}$ is an operator, define the multiplicative product $\mathbf{T} = u\mathbf{S}$ by the same formula

$$(\mathbf{T}\cdot\phi)(x) = \big(\mathbf{S}\cdot((\tau_x\check{u})\phi)\big)(x).$$

One easily sees that the $\mathbf{T}$ defined this way is indeed an operator. This formula is not beautiful, but what could I do about it? It left me with a bitter taste.

## After three months, finally distributions

I then tried to define Fourier transforms of operators, and that is where I encountered a total obstacle. The very special role of convolution in the definition of operators leaves absolutely no room for a Fourier transform. So I remained with a completely open problem, and a lingering feeling of dissatisfaction with the multiplication formula for $u\mathbf{T}$ above. I left for Grenoble in 1944 and continued feverishly to study operators. More particularly, I searched for a definition of the Fourier transform, but in vain. Suddenly, I had the idea that I should have had from the beginning: it is really unacceptable that I didn't have it earlier! If I'm telling this whole story in such detail, it is not in order to display my personal imbecility, but the defects of thought, because I'm sure I'm not the only one to make such mistakes. Furthermore, I had already told my ideas to Cartan and other members of Bourbaki. When circumstances lead one to formulate new objects, even if they look very promising, the first definition one finds is not always the best, and yet at that stage one cannot possibly imagine any other. One must be able to accept a revolution in one's ideas, and to start again from the beginning, rediscover all the properties one has already discovered and then penetrate even further, sometimes months after the initial discovery. It can happen that such a revolution needs to take place several times. Maybe it's necessary to be young for this to be able to happen, although I don't really think so, as I myself have had this experience many times, at all ages, which makes me think it's more a question of character. The discovery of almost every theorem proceeds in zigzags, and the final result is often quite near the starting point. When I obtain a new result, I then search for the most direct road leading there; others are contented with the result alone and publish their zigzags, it's a question of temperament. However, in the case of distributions, I was not pushed to change my definition by natural purism, but by the obstacles which arose in my path using the initial definition. I understood convolution perfectly; thanks to André Weil's book, I swam among convolutions like a fish in water. But I was also familiar with duality, since Weil defined Radon measures on a locally compact group $G$ as linear forms on the space $C_{\text{comp}}(G)$, equipped with the pseudo-topology defined by the $C_K$ for compact subsets $K$ of $G$. Better, I had studied duality of topological vector spaces in Clermont, from the course taught by Dieudonné and taken by Marie-Hélène, and this notion was the basis of my thesis; in a few weeks in Saint-Pierre-de-Paladru in 1943, I had even rendered duality general! Why on earth, then, in order to generalize functions, did I think of operators and of the space $\mathcal{L}_b(\mathcal{D};\mathcal{E})$, instead of duality and the spae $\mathcal{D}'_b$? It was enough to take Radon measures, and replace $\phi \in C_{\text{comp}}$ and $\mu$ a

Radon measure, with the scalar product $\mu(\phi) = \langle \mu, \phi \rangle$, by $\phi \in C^{\infty}_{\text{comp}} = \mathscr{D}$, $T$ a distribution in $\mathscr{D}'$, and the scalar product $T(\phi) = \langle T, \phi \rangle$. Just a little substitution was all that was needed. I could have generalized functions by distributions just as well as by operators.

Thus, it was the ideas of Dieudonné and Weil, which had already played a fundamental role in my thesis, which also essentially inspired the invention of distributions. Dieudonné introduced the importance of duality in topological vector spaces into my work, and André Weil introduced both duality and convolution on a locally compact abelian group, by his 1940 book. I already recounted how André Weil and Henri Cartan were the two first founders of Bourbaki. The influence of André Weil on French and international mathematics was considerable. Whether or not he is aware of it, every French mathematician owes something to Weil. Weil emigrated to the United States during the war, and spend the rest of his life at the Institute for Advanced Studies in Princeton. But he often came to the Bourbaki Seminars in France, and always had a lot of influence over Bourbaki. Everybody admired him enormously, and Dieudonné used to say humorously: "I won't drink my café au lait as long as André Weil hasn't drunk his." But admiration never engendered a submissive spirit within Bourbaki, and all members were equal during discussions.

During the last few years, Weil used to spend a few weeks in Paris at the end of spring; he always came to have dinner with us at least once each time, and we would spend an agreeable evening discussing various things. He played an important role in Marie-Hél'ene's training, visible in all her work, and also in mine, not only for the ideas I mentioned above, but also for example in increasing my knowledge of Kähler manifolds and complex analysis in general, and Chern-Weil forms. After the death of his wife Evelyne a few years ago, he lost all happiness and desire to live. He became more and more infirm, and lost his eyesight, and lately desired only to die. He died in August 1998, very recently, between the first and second editions of this book, at the age of 92. This is not the place to give an analysis of his work, but it is certain that his role in twentieth-century mathematics was of the highest importance.

Of course, my definition via operators came directly from the theorem of Choquet-Deny and my little article following theirs, and convolutions appeared everywhere. I paid for my mistake with several months of complications. Contrarily, Sobolev lived in the USSR where there were numerous specialists in partial differential equations (Petrowsky was still young in 1936, and his first big articles dated from before the war), and he was familiar with convolution, which he used for example in his famous theorem on convolutions with powers of the distance from the origin, but he was less familiar with it than I was.

Schwartz, Cartan and Weil

His 1936 article came long before Weil's 1940 book. So he oriented himself immediately towards distributions rather than operators, and did not stop with convolutions of distributions, although they play a major role in the entire theory. Thus, Sobolev and I and all the others who came before us were influenced by our time, our environment and our own previous work. It makes it less glorious, but since we were both ignorant of the work of many other people, we still had to develop plenty of originality.

I thus abandoned the operators of my marvelous night, and considered distributions instead. In this way, one naturally discovers Dirac's distribution, which was just the Dirac measure defined by the formula $\langle \delta, \phi \rangle = \phi(0)$. The space $\mathcal{D}'$ is then the topological dual of $\mathcal{D}$, so we can define an operation of transposition. If $f$ is a $C^1$ function on $\mathbb{R}$, how can we define its derivative $f'$ as a continuous linear form on $\mathcal{D}$? We do it by the following formula:

$$\int f'(x)\phi(x)\,dx = -\int f(x)\phi'(x)\,dx,$$

obtained by integrating by parts, where the integrated terms disappear because $\phi$ has compact support; this formula can be written simply as $\langle f', \phi \rangle = -\langle f, \phi' \rangle$.

The minus sign which is introduced here did not occur in the theory of operators. We will say that a distribution $T$ has derivative $T'$ if we have the analogous formula $\langle T', \phi \rangle = -\langle T, \phi' \rangle$. Differentiation on $\mathcal{D}'$ is the opposite of the transpose of differentiation on $\mathcal{D}$. On $\mathbb{R}^N$, we have

$$\langle D^p T, \phi \rangle = (-1)^{|p|} \langle T, D^p \phi \rangle.$$

Thus, the differentiation of distributions generalizes the differentiation of functions, and we rediscover the fact that every distribution on $\mathbb{R}^N$ has successive derivatives of every order, and that we can invert the order of differentiation. We did not use any other ingredients than integration by parts and transposition! Convolution no longer plays any particular role. This makes it just a little more complicated to define than for operators. Everything is written in my book on distributions, so I will just give the formula here:

$$\langle S * T, \phi \rangle = \langle S_x \otimes T_y, \phi(x + y) \rangle.$$

It is necessary to give a certain number of properties which are less obvious than for operators, but still very natural. I won't detail them here since they are completely described in my book and all other books on distributions. We see that convolution still has meaning if one of the distributions $S$ and $T$ has compact support. We can define the convolution product of several distributions as long as all of them except at most one has compact support. This convolution product is associative and commutative. Moreover, to differentiate a convolution product, we simply differentiate any one of the factors. With only slightly more trouble, we obtain all the magnificent properties of convolution. The notion of support is easily defined, but instead of using a finer subcover, one needs to use a finer $C^\infty$ partition of unity.

The finiteness theorem which I had not been able to obtain with operators, and which I hadn't even thought about, became altogether simple with the continuation of Hahn-Banach continuous linear forms, whereas there is no such continuation for continuous linear maps with values in a topological vector space. Finally, multiplication became obvious: if $T$ is a distribution and $\alpha$ a $C^\infty$ function, then $\langle \alpha T, \phi \rangle = \langle T, \alpha\phi \rangle$.

Operators and distributions each have their own advantages, but finally everyone agrees that distributions are the more useful version.

Having discovered the main results on operators in a very short time, I was loaded with a bad definition and only found the right one two or three months later, in Grenoble. When I communicated it to Cartan, he answered "Oh, it's easy as pie! So there's nothing more to differentiation than integration by parts. How could we not have seen that when we talked about it together? It was just because we didn't want to change a definition we'd already worked on so much." The correspondence between distributions and operators is obvious: if $T$ is a distribution, it defines the operator $\mathbf{T}$ by $\mathbf{T} \cdot \phi = T * \phi$, and if $\mathbf{T}$ is an operator, it defines the distribution $T$ by $\langle T, \phi \rangle = (\mathbf{T} \cdot \check{\phi})(0)$.

## Fourier, and tempered distributions

But I was still not finished. The Fourier transform remained a major obstacle, and it was not until nearly the end of the spring of 1945 that I figured out how to overcome it. Slowly, I came to realize that it would not be possible to define the Fourier transform of a distribution to be another distribution. It was necessary to make some restrictions. I was led to replace the space $\mathscr{D}$ by the space $\mathscr{S}$ of $C^\infty$ functions on $\mathbb{R}^N$, without compact support, but rapidly decreasing at infinity; I also required this condition of all of their derivatives. Here, rapidly decreasing means decreasing more rapidly than any power of $1/|x|$. I put an obvious topology on this space, which made it into a Fréchet space (which $\mathscr{D}$ was not, since it possessed only a pseudo-topology), and I defined a tempered distribution to be a continuous linear form on the space $\mathscr{S}$. The dual space $\mathscr{S}'$ of $\mathscr{S}$ became the vector space of tempered distributions, a Fréchet dual, all of whose properties I already knew well since my stint in Saint-Pierre-de-Paladru. In Grenoble, I gave an exact definition of the real topology corresponding to the pseudo-topology on $\mathscr{D}$, which later, in 1946, Dieudonné and I took to calling an inductive limit topology. The pseudo-topology is not enough; in order to apply the Hahn-Banach theorem and to study the subspaces of $\mathscr{D}$, you need to work with a real topology.

Why did we choose the name distribution? Because, if $\mu$ is a measure, i.e. a particular kind of distribution, it can be considered as a distribution of electric charges in the universe. Distributions give more general types of electric charges, for example dipoles and magnetic distributions. If we consider the dipole placed at the point $a$ having magnetic moment $M$, we easily see that it is defined by the distribution $-D_M \delta_{(a)}$. These objects occur in physics. Deny's thesis, which he defended shortly after, introduced electric distributions of finite energy, the only ones which really occur in practice; these objects really are distributions, and do not correspond to measures. Thus, distributions have two very different aspects: they are a generalization of the notion of function, and a generalization of the notion of distribution of electric charges in space. The total charge is defined by $\langle T, 1 \rangle$, which has no meaning in general, but which always has a meaning if $T$ has compact support. The charge of a dipole is zero. The support of a distribution becomes the support of the charge-distribution, i.e. the smallest closed set containing all the masses. Both these interpretations of distributions are used currently.

## Currents

The next step was to make a general study of de Rham currents. It is a curious fact that I did not work on this myself, although currents had been one of my strongest motivations. But I was too busy looking for a definition of the Fourier transform to think about generalizing currents at the same time. When de Rham heard about my definition of distributions, which I quickly sent him, he immediately answered by redefining currents and giving their fundamental properties. In fact, I knew these things implicitly already, but not explicitly, and the discovery truly belongs to him. I soon defined distribution-sections of a vector bundle of finite rank, which occurred in the study of differential or pseudo-differential operators on manifolds and in the great index theorem of Atiyah-Singer. But currents, which corresponded to the scalar case, had a specific definition which de Rham immediately indicated. In 1955 he published his famous book *Formes, Courants, Formes Harmoniques*, presenting his complete theory. I only published *Sections distributions d'un espace fibré à fibres vectorielles* in 1958, in the Annals of the Institut Fourier; I had actually discovered these distribution sections very quickly, but the fiber was infinite-dimensional.

## Yet more spaces

My first publications on distributions are in the Annals of the Institut Fourier in 1945 and 1947. My article of 1945 is often referred to as my propaganda tract. It corresponds to part of what I talked about in the harmonic analysis seminar in Nancy; the 1947 article contains the Fourier transforms. To define multiplication and convolution of tempered distributions in such a way that everything is compatible with Fourier transforms, it was necessary to introduce still more spaces, called $\mathcal{O}_M$ and $\mathcal{O}'_C$. After that, I did not encounter any more serious obstacles; the most serious problem had been solved by the introduction of $\mathcal{S}$ and $\mathcal{S}'$, which was contrary to all my acquired habits. Here, also, I needed to vanquish some inhibitions. The spaces previously used in functional analysis were vector spaces of functions of a type known beforehand and fairly general. We had $C^m$ spaces and $L^p$ spaces. But it was altogether unusual to define new elements as being those of a dual space. It was done in the André Weil's book since measures lived in a dual space. But measures had no pretention to be generalized functions and they were not differentiable. They were only really a more correct definition of certain simple charge distributions. On the other hand, the space $\mathscr{D}'$ was totally new. It was a priori vexing to have to use more particular spaces like $\mathscr{S}$ and $\mathscr{S}'$ in order to have a Fourier transform, and even then to have to introduce $\mathscr{O}_M$ and $\mathscr{O}'_C$. If each new problem was going to require the introduction of new function spaces, how complicated was mathematics going to be? I was not at all inclined to adopt this view. At each step, I needed to overcome some inhibition, but once the result was proved, it became clear that it would be easy to generalize. All mathematicians working today in the subject of partial differential equations consider this a common practice, which considerably augments the number of classes of topological vector spaces used in analysis, but it doesn't bother anyone any more.

The book on distributions was not completely finished until 1950 for the first volume, and 1951 for the second. Seventeen years separated my first reflections in hypotaupe, in 1933, from the appearance of the first volume in 1950. When I received the Fields Medal in 1950, only the first volume was written. Of course, the second volume was not far from being ready, and a short version had already been published. In any case, my talk at the 1950 congress, relative to the theorem of kernels which uses distributions, went even farther than the book, and was published in a separate article. This kernel theorem has become essential in the study of pseudo-differential operators and Fourier integrals. It had been expressed in a vague and intuitive manner by Dirac, much earlier. It led me very far in the following years: my 1958 book on vector-valued distributions is a fruit of it, as is the theory of topological tensor products of Grothendieck. The two volumes of 1950–51 together make four hundred pages. It was a long way from the initial "propaganda tract". And I even think the book is rather too short. I was impregnated with Weil's attitude of rejecting books and long articles. The theoretical part of the book is well done, but practical computations are missing for the most part. The result is that a reader who needs to understand certain formulas is obliged to prove them for himself. When I disappear, certain formulas will not be written down anywhere at all. Fortunately, the Swiss mathematician Méthée, a student of de Rham, published all the necessary complements for a complete understanding of the finite parts of singular functions of relativistic quantum mechanics.

I gave my first lectures on distributions in Paris, in the Peccot course, just after going to live in Nancy. The twenty-five or so auditors who actually followed the course from beginning to end were delighted. About half of them were physicists, and in particular electricians. In 1947, I gave a general talk in the harmonic analysis seminar in Nancy.

## A difficult success

Modern mathematicians have the impression that as soon as the theory of distributions was invented, it was universally accepted. This is not true at all. I bring to witness the testimony of mathematicians who were there in the early stages, such as Lars Gårding in Sweden. I received two kinds of criticism: some people found that the whole idea was so simple that it couldn't possibly be useful, and others that the generalization of functions by continuous linear forms on a topological space was so complicated that it couldn't possibly be useful. It goes without saying that young people were the main critics of the first type, whereas older mathematicians had reservations of the second type. Both criticisms corresponded to doubts of my own. I really wondered whether distributions would turn out to be useful. The people who ended up making distributions acceptable were the young mathematicians who devoted their theses to them because they had been raised on them: Malgrange, Lions and above all Lars Hörmander. The work of these young people gave rise to a general impulse in favor of distributions. In the end, they became not just acceptable but I really believe (without undue modesty!) that they deeply changed the whole nature of analysis. They are now taught currently in universities, and I taught them myself in a course on "Mathematical methods in physics", at the University of Paris from 1952 to 1962, and at the Ecole Polytechnique. The book on generalized functions by Gel'fand and Shilov appeared in 1955. A few years later, Sato introduced hyperfunctions. These were an extension of distributions, and the notions are more difficult in this situation, but they are essential for many applications to the theory of partial differential equations.

When I introduced distributions to physicists in 1953, I started with the Dirac function which they knew, and showed them that it did not really exist and that distributions were a more or less inevitable generalization of the Dirac function and its successive derivatives. Thus, the students were led to understand the necessity for distributions. When I began my course on distributions at the University of Paris, I proceeded in the same manner. But those young people had never heard of the Dirac function and its derivatives, and they found my whole approach rather curious. However, when I tried approaching the question via continuous linear forms on the space $\mathfrak{D}$, they breathed again – "Finally we understand what he's talking about". I hastily abandoned the idea of introducing distributions starting from the Dirac function to these students – it was not only useless, but confusing! Later, once they had understood distributions, I could talk to them about the functional notation for distributions, where $\int T(x)\phi(x)\,dx$ stands for $\langle T, \phi\rangle$, but $T(x)$ doesn't mean anything.

## Other distributions

Other attemps were made to introduce distributions in a clear and direct way for engineers. I'm not sure that this is so important, or even possible. The other definitions run into various obstacles which usually end up making them more difficult instead of easier. Lighthill and Mikusinsky introduced distributions as limits of sequences of continuous functions, giving a suitable equivalence relation on such sequences: two sequences of continuous functions $(f_n)$ and $(g_n)$ are equivalent, i.e. have the same distribution as limit, if for every $C^\infty$ function $\phi$ with compact support, the integrals $\int f_n(x)\phi(x)\,dx$ and $\int g_n(x)\phi(x)\,dx$ have the same limit for infinite $n$. But then what is the difference between this notion and the notion of a continuous linear form on $\mathscr{D}$? Maybe, after all, the mind of an engineer is more adapted to understanding limits of sequences. But I think this kind of obstacle no longer exists, and engineers are used to continuous linear functionals; the abstract has become concrete. Another aspect of Mikusinsky's theory could have gone much farther: he considered the space of functions which are continuous on the right, with support contained in $[0, +\infty[$. He considered the space of these functions as an algebra under convolution, which actually takes a special form in this situation. This algebra, as he showed, contains no zero divisors. So, it can be extended to a field. For example, $\delta$ is the unit element of this field, and $\delta'$ is simply the inverse of the function equal to 1 on $[0, +\infty[$ and 0 elsewhere, which is just Heaviside's function. However, this method does not allow one to obtain distributions along the whole real line, and ended up being abandoned.

Hyperfunctions, an important extension of distributions, have played a specially important role in recent years. They were introduced using cohomological methods by Mikio Sato. Another new notion, the "wave front set" was created by Lars Hörmander. A distribution is singular at a point $a$ if it is not a $C^\infty$ function in any neighborhood of $a$. When we know that a distribution is singular at $a$, we can study "the hyperplane directions in which it is singular", and these directions form the wave front set at $a$. One can also consider the wave front set as the singular support in the product of the configuration space and the phase space. Its definition uses the Fourier transform. I had all the elements to discover this wave front set and its many important applications, but I never thought of it. Pseudo-differential operators are a remarkable generalization of differential operators with $C^\infty$ coefficients. For such an operator, convolution and the Fourier transform play a fundamental role: a pseudo-differential operator is a complicated mixture of multiplication and convolution. The beginnings of these operators were the singular integral operators of Calderon-Zygmund, which dated back several decades, then those of Seeley, and their final, formal introduction was due simultaneously to Kohn and Nirenberg in the US and the Unterberger couple in France (in their thesis written under my direction). They were used in the Cartan-Schwartz seminar on the Atiyah-Singer index formula in 1963–64, and in a large article, Hörmander reintroduced them and gave remarkable new applications. They were then extended by the invention of Fourier-Integral operators on manifolds, and came to play a greater and greater role in all the theories making up what is known as microlocal analysis. These are quite extraordinary generalizations with immense consequences. It is what I call the "triumph" of Fourier and convolution. In my book on distributions, I had conjectured as an obvious fact that the Fourier transformation would play an essential role in the theory of differential operators with constant coefficients, but no role at all for differential operators with variable coefficients, and mathematicians such as Lions, Malgrange, Hörmander and Trèves agreed. But the situation changed radically. Now the Fourier transform is everywhere, throughout microlocal theory; it has become a universal tool.

Moreover, it gives rise to a typical case, the prototype of the spectral decomposition of operators. Taking the spectral decomposition of an operator means decomposing the space into a sum (or integral) of subspaces on each of which the operator acts as multiplication by a scalar. The Fourier series and the Fourier integral give the spectral decomposition of the differentiation operator, since the derivative of $e^{i\alpha x}$ is equal to $i\alpha e^{i\alpha x}$: if a function of $x$ is written $\int_{-\infty}^{+\infty} f(\alpha)e^{i\alpha x}\,d\alpha$, its derivative is given by $\int_{-\infty}^{+\infty} i\alpha f(\alpha)e^{i\alpha x}\,d\alpha$.

The wavelets invented by Yves Meyer now replace Fourier analysis in a certain number of questions, but they also use Fourier analysis and exist more or less within the same framework.

## The zigzags of discovery

The image of discovery is quite different from what the public imagines it to be: according to the public, one progresses from beginning to end by rigorous linear steps, in a well-determined and unique order corresponding to perfect logic. Zigzags are not recognized. It's a pity. It makes mathematics and all sciences appear too rigid, les human, more inaccessible, giving no rights to hesitations and errors. My own testimony is probably valid for all mathematicians and all of scientific research. Research in physics, chemistry or biology is not very different. One passes right by obvious things, searching for things which are much too complicated. Only afterwards does one arrive at a simple result. *La Statue intérieure* (*The Inner Statue*) by François Jacob shows the difficulty of the discovery of RNA. There are theorems which are difficult at first but become simple once you clean them up. Those who read the final result find it almost obvious and have a tendency not to attribute much merit to the mathematician who introduced it. Yet nobody discoverd the result until he did, probably because of a collective inhibition on the part of the scientific community which made the discovery impossible. On the other hand, certain very complicated theorems remain complicated. The theorem of the distribution of prime numbers, for instance, has been improved, but it remains just about as difficult as it was at first.

Here is an example of a theorem which is very short but which took me a very long time to discover, although I proved some theorems which appeared more difficult quite quickly. The reason is that this theorem deals with a counterexample, and counterexamples are always delicate. I believed that I could prove that every distribution supported on a compact set is a finite sum of derivatives of measures supported on the same compact set. One evening when I really thought I had proved the result, and felt both tired and tense, I took a sleeping pill to get some rest. The following morning, on waking, I had to admit that the proof I had found the previous evening was wrong. I also searched for counterexamples, in vain. This laborious research lasted for a whole week. Finally, it turned out that the theorem was false, and on the morning of the eighth day I finally got hold of the counterexample. The theorem held only if the compact set $K$ possessed certain properties which I called regularity in the sense of Whitney. A whole week of suffering. Research is not always enjoyable; it is when you finally find a result, but sometimes you have to labor for a long time, so that happiness is born of suffering. I expressed this feeling later in a kind of self-mockery in the first version of my book, writing "One might think that the following theorem is true" – followed by the statement – "but it is false, as the following counterexample shows" – followed by the counterexample. Nobody reading that passage could guess how much suffering that counterexample cost me. In fact, that passage disappeared from the final version of my book on distributions, where the counterexample is not specifically given, but occurs in the midst of a sequence of theorems which show in particular that the simplified conjecture is false.

## Drying up and its consequences

It happens that while one is engaged in a piece of research, one asks oneself a question (you have to be curious to be a researcher!) whose answer is not forthcoming. Most of the time, after a brief period of investigation, you move on to something else. Sometimes, you continue, but you still don't find the answer, and you keep the problem in a corner of your mind, to think about it again later. It happens that unexpectedly, you actually discover even more. But sometimes, that extra thing is not interesting, and perhaps it doesn't deserve to be developed or published. Yet, one continues asking oneself related or different questions, which constitute a stock of open questions. Very often, the solutions to many of these questions come simultaneously.

This "drying up" is an integral part of research. It can become painful if it lasts too long. But to a certain extent, I actually like it. When everything moves along too easily, one has a tendency not to make enough effort. But during the drying up periods, one wants to overcome the difficulty at any price, and strains one's whole being to that end; often, at the end of a day of this kind, one has discovered something. Not necessarily what one was looking for, but perhaps something interesting anyway. I'm always pleased when I reach one of these periods, but I become unhappy if it lasts too long. After the enthusiasm of the sudden discovery, one always encounters some obstacles. I gave some examples above. Then, one finds that one has asked oneself a bad question, or what is worse, one is using a definition which is ill-suited to further developments. One has to be ready to make painful changes. When one has elaborated a whole theory from a definition, it's not easy to change it. And yet, as I said above, it is often quite simply indispensable, because there's no reason that the first definition one finds is the one best adapted to further developments. At this point, it takes real courage to start again. Fortunately, it's not as difficult as you might think, because one has thought so much about all the different parts and their connections to each other, that it feels natural to readapt them all to the new definition. This alternation of joy and suffering is at the heart of research. Young people need to get used to it. High school students all too often think that it isn't normal to think about a problem for more than one hour, whereas to find something really important, it's necessary to think for several days, if not months, and even years. Naturally, one is not forced to think about only one problem at a time. When you read a well-written book, you don't think about the joys and sufferings of the author who wrote it. It can be quite instructive to learn about them. In high school, there is no time to study science together with its history. Most of the time, teachers are obliged to teach in an imperative, dogmatic manner. But from time to time, we should encourage students to search things out for themselves, and teach them something about the history of science. The collaboration of some historians of science would be welcome. This subject has developed remarkably in recent years. A few decades ago, people studied mostly the development of science in antiquity, but progressively, it has turned its attention towards modern times. It's very useful to insert a little history when teaching a new theory. It just isn't done enough in high school mathematics classes, to let the students understand the enormous distances which our predecessors had to traverse in order to obtain today's perfected results. They should also know that if a theory seems well-done but some of its aspects remain incomplete, those are probably the most interesting ones for future research. The goal of science is not to ingurgitate complete, ready-made ideas, but to imagine new concepts. And this is generally achieved by overcoming inner obstacles.

## A complete change of research subject

Sometimes, during a period of research, an accident happens and disturbs everything. Simply allowing oneself to be disturbed by it is not enough. One must hunt down the true source of the problem, and this, again, demands a sense of curiosity. It happens that while doing this, one notices a fact which is actually much more important than all the research one previously did. At that point, one should abandon one's previous train of thought to follow the new one, which often gives much more fruitful results. This happens frequently. A researcher needs to have a mobile, original mind, always ready to accept revolutions, and very stubborn. It demands courage and curiosity. I often mention the example of the biologist Fleming who studied bacillary cultures. One day, he noticed that one of his cultures had been invaded by mold. Instead of throwing it away, he had the curiosity to examine it, and he saw that the bacilla were all gone. He had the courage to abandon his previous line of research in order to think about this new phenomenon. And that is how penicillin was discovered: it is a product of the mold which killed the bacilla, it is a defense method of the mold itself. The discovery of penicillin was far more important than the work Fleming was previously engaged in. Everyone is not Fleming, but it is a good example. I also like to mention Christopher Columbus, obsessed by the idea of finding the occidental route to India, and discovering America instead. The Vikings, who overran and conquered part of Europe, had actually also landed in North America. They even had a woman on board, who had a baby there! But instead of exploring further south, they left the new country and returned home, and the woman explained that her baby had been born in some far away place like Greenland. Through lack of curiosity, they did not discover America! The story of distributions is a similar one: I had written a little article to generalize a property discovered by Choquet and Deny, and for that I had introduced regularization by $C^\infty$ functions with compact support, and I saw immediately that this could be useful for many other, very general things.

## A lion in the sense of distributions

When I was writing up the convolution product in Nancy, the wonderment its numerous possibilities aroused in me inspired me with the following dream. I was telling someone that it was possible to convole two arbitrary distributions, but naturally, the convolution product was also a distribution. For example, I told him, we can take the convolution product Mozart$*$Beethoven, but naturally the result is not a musician, it is a distribution. We can also multiply Nancy$*$Strasbourg, or Tomblaine$*$Tomblevent. Tomblaine is a town near Nancy, and Tomblevent does not exist. Naturally, however, all the products are not towns but distributions, or rather, I said, towns in the sense of distributions. This repeated use of the expression "It is not a ... but a ... in the sense of distributions" was amusingly caricatured by Tuckey. He published a long article full of humor on forty mathematical methods for hunting a lion; it was so funny that it was published in a very serious mathematical journal, the *Duke Mathematical Journal*. One of the methods for hunting lions was the inversion method. You want to capture lions in the desert: you construct a metal sphere and place yourself inside it, but not in the center. Naturally, the lions in the desert all lie outside the sphere, as does most of the universe. You then invert the sphere (sic). This way, you are outside the sphere, but not too far, since only the center goes to infinity. But everything which was previously outside the sphere is now inside it. The whole universe has passed inside the sphere, in distorted form, to be sure, but the lions of the desert are nonetheless inside, albeit somewhat demolished. He then went on to define a method for hunting lions using distributions. Anything in the desert is a lion, but in the sense of distributions. For example, a stone is a lion in the sense of distributions. Thus, it suffices to regularize it by a $C^\infty$ function with compact support, and it becomes a real lion, which fiurthermore is infinitely differentiable!

## My interior palace

Distributions, like all the mathematical knowledge which I accumulated, live inside my brain in a well-structured manner. Each part is connected to other parts, each part is preceded and followed by other parts. The whole forms a beautifully ordered set. This structure is as beautiful to me as a palace. It has a rigid structure. If I want to take a certain path, I can only follow it in a given direction. If someone explains it to me in a different order, taking as a definition what for me is an example or a final result, and inverting the usual direction, I feel deeply troubled. It's even worse if someone tries to introduce a new notion to me. Every time I try to read mathematics or listen to a talk, I feel as though it is an assault. It's as though they were trying to destroy my castle. Usually I'm just not able to understand what they're talking about at once; I take notes and reflect over them alone at home. Of course, I could cede to temptation and simply ignore them, in order to preserve my castle intact. It almost seems as though my castle is an obstacle to development. But I don't believe so; it seems to me to be an indispensable step. I can't see how I could do new mathematics if my internal mathematics were not so well-organized. I feel a kind of imperialist desire for total knowledge, not only for mathematics, but also for sciences and everything to do with life and society. For me, everything ought to be perfectly logical. I can't tolerate fuzziness. If I don't know a theory well, I feel as though I don't know it at all: I have difficulty accepting half-measures. And since new results feel like assaults, I resist them and it takes time for me to assimilate them. But once I have assimilated them, I remember them forever, thanks to my excellent memory. When I receive an impression from outside, I have to put a whole series of phenomena back into their places and include the new idea into my own scheme of things. My castle is then even more perfect than before. Maybe some parts of it have been rejected as useless from time to time, but it's necessary to clean up every now and then. Other elements have been incorporated into it. And the castle slowly becomes modified, without actually seeming to get larger.

When I was young, I had a powerful memory. Until the age of about fifty-five, I think I remembered every single thing I had ever thought about. I never took any notes. Only in 1972 did I begin to take notes, following Grothendieck's advice. Every time I learned something new or read a book, I wrote a memo for myself, mostly in a way that made sense only to me. I wrote just a few pages, and didn't publish them, but gave them a number. There was no going backwards: every new reflection got a new number. The texts are not in any logical order. I have a table of them in my computer now. This pile of memos is my written memory. I'm up to number 380, and since the average length of each memo is about ten pages, I guess I've written somewhere between 2500 and 4500 pages. Unfortunately, my "living" memory is weakening with time; it used to be excellent, but it isn't any more. I don't even remember the statements of many of the theorems I've proved. Naturally, if I look into my written memory, I find it again, but it gets more and more difficult. A few years ago, it took me ten minutes to find a memo, look over it and remember what was in it, but nowadays I sometimes need an hour or even several days. This is partly why

I have such difficulty now with mathematical research. I'm more than eighty years old, and it's very rare for a mathematician to do original research at such an age. It's the lot of every mathematician.

I know many mathematicians who feel about new ideas the way I described above, and others who have an incredibly rapid mind. The latter kind must have an interior castle with a different kind of structure. For example, Lions, Dieudonné, Dixmier are capable of reading articles very quickly and understanding them immediately, without feeling bothered at all, whereas Henri Cartan and Alain Connes are more like me. I wrote an article about this once, and Alain Connes told me he felt as though he could have signed it himself. Of course, there are often many ways to get from one point to another, and a well-constructed castle contains them all. It's more or less the same system as neurons. There are theorems I remember so well that I can give a talk on them with no preparation at all. There are others whose proofs I always forget, even in my university courses. For those courses I needed to prepare very carefully. For example, the Hahn-Banach theorem is so deeply ingrained in me that I could prove it asleep. But Banach's closed graph theorem has always seemed somehow difficult to me, even though I always proved it for myself before teaching it to students.

As I've already said, mathematicians are sometimes happy with proofs that go in zigzags, and write them up in this form. One might think that this method is didactically better for the reader than the well-constructed proof in the shortest line. But it's not certain, because your zigzags may not be the right ones for another mathematician!

## The age of a child

My son was born on March 17, 1943. The day of his seventh birthday, on March 17, 1950, I asked him "Can you tell me what day you were born on?" He thought, subtracted and answered "March 17, 1943". "Good," I said. However, he added "If you had asked me yesterday, I would have had to say March 17, 1944 since yesterday I was only six." I was quite taken aback by this reasoning: he had changed his date of birth! I went into an explanation which was not so easy for a seven-year-old to understand, about how yesterday he was seven years minus one day old, and that counting in years alone was not sufficient to measure age, you needed to count days too. When he understood that, I asked him "But isn't there some simpler reason for which the day of your birth is always the same, no matter when someone asks you about it, and always will be, even after you're dead?" The fact that the day of his birth was a historical fact whose date could not be changed was not at all obvious to him. Birth had no meaning for him; he hadn't seen it, he hadn't lived it, he could only obtain a birth date by subtracting from the day of his birthday. The date of his birth had no reason to be fixed. He understood it only by an inversion of cause and effect.

I tried the same experiment with my daughter on her seventh birthday. She was born on July 30, 1947, and on July 30, 1954 I asked her what the date of her birth was. Without hesitation (she was very talented in mathematics and became a mathematician and a professor at the University of Grenoble), she answered that she was born on July 30, 1947. I said to myself "Now, I'll ask her what she would have said if I had asked her the question yesterday." But the astute little creature got ahead of me and announced "When you'll ask me on my next birthday, I'll have to answer July 30, 1946 since next year I'll be eight years old." I told her to think for a minute, and she soon realized that next year would be 1955. Triumphantly she announced her conclusion: "Today, I'm seven years old, and it's July 30, 1954, so I was born on July 30, 1947. Next year, I'll be eight years old, and it will be July 30, 1955, so I'll still be born on July 30, 1947. So every year, I'll give the same answer." She had proved a theorem, namely that the difference is constant. It's a typical case of the zigzags via which she reached her conclusion: the date of birth is a constant. I then asked her the direct question: "Don't you see any natural, simple reason for which your date of birth is always the same and can't change?" Claudine didn't find the answer any more than Marc-André; birth is just not an intuitive thing, it can only be obtained by taking differences, or by proving a theorem.

If I'm telling this story, it is not only because it's adorable, but to show that all scientific discoveries take place this way. By a succession of reasonings starting from things we already know, we find that a certain quantity is constant (for example, in relativity, we find that time and space are not constant, but the speed of light is a constant, and Maxwell's equations are relativistically invariant. These are things which are not intuitive, which appear almost shocking. That is why it took so long to discover them, and Einstein had to do a lot of propaganda and demolition of previous ideas before his theory became universally accepted. Now, of course, we know that we should begin at the other end. Space and time change, but Maxwell's equations and the speed of light are absolute constants.

## Mathematics and experimental sciences

As we see, contrarily to what everybody thinks, mathematics is not a rigorous science, in opposition to the experimental sciences which are perceived as intuitive sciences. There's exactly as much rigor and as much intuition in

Schwartz with small Marc-André

mathematics as in physics. But the rigor is not exactly the same. In mathematics, rigor requires a formal proof by the rules of reasoning, whereas rigor for a physicist means experimental verification. A physicist who wants to verify a theory does not prove it, he verifies it experimentally. In mathematics, we also make experiments, since in order to discover a property, we make several efforts with approximative models before finding the best result. For example, in twelfth grade, to study the curve described by a point, we study a particular situation and the particular points of that curve, which helps tremendously to find the final correct result. Intuition is a fundamental part of our work. We check things rigorously, but we find them using intuition. It is intuition which pushes us forward, and I have a tendency to go very fast and farther and farther, without really proving everything as I go along. Then comes a point where I finally get mixed up, and intuition is no longer sufficient. At that stage, I go back and rigorously prove all the steps, and even write it all down. From then on, everything is well-established; the consequences I hastily discovered at first may not always be correct, but I can continue on a solid basis.

I'd also like to add a word to the published controversy between Alain Connes and Jean-Pierre Changeux about the nature of mathematics. Alain Connes considers – and I completely agree with him – that mathematics correspond to a deep and independent reality, and Changeux considers that unlike other sciences, mathematics are a pure construction of the imagination.

Which of our findings are discoveries and which are inventions? Discovery is the finding of an object outside of ourselves, which has always existed and will exist after us, even if we only learned about it recently, and which leaves us no possibility of choice. But invention is the finding of a new object which did not exist before it was found, and which we are perfectly at liberty to change.

The sciences of the earth and of the universe are uniquely sciences of discovery: geography, geology, astronomy are the study of outside objects, and so is history. One can occasionally discover inventive parts of these subjects, for instance the conception and the construction of a bridge or a dam or a town, or the diversion of a river. Modern reflections on geography and history are already closer to inventions: geography does not just create maps, and history does not just relate facts; every decision of a king or a people, every election of a president is an invention. At the other extreme, music, painting, sculpture are pure inventions of the artist, who is entirely master of his choice. But in those domains, discoveries also exist. For almost every other activity of the mind, or of the hand guided by the mind, there are parts having to do with discovery and others having to do with invention. The knowledge of the laws of biology, DNA and RNA are essentially discoveries, as are the laws of immunity, the HLA system, the AIDS virus. But the vaccination against scarlet fever, the antitetanus serum and the utilization of corticoids and morphine, general and local anaesthesia, contraceptives, in vitro fertilization: all these things are inventions. Everything we know about the laws of chemistry had to be discovered, but the synthesis of new bodies, medicines, colorings and explosives are inventions. As materials, wool and silk were discovered, but to weave them was an invention. In physics, the laws of statics and dynamics, universal attractyion, the motion of a gyroscope, protons and electrons, electromagnetic induction, Maxwell's equations, radioactivity, superconductivity, relativity, quantum theory are all discoveries; the last two have become accepted only recently and after encountering a certain resistance. But a continuous or alternating electric current, the telephone, the television, cars, railroads, airplanes, the atomic bomb and electronuclear energy are inventions. Continuous current is an invention, but once it was found, Ohm's law, Joule's law, and electrolysis are discoveries.

After all, the situation is exactly the same in mathematics: integers and fractions, prime numbers and their distribution law, the numbers $e$ and $\pi$, the line, the plane, 3-dimensional space, circles and conics (Kepler's trajectories), the (caustic) cardioid, Cassini's oval and Bernoulli's lemniscate, minimal surfaces, Betti numbers, curvature, derivatives and integrals, differential equations and partial differental equations (harmonic functions for the electrostatic potential and the equation of propagation of sound or light waves) are all discoveries. Complex numbers may be called an invention, if you introduce them by $x+iy$, but they are a discovery if you prove that the ring of real polynomials modulo $(x^2+1)$ is a field, which is denoted by $\mathbb{C}$. It's a discussion about the sex of angels .... But once the complex numbers are defined, the theories of Weierstrass and Cauchy on holomorphic functions, the residue theorem , Picard's theorem are discoveries. Moreover, the invention of complex numbers came at a certain moment in the history of the Western world, but if it had not occurred at that moment, it would certainly have occurred later. Vector spaces of dimension $\geq 4$, Banach and Hilbert spaces, inversion and transformation by reciprocal polars, distributions, wavelets, the computer, these are all inventions. But once Hilbert spaces were found, spectral theory is a discovery. Once distributions were found, their Fourier transform is a discovery. A cave is discovered, a hut is invented, wool is discovered, weaving is invented. This little game is easy to play and not very deep, and everything I've just said is controversial.

It shows that discovery and invention are deeply interwoven, and can occur relative to the simplest and to the most complicated objects; there's no real difference between mathematics and the other sciences. Complex numbers may be the fruit of the imagination, but then they acquire a deep reality, they behave according to laws which have to be discovered (I already mentioned holomorphic functions) and which are written in thousands of books (thanks to the art of printing which is an invention); they are the same for every mathematician in the world, and they could be communicated to intelligent extraterrestrians. The Dutch mathematician Freudenthal wrote an excellent book to explain how to communicate all of mathematics to extraterrestrians. A sculpture by Rodin is the pure product of imagination. He would not have been able to communicate it to anyone before sculpting it, least of all to an extraterrestrian. But once finished, the statue becomes a beautiful reality, just like complex numbers! The famous mathematician Charles Hermite, who proved (among many other things) that $e$

is a transcendental number, wrote "I believe that the numbers and functions of analysis are not arbitrary products of our minds: I think they exist outside of us, with the same character of necessity as the things of objective reality, and we encounter or discover them and study them like physicist and zoologists" (this comes from the correspondence between Hermite and Stieltjes, 1905). When I sent my text about the inner castle to Alain Connes, he wrote to me that he could have signed it himself; when I showed it to Changeux, he said he found in it a total confirmation of everything he had written about mathematics. But I also have an inner castle for all the other sciences and their mutual relations ... and for the whole of the one thousand nine hundred species of my butterfly collection, which are not at all the fruit of my imagination, but whose names are inventions.

A recent book by Claude Allègre, called *La Défaite de Platon*, goes farther than Changeux and declares that mathematics "are not really a science", not "a science like the others". If he means "not an experimental science", then it's all right, but to say "not really a science" is totally absurd and unacceptable; I say this in spite of my friendship for him and the admiration I have always felt for his work and his books on the physics of the globe. Science is the queen of the universe (if this phrase means anything) and mathematics are – not the queen of the sciences as has been said too often; there is no queen of the sciences – but a very great, true and magnificent science.

# Chapter VII
# Militating, Teaching, Research

The invention of distributions coincided with the return to life, the end of the war. Our existence became normal once again. At the university, my position came to correspond to my mathematical level. My political activities also took a new turn.

## A "legal" Trotskyist

As soon as I returned from Saint-Pierre-de-Paladru to Grenoble, in the summer of 1944, I took up my political activities again. For a while, I remained "clandestine", and according to the rules of the party, I concealed my Trotskyist affiliation. This did not stop me from giving talks in which I expounded on the Trotskyist analyses of the international political situation and denounced the enormous concessions to right-wing ideas in view of a so-called reconciliation. Franco remained all-powerful in Spain, and we justly feared that his regime would become permanent. Of course, the Allies owed him a certain debt of "gratitude"; he had allowed many Frenchmen to join the Free French Forces (my brother Bertrand was among them), and above all, during a solemn interview with Hitler, he had refused to allow him to let his troops march through Spain. And yet, this did not justify the yoke he imposed upon the Spanish people until 1975. Moreover, it was not Franco's attitude which motivated the Allies; they simply preferred his dictatorship to the specter of Communism or even a left-wing democracy. In Greece, the English had brutally disarmed the forces of the Communist resistance, which had battled valiantly against Hitler and Mussolini, and established a puppet regime in their stead.

General de Gaulle worked at restoring order to the country. The Communist Party did not present itself as a revolutionary party, and de Gaulle allowed it to share power, within the framework that he constructed. The patriotic militias of the resistance, in the hands of the Communist Party, could have become the instrument of an opposition power, or even a revolution. Maurice Thorez, who had declared in 1936 that one must know how to terminate a strike, confirmed de Gaulle's order that they be dissolved.

My listeners were always a varied lot. There was a striking contrast between the attitude of the leaders of the Communist Party and the mass of members who often listened to me with interest and even enthusiasm. It even happened that a union member proposed to distribute the text of my lecture to members of the CGT while the leaders, guessing how deeply "subversive" I really was, worked hard to prevent anyone from listening to me. Hostilities never went farther than that, however. Little by little, as the months passed and life became normal, the leaders of our party decided that the time had come to "legalize" ourselves, to openly declare ourselves as Trotskyists and to act in consequence. It was reasonable.

At the end of the spring of 1945, just after the war, the mathematician Küntzmann returned from the camp where he had been a prisoner and took possession of his position at the university which I had temporarily occupied, so I needed to look for a position as assistant professor in a university. Dieudonné, who had passed the academic year 1944–1945 in Nancy where he had obtained a position before the war, and Delsarte who was also there, battled energetically against mathematicians' tendency to want to go to Paris, and decided to turn Nancy into a great center of mathematics. They soon asked me to join them there. Dieudonné asked for only one condition: that we really come to live in Nancy, instead of living in Paris and commuting to Nancy. I was quite simply enchanted by his proposal, and we moved to Nancy at the beginning of the autumn of 1945.

In 1945 and 1946, we decided to put up some candidates in the legislative elections in Grenoble. The first time, I was still living in Grenoble, and the second time, I returned there from Nancy. There were six names on our list of candidates. Apart from myself and another member of our group, we had convinced four former Communists who disagreed with the Communist regime after 1944–1945 to join the reborn Trotskyist Party. When Raffin-Dugens, a former Communist more than eighty years old, joined the IVth International, it created a certain sensation. There was another very old gentleman, Horace Martin, a former railway worker well-known in the Communist hierarchy. The year before, we had also met a factory worker and long-time militant, whose name was Charles Martel[35) and who was very proud of it. His father, his son and his grandson all bore the same name, and like kings, they had numbers to distinguish them from each other. He earned his living with difficulty as a night watchman and possessed a solid political culture. He had been the first federal secretary of the Communist Party of the Isère region, and the secretary

35) Charles Martel is the name of an early king of Gaul (714–741), the grandfather of Charlemagne.

of the Krestintern, the Red Peasant International of Latin countries in Moscow. He was a strong believer in free thought, and played an active and original role in our campaign. But the whole list was still a little too "golden-oldie".

The fourth member was a very interesting person, the father of Henri Fabre. The latter was a young doctor who had been one of our militants the year before, a contact during the war and responsible for the family planning movement of the Isère. His father was a retired schoolteacher from La Mure, a little mining town with a population of seven thousand, where he had been a dynamic member of the Communist Party, and had a lot of influence on the population. The letter he wrote when he left the Communist Party and joined ours criticized the nationalism and the chauvinism of the French Communist Party, and encouraged our internationalism. He was younger than the other three members of our list, and was really a choice recruit for us. His whole family, including Henri's younger sister Paulette, became militants for our party. The family was very politically involved and we were close to them. About fifteen people helped us with our campaign. We all signed a profession of faith of which I have forgotten every word, but which was totally sectarian. Just to give an example, the black market was a considerable problem after the war; not the minor black market transactions that all of France including myself engaged in, but the grand organization. Pierre Mendès France (like Yves Farce, the Minister of Food Supplies) asked that exemplary sanctions be meted out to the organizers. Our group quite simply required that they receive capital punishment. We sent our vast campaign out to the voters, and I talked about it on the radio; all candidates were given an equal amount of time on the air.

The team was completely devoted, showing up at every right-wing and left-wing meeting and generally being so active that people thought there were at least a hundred of us. At our main meeting in the big hall of Grenoble, Maurice Laval took on the French Communist Party. Maurice and Renée Laval were young members of the party; they had been deported, separately, but had both returned. For some time, they had lived in our house in Meylan, as had Paul Parisot. Maurice gave an impressive lecture, moderate and reasonable. The Communist candidate was taken aback. The anti-Trotskyist campaign led by the Communist Party, which would soon become violent, had not really begun yet and Maurice Laval was a returnee from deportation. So he was answered with serious consideration, and a truly exciting discussion began. One of the things we demanded was the mobile scale of salaries, which was generally considered as demagogic and exaggerated at the time. Aimed at maintaining the purchasing power of the population, it made it possible to control prices effectively, and in fact it ended up being adopted even by right-wing regimes. But some of our other slogans like "government by committees" and "permanent armed workers' militias", inspired by the Russian revolution, now appear to me to be essentially unrealistic. We had not yet understood one fundamental fact: revolutions can be indispensable to modify certain situations, and armed groups certainly play a crucial role in them, but in a postwar period in a country which has just undergone an occupation, when the main concern is the re-establishment of a normal life, what is needed is a set of stable institutions: justice, police, an army and an elected parliament, a free press and everything else which contributes to a democratic state. This is what we refused to accept, by taking the Russian revolution of 1917 as a model. But that revolution had degenerated, the power of the Soviets had turned into a dictatorship, not even of the proletariat but of a single party, itself under the tyranny of a single man. I didn't see then that this degeneration was unavoidable. We continued to insist that the degeneration of the Russian revolution was not inevitable and that it was possible to establish Soviets in France.

While waiting to realize this objective, we wanted a Blum-Thorez government which would exclude all bourgeois Ministers, according to the formula which we had unsuccessfully supported during the formation of the Front Populaire. Legality had been re-established by General de Gaulle, with the modest but clear participation of the Communist Party. Of course, this could change. The General was not eternal, and his regime could perfectly well have been followed by a left-wing regime with Socialists allied to Communists.

During the second electoral campaign, I stayed with my cousins the Smoukovitches (Jacqueline Smoukovitch was a first cousin of my mother). They were both ardent Communists, but we liked each other and they certainly took very good care of me during this tiring period (we often didn't get home until late at night). They didn't mind that I was a Trotskyist and I told them everything that happened day by day without any embarrassment. Several other Communist friends later assured me that they would have put me up at any time I wanted, except during my electoral campaign.

We toured practically all the big and medium-sized towns in the Isère region during our campaign; the election was determined uniquely by the number of votes. Some of our meetings were fiascos, like the one we held in Bourgoin; others attracted an audience of friendly and receptive workers, for example in the small industrial town of Vizille. Thanks to the schoolteacher Fabre, we pulled off a real success in La Mure; we were listened to attentively and earned seven hundred votes. The total number of votes we received in that first election was two thousand seven hundred and ten. This was a bit ridiculous compared to the big parties which earned more than sixty thousand votes, but for a party like ours, it was relatively successful and better than we had hoped for. The detailed vote count showed us that we had earned a few votes in every single community. At that time, the anti-Trotskyist campaign was in full swing: in the Communist newspaper of the Isère, the detailed vote count was published, and our party was listed among the extreme right-wing parties, under the name "Hitler-Trotskyists". This was the newspaper which referred to me as the Hitlerite Schwartz, a title which the French Communist Party press gratified me with for years afterwards. It is scandalous that the many Communist or Communist-sympathizing intellectuals who knew and esteemed us never protested against such proceedings, whether directed against me or against anyone else. They paid for it later, by participating in the slow but inexorable decomposition which the Communist Party has undergone since 1945, until they themselves either left it or were excluded.

The result of the second campaign was not very different from the first. Other Trotskyist candidates campaigned in diverse parts of France. Ivan Craipeau was the main candidate of a list which obtained fourteen thousand votes in Seine-et-Oise.

In Nancy, as in Grenoble, we formed a little group of less than ten people. For a long time, I believed that we could conquer the world starting with the one or two hundred Trotskyists in France (after all, the Bolshevik Party had also started out as a tiny island). Apart from a few intellectuals, we had recruited a government tax worker, a taxi driver named Telesphore Fleurentin and two real workers who were very active in politics and in the union: Georges Paget, an iron miner, and Raymond Florence, a metallurgist. Both of them worked in the metallurgy factories in Neuves-Maisons, a few kilometers from Nancy. Florence (1908–1981), who was a well-known Trotskyist, was an inflexible yet subtle man, who was held in high esteem even among Communists. He confided to me once "When salaries get low, lunchboxes are practically empty." He was a qualified worker who joined the Trotskyists in 1931, at the age of 23. He met Trotsky in 1933 and became the secretary of his local section of the Socialist Youth. Georges Paget (1904–1984) lived all his life in Chaligny, near Neuves-Maisons. He also had a terrific revolutionary past: he had left school at the age of ten because of the First World War which he soon protested, and at the age of fifteen, in 1919, he joined the Socialist Youth. He became a Trotskyist in 1927 and met Trotsky in 1934. He was also a well-known Trotskyist and even managed to get himself elected as municipal adviser on a SFIO list in Chaligny in 1935. He was a prisoner from 1940 to 1945. Florence and Paget had recruited seven Trotskyists in Neuves-Maisons. Our very "proletarian" activities consisted in selling the party newspaper *La Vérité* every Sunday, calling out "Buy *La*

*Vérité*, the newspaper of the International Workers' Party, French section of the IVth International!" It took courage to sell the newspaper this way, especially at first. But in every street in Neuves-Maisons, some people came out to buy our newspaper. We sometimes sold as many as fifty in one morning. The party meetings took place in my home, once every two weeks. I distanced myself from Trotskyism in 1947, but Florence and Paget remained members of the central committee till 1949, after which they also left, for reasons similar to mine. This long contact with two first-class militant workers made a deep impression on me.

The two last years, I was elected to the central committee. Some of my comrades even tried to persuade me to abandon my professional activities to become the secretary general of the party – in a word, a miniature Trotsky! I admit that I did not immediately reject the possibility. However, after some reflection, I could not hide the absurdity of the situation from myself, and my closest friends in the party, particularly the Filiatres, did not disagree with me. The idea of becoming an apparatchik was not exalting, but even though this may seem surprising today, I did not radically reject it until after a couple of weeks of reflection. At that time, I was a firmly convinced political militant. What I really wanted was to help the party to emerge from a difficult phase and come into its own.

## Goodbye to Trotskyism

I only began to distance myself from Trotskyism progressively, over the year 1947. First of all, I saw that neither the party nor its ideas were progressing at all in the eyes of the "masses". How much time could I go on fooling myself with arguments such as "We are surrounded by enemies, nobody knows about us, but if the masses of workers could read our newspaper, there is no doubt that they would quickly join our revolutionary party." I believed this sincerely. But I was soon forced to accept decisive proof to the contrary. Our party was the only one to support the famous strike of postal workers of 1947, initiated by the FO (Force Ouvrière)[36] and condemned by the press and all other political parties. Thus *La Vérité* was the only newspaper which the postal workers delivered, and for a brief period, it was the only available newspaper. I had no doubt that our ideas would spread among the populace like wildfire. But after the strike was over, we found ourselves with quite exactly the same number of party members

36) FO: The Workers' Force, a relatively moderate union which broke away from the CGT in 1947, wishing to escape excessive Communist influence.

as before. After that, I could no longer avoid realizing that our ideas were not reaching the masses, not because of the ignorance of the proletariat, but simply because they were not credible.

The objectives which had filled me with hope before the war now came to seem ridiculously unrealistic to me. In Paris, things slowly degenerated. The struggles between different currents within the party became exacerbated and ended up by consuming the whole of our activity and becoming unendurable. We were not immune to the sclerosis from which every party eventually suffers.

Other facts contributed to this evolution. At the ENS, I had lived in a protected world where the most eccentric opinions could be expressed. I already talked about our isolation during the war, apart from the period spent at the university from 1940 to 1942. We didn't have a social life worthy of the name until we moved to Grenoble in 1944. My lectures were always successful, thanks to my own conviction in delivering them, but the confrontation with a wider circle soon revealed contradictions. A Trotskyist was supposed to display an aggressive attitude, especially towards Socialist or Communist militants, and even when the parties were supposed to form a united front. But university tradition holds that political views should play no role in the contacts among the members of the university, be they professors or students. Because of this, a dichotomy or even a slight schizophrenia infiltrated my existence. I was on the best of terms with reactionary professors and students, but I had to fiercely criticize militants of other parties, even left-wing parties, which was an absurd and caricatural situation. In Grenoble, and then in Nancy, I had relationships similar to those one finds in all cultivated circles. Trotskyism did not go together with this reality. I came out of Trotskyism to enter into civil society, where in fact I already was. I continued to feel myself as politically left-wing, even extreme left-wing, but intransigence gave way to tolerance and increased comprehension for ideas different from mine. Little by little, slowly but surely, I became detached from the IVth International, which I found absolutely arteriosclerotic and outside of reality.

I was compelled to reconsider the whole system of Marx, Lenin and Trotsky. Already at the ENS, Ruff had sown the seeds of doubt about Marxism in my mind, because the proletariat had insufficient economic and cultural power. Moreover, as I more recently came to understand, the proletariat never really tries to acquire this power. Even nowadays, the proletariat and the peasantry do not progress sufficiently in the educational system, not only because of defects in the structure of the system, but because of their own lack of ambition, and the kind of "stimulants" that there used to be under Jules Ferry[37]. Jules Ferry had encouraged the desire for social promotion by making schooling mandatory and spreading the love of knowledge and the value of success. But today's egalitarian system equalizes by cutting off everything which stands out, and thus tends to level things from below, which strongly privileges the children of intellectual families. Marx considered that religion was a promise of future happiness which kept the people passive, and called it "the opium of the masses". But since he did not impose the education of the masses, it is perfectly possible to consider Socialism, which dangles a shining image of liberation following inexorably from the contradictions of capitalism, in the same light. Communist countries have certainly taken large steps in the right direction, but only after taking power. It is true that education of the masses, sometimes at an impressive level, is one of the major accomplishments of Communist regimes. But the effort ought rightly to be accomplished before the seizure of power. Because of this, the whole foundation of Marxist Socialist theory is false.

In the USA in 1948, Van Heijenoort wrote a sensational article called "A century's balance sheet" published in the *Partisan Review*, in which he distanced himself from Trotskyism by formulating an original criticism of Marxism and Leninism. He carefully studied all of Marx's mathematical manuscripts and found that he had no more knowledge than a reasonably alert student in the first couple of years of university; he was surprised that Marx could have seriously taken himself for a mathematician. If he was so arrogant, Van Heijenoort asked, then how could one trust his judgments on other questions? This verdict was perhaps hasty, but it was based on an intuition which cannot simply be ignored. He cites many mathematical errors of Marx and even more by Engels, together with signs of excessive pride. Glaeser had also noticed some of them. For instance, in a letter to Engels, Marx notes that he had found a new definition of the derivative, without using infinitesimals, which was simpler than the one mathematicians use. It was in fact Lagrange's definition. Of course, one cannot reproach Marx with rediscovering a formula known to Lagrange! The trouble is, that that formula was the only major error of Lagrange, whose extraordinary mathematical genius forces respect. Engels' answer to Marx reveals the megalomaniac nature of the whole dialogue: "Bravo, you've finally explained what those Gentlemen of the University tried so hard to hide from us." Unfortunately for Engels, time eventually consecrated the University Gentlemen's definition. Engels made things even worse by trying to justify Hegel's principle

37) Jules Ferry, Minister of Education in the 1880's, passed controversial laws rendering elementary school free, mandatory and secular, and opened high schools for girls.

that negation of a negation is a superior affirmation by the following argument: the negation of a real number $a$ is $-a$ and the negation of the negation of $-a$ is $(-a)(-a) = a^2$ which is greater than $a$! Absurd. Firstly, he confuses negation of a real number with negation of a logical proposition. If you really wanted to do this, then you should at least say that the negation of the negation of $a$ is $-(-a) = a$, not $a^2$. And even if you insist on $a^2$, it may very well not be greater than $a$, if $a$ itself is less than 1! After all, if $a = 1/2$ then $a^2 = 1/4$. Knowing all this, a mathematician reads the *Anti-Dührung* with a different pair of spectacles.

Lenin's original sin is his theory, explained in *What Should We Do?* published in 1901, of the conquest of power by a unique party of professional revolutionaries and organized on a model of democratic centralism. This theory was successfully applied in 1917 (over the objections of many members of the party). Undoubtedly, it generated within the USSR the dictatorship of a single party, and then of a single man, the secretary general of the party, whose transformation into a tyrant no one had predicted. Lenin's interpretation of Marx's ideas was rather forced. Marx had justified the dictatorship of the proletariat, a class destined to become "the crushing majority" of the population, facing the "exploiting minority" of the bourgeoisie. Marx's prediction was not realized anywhere in the world, and certainly not in the USSR, where the proletariat counted for 2 percent of the nation in 1917. The proletariat may have been the motivation of the revolution, but power was seized by the single party; for some time the party functioned democratically like the Soviets, but it eventually degenerated. Life within the Soviet Communist Party became intolerable as soon as Lenin fell ill in 1922, and indeed even earlier. Marx gave rise to Lenin and Lenin gave rise to Stalin. The power of Stalin's apparatus combined with the world's reaction was too much for Trotsky. His errors have been analyzed many times, and there is no need to re-analyze them here. I understood some of them very quickly, and others only much later.

Perhaps one of Trotsky's errors has not been sufficiently emphasized. In spite of his original thoughts and his views which were sometimes almost prophetic, he never ceased to prepare a revolution like that of October 1917 and a war on the model of the 1914–1918 war. He attempted to form a IVth International, after 1933, consisting of new parties of the Bolshevik type made up of small numbers of professional militants. Like the Bolsheviks in Russia, these parties were supposed to seize power while remaining small minorities, in all the bourgeois democracies of Western Europe; the whole program was destined to fail. He wrote in an article that "the next war will begin where the last one left off, with tanks and airplanes, machine guns and automatic weapons," and he also thought that the armies would have a horse for every two men, as in the First World War. He always claimed to be a specialist on military matters, and he certainly displayed true talent when he created and organized the Red Army. In Coyoacan, he read military reviews from several countries, but as Jean Van Heijenoort showed, he did not evolve towards a comprehension of modern warfare with tanks, airplanes and the necessity for command of the air. His most tragic confusion really concerns the fundamental nature of the war. The Second World War was an imperialist conflict, but its essential aspect arose from the very nature of Nazism and the participation of the USSR. In order to resist, it was absolutely indispensable to give up the notion of revolutionary defeatism. Trotsky only came to realize this in 1940 (he asked the American proletariat to push Roosevelt to enter the war), but only timidly, and his change was not known in Europe. He should have come to this conclusion much earlier, and proclaimed it loudly. In 1940, it was already too late. This blindness disarmed the IVth International politically with respect to the war and paralyzed it for the whole of its duration. I felt this contradiction very painfully.

I no longer really consider Engels as a great man. But I persist in considering Marx to be an essential thinker, even though the bases of his theory have turned out to be false. Could he really predict the future for the rest of eternity? However, thousands of intellectuals, myself among them, dozens of revolutionary or reformist movements (like the Socialist Party of today) who for a century and a half raised Marxism to the level of a revealed religion, really went astray. And even if I support the Socialist Party (not with enthusiasm but just because there is nothing better), even if I voted for Lionel Jospin in the presidential elections of 1995, I do not consider myself a Socialist, at least as long as the word has not been carefully redefined. I still admire Lenin in spite of his tragic errors. There is an unquestionable affiliation between Marx and Lenin, and between Lenin and Stalin. But Lenin died in 1924 and the horrors of Stalinism cannot reasonably be imputed to him. His well-known testament, dated December 25, 1922, was carefully balanced between Stalin and Trotsky, but he modified it on January 4, 1923 with a postscript in which he asks that Stalin not be entrusted with the post of secretary general of the party. Why did Trotsky not make this testament public until his exile, much later, even though Lenin could not do it himself once he became ill? Again, I repeat that the event which is at the origin of the sufferings of this century is not the Russian revolution of 1917 but the First World War, which Lenin opposed fiercely in his text *Against the Current*. The 1917 revolution was precisely a revolt against the war, and suffered its after-effects.

Stalin remains one of history's greatest criminals, but it is true that he directed the war against Nazism, which is no small feat. One can ask another question. What should or could Lenin and Trotsky have done to rectify the situation before it became desperate? Considering that they did not believe it was possible to realize Socialism in a single country, should they have tried it anyway? Should the Bolsheviks have become involved in the October Revolution? Of course, the most advanced groups (such as the workers of the Putilov factory and the Cronstadt sailors) were bursting with impatience, but the whole of the country did not trust the Bolsheviks, as became apparent later. During the Carnation Revolution in Portugal in 1974, a victorious "Lisbon Commune" would have been bloodily crushed because of the peasant masses; the colonels wisely gave up beforehand. Portugal became a democratic country led by the Social-Democratic Party of Mario Soares. The aesthetics of the revolutionary dream were lost, but the solution is not a dishonorable one. Kerensky wanted to continue the war on the side of the Allies; the armed intervention of the Bolsheviks was certainly necessary to make peace with Germany and give land to the peasants, but they should have withdrawn afterwards. In a republic, the Bolsheviks could have continued to play the role of a very dynamic opposition. After the seizure of power and their defeat at the constituent Assembly (November 25, 1917), instead of dissolving the Assembly on the 19th of January, they should have governed democratically together with the winners of the elections. Having conserved power, they should then have deepened the NEP[38) after 1921 and negotiated with the revolting Cronstadt sailors instead of crushing them. If the NEP had continued, it would have re-established a partially capitalist economy or at least a market, especially in the countryside. The only alternative to the "dekulakization of the countryside" was contained in Bukharin's order to the peasants: "Get rich!" (kulaks was the word used to designate rich peasants). On December 27, 1929 Stalin decreed "the liquidation of the class of kulaks", which soon turned into a liquidation pure and simple: millions of peasants were deported, both rich and poor, and the inevitable result was a tremendous solidarity of all classes of peasants against the power of the towns. When the Bolshevik commissaries came to the countryside, they were massacred by peasants. Rather than sending their livestock to the town, they slaughtered it and ate meat for weeks. The quantity of livestock in Russia was lower than in 1913 for decades.

Bukharin upheld a so-called right-wing position at that time, but was nonetheless faithful to Bolshevism. The left-wing opposition had been defeated by a

38) NEP: New Economic Policy. Lenin created this policy to slow down his own reforms because they were causing massive death and deportation.

Stalin-Bukharin coalition, but Stalin then turned against Bukharin and the peasants. Obsessed by classical left/right divisions, Trotsky underestimated the dangers of implacable dictatorship which Stalin represented. The main problem at that point was the elimination of this danger, which would have meant allying himself with Bukharin against Stalin, but instead Trotsky decided to ally himself with Stalin against Bukharin because Stalin was supposedly more to the left than Bukharin: his decision was the exact inverse of what should have been done to eliminate the danger of dictatorship. If he had tried this, the left-wing opposition would not have followed him, and in any case, he would have been unable to change past political patterns. When Bukharin was defeated in 1929, the die was cast. Of course, allying himself with Bukharin (which Bukharin would have accepted because he already feared for his life) would have sealed an alliance with the peasantry, and signified a "backward step" towards a market economy. "There's no need for a revolution if it is to end up so," Trotsky said. But that is exactly what ended up happening under Gorbachev in 1987, and it would have been better to do it earlier, with less cost in human lives. Stalin was strong, but not yet all-powerful. Trotsky and Zinoviev were excluded from the central committee only in 1927, and they should have been concentrating on eliminating Stalin much earlier than that, as Lenin had demanded in his uncirculated testament of 1923. From that year, the threat of dictatorship was perfectly visible. Trotsky addressed plentiful reproaches to the German Communist Party for having ignored the threat of fascism until 1933, but he himself ignored the deep reality of the Stalinist threat until 1929, the year of his exile to Alma-Ata. It was like the farce of the resistible ascension of Arturo Ui in Germany.

In conclusion, it is not always possible to have recourse to democratic methods. Fidel Castro could not eliminate Batista otherwise than by force. Without force, Algeria and Viet-Nam could not have conquered their independence. But when one has used force to obtain power, one must not take avantage of it to stick to power, muzzling one's opponents. The absolute power of a single party is always illegitimate: one must return to democracy, share power, or know how to lose it. I learned about an edifying alternative during a long discussion with my friend Chandrasekharan in Bombay, in 1955. Gandhi had mobilized the Indian people in a non-violent struggle against the British, the *satiagraha*. When the struggle intensified to the point of violence, Gandhi had the courage to slow down the movement, taking the risk of generating a deep sense of disillusionment within the people and postponing decolonization for several years. As a Trotskyist, I had judged Gandhi's political attitude to be timorous. Chandrasekharan pointed out to me that the struggle not only amplified itself, but ended up obtaining a pacific decolonization of India (although it did not go so far as to establish equitable relations between Hindus and Muslims, but that is another story). India remained a democracy, not without many defects, not without corruption, but still a democracy. In July 1974, Vidal-Naquet and I published a manifesto in *Le Monde*, called "Truth and morality in politics" which collected four hundred signatures, in which we asserted that every party or organization which lays claims to an absolute truth supposed to save the world was a prime candidate for dictatorship complete with concentration camps and torture.

Having broken with Trotskyism, I feebly joined the RDR (Rassemblement Démocratique Révolutionnaire)[39], a committee without much future started up by Jean-Paul Sartre and David Rousset in 1948. I was immediately excluded from the Trotskyist Party which I had actually left in 1947, shortly after the elections, with some of my closest friends, Parisot, Demazière and Craipeau. From the end of the fifties to the beginning of the New Left, I did not belong to any party; later I joined the UGS (Union de la Gauche Socialiste)[40], and finally the PSU (Parti Socialiste Unifié)[41]. My political experiences, my years of thinking about the various theoretical aspects of Trotskyism, Marxism-Leninism and the revolution, left me with a particular form of political reasoning and analysis, whose rigor is close to mathematical rigor. Many people continue to consider me a Trotskyist, which is not true any more. But I do not deny my past and I have kept friendly relations with many Trotskyists.

## First postwar engagements

In any case, there are two subjects on which my Trotskyist ideas have not changed at all: internationalism and anti-colonialism. I threw myself into the combat against the Indochina war with no hesitation. My possibilities for action were limited because Nancy was a conservative little town, and politically more or less asleep. There were also the affairs of Tunisia, Morocco and Madagascar. France cared about its empire, and de Gaulle had mobilized the country in the war against Nazism. But after the war, the countries of the French empire aspired to a certain liberty, and France refused to grant it to them. In Tunisia, tension began to grow from the beginning of the fifties. I had contacts with the freedom movement of Habib Bourguiba, the Neo-Destour. I still remember one trip to Tunisia in October 1953. I was invited to give a lecture on distributions at the Scientific University of Tunis by the associate professor Jean Riss. Then

39) RDR: Democratic Revolutionary Rally
40) UGS: Union of the Socialist Left
41) PSU: Unified Socialist Party

he took me traveling around Tunisia. On the road, children sometimes threw stones at our car. Once, I got out of the car and they shouted at me "Ya ya Bourguiba!" I told them that I, also, was on the side of Bourguiba, and they became intrigued and started a discussion. I left them quite amicable. Shortly after the Geneva conventions on Indochina (July 27, 1954), Mendès France undertook an improvised trip to Carthage, accompanied by Marshal Juin and by Fouchet, in order to grant the bey (to his great surprise) internal autonomy for Tunisia, which unblocked the situation. This was on July 3, before the beginning of the war with Algeria; later, the bey gave up his place to Bourguiba, who became the president of the Republic in July 1957.

I took another trip to Tunis in April 1956, invited by Pallu de la Barrière. Again, I was taken around the country, which was covered in multicolored spring flowers. After several very agreeable days in Djerba, my friends took me to the edge of the Sahara, in a desert with little bushes where I picked some sand roses. Always hunting for insects, I amused myself particularly by lifting large stones underneath which I was almost sure to find relatively dangerous white scorpions. I also took advantage of this stay to try to sort out a thorny problem with the Minister of National Education. The young Tunisian mathematician Mohammed Salah Baouendi had brilliantly passed his exams for the degree of Licence, thanks to a scholarship from the Tunisian government. I had been his professor and knew that he could become a very good mathematician if he was able to do a Ph.D. thesis in France. But when his scholarship expired, the Tunisian government ordered him to return to Tunisia immediately and take up a job as a high school teacher. I wanted to convince the Minister that Baouendi simply would not reach a sufficiently high level if he did not spend time in a research department in favorable conditions. I had met the Minister in France during the combat for Tunisian independence. He thought that Baouendi could do his thesis in Tunisia and then maybe leave for France on another scholarship. I explained to him that this was not a good method, and he said this sentence, very surprising coming from a Minister: "We don't have enough experience; I can see that I made a mistake again." Finally, Baouendi received permission to return to France the following year, and did an excellent thesis under Malgrange. I immediately wrote a letter to the same Tunisian Minister asking that he be given a position in which he could direct research at the University of Tunis immediately, in spite of his youth, and that furthermore he be given opportunities of traveling and of inviting foreign mathematicians: all this was granted. We had judged the situation well, but finally, there were so many administrative difficulties that Baouendi gave up his Tunisian nationality and went to the USA, where he became one of the great American analysts.

I am telling this story in detail, although it does not have much to do with the struggles for independence of the 1950's, because it's typical of the tragic situation of research in developing countries. Even if they are able to train scholars, they cannot offer them suitable working conditions.

I also became involved in the struggles for independence of Morocco and Madagascar, but my involvement remained low-level until I came to Paris.

I was also deeply interested in the Yugoslav crisis. On June 28, 1948, Tito broke with Stalin. For the USSR, Yugoslavia had been "the eldest daughter of the church". Tito was even more Communist than Stalin. Yugoslavia was the only country in Europe to have liberated itself from Nazism by itself. Tito's army consisted of one million seven hundred thousand men and had struggled valiantly against the Nazi troops. This revolt of a satellite country against the Soviet Union came like a thunderclap. In Poland, Gomulka had tried to raise his head, but he was imprisoned and order was re-established. Tito, who had expressed divergences, was summoned to Warsaw to explain himself before the Kominform, the Communist leaders of all the Communist popular democracies and Stalin. He refused to go, and the Kominform publicly condemned and "excommunicated" him. We were astounded when we read this in the news. Tito then broke radically with the USSR, followed by his party and the population, insulted by Communist newspapers the world over who called him a fascist and an agent of American imperialism. It is obvious that without the implicit protection of the US and the certitude that the Yugoslav people would defend itself tooth and nail, the Soviet troops would have invaded Yugoslavia. In the neighboring Communist popular democracies, Stalin set up large-scale trials like the famous Moscow trials, in which authentic Communist leaders (like László Rajk in Hungary in 1949) accused themselves of every crime, the worst being to have conspired with the fascist traitor Tito. They served as large scale intimidation maneuvers. The case of the Bulgarian Communist leader Kostov, a former member of the International Brigade in Spain, was very particular. During his trial, he refused to admit having betrayed his country for Tito, and furthermore revealed that he had been tortured. He was taken away secretly and tortured even more, until finally he ended up accusing Yugoslavia. Then the Yugoslav press accused him of being a criminal. Quite soon, however, a protest against this horrific judgment arose within the ranks of the Yugoslav Communist Party, so that the press soon had to make a public self-criticism (the original error was mostly due to Tito himself) and declare Kostov a hero. This revolt of a Communist Party against its leaders, leading to a declaration of error, is unique in the history of Communism.

In spite of all the parodies which we had already witnessed, the Slansky trial which occurred shortly after (1952) managed to reach an exceptionally high level of ignominy. It was a typically Stalinist trial, with openly anti-Semitic tendencies; eleven out of fourteen of the accused were Jewish and had been denounced as Jews. Slansky had been one of the perpetrators of base missions for the Soviets in Yugoslavia. The accused were systematically tortured and all "admitted" having conspired with Tito. All except Arthur London and three others were executed. London was a man of exceptional destiny. He became a Communist at the age of fourteen, fought in the International Brigades and in the French Resistance, was arrested there by the Gestapo in 1942 and deported to Mauthausen in 1943. He was accused of spying in France after the liberation and was able to return to Prague only in 1949, with the help of Clementis. There he was Vice-Minister of Foreign Affairs when the Stalinist accusation fell, and he survived the trial only because his wife was a relation of the French Communist Guyot, an extreme Stalinist with a lot of influence in the party, who succeeded in obtaining that London simply be sent to prison. He came to France only many years later. He wrote a book called *L'aveu* (*The Confession*), in which he described the whole process of the trial in detail, and which created a sensation. In 1970, the book was made into a remarkable film by Costa Gavras in which London and his wife were played by Yves Montand and Simone Signoret. When the film was broadcast on television, it was followed by a debate between the television panel and the public who asked questions by telephone (the number of spectators was estimated at four million). Among the members of the panel were London and his wife, some Communists for instance Jean Kanapa, some non-Communist left-wingers like Jiri Pelikan (a former Czech Communist) and myself. Kanapa was told by the French Communist Party to emphasize the modernist evolution of the party, and stole the show. The panel was overwhelmed by telephone calls from the public, many of whom believed that the whole thing was a setup by the Pompidou government to discredit Communism, and pleaded with London to tell the truth. He told the truth, confirming the veracity of everything in the film. I played the role of a left-winger who had never joined the Communist Party, and intervened mostly on two subjects: Yugoslavia and the Khrushchev Report[42] of 1956. Exasperated, Kanapa told me that the French Communist Party had never heard of the Khrushchev Report! It had been communicated to Maurice Thorez who had decided not to make it public, but obviously it had been read by every member of the central committee. This question was abundantly commented by the press in the following days and

42) This refers to the text of Khrushchev's secret speech to the 20th Party Congress in the USSR in 1956, denouncing Stalin.

seriously embarrassed the leaders of the French Communist Party. *L'Humanité* ended up by publishing a grotesque communiqué essentially claiming that: "we have tried to find out if the central committee of the French Communist Party was informed by Maurice Thorez of the contents of the Khrushchev Report. A delegation was sent to Georges Cogniot, the only member of the central committee of 1956 still alive, and we asked him; he answered positively." This was a new, unforeseen step on the part of the Communist Party. London worked intensively to inform the world public about Stalin's crimes.

We were full of sympathy for the new Yugoslavia, the more so as the regime established a new system of self-management in industry; it was not absolutely perfect but it was still very serious. The Trotskyist Party sent brigades of young volunteers to work in Yugoslavia. I was invited to give mathematical lectures in Belgrade, Zagreb and Liubliana, over a three-week period in 1951. I met the principal Yugoslav mathematicians, in particular Kurepa of Zagreb who had spent a long time in France before the war. The mathematician Karamata, from Belgrade, a mathematician of international level who specialized in tauberian theorems, was absent, but I met his wife to enjoy a friendly lunch topped off with an excellent Turkish coffee. We knew of each other by reputation, and she knew that she could talk to me with confidence. Our long discussion gave me a much better idea of life in Yugoslavia and of the history of the country than my many meetings with officials.

It is regrettable that mathematics never really developed in Yugoslavia. In spite of the presence of Karamata, there was no real Yugoslav school of mathematics.

My guide was very nice, but he tended to speak in the usual Communist language of formulas. I visited museums, touristic sites such as the famous Postunia grottos, which are several kilometers long and shelter proteans, blind amphibians; I also went to see a traditional and very sentimental play, the *Legend of the Lake of Okhrid*. It was obvious that the regime was still absolutely "Socialist" and that Tito's politics were solidly supported. The rupture with Stalin had not had any effect on the mistrust with which the Yugoslav politicians regarded western capitalism. "You in the West . . . we here . . . ."

The personality cult of Tito also continued to exist. When he entered an assembly, as I personally witnessed, the whole public stood up and applauded for five good minutes. The Communist League preserved the same privileges as the former Communist Party, and their powers were similar. But the country did not undergo a bloody purge. Stalin was a great criminal, but Tito was not; he was on a completely different level.

One must remember that when the regent Paul of Yugoslavia concluded a pact with Hitler in April 1941, the Serbs of Belgrade took to the streets in massive protest and obtained its retraction. Hitler had to invade Yugoslavia; it took him three weeks, which were fundamentally lacking during the summer of 1941 when he tried to take Moscow. The Serbs indirectly saved the USSR. We know of the successes of Tito's partisans during the war. The Serbs were extraordinarily proud. During one conversation when I happened to say "When you were liberated . . ." my interlocutor stopped me immediately to rectify "You mean when we liberated ourselves!" which is perfectly accurate. In 1941, the Croat fascist Ante Pavelic obtained the independence of Catholic Croatia from Hitler, and had hundreds of thousands of orthodox Serbs assassinated, and others forcibly converted to Catholicism. The Serbs massacred just as many Croats after peace was established. Tito was able to guarantee the unity of the country in spite of the powerful nationalist movements. Not only did he stand strong against Stalin, but on May 26, 1955, when Khrushchev was forced to change his politics and visit Belgrade, and tried to impute all past errors to Beria, Tito asserted directly that they were due to Stalin. The Serb resistance against the Nazis inspired our admiration. The massacres, rapes and ethnic purging exercised by the Serbs in Yugoslavia today are revolting to the human conscience. Are we just seeing the other face of their nationalistic passion?

On my return to Nancy, I gave a lecture about my trip. Suddenly, Communists burst into the hall and expelled us from it. One particularly virulent fellow called me a fascist who received money from the prefecture every week. I finished my lecture, but in my house. In a similar vein, some (not all) of my Communist students distributed tracts against me in my lecture halls using the same terms. I knew who was leading them, and anyway it did not have much influence among the students, with whom I tended to be popular. I did however eventually counter-attack during a course. I told the class that it was their duty to remain neutral, and that even if they disapproved of me politically, our cordial relations should prevent them from accepting that such insults be showered on me. They burst into applause and the leader calmed down a bit. The day of the final exam arrived. He passed the written test reasonably well; he was a mediocre but serious student from the Institute of Mechanics. But at the oral exam, faced with my glacial attitude, he became so meek that he couldn't pronounce a single word. I passed him anyway on the grounds of the written test, but he was mortified; perhaps it made him think a bit.

## My union experience

Professors and associate professors at the university met regularly in faculty meetings, but also at the Union of Higher Education.

In Nancy, the professors, associate professors and assistant professors belonged to a union directed by Marcel Roubaud, the director of the Institute of Geology and Mining Prospection; he was a Socialist and a firm believer in the unionist movement. We had the habit, which was quite unusual in other universities, of discussing our professional problems from a unionist point of view. I was the most political among my colleagues, and they elected me secretary of the union. This was the beginning of an intense period of activity during which I was led to deal with problems I had never previously thought about, for instance questions of retirement (I was thirty). We discussed the status of a public functionary established by Maurice Thorez, and problems of salaries and credits. For me, it was a whole new training, which led me to meet colleagues from almost every university department.

Teachers' salaries were very low at that time. To increase them, the union started a one-day strike, which was massively followed, and they organized a public meeting which was attended by an immense crowd. It was most unusual to see university professors on strike. Roubaud gave an excellent talk, there were talks by schoolteachers, and I gave the final talk, as the departmental secretary of the FEN (Fédération de l'Education Nationale)[43]. A contradictor was present, who asserted that schoolteachers were well paid, that in most communities they also had free lodgings, and that in fact, they often benefited from a double salary, seeing that schoolteachers often married other schoolteachers! I don't have much presence of mind, but I answered him that anyone in France has the right to marry a schoolteacher. I believe that our teachers' strike in Nancy was one of the most successful in France.

I also had more political contacts with other circles. At first, as the secretary of the Union of Higher Education, I became interested in the other teachers' unions. I came to know many schoolteachers and professors; I was one of the most active among them. This perhaps explains why one of them, named Mougenot, came one fine day to tell me that in my absence, they had elected me secretary of the departmental Federation of National Education, which grouped all the teachers' unions. I accepted the task, and made many contacts which I never regretted. At the time of the grand schism between the CGT and the Force Ouvrière, in 1947, discussions everywhere were very lively; the huge majority of the members of

43) FEN: Federation of National Education

the unions of Meurthe-et-Moselle, from elementary school teachers through to university professors, were against the schism and remained with the CGT. I was the delegate of the Federation to the meeting of the FEN in Paris, during which we declared the FEN independent. This decision was necessary, I believe, but the citizens of Nancy were not enthusiastic. I suggested a motion which was neither for the FO nor for the CGT, nor even for the independent FEN, called "For a Federation of National Education joining a democratized CGT". The adjective "democratized" was obviously essential. The success of my motion was very limited; I believe I obtained two votes, one of which was mine. At that point, a FEN-CGT was formed; it was a grouping of unions which belonged to the independent FEN but also maintained their membership in the CGT. Even some non-Communist union members from Nancy, including myself, joined it.

Things did not stop there. As the secretary of the union and of the federation, I had regular contacts with the Bourse du Travail[44] where union members from every branch of work met regularly, and the departmental conferences of the CGT were held. Naturally, the schism extended to the Bourse. The relations I formed with departmental secretaries of unions of metallurgy, shoemaking, chemistry etc. were quite interesting. They knew about my Trotskyist leanings, but it never caused any difficulties in our relations. I also came to know many workers, some of whom visited us at home. Not many professors actually came to know workers. Many of the workers were Stalinists, others were Socialists, some joined the FO, others stayed with the CGT, but none of these things disturbed our personal relations. I still remember the naive enthusiasm of one shoemaking union member who had benefited from a paid trip to Hungary to make contact with the unions of that country. On his return, he described with fervor the political education of the Hungarian workers, compared to that of French workers. It was hard to get French workers to sign the Stockholm Appeal against atomic weapons, he said, whereas in Hungary whole factories signed it.

Later, I did not occupy any directorial position. The SNESup (Syndicat National des Enseignants du Supérieur)[45] disappointed me, although I remained a member for some time, even when it defended positions which were the opposite of mine. I haven't belonged to any union for over twenty years. Former Trotskyist, former unionist, always left-wing, even extreme left-wing, I feel better and more useful this way. Anyway, the SNESup degenerated completely and practically doesn't represent anything any more. More generally, all the

44) a location where the unions have their offices

45) SNESup: National Syndicate of Teachers in Higher Education

teachers' and students' unions of today (except for elementary school teachers) consist of ridiculously small numbers of members, and only a tiny proportion of workers belong to unions, compared to the numbers in England and Germany. Only the CFDT (Confédération Française Démocratique du Travail)[46] still has a positive image in my eyes, but not its teaching union the SGEN (Syndicat Général de l'Education Nationale)[47]. I am a strong believer in unions, and even in powerful unions which represent the working classes, but I cannot belong to today's teaching unions.

## Teachers, students, ENS students, researchers

The first courses I taught in Grenoble in the first year of university mathematics were fairly modern. I taught linear algebra, i.e. vector spaces of arbitrary dimension. The students took the bait, and the rumor soon circulated that I was teaching spaces of dimension greater than 3, so I was teaching relativity .... Several students offered to write up notes for the course. On the whole, the level was not too high but still reasonable; Grenoble has always been one of the great French universities.

That same year, I taught the agrégation preparation courses to students of the ENSJF (Ecole Normale Supérieure de Jeunes Filles)[48]; I traveled regularly up to Paris to teach there. It was by no means disagreeable to teach a class of twenty attentive and intelligent young ladies; I shouldn't think any young man would remain insensitive to the charm of the situation. A strong current of sympathy soon established itself between us. They prepared their lessons solidly and conscientiously. Some of them were so enthusiastic that they proposed to contact people from the Resistance in order to have me become a professor at their Ecole Normale! They were amazed when I rejected such methods and appeared altogether taken aback when I explained to them that university promotion was a serious and codified procedure having nothing to do with the Resistance.

The students of the University of Nancy were more varied. I taught "general mathematics", the first year curriculum, and the students were quite bad. I had about fifty of them. I remember a discussion with one of them, an abbot who taught in a private school, and who had just obtained 2 out of 20 on his exam for the third time. "It's strange," he told me, "this time, I really thought I had succeeded." Dieudonné taught differential and integral calculus, and I taught rational mechanics. He had about ten students, and passed two of them on the

46) CFDT: French Democratic Confederation for Work

47) SGEN: General Syndicate of National Education

48) ENSJF: formerly the feminine counterpart of the ENS (which has now become co-educational)

average. The situation was not comparable to what happens today! Dieudonné was filled with bounty and generosity, but the students feared his bursts of anger. He was an agreeable man, but subject to brusk, ephemeral outbursts, some of which remained legendary. One more or less hopeless young girl found herself threatened, during her oral exam, by "If you go on like this, I'll throw you out the window!" upon which Dieudonné crossed to the window and really opened it! Se non è vero, è ben trovato. He was an exceptional man altogether. He got up every morning at six and went to bed at midnight; most of his waking hours were spent doing mathematics. His pedagogy was triumphant; he lectured with communicative conviction and a radiant smile. He was an excellent musician; he could direct a concert of chamber music. He also played bridge regularly and attended all important cultural events. He was a sophisticated cook. His wife took care of the daily meals, but handed the kitchen over to him whenever guests were invited, and he was to be seen bustling there, dressed in his cook's cap and ample white apron. I consider myself as a student of Dieudonné (in mathematics, not in cooking). He trained me in Bourbakism in Clermont-Ferrand, and continued to have a lot of influence over me in Nancy, where we worked on topological vector spaces together. He taught me a great deal of algebra.

Delsarte was just as interesting. He was older than the rest of us, and was dean of the faculty of sciences for a long time. He was conservative, a practicing Catholic with links to the bishopric of Nancy, but he was capable of breaking free and evolving. We did not frequent the same circles in Nancy. He lived for several years in Japan where he directed the French-Japanese House, and he was so impressed by the religiosity of the country that his Catholicism took on a quite unorthodox coloring. He was a deeply righteous man. If it was necessary to intervene with the authorities to straighten out injustice (I remember one particular trial where there were some echoes of anti-Semitism), or cases of unjustly fired employees, Delsarte and I undertook action together. We got along very well, and our undertakings were generally approved. Every fall, he taught a new graduate course on a subject which he had carefully studied the year before and over vacation: analysis, geometry, algebra, astronomy. Many of them ought to have been published. He was very short-sighted and never looked at the class, but gazed vaguely at the ceiling. His courses were luminous. He also taught a very funny, naughty song called *The Wife of the Ro-Ro*, to the members of Bourbaki, and we sang it together at our meetings. His seriousness when faced with difficult situations and his absolute integrity got him elected to the position of dean of the faculty for many years.

The analyst Roger Godement and the algebrist Luc Gauthier obtained posi-

Delsarte and Schwartz

tions in Nancy after I did. Godement had done beautiful work on group representations and functions of positive type, essentially in order to generalize to non-commutative groups the results on commutative groups which can be obtained using the Fourier transform. Convolution still exists in this situation, but there is no Fourier transform. He wrote a beautiful paper on spherical functions. He was our best expert on the works of Gel'fand (USSR) and Harish-Chandra (USA). Godement struggled throughout his life against the international military-industrial complex, with a rage and a passion which rendered any discussion with him difficult. For this reason, he was rather isolated. It is a pity, because he had a lot to say. In spite of his irascibility, he was liked and esteemed by his colleagues for his intellectual qualities and his moral rectitude. He never pardoned von Neumann for having abandoned mathematics to create computer science.

I was close to Luc Gauthier, a well-known algebrist who replaced Capelle in applied mechanics. He died young after a serious illness. We were all on excellent terms; even though we lived near each other and could have met easily, we used to telephone each other interminably just as though kilometers separated us.

Godement, writing on outdoor blackboard (1954).

Starting in 1947, a unique position of assistant professor was created in the department. It was occupied by Georges Glaeser. He was an old friend of mine and a schoolmate of Bertrand, and had become a high school teacher after passing his agrégation. He lived in Vichy during the Occupation, with his wife Pauline, a philosopher, and he belonged to our Trotskyist group when I was in Clermont-Ferrand. Then I lost sight of him until after the war. When we saw each other again, he was looking for a university position. We gave him the job. He was an assistant for every course and had to work incredibly hard: he had full responsibility for all written and oral exercises in general mathematics, mechanics, and differential and integral calculus. He did it all at the price of terrific overwork, and succeeded in working on Banach algebras with me as well. He finished his thesis with Bruhat in Nancy and obtained a position at the Scientific University of Strasbourg, where he spent the rest of his life. He proved a remarkable and difficult theorem about differentiable symmetric functions of a finite number of variables, and later devoted himself to pedagogy and didactics.

Unlike Dieudonné, I was lucky enough to have very good students in mechanics. There were students in Nancy from the ENSI, the Ecoles Nationales Supérieures for engineers, for mechanics, chemistry, geology, mining, brewery, dairy. Each of these institutes was staffed with excellent university professors whose students had to take courses and pass the same exams as the students at the university. The students I had from the Mechanics Institute gave me considerable satisfaction. It was odd that I should have had to teach them, because I am not at all a specialist in mechanics, which I taught in a most conventional manner, without bringing in anything new. Nowadays, mechanics courses are much more modern in any university.

Sometimes I introduced variations on the usual themes: a little hydrodynamics, or the equation of vibrating strings or sonorous pipes. Once or twice I taught restricted relativity. Anyway, I was happy and the students were too.

Apart from the low-level teaching, all the teachers shared the task of teaching the preparatory courses for the agrégation; each took his turn for one month. Teaching in Nancy was less interesting than teaching the young ladies in the ENSJF. For one thing, the level was not comparable. We had to give courses preparing both the written exams and the oral lesson of the agrégation, which took the form of a class taught by the candidate; in the ENSJF we taught only preparation for the oral lesson. Delsarte was the director of the agrégation classes in Nancy. The duty of correcting the written homework was shared by all the professors and I had to give and correct an agrégation homework about once a month. The subject of the homework was always chosen from a past agrégation exam, and each one lasted seven hours: it took me that much time to prepare them and correct them. I have a rather bad memory of the whole thing.

The best students in France went to the Grandes Ecoles, or at least to a few prestigious universities such as Paris, Strasbourg and Grenoble. This left Nancy very few students actually intending to pursue research. Every year, we had one foreign student who received a position in the CNRS. There was an American interested in harmonic analysis and a Brazilian, Paolo Ribenboim, an algebrist who wanted to work with Dieudonné and with whom we remained friendly. He fit into Nancy so well that he got married there. He is a great specialist of Fermat's last theorem, which he proved for a large class of exponents (the general statement was recently proved by Andrew Wiles using a most remarkable procedure). One thing about Ribenboim always struck me. He was a well-trained algebrist, but he knew next to nothing about analysis: when I asked him what $\int_{-\infty}^{+\infty} dx/x$ made him think of, he answered that it didn't make him thing of anything at all. He later returned to Brazil, then went to Canada. The arrival of these foreigners were like an ephemeral and short-lived breath of fresh air.

The situation changed interestingly in 1948, thanks to Henri Cartan. He was one of the professors at the University of Strasbourg who had taken refuge in Clermont-Ferrand, and he had been given a position at the ENS in 1940. He taught there until the beginning of the academic year in 1945. At that point, he spent two years in Strasbourg and then a year in the USA, after which he returned to Paris in 1948, at the beginning of the academic year. He devoted all his energies to the students of the ENS and was a first-class professor. He took careful notes on each student which he has kept until today. He was really a mentor for them, in the noblest sense of the word, and a large number of today's French mathematicians owe their entire training to him. If one considers that on a world scale he is one of the most eminent living mathematicians, one has some idea of his role in the destiny of modern mathematics. He was also one of my best friends. We talked about the most diverse subjects, and during some periods we telephoned each other every day.

Apart from his teaching the different classes at the ENS, he organized a seminar for "adults" every year with his most brilliant students, particularly Jean-Pierre Serre. Every lecture was minutiously prepared. The seminar dealt with a different subject each year and attracted mathematicians from all over the world. I went several times. Algebraic topology and analytic and algebraic geometry were the principal themes. The volumes of notes of these seminars form a monumental publication. But it was difficult for him to take care of all the ENS students from three successive years. There were two other professors from the Sorbonne who also gave seminars at the ENS, Maurice Janet and Georges Valiron (just as in my time, he taught functions of one complex variable, a rather worn-out subject); it was a little depressing. Many students felt like "orphans" and Cartan worried about them.

And what happened in Nancy? In 1947, while Cartan was still in Strasbourg, but organizing the training of the ENS students from a distance, he worried a lot about the fate of his "orphans". In particular, he had trouble training future analysts. Dieudonné advised him to send a few of them to Nancy, the capital of analysis. Thus, André Blanchard and Bernard Malgrange spent the first semester of their second year at the ENS in Nancy in 1948–1949, and the following year we welcomed François Bruhat and Marcel Berger. Alas, that was the end of it: Maurice Janet, another professor at the ENS, objected to the system, saying "There is only one ENS, in the rue d'Ulm, and the students should be spending their time there." So we had no more students from the ENS.

Fortunately, Malgrange and Jacques-Louis Lions came to Nancy in 1950–1951, as soon as they left the ENS and received scholarships from the CNRS. The breath of fresh air lasted three years. Malgrange, Lions and Bruhat did their theses with me in 1954 and 1955. All three then became university professors, among the best mathematicians in France. The young Paul Malliavin, who was not from the ENS, came by choice to spend several years in Nancy.

He is known worldwide. Lions, Malgrange and Malliavin are members of the Academy of Sciences. Lions, who is the founder of applied mathematics in France, is a professor at the Collège de France and used to be a professor at the Ecole Polytechnique. Bruhat was an excellent specialist of Lie groups, and collaborated with Jacques Tits (another member of the Academy). He taught in Nancy for several years before moving to Paris, then he sacrified part of his life as a researcher to tasks of general interest: he became Vice-President of the University of Paris 7 after 1968. Only Berger was not happy in Nancy. We did mostly analysis and he was a geometer. Later, he became one of the best French geometers. He trained many students and for fifteen years he was director of the IHES (Institut des Hautes Etudes Scientifiques) in Bures-sur-Yvette. He is a corresponding member of the Academy of Sciences. Even if our breath of fresh air only lasted three years, it produced a miracle. A few good students, even just two a year, among five professors – everyone was enthusiastic.

I would like to mention two other students whom I met a little later, in Paris, but who belong to the same generation as those I've just spoken of: Martineau and Trèves. André Martineau was a student of the ENS. He worked with me for several years and defended a beautiful thesis on analytic functionals. He was intelligent, subtle and very original, with a sweet and generous temperament. His work was the basis for a lot of later work, and his name is well-known in the mathematical world. In 1972, he died of cancer in the prime of his life, deeply regretted by his numerous friends and colleagues. François Trèves was born in Italy but raised in Paris. From the beginning of his graduate studies he worked with me and passed his thesis on partial differential equations, like Malgrange and Lions; it is an immense subject which he spent his life exploring. Very soon, we became intimate, almost inseparable friends. Even though our political and moral ideas are very close, we are never tired of discussing ways to change the world or to foresee its end. He likes to laugh and has a lot of humor. Sometimes even involuntary humor: once he introduced himself quite unexpectedly and involuntarily as "François Premier". He himself never understood what provoked this hilarious lapsus. A frequent visitor in Autouillet, he took part in many of the important episodes of my life. He was not only my student in mathematics, but also in entomology, as he started a butterfly collection after our stay at an international conference in Boulder, Colorado. Slowly, he came to specialize in Morphidae, and has a beautiful collection of them. For many years, we traveled together and went butterfly hunting together. He left France for the US and now works in Rutgers and lives in Princeton. Our families are close, particularly Claudine who loves to see François. He is married and his two children are now grown up. Of course, we telephone each other frequently and never lose an occasion to make a (possibly entomological) detour when traveling. He works enormously and produces remarkable articles and very pedagogical and reputed books, for example his book on pseudo-differential operators and Fourier integrals, which I use frequently.

## Alexandre Grothendieck

The most fantastic gift to Nancy came in the person of Alexandre Grothendieck. Born in 1928 of a German mother, a journalist living as a refugee in France, and a Russian anarchist, he came to France at the age of nine, shortly before the war. His father was deported and died during the war, and his mother, to whom he was deeply attached, worked cleaning houses in order to survive; during the war they were interned several times in the Gurs refugee camp. To say his secondary education was perturbed by these circumstances is of course an understatement. However, he was able to pursue them thanks to Magnier, a high school teacher who worked for an association helping refugees. He passed his Licence at the University of Montpellier, without being noticed by his professors (whom he did not particularly notice either!) He then traveled to Paris in order to become a mathematician: he wanted to work in analysis. After having looked around a little, he went to see Cartan who advised him to come to Nancy. We received him in October 1951. He first presented Dieudonné with an article of about fifty pages called "Integration with values in a topological group". It was very precise, but not at all interesting. Dieudonné displayed his characteristic – ephemeral – aggressivity and scolded him energetically, telling him that generalizing for the pure pleasure of generalizing was no way to work. One should attack difficult problems which could then have applications in the rest of mathematics, or in other sciences: the results in his article would never be of any use to anyone. Dieudonné was right, although Grothendieck never admitted it. However, Dieudonné did not give up on him. He and I had just published an article on "The spaces $\mathscr{F}$ and $\mathscr{LF}$". Distributions had started up a whole series of new topological vector spaces, above all the space $\mathscr{D}$ of $C^\infty$ functions with compact support. I had put a very special kind of topology on this space; its general type has not yet been understood. We had more or less (rather less than more) handled topological projective limits for some time, and André Weil had studied them well in the framework of projective limits of topological groups, in his 1940 book on integration in topological groups. We used it more and more, but still parsimoniously; since then it has become a very familiar notion. I believe that it had not yet been stated that a Fréchet space is the projective limit of a sequence of Banach spaces. As for the notion of the dual of an inductive limit, it was not known at all! Now, the topology on $\mathcal{D}$, which I had introduced carefully and prudently after working with pseudo-topologies, was exactly the inductive limit of the topologies on the $\mathcal{D}_K$ for compact subsets $K$ of $\mathbb{R}^N$. I carefully defined the neighborhoods of the origin in $\mathcal{D}$, then gave the characteristic property which was precisely that of being an inductive limit, without giving it a name. I only did this for the particular object $\mathcal{D}$, without daring to introduce a general category of objects. Mathematical discovery often takes place in this way. One hesitates to introduce a new class of objects because one needs only one particular one, and one hesitates even more before naming it. It's only later, when the same procedure has to be repeated, that one introduces a class and a name, and then mathematics takes a step forwards. Other inductive limits were introduced, then the theory of sheaves used them massively and homological algebra showed the symmetry of inductive and projective limits. Dieudonné made the effort of doing this for $\mathcal{D}$, without my having thought of it. We then worked in close collaboration on the $\mathcal{F} - \mathcal{LF}$ article: I gave all the theorems I needed for distributions, which I had just finished writing up, and together (he more than I) we proved them in the general framework. The proofs closely followed those I had developed for distributions, but they were more complicated. I owe Dieudonné a lot for all the algebra I learned from him.

To return to Grothendieck, our article ended with fourteen questions which we had not been able to solve. Dieudonné proposed to him to think about any of them he liked. We didn't see him any more for several weeks. When he reappeared, he had solved more than half of them! The solutions were deep and difficult and introduced new notions. We marveled. Obviously, we were dealing with a first-class mathematician. He immediately published this manna of results and became our collaborator. We set to finding him a thesis subject. In 1952, just before leaving to spend the summer in Brazil, I proposed the following subject to him: put a good topology on a tensor product $E \otimes F$ of locally convex spaces $E$ and $F$. I was then beginning to write up the theory of vector-valued distributions, where a good topology was obvious for $\mathcal{D}'(F) = \mathcal{L}(\mathcal{D}, F)$ (the space of continuous linear maps from $\mathcal{D}$ to $F$), so that it induced a good topology on its dense subspace $\mathcal{D}' \otimes F$ and $\mathcal{D}'(F)$ became the completion of $\mathcal{D}' \otimes F$ equipped with this topology. But I wasn't succeeding completely, a proof that I did not yet manipulate topological vector spaces well enough. Grothendieck was exactly the man to discover such a tensor topology. At the end of July, in Brazil, I received a very disappointed letter from him: there were two locally convex topologies on $E \otimes F$, both equally natural and yet different! So there was nothing interesting to say. I didn't know what to answer him, and yet, on $\mathcal{D}' \otimes F$ there was clearly just one good topology. But difficulties and defeats can be the source of victories. Two weeks later, I received a new, triumphant letter: the two topologies coincided in the case of $\mathcal{D}' \otimes F$. There is a type of locally convex space $E$, which he called "nuclear", such that for every $F$, the two topologies on $E \otimes F$ coincide. Everything became clear. The space $\mathcal{D}'$ is nuclear. The word "nuclear" comes from the fact that if $F$ is another $\mathcal{D}'$ space, one rediscovers the kernel theorem which I had lectured on at the International Congress of Mathematicians in 1950.

He rapidly discovered the fundamental theorems of the theory of topological tensor products and published a summary at the end of 1952. At the beginning of 1953, his thesis was entirely written. It was a monument of more than three hundred pages, a masterpiece of immense value. I had him defend it towards the end of the year. It was necessary to read it, to learn it, to understand it, because everything in it was deep and difficult. I spent six months full time on it. What a job, but what a pleasure! The statements of the theorems were miles long, because the reader was not spared the slightest detail. I learned a large quantity of new things. It was the most beautiful of "my" theses. I had known the joys of teaching and personal research, but the collaboration with this talented young man was a new, fascinating and enriching experience. Grothendieck was immediately recognized as the world's grand master of topological vector spaces. In some sense, he played a mean trick on me: encouraged by his thesis, possessing topological tensor products, I decided to attack the problem of vector-valued distributions in a general framework, and spent too much time and energy on it. In that year 1953 he was recruited by Bourbaki. But he only remained for a few years: he lacked humor and had difficulty accepting Bourbaki's criticism which is customarily rather virulent, and he left.

Our friendship was deep. He often came to play the piano at my house. He was naturally very kind, and was particularly affectionate with our housekeeper Alice, who reminded him of his mother. He moved to Paris in 1953, then spent a year in Brazil where he wrote something on topological vector spaces which appeared in 1958. He lived alone and worked twenty-five or twenty-six hour days, which had the effect of shifting his daytimes throughout the usual cycle, so that sometimes he worked in the daytime and other times at night. It is indeed a little-known peculiarity of some mammals to have a natural cycle which is actually slightly longer than 24 hours. If you keep them in a lighted underground room, their circadian rhythm reaches cycles from 25 to 27 hours. Their usual 24-hour-a-day life is just an adaptation to the solar rhythm. This is why an airplane trip to the west is not so difficult for most people, whereas a trip to the east creates longer and deeper problems. Grothendieck's underground room was the hotel room where he worked.

He soon changed subjects. After attending Cartan's seminar, he began to work in algebraic geometry with the modern tools of the 1960's, and he renewed and completely changed that branch of mathematics. One of his most beautiful theorems is the following. The famous Riemann-Roch theorem dates back to the last century; the German mathematician Hirzebruch generalized it from dimension 1 to higher dimensions. Grothendieck transformed it completely in 1957, turning it into an extremely general theorem of algebraic geometry. His theorem, known as the Riemann-Roch-Grothendieck theorem, was lectured on by Serre at the International Con gress of Mathematicians in Stockholm in 1962. For his sensational work, he was given the Field's medal at the International Congress of Mathematicians in 1966. As a protest against the repression of the Hungarian insurrection in 1956, he refused to go to the USSR to receive it. He wrote a letter on the subject and demanded that it be read, but the president of the Fields committee, Georges de Rham, refused to read it. The actual medal was given to him later, and he offered it to Viet-Nam during the war with the US, to use as gold. It is not irrelevant to add that he spent several weeks teaching mathematics in a Vietnamese university which had fled to the countryside during the American bombings.

His work constitutes one of the great scientific realizations of this century. A little later, he spent some years as a member of the IHES (Institut des Hautes Etudes Scientifiques)[49] in Bures-sur-Yvette, and while there he was awarded the Swedish Crafoord Prize, which he refused for moral reasons which I did not really find convincing. After 1968–1970, he devoted himself to ecology, and then took to living as a recluse in the countryside, devoting himself to solitary meditation. In 1983, he decided that he no longer wanted to teach in university, and asked for a job in the CNRS, submitting a research program called *Esquisse d'un Programme* although he had not published anything for over ten years. This caused a long discussion within the CNRS committee, with ardent opinions for and against. Finally, thanks to the influence of Jacques Dixmier, the reasonable solution was adopted. After all, many foreign scholars had obtained positions where no particular activity was required of them, for example Dirac. It's not that it was advantageous for Dirac or for science, because he ceased to work very soon. But Sir Michael Atiyah has an analogous position at Oxford (Royal Society Professor) and he remains very active. Grothendieck had already done enough work for several mathematicians, and they gave him the job. However, he devoted himself to meditation more than to research. Thus we lost the benefit of his mathematical ideas, and I regret it, but he amply deserves homage and

49) IHES: Institute for High Scientific Studies

no one can oblige him to devote himself to science. There was another factor which probably had a negative influence. With his usual modesty, Dieudonné proposed to write up Grothendieck's results. Grothendieck thought, Dieudonné wrote. The situation was quite ambiguous, as Grothendieck had a slight feeling of contempt for Dieudonné. But Grothendieck himself lost any capacity for writing, and especially towards 1970, in the final years of his seminar which was at an incredibly high level, his lectures grew obscure and the audience, which consisted in several of the most brilliant young mathematicians of the time, had difficulty following them.

Grothendieck was furious with them for having distorted his ideas, and attacked them violently in a book called *Reaping and Sowing*, which no editor accepted to publish. The editors were right not to publish the public attacks, but it is a pity that the scientific community never profited from the deeply justified reflections on research and ethics contained in the book. Unfortunately, his attitude made any kind of compromise difficult. I tried to persuade him that nobody in the mathematical community was angry with him, and that if he didn't want his ideas to be distorted, he should explain them himself. It was not too late, and his lecture halls would certainly have been packed full. But he answered me that he had given up such ideas long ago, and did not want to abandon meditation, which was his most essential activity in his eyes. It is a pity. I doubt that Grothendieck will ever write his autobiography or the results of his meditations, so in these pages I tried to give an idea of the gigantic scientific, intellectual and human dimensions of the man.

## Research

The discovery of distributions was one thing, writing it up was another. The task I gave myself from the year I spent in Grenoble until the appearance of the two books on distributions in 1950 and 1951 was to write up everything completely. The two books together contain about four hundred pages. Naturally, they contain some grand ideas, but also a large number of technical parts. In fact, as I said before, I find the book too short. On the whole, though, it is well-written. Generally, publications are not very time-resistant, and it is necessary to rewrite them or have them rewritten. But the book on distributions has withstood the test of time. I believe that for the most part, I introduced the best possible definitions and proofs.

This was my working schedule at the University of Nancy. Every morning, I worked on pure research – not distributions – on all kinds of subjects, and in the afternoons, I wrote up distributions. The combination of these two tasks and the need for scrupulous technical rigor caused me to advance rather slowly. Gel'fand's book on generalized functions only appeared in 1955. I devoted so much time to research that if I did not pay attention, I started to suffer from overwork. To avoid that, I finished every afternoon with a non-mathematical activity. I also thought about varied subjects, often inspired by the Saturday seminars in Nancy. Before Nancy, I had thought about potential theory, with Brelot in Grenoble and thanks to an article of Cartan which I read. In our Saturday seminar, we had talks on spectral decomposition, representations of compact groups (Peter-Weyl), a few rather poor ideas on representations of locally compact groups, without Harish-Chandra's theory. We studied Dixmier's direct sums of Hilbert spaces and his trace theorems, and von Neumann's classification of algebras into types I, II and III. Lie algebras and Lie groups were also studied. I was very interested in them and wrote up the version zero of the book on Lie groups for Bourbaki in the fifties. Once, I was dictating a letter to Dieudonné to my secretary and I said "Lie groups (groupes de Lie) prevent me from sleeping" which she transcribed "Groups of beds (groupes de lits) prevent me from sleeping", which is, after all, no less logical. (Another time, when I dictated "théorie quantique des champs" she wrote, always with logic, "théorie cantique des chants"). However, I never studied their classification and know almost nothing about exceptional Lie groups. Dieudonné worked on Dedekind number fields and $p$-adic fields, adeles and ideles. We also studied algebraic topology. After the publication of a famous little book by Cartan and Eilenberg called *New Methods in Algebraic Topology*, the use of exact sequences became a dominating technique in the subject. Our study of cohomology was rocked in the cradle of short exact sequences and long exact sequences with coboundary operators. I looked at Cartan and Eilenberg's book on homological algebra, but not very deeply. I studied Godement's book on sheaves more carefully, but avoided spectral sequences, which I have never used. I wrote up the version zero of the Bourbaki book on differential manifolds, a subject which was very familiar to me. I used it in analysis and even in probability (semi-martingales on differential manifolds in the seventies and eighties). Jean-Pierre Serre told me that when he read the version zero of the theory of manifolds – he was still a guinea pig in Bourbaki – he was astounded to learn that the tangent bundle of a variety was not always trivial! Since then, and this is an understatement, he has done first-class work on fibre spaces and received the Fields medal in 1954! He is one of the very best mathematicians of our time!

I brought an unfortunate modification to the definition of the word "elliptic" for partial differential equations. I said that an operator $P$ was elliptic if in every open set where $Pf$ is $C^\infty$, $f$ is also $C^\infty$. This is one of the properties of operators which are elliptic in the usual sense, and I considered that it was one of their main properties and could be used as a definition. But it was an error. The grand classification of differential operators into elliptic, parabolic and hyperbolic operators is more than a century old and should remain as it is. At a colloquium on harmonic analysis in Nancy in 1954 (I was living in Paris) I lectured on the work of Mizohata, a Japanese mathematician who had written his thesis with me, and called it "Ellipticity of parabolic operators". And indeed, according to my definition, a parabolic operator was elliptic. It provoked murmurs of scandal in the audience, who decided unanimously that the word elliptic should be used in the classical way, and that the property I mentioned above should be called hypo-elliptic. This was universally adopted.

I read Jacques Hadamard's magnificent book on hyperbolic partial differential equations of the second order, in 1951, and it was extremely difficult for me. It uses finite parts in a deep way, because the elementary solution is expressed by a distribution-finite part. I read it, naturally, applying distributions automatically everywhere, which gave me great pleasure. It is probably Hadamard's most important book. In my book on distributions and in all the research I worked on at the same time, convolution, Fourier transforms and Laplace transforms played a fundamental role. The theory of group representations is impregnated with it.

Naturally, we were enthusiastic about Gel'fand's discoveries in spectral theory and operators on $\mathbb{C}^*$ Banach algebras. Gel'fand was a Soviet mathematician, and because he was Jewish, he was very late in entering the Academy of Sciences. Gel'fand's theorem stating that every commutative $\mathbb{C}^*$ Banach algebra is isomorphic to the algebra of continuous functions of a compact space is a classic today, taught in many courses at Licence level. He has another equivalent, analogous but much more complicated theorem on non-commutative Banach $\mathbb{C}^*$-algebras.

I also studied complex analytic manifolds and their singular sets, Stein varieties and Cartan's A and B theorem, as well as Kähler manifolds. The Weierstrass preparation theorem and its application to the structure of analytic sets became familiar to me. It goes without saying that I also studied Hille-Yosida's theory of semi-groups, which I pushed quite far since I extended it to semi-group distributions. It was further generalized by Lions; I don't think these generalizations are well-known enough. Topological vector spaces is one of the subjects I studied most closely. Dieudonné extended a certain number of Banach's theorems to them. Distributions naturally introduced topological vector spaces rather than Banach spaces. Thanks to Grothendieck's work, they were extended even further. I believe that by the end of my stay in Nancy, I knew everything which could be known about topological vector spaces, spending even too much time on certain details. In particular, I continued my work on distributions by setting to work on vector-valued distributions after reading Grothendieck's thesis, and they occupied me from 1954 to 1958. I exhaustively considered the most general cases; I told myself that if I did not write up everything in detail, nobody ever would. Vector-valued distributions are used frequently and theorems from the two books are often cited. However, what people usually use is a set of relatively elementary theorems on vector-valued distributions. The four hundred pages could probably have been written in one hundred. With hindsight, I regret not having known how to economize and organize my time more reasonably. It would have been much wiser to leave certain cases aside, or let them be considered by other authors. The difficult parts of this vast and complete book are not much used. And it had the effect of preventing me from becoming involved in other research and reducing my breadth of activities. This is one of my major errors in the organization of my research. I made it again several times in later life.

For a while, the surge of enthusiasm for topological vector spaces led people to believe that Banach spaces had finished their career. It is true that the spaces which occur in the theory of distributions and their applications are not Banach spaces but general locally convex topological vector spaces, sometimes nuclear. So the theory of Banach spaces stopped progressing for some time. But Israeli mathematicians took it up again and today, it is the theory of topological vector spaces which can be considered as finished – indeed, it is enough to know some of the major theorems of the theory, the rest having become a little obsolete – whereas Banach spaces have returned in triumph, in the form of geometry and probability theory in Banach spaces. After the theorems of Pietsch and other Polish and Israeli mathematicians, I devoted myself to the subject, while I was at the Ecole Polytechnique in the 1970's. Today, this theory is more flourishing than the theory of general topological vector spaces. Naturally, metric questions play a fundamental role, and seminars were organized in France (with Pisier, Maurey and Bourgain) as well as in the USA, in the USSR, in Poland, in Israel and elsewhere. There are deep and difficult areas within this theory. Such returns of partly abandoned theories are quite frequent in mathematics.

One of my main subjects in Nancy, spectral analysis and synthesis (we also say harmonic analysis and synthesis, or even just harmonic analysis), is still linked to convolution and to Fourier. The general problem is the following. If $E$ is a space of functions or distributions on a commutative group, and if $F$ is a closed vector subspace of $E$ invariant under translations, then what are the exponentials (or characters) or more generally the exponential-polynomials contained in it? (An exponential polynomial is the product of an exponential by a polynomial.) They constitute the spectrum of $F$. Is there always at least one? This is spectral analysis. Spectral synthesis treats the following problem: do the linear combinations of the exponential-polynomials in the spectrum of the topological vector space $F$ topologically generate the space? This is an extension of the theory of mean-periodic functions introduced by Jean Delsarte. Letting the topological vector subspace $E$ invariant under translation vary, we obtain a series of problems whose answers also vary. In certain cases, spectral synthesis is possible and in other cases it is not. At first, in Nancy, I studied one of the most difficult cases, where the space $E$ is the space of continuous functions on the real line $\mathbb{R}$ equipped with the topology of compact convergence, or the space of distributions on $\mathbb{R}$. I proved that spectral synthesis is possible. This is a generalization of one of the problems posed in my thesis and the article which followed it. Moreover, the convergence theorem as it is written up in my thesis, by groups of terms and Abel exponential factors, remains valid. The proof is difficult.

In his thesis, Jean-Pierre Kahane simplified everything to do with spectral synthesis remarkably, although his method does not prove the nature of the convergence. Logically, I was in his thesis jury; the president was Szolem Mandelbrojt. Kahane later became an absolutely first-class mathematician. He continued to work in harmonic analysis and the theory of Fourier series, using probabilistic methods. In fact, our work often converged. I tried to study the case of other topological vector spaces $E$. First of all, I tried replacing $C(\mathbb{R})$ with $C(\mathbb{R}^N)$ for $N \geq 2$. But I had no success: the problem is a priori much more difficult, because the case $N = 1$ uses holomorphic functions of a single variable, whereas the higher dimensional cases use holomorphic functions of several complex variables. In his thesis, Malgrange tried to generalize the problem to higher dimensions, but he only obtained some very partial results. The problem was only solved, in the negative, in 1971 by the Soviet mathematician Gurevich: spectral synthesis is possible in $C(\mathbb{R})$ but impossible in $C(\mathbb{R}^N)$ for $N \geq 2$. There are even subspaces with empty spectra, containing no exponential at all. We had searched for the result for so many years without success that we had become discouraged. Nobody thought about the problem any more, and Gurevich's negative result had almost no repercussions. Ingratitude of mathematics and of mathematicians! Actually, this result whose proof is extremely ingenious is strange and quite disconcerting.

I then attacked the case where $E$ is the space $L^\infty(\mathbb{R}^N)$. Wiener's grand tauberian theorem (130 pages, 1982) had started up the dual theorem in $L^1$. In modern terms, he showed that every closed ideal of the convolution algebra $L^1$

is contained in the maximal ideal formed by all the functions whose Fourier image vanishes at one given point. One deduces the spectral analysis of $L^\infty(\mathbb{R}^N)$ (equipped with the $*$-weak topology) from this by duality: every closed subspace of this space, invariant under translations of $\mathbb{R}^N$, contains at least one exponential, purely imaginary and without a polynomial; the spectrum is never empty. There are no analytic functions in the proof, only Fourier transforms. Wiener's theorem not only had remarkable applications in tauberian theorems, including for the proof of the distribution of prime numbers by the Riemann $\zeta$-function, but it is the origin of Gel'fand's theory of Banach $\mathbb{C}^*$-algebras. It is one of the grand theorems of analysis. I don't think Wiener saw that the dual problem, for $L^\infty(\mathbb{R}^N)$, was interesting: he missed a beautiful theory there. Since, in the case of $L^\infty(\mathbb{R}^N)$, spectral analysis is possible, what about spectral synthesis? Is a closed subspace invariant under translation generated by the exponentials of the spectrum? I believe that the Norwegian mathematician Arne Beurling and I were the first to ask this question, in 1947. Many mathematicians worked on it. One day, while I was suffering from the flu, I suddenly realized that the work I had just done on Bessel functions provided me with a counterexample: spectral synthesis is not possible in $L^\infty(\mathbb{R}^N)$ for $N \geq 3$. The proof fitted into the four pages of a Note in the *Comptes Rendus de l'Académie des Sciences.* A marvelous occurrence of the Bessel functions! The cases $N = 1$ and $N = 2$ remained absolutely open until Malliavin discovered that the answer is also negative in those two cases; his proof is much more difficult than mine for $N \geq 3$. Thus, spectral synthesis is never possible in $L^\infty$ spaces. Varopoulos found a much simpler proof, introducing tensor products: he uses the tensor product to reduce the case of all $N$ down to the case of one single value of $N$, and then it suffices to use the negative result for $N = 3$. Thus, although I gave only a partial solution to the problem, it indicated the right direction and was useful in the final proof. Naturally, I also considered the case where $E$ is the space $\mathcal{S}'$ of tempered distributions. A Fourier transform obviously transports this problem to another one which is a pure problem of differentiable functions. Consider the multiplicative algebra $\mathcal{A}$ of $C^\infty$ functions on $\mathbb{R}^N$, and let us study the closed ideals of this algebra. To each element $f$ of $\mathcal{A}$, and to each point $a$ of $\mathbb{R}^N$, we can associate the formal Taylor expansion $f_a$ of $f$ at the point $a$. Thus, we obtain a homomorphism from the algebra $\mathcal{A}$ to the algebra $\mathcal{A}_a$ of formal series in $a$. An ideal $\mathcal{I}$ of the algebra $\mathcal{A}$ then gives an ideal $\mathcal{I}_a$ of $\mathcal{A}_a$, which is always closed. The problem of spectral synthesis can be stated this way: is a closed ideal $\mathcal{I}$ of $\mathcal{A}_a$ generated by the set of ideals of formal series of the form $\mathcal{I}_a$ for $a \in \mathbb{R}^N$? Or is it the intersection of all these ideals? This problem appeared very complex to me at first and I couldn't solve it. It was finally solved by Hassler Whitney of Princeton, a mathematician who was extremely competent on the subject of algebras of differentiable functions.

The Scientific University of Nancy had became one of the best centers of mathematics in the world, so it was natural to invite foreign mathematicians. In 1946, Delsarte organized a symposium of harmonic analysis in Nancy. Apart from mathematicians from Nancy and from Paris, such as Henri Cartan, we invited Harald Bohr and Börge Jessen from Denmark, Michaël Plancherel, a Swiss mathematician well-known for Parseval-Plancherel's equality in the theory of Fourier series and very energetic in spite of his sixty years (he went swimming every day at the town swimming pool in Nancy), Norbert Wiener, a prestigious American mathematician well-known for his intransigent pacifist views but also for sleeping and even loudly snoring during lectures and yet asking pertinent questions (prepared beforehand) at the end, Arne Beurling who was interested in the problem of spectral synthesis in $L^\infty$ I discussed earlier, Michel Loève, an American probabilist, and Szolem Mandelbrojt. This symposium remained famous among mathematicians. I gave a lecture on distributions, followed by my first publication on the subject, a little pamphlet in the *Annals of the Fourier Institute*, which is still called my "propaganda tract" and which is only about fifteen pages long.

Whitney also came, on the occasion of a conference on algebraic topology which was being held in Paris at the torrid beginning of the month of July. He came to Nancy near the end of our symposium. We waited for him during the unique dinner organized in our home for the members of the conference. Nobody knew him personally except Wiener, who said to me with humor: "There is an infallible criterion for recognizing him: he looks older than he is." Equipped with this invaluable piece of information, I went to the station and found him: two men looking for each other in a provincial train station can hardly miss each other. At the end of the dinner, I gave him my problem of differentiable functions coming from Fourier transform on a problem of spectral synthesis. "I'll give you the answer in a quarter of an hour," he said. In fact, I waited six months before he sent me the solution.

The study of the subject lasted several years. Among other important theorems, Bernard Malgrange discovered the theorem of differentiable preparation, and in 1966 he published a book called *Ideals of Differentiable Functions*. This proved my spectral synthesis in $\mathcal{S}'$. Thus, I personally only solved some parts of the large set of problems of spectral analysis and synthesis, which were worked on for a long time: the part concerning mean-periodic functions on the space of continuous functions in dimension 1, the counterexample on $L^\infty(\mathbb{R}^N)$ in dimension $N \geq 3$ and the passage via the Fourier transform to a property of differentiable functions. Gurevich, Malliavin, Varopoulos and Whitney solved the other parts. The theorem of mean-periodic functions in dimension 1 is one of the most difficult. Some ideas are due to me, because although the problem of spectral analysis existed in the sequence of Wiener's tauberian theorems of 1932, I stated the general problem of spectral synthesis and then gave partial solutions in several different directions. It is possible to give a physical interpretation of spectral theory. A wave can be defined as a function $f$ of time; dephasing it would be having it undergo translation in time, $t \mapsto f(t - t_0)$. A monochromatic wave would be of the form $t \mapsto ae^{i(\omega t+\phi)}$, an imaginary exponential. The problem of spectral analysis is the following: among the combinations of the dephasings of a wave, is there always at least one monochromatic wave, which is said to belong to the spectrum of the wave? And can a wave be reconstituted as a combination of the monochromatic waves in its spectrum? In this vague and general form, it would be possible to think that the answer is always positive. We just saw that these are questions which should be posed in terms of topological vector spaces, that the answer is different in different cases and that the proof is always difficult.

Marie-Hélène wrote her thesis in Nancy. She was working in very favorable conditions, leading a happy and stable life with her children, and receiving a CNRS scholarship. Starting in 1947, we had two salaries and no more financial problems. Thus, we were able to hire a domestic help who was very precious to us, allowing us both to work full time on mathematics even while having children. We are deeply grateful to all the people who successively participated in our family life.

The idea of Marie-Hélène's thesis was born from her mathematical reveries during the war. In the sanatorium, Marie-Hélène had more or less hoped to catch up on her lost time and pass the agrégation; she didn't think of doing research. And yet, without her ever really talking about it, the desire to do research worked inside her. Progressively, she was led towards one of the grand problems of the 1930's: generalizing Ahlfors' asymptotic theory to complex dimensions greater than 1. In fact, she studied a slightly different situation. In 1944, Chern's first paper appeared, in which he defined curvature and transgression forms. She deduced a finite formula from it, which served as an intermediary for the asymptotic computations she needed to make. She defended her thesis in May 1953, with a jury consisting of Georges Valiron as president, Henri Cartan and André Lichnerowicz. If someone had told her in 1935, as she was entering the sanatorium, that she would overcome so many obstacles including even a world war, she would have thought it a joke.

## An open life

The teachers I met, of varying horizons, introduced me to other circles in Nancy; I met many interesting people through Dieudonné and his wife Odette. We lived in the cultural heart of the city, a few minutes from each other, from the university in the rue de la Craffe, from the cinemas, the theater, the concert hall, the museum, the main post office and the station. We spent our leisure time going out assiduously. The salle Poirel was a remarkable concert hall. Great interpreters such as Gieseking came to play there. At concerts, as everywhere else, it was perfectly possible to find excellent seats even if you arrived with just five minutes to spare. Through faculty and union meetings, I knew almost all the teachers in Nancy, in every subject. Dinners at their homes were frequent. Our simple but intense social life was an incredible breath of fresh air after our cloistered life during the war.

Nancy was quite charming. Louis XV had married Marie Leszczyńska, the daughter of Stanislas Leszczyński, the king of Poland, who had brought Lorraine to France. He had a square constructed called Stanislas Place, which together with the Carrière Place is one of the most harmonious in Nancy and one of the most beautiful in France.

At that time there were not many students, and professors, who were also not very numerous, inspired a certain deference. Once I was riding around on my bicycle, and I was stopped for not having renewed my license plate. The sergeant who stopped me took down my name, intending to fine me. But when I told him that I was a professor at the university, he closed his notebook, saluted smartly and said "At your service, Mr. Professor"!

After the difficulties of the war, our seven years in Nancy were among the happiest of our existence. Never again did we recapture the joys of our simple, stressless life, in which we met so many different people. There were interesting paintings by Claude Gellée, called the Lorrain, in the museum of Nancy, and powerful engravings by Jacques Callot, particularly on the subject of the Thirty Years' War. Apart from the period just preceding our marriage, I had never gone to museums much. But when I became an assistant professor at the Ecole Polytechnique for a temporary period of two years after 1945, I went up to Paris regularly to spend a day giving problem sessions; I must have gone about fifteen times over the two years. Each time, I spent two days in Paris at my parents' house: the first day, I taught, and the second I spent in museums.

I never neglected to explore new ones whenever I visited foreign countries. Without being very knowledgeable in painting or sculpture, I became a modest amateur of the fine arts. I loved the Rodin museum more than any other, and was never tired of admiring the series of *Kisses* which are authentic masterpieces, but also the *Bronze Age*, the *Gates of Hell*, the *Thinker* and the *Bourgeois de Calais* group, which moved me even more. Later in Yugoslavia, I encountered a student of Rodin called Meštrovitć, whose work I also appreciated.

For the first six months after our arrival in October 1945, we lived in a little room on the boulevard Lobeau. I worked high up in a freezing cold nest. Then the Dieudonnés left for Rio de Janeiro and São Paulo from May 1946 to December 1947 (the French members of Bourbaki are at the origin of Brazilian mathematics: Dieudonné, Delsarte and Weil went there. The Brazilian mathematician Leopoldo Nachbin, who died recently, continued the tradition by regularly inviting French mathematicians.) During their absence, we stayed in their apartment at 26, rue Saint-Michel. We lived very modestly. Previously, my salary came from the CNRS and the ARS. When I received a position as assistant professor in the university, I was attributed the very lowest possible salary. Similarly, when I became a full professor, I was given only the lowest possible raise, up to the second lowest level of possible salaries. When I became an assistant professor in Paris, I progressed very slightly, with the lowest possibly salary of assistant professors in Paris. My nomination to Paris in 1952 reads: "Monsieur SCHWARTZ, professor at the scientific university of Nancy, is hired as assistant professor at the scientific university of Paris; he will receive the salary of the third class of assistant professors of the Scientific University of Paris, in which category he is placed." This was the rule at the time. It was even more scandalous for Grothendieck, who was paid for some time by the CNRS, because at that time foreigners could not teach at universities. When he was finally allowed to, he was a director of research at the CNRS and a Fields Medalist. He was given a job as assistant professor at the University of Montpellier, with the lowest possible salary. Today's salary calculations including previous positions did not exist. This explains why I was unable to buy myself a new typewriter. Thanks to the advice of the counselor of our university, M. Lemoine, with whom I was on good terms and whose daughters were students of mine preparing the agrégation, I bought an old, very heavy typewriter at an auction; it made a sound like a truck and Godement called it my tank. Naturally, we had no offices nor secretaries at the university, and I often spent my Sunday mornings typing my voluminous correspondence. I typed with three fingers, staring at the keyboard, and it took an incredible amount of time.

When Dieudonné returned in 1948, we moved to 30 cours Leopold, to a spacious apartment, with a garden giving onto the street, which is really more of a wide, long square of monumental proportions. Sometimes, Nancy was invaded by a fog you could cut with a knife. One day when you couldn't see a thing, I asked my way to the cours Leopold, only to find that I was already in it. My office, an enormous room with windows on two sides which gave onto the rue Saint-Michel on one side and onto the cours Leopold on the other, was decorated with a blue chimney covered with white fleur-de-lys and fortunately heated by an excellent stove. A little farther on was a bathroom, which was not heated but contained a magnificent bathtub in which, alas, one was not inclined to spend much time in winter. Sometimes the water we kept in a glass on the nighttable froze overnight. But as soon as we got up, we lit the stoves and everything was for the best in the best of worlds. And indeed, we were happy in that house. In Paris, in 1953, we did not have central heating immediately either; we had to wait until we moved to the rue Pierre Nicole in 1958.

We held many receptions in our house on the cours Leopold, particularly around a Christmas tree. Draped in an enormous red bathrobe, hidden behind an immense cotton beard, disguising my voice, I was unrecognizable. Marc-André, who was six years old in 1949, didn't recognize me. That year, he had invited some school friends, among which was a little girl called Danielle from a modest family, with whom he was quite in love. She embarrassed me greatly by asking in front of her mother why I had not put all the presents she had asked for in her stocking. I answered that I didn't have enough presents to satisfy all requests; her mother had told her the same thing. The Christmas presents were spread out in front of the large closed chimney in my office; one wondered where Santa Claus was supposed to come down. Unlike parents from the generation of my mother and father, we played a lot with our children, and the animated atmosphere that reigned in our household at that Christmas celebration remains a moving memory.

The garden was also a non-negligible resource for the children. Before Marc-André went to school, they spent a large part of the day playing outside. Madame Delsarte came often and brought candy to Marc-André, who developed the unpleasant habit of asking for it. After we scolded him for such effrontery, he changed to asking her "Do you have anything interesting to tell me?" which was merely a discreet manner of saying the same thing. Madame Delsarte was delighted.

Towards the age of six, Marc-André liked to count the trams which passed in front of the house. Naturally he had to stop counting during meals, and he complained bitterly, saying "But then my whole counting for the day doesn't make any sense." In school, he was embarrassed to be the only one in his class who did not go to catechism. The parish priest was very nice to him, but that only increased his embarrassment. He received a secular education, which was a rare thing in Nancy.

The cours Leopold had just one defect: the annual fair took place there in the month of June. The street was filled with stands, and nothing was lacking: there was every kind of merry-go-round and wooden horse, huge Ferris wheels, bumper cars, roller coasters, cyclotrons (rapidly spinning cylinders on a vertical axis), wild animals ("Professor Lambert" had his crocodiles; I was always tempted to address him as "dear colleague"), tarot readers and so on. The incredible racket went on from nine in the morning until midnight, and loud popular tunes that could be heard from one end to the other of the street (do-ti-la-sol fa-mi-re-mi do-ti la-sol sol-la-ti-do sol do) prevented us from sleeping. The Gauthiers shared our misfortune of living on the cours Leopold. The children, of course, were absolutely delighted from beginning to end, and I had to sacrifice to the god of paternity by climbing onto all kinds of dizzying machines. The two dark spots in our life in Nancy were grading the agrégation homework and the annual fair.

During the vacations, we usually went to Autouillet. The children soon enjoyed themselves there as much as we had. Marc-André was always very busy around his grandfather who took care of the trees at the bottom of the field. Starting at the age of five, he would go there by himself and from far off, he would start to call out "Cuckoo, my grandpa" and my father would answer "Cuckoo, little cannon". Marc-André chatted amiably with my father while he worked, then came home all proud to tell us that he had "helped grandpa".

The birth of Claudine, which was expected in July 1947, was an exciting moment. One week beforehand, we sent Marc-André to Autouillet. He sang at the top of his voice on the way to the station, but as soon as we started moving, he realized that Marie-Hélène had remained behind, and reproached me with having tricked him. In Autouillet, he was gay as a lark, so that I was able to leave him behind serenely. My parents took great care of him, but he kept repeating "I'm sad for my Papa and my Mama." Finally, my mother asked him to make this statement no more than once a day, at bedtime, which he did very politely. To amuse him, they let him drink in a little pharmaceutical bottle instead of a glass. They also talked to him frequently about the little brother or sister that was soon to arrive. Marie-Hélène was supposed to have the baby in Paris. The day before, we had a friend to dinner. We had reserved places in the train for the next day, which took five hours to get to Paris, stopping in Bar-le-Duc. What had happened for Marc-André happened again: the agitation of the departure hastened the birth. As soon as our friend had left, Marie-Hélène rushed to the hospital in Nancy and gave birth the following morning, on July 30, 1947, the hottest day in a century. I telephoned my mother: "Tell Marc-André that he has a little sister!" My mother made me repeat it again and again. She was deeply moved, because she had had only boys in her family: three brothers, three sons and three grandsons (my son and Daniel's two sons Maxime and Yves; Bertrand was not married yet). We had just made the first break in the cycle: a female after nine males. Daniel immediately started up a new baby, since it seemed possible to have a girl after all – and it turned out to be his daughter Irene! After that, the equilibrium between the sexes was re-established in the family. My mother announced the news to Marc-André while I was still on the telephone, and he immediately and relevantly replied "You have to tell Mama that I drink in a little bottle!" One week later, we were all united in Autouillet for the rest of the summer.

To our great joy, Bertrand and his wife Antoinette came to live in Nancy. Bertrand married in 1950, at the age of 31. They soon had four children, Olivier, Alain, Isabelle and Jean-Luc, all born in Nancy. Bertrand became director of the School of Mining, and undertook important reforms there. In 1960, he became director of the CUCES, the University Center of Economic and Social Cooperation, which among other things organized continuing education courses for the mining workers of Briey. Some of them had forgotten how to read properly, even though they had been to school. Bertrand's pedagogical methods were the fruit of much reflection, and they were fundamentally new. Now, he works in the education of young children who are failing school; he is convinced that it is possible and necessary to give them further training as adults, in order for them to be able to have professions. Recently he wrote an extremely interesting book, *Moderniser sans exclure* (*Modernizing without Excluding*). Bertrand was an educator, Daniel was a doctor of epidemiology and I was a mathematician – all three of us are deeply attached to teaching ... but we address three completely disjoint publics.

I've always been very close to Bertrand's family, the more so because we both lived in Nancy. We spent summers together under the same roof in Autouillet. In Paris, they lived very near us, but they moved in 1996. We also shared Autouillet with Daniel. But we saw each other less in Paris, though we spent hours on the telephone. My mother, who never ceased to weave powerful webs attaching all the members of her descendance, direct and indirect, wrote long letters which were almost illegible because she broke her thumb, detailing the health and activities of every member of the extensive family. All by herself she formed an efficient liaison and information center.

Shortly after the war, my father, who was tired by old age (he was seventy-three) and affected by the extreme difficulties he had traversed, suffered from various health problems. But he survived them, and when he retired he worked as a consulting surgeon for Social Security. I can't help wanting to recount an

Schwartz' mother with Claudine

amazing coincidence. We were living in Paris, avenue Daumesnil. One day, in the metro near the Daumesnil station, an Arab stopped to ask me something. He did not know a word of French and could only show me a paper on which to my surprise I saw written the word "Schwartz". Surprised, I looked closer and read my father's name, "Anselme Schwartz". The Arab had an appointment with my father and was lost in the metro. I explained to him as best I could how to get to his destination. It gave me acute pleasure to pretend witchcraft with my father, who couldn't understand how I could possibly know who he had seen in his office. Another time, by pure chance, my father was called to a man who had just fallen from the sixth floor of an apartment building. He had

Schwartz' father with Claudine

caught onto several balconies while falling and astoundingly, he had no broken bones nor even bruises! He was simply in shock. In the midst of the people standing around him, my father examined him and said "At least bring him a glass of water!" And the gentleman showed that the fall had not affected his lucidity by responding "Good God, what floor do I have to fall from to have the right to a glass of wine?!"

My father died suddenly in April 1957, at the age of eighty-five, from an inflammation of the arteries he had suffered from for several years. It was in Paris. He felt ill, lay down, wandered a little and suddenly he was gone. My mother telephoned us; we were all in Paris, but when we arrived it was too late. It took me months to get over mourning for him.

# Chapter VIII
# International Recognition

## A vaster world

The symposium on harmonic analysis organised in Nancy had numerous and unpredictable consequences for my existence. Harald Bohr became enthusiastic about the theory of distributions, which he studied in detail, exchanging many letters with me. Soon after, I received an invitation to Copenhagen to give a series of lectures in October 1947. At that time, I had also solved the problem of the Fourier transform by tempered distributions. It was my first really important university trip, and also one of my first escapades. I stayed in Copenhagen for three weeks, and was received there with the famous Danish hospitality. Memories of the war were still vivid in 1947, and everything appeared wonderfully fresh to me. Naturally, I am a little more blasé now.

My contacts with the Danish mathematicians were free and easy and overflowed into a social life. I met Niels Bohr at the Academy of Sciences. Harald Bohr gave a talk addressed to all Academicians, in which, to simplify, he defined $\zeta(s)$ as $\sum n^s$ and considered the cases $s = 2$ and 3 explicitly. Apart from the mathematicians, it seems that no one was taken aback. Learning that I was in Copenhagen, Lars Gårding invited me to spend three days in Lund, one of the best mathematics departments in Sweden. He initiated me to the very important work of the Soviet mathematician Petrowski; "Petrowski is my God, and I am his prophet," he said laughing. I returned to Lund several times later on, and in 1981 I was awarded a doctorate *honoris causa* there. I had many conversations with Lars Gårding, a devotee of distributions, and with Marcel Riesz. Marcel and Frederic Riesz were two remarkable Hungarian mathematicians: Frederic was tall and thin, Marcel small and fat (Laurel and Hardy). Frederic lived in Hungary, but Marcel was in Lund. He discovered results on partial differential equations, in particular the solution of hyperbolic equations of non-integral order, via symbolic calculation on Hadamard's finite parts. This problem was adapted to the use of distributions, so he was interested in my talks. Marcel

Riesz was a solid drinker, occasionally pushing the limits. One night, we stayed in his house until two in the morning, with a glass of liquor in front of each of us. His glass kept refilling itself. I didn't touch mine, so every time he looked at it, he saw it was full and thought I had served myself. Later he told Gårding that I was an "excellent drinker". I remained very close to them. When Riesz came to dinner at our house, we served a chocolate cake, decorated with the wave equation

$$\left(\frac{1}{v^2}\frac{\partial^2}{\partial t^2} - \Delta\right)u = 0,$$

written in white cream – he was enchanted! The young Lars Hörmander was at that time still in high school. Some years later, this future Fields Medalist, trained by Gårding, stood out at the university and became one of the world specialists in partial differential equations. He profited indirectly from my stay, because he made extensive use of distributions. We also became close friends.

The same year, Michel Loève, who had been at the colloquium on harmonic analysis, invited me to London in January 1948, while one of his colleagues, J.H.C. Whitehead, invited me to Oxford. Naturally, it was out of the question to give my talks in any other language than English. However, although I had studied Latin, Greek and German in high school, I did not know a word of English. I had to prepare myself in a rather unusual way. Norbert Wiener's book *Fourier Transforms in the Complex Domain* introduced me to the English mathematical vocabulary, which is very small. Then I wrote the text of my talk in French – a very exceptional procedure for me since generally I just scribble down a few notes – and finally, I translated it into English with the help of a dictionary and Marie-Hélène who had studied English in high school. In this way, I learned some elements of English grammar and syntax, which are not too difficult. All this was quite indispensable, since I didn't want to give a talk in some kind of approximative English. To express a mathematical reasoning, it's necessary to use subjunctives and conditionals, for instance: "if this function were discontinuous, we would have a contradiction; it is necessary that this series converge ..." I needed to learn the names of digits and numbers in English, and the letters of the alphabet, since you're not supposed to be mute as an oyster when writing down a formula on the blackboard. Finally, I needed to know all the vocabulary necessary to my limited subject. Jacques Hadamard and a friend Anne Nordon corrected my text, and Anne helped me with my accent. I learned to speak correctly with a more or less British accent; although my French accent can still be heard, my English is good enough. Dieudonné and Ehresmann used to speak with an excellent British accent! I read and reread my talk, but left for England with the firm intention of reading it rather than improvising as usual.

When I arrived there, I was surprised by the bitter cold. It was winter, and the houses had almost no heating. People lit the heating in bedrooms only at bedtime, and warm covers were indispensable. Later, I returned to England many times, visiting London, Manchester, Oxford and Cambridge. The latter two towns are extremely beautiful, but I discovered them around Easter. In Cambridge, I was lodged in an ancient, church-like building. The walls were a metre thick and had retained all the cold and humidity of winter. The sheets were humid. The little gas pipe in the fireplace had no effect. I think I was never so cold in my life. Finally, I decided to keep all my clothes on, including my big winter coat, underneath the covers and the quilt; I placed a table upside down on top of it all to diminish caloric loss .... I was like an ice cube when I woke up.

The traditionalism of British universities is not a myth; when not taken literally, it was very pleasing to me. To penetrate into the unique collective dining room of Magdalen College in Oxford, the professors line up in order of seniority. Then they all sit down at a table on the podium, while about a hundred students eat at long tables lined up below. After dinner, the professors meet in a long, dark room where a minuscule table carrying cookies and liquors circulates on wheels. A muffled conversation takes place in the half-darkness.

My first talk in London went perfectly. I barely needed to read my text, which I practically knew by heart. Everyone understood me perfectly! Alas, at the question session afterwards, I could not understand a single thing that was said to me. This difficulty has pursued me all my life. I can talk adequately in English – but I can't understand it! Loève did me the honors of the Tate Gallery.

The trip to Oxford to meet Whitehead was much more of an epic. I warned him that I knew practically no English. This worried him somewhat. He met me at the station, accompanied me to his house, and introduced me to his wife. Since I obviously didn't understand a single word he was saying, he was absolutely convinced that I could never give a talk.

I showed him my paper and reassured him. Then he took me around to visit the town with its old churches. We arrived at the university, but he had forgotten to go back home to pick up my briefcase, and I didn't have my text when we got to the conference room! I just had to go back to my usual method of improvising. My vocabulary was restricted to barely a few hundred mathematical words, but I had picked up enough grammar and syntax to use them without referring to my paper. Afterwards, I was asked some questions which had to be translated to me; I answered them in English. Whitehead was indignant. "You're a crook and a liar – you tricked me, your English is perfectly good!" I explained to him that apart from mathematical words I was ignorant of even the most common expressions. He didn't appear convinced, but remained friendly. He himself was deeply British; slow and awkward in his talks. He couldn't believe that a foreigner could easily do something which was so difficult for himself, and yet be incapable of a simple conversation in English. The words for trip, street, bread, milk, bed, house do not belong to the mathematical vocabulary. Only during my next trip, to Canada in 1949, where I remained for several weeks, did I learn some of the "rest" of the English language. I worked on it conscientiously, reading books and newspapers, not only to understand them, but to grasp the structure of words and phrases.

I learned several foreign languages this way – I recommend it to all scientists – starting by reading a mathematical paper, whose vocabulary is very restricted. After English, I learned Spanish and Portuguese which I speak more or less fluently. I've also given talks in Italian, but I've forgotten it; I studied Russian grammar and syntax, but since I could never travel to the USSR for political reasons, I forgot everything. But in all I've "touched" nine languages: French, English, German, Spanish and Portuguese which I really speak, and Latin, ancient Greek, Italian and Russian. My passion for all languages, dead or living, goes back to my earliest childhood.

During a brief trip to Germany in 1950, I was struck by the misery which still reigned there. Many houses had been destroyed, families lived in basements where air entered only through a vent. Small children begged at train stations. It was impossible to feel any real consciousness of what had happened during the war; it was not at all like today's Germany. At the university, I exchanged views with some colleagues, in particular the mathematician Gottfried Köthe, a specialist in topological vector spaces defined from sequences of numbers. He surely did nothing wrong during the war. He was a nice, agreeable man; we kept in touch. But like the majority of his countrymen, he didn't seem to attach any importance to Germany's past. I owe him some good theorems on distributions. Some of his colleagues had been loyal servants of the Nazi regime. One of the faculty members who had more or less flirted with the Nazis talked to me at length about his fear of the rise of the Catholic church. That was his main worry for the present and the future! On the other hand, I met some young people who had been real anti-Nazis. One of them, Maak, told me how much Germany had suffered under the internal dictatorship of Hitler. Starting around the end of 1943, he had begun to hope for the military defeat of his own country; he admitted not having actually done anything toward this result, because it was beyond him, but at least he had hoped for it. He applied to the letter (though not in his acts) the revolutionary defeatism of Lenin. I remained in touch with him for some time. Desertion, in Germany like in every country, was reproved by the entire population. A young and very brilliant German-Jewish probabilist, Döblin, who was a refugee in France with his mother and who had fought in the French army against the Nazis, died during the war. His mother, returned from Paris to Germany, but there she met with no indulgence at all for the so-called "treason" of her (Jewish!) son from her old friends, even those from quite liberal circles. Desperate, she committed suicide.

My talks in those first years all dealt with the same subject: distributions. Certainly, I like to give talks on anything I find beautiful and interesting, but I ended up having enough of the ceaseless repetition of the same thing all over the world. I tried to give talks on other topics, but that was not what people wanted of me. One non-negligible result was my introduction of the whole of Latin America to the use of distributions. At the end, I actually became a sort of depersonalized double of myself. I recited my talk on distributions mechanically, while scolding myself inwardly at the same time in the third person: "God, he's boring! God, he's monotonous!" At any rate, these tours were always triumphant successes, so I can't really complain. The audience generally asked quite pertinent questions. Later, I read a report on Niels Bohr's tour around Europe to talk about his atomic model, which had received an enthusiastic welcome everywhere. That must be nice, I said to myself, before realizing that I had experienced much the same thing! I don't know if Niels Bohr also ended up having enough of it after a while.

My next trip, which was full of consequences for my future, brought me to Canada, where I was invited to the Canadian Mathematical Conference which lasted a whole month, in 1949. It took place every four years, between two International Congresses, and its goal was to unite people who lived scattered apart along the thin, six-thousand kilometer band of land touching the United States from Newfoundland to Vancouver. The Canadian mathematicians, who were in close contact with their American counterparts, rarely met the Canadian mathematicians from the other end of the band.

The Vancouver conference was the occasion of my first, unforgettable trip in an airplane. The propeller airplanes used at the time shook a lot in the air. First, a plane of the BEA took me to London. Then a BOAC flew for fourteen hours (there were no jets back then) to Montreal. We had plenty of time to get tired. My longest trip in a propellor airplane, from Montevideo to Paris, took twenty-eight hours (with some one-hour stopovers), on a plane on which I was the unique passenger. The stewardesses took care of me so intensely that one of them spilled a glass of tomato juice all over me. Perhaps I flustered her . . . .

Shannon, the farthest point of Ireland, Iceland, Greenland, then Labrador, and finally Montreal revealed to me their varied splendors. In Iceland, at our stopover in Keflavik, near the capital Reykjavik, it was cold even at the beginning of summer. The wind blew so strongly that a whole series of projectiles rolled along the airport runways. In Greenland, the northernmost point of the voyage, the magical vision of luminous, floating icebergs awaited me. There were two possible stopovers for planes flying to Canada: Gander in Newfoundland or via the northern itinerary, Goose Bay in Labrador. The view in Labrador is phenomenal, frozen in perpetual cold, scattered with tiny trees. We flew over myriads of lakes sparkling in the sunshine. Traces of snow drew mysterious signs. We stopped at Goose Bay. In Montreal, we were met by torrid and humid heat. I had traveled first class, as university professors generally did, and I continued to do so for many years. Now, we travel like everybody else, in economy class. Naturally, I gave a talk in Montreal in English. At that time, English still dominated. Today, it would be insulting not to speak French for a public address there. For several days, I visited the town. My brother Bertrand was also to be in Canada for six months, for his job, but we didn't think we would be able to get together. Suddenly, we met each other face to face in a street in Montreal! But we couldn't continue our travels together. I joined a group of a dozen young Canadian mathematicians who were traveling to Vancouver by train. The trip lasted a week, with stops from time to time. We passed through the regions of the Great Lakes, Prairie, the Rocky Mountains. I discovered storks, bears and beavers in their natural habitat. My meeting with a bear caused tremendous delight to my six-year old son Marc-André. Canadian bears were so tame that the mother bears would send the little ones to beg in front of cafés. Marie-Hélène and the children were with my parents in Autouillet. I sent my letters to her there, and she read them to the family. When I came back, I realized that I had vexed my parents by not sending them one or two letters addressed to them personally, which they could then have read to the family. My mother was sad and said "I can see that you don't even think about your old parents any more." It was the first time she had ever used that expression. In 1949, she was sixty-one and my father was seventy-seven. From that time on, I had to admit that they were indeed my old parents. I never left again for a trip without writing to each of them personally.

It was during this trip that I learned most of the English vocabulary of daily life. When I finally arrived in Vancouver, I knew English more or less. The Great Lakes were majestic; they carried great rows of floating logs which were sent this way from one end to the other. The Prairie seemed to be an interminable wheat field. Even more than today, Canada was an inexhaustible producer of cereals. Canadian wheat is not the same as the European variety. Its stalk is short and it matures early. Introducing wheat to the rough Canadian climate was no child's play. For years, the farmers who sowed in autumn saw no result in spring, because the seeds had frozen over the winter, in those prairies where the temperature often descended to minus thirty or forty degrees. It took one farmer who found one miserable stalk of wheat which had survived the winter, and had the idea of planting the grains of wheat growing on it. They all grew the next spring, and all the wheat in the Prairie grew from that one modest stalk. At least, so they say. Apart from a giant silo here and there, the countryside is empty, almost without villages, just fields and fields with no separations, as far as the eye can see. For more than a full day's train journey, you sail on an ocean of wheat. Crossing the mountains (Mount Eisenhower) is much more varied. The view is so astounding that sometimes the train stopped for ten minutes just to let the passengers admire it. The coolness – we were quite high – was a relief after the heat of Montreal. I'll never forget Lake Louise. In its pure water, you can see the reflection of every tree of the giant forest surrounding it, and its bank is covered with multicolored wild flowers. I encountered totem poles, those huge obelisks of sculpted wood made by the Indians. I met no Indians, but they were still the object of a revolting racism. In Montreal, I saw a lot of restaurants marked "Forbidden to Indians". At that time, there was segregation in the United States. I wasn't a very good walker, but nevertheless I undertook some hikes of several hours with the young mathematicians; the air was so pure that I was able to manage them. They also took me in a canoe. I didn't know how to swim and found it far from reassuring. However, that is where I met Dirac and Bhabha, who were invited to the same mathematical congress.

Each of the four principal invited speakers gave a cycle of talks. Paul Dirac, one of the great English modern physicists, was the inventor of spin and of the theory of the electron. Homi J. Bhabha was an Indian nuclear physicist (who probably bears some responsibility for the Indian atomic bomb). Antoni Zygmund from Chicago, of Polish origin, was the world specialist on Fourier series, but also, together with Calderon, the creator of singular integral operators. Other important mathematicians such as the Hungarian Gabor Szegö were also at the conference.

Vancouver is a magnificent region. The climate is warm but maritime, so quite bearable. There, as I later saw in Berkeley too, a lot of animals come into the town, particularly deer, does and fawns who approached fearlessly to eat the leaves from trees or the tufts of grass we held out to them. Just as tame, American squirrels, chipmunks and swiss came to eat from our hands. We find the same animals in the reserves in the eastern United States.

The foreign invited speakers were lodged in the mansion of a very rich family, the Grahams, who were the sponsors of the university and watched over us. We took certain meals with them. They had a magnificent garden and a small park full of wild birds. It was there that I first saw hummingbirds, who use their long, sticky tongue to search for nectar in flowers, and whose flight resembles that of certain butterflies, the hawkmoths, who hover in place while they take in nectar through their proboscis just as we drink juice with a straw. For a long time, I actually thought that hummingbirds were hawkmoths, and I tried to capture some, but they were much too fast for me. I saw hummingbirds on all my trips to America and Canada, and also to South America. They are not fearful, and ready to feed from anything at all, as long as it is sweet. It's easy to attract them by attaching a long tube to a bottle of fruit juice, so that they can plunge their long, sticky tongue into it. During our summer in Berkeley in 1960, we had a little garden where a couple of hummingbirds lived; they liked fuchsia nectar. They knew us well, and when I was working as usual in the garden, they came regularly once a day to hover right in front of me and inspect me, so as to make sure I was still the same. If ever another hummingbird came there, one of ours flung itself at the newcomer and chased it away in a wink.

But let us return to the Grahams' house. We took our meals together, all five of us, sometimes alone, but mostly with our hosts, or at least with Mrs. Graham. My English continued making progress.

My cycle of talks went well enough. I met other participants of the conference, in particular Canadian professors and students from different provinces. For them, Vancouver was the ideal spot to hold a conference. They learned a lot about European mathematics. We gave our talks at one of the good Canadian universities, the UBC (University of British Columbia). Some of the Canadian participants made up a rather ironic report of my English mistakes. I had assimilated the vocabulary rather too hastily and made a lot of mistakes with "false friends", for instance "I shall now resume my lecture", giving resume the French meaning of "summarize", instead of the English meaning of starting again. I made several similar mistakes at dinner. For instance, I excused myself to Mrs. Graham by saying "I excuse me, I am in retard." I don't know what she thought I was saying, but she answered "It doesn't matter, we don't dress here in the evening", a reply which I didn't understand either. Another time, someone asked me if we had a lot of destruction in Nancy. I thought he said "distraction" (i.e. entertainment), and replied that there was not a lot, but that I was not very demanding on that score. Everyone looked extremely surprised at my answer, and we ended up understanding each other. Mrs. Graham took wonderful care of her guests. She took us to the Malibu Club in a hydroplane;

the strange sea-landing takes place in a rainbow. After some fishing, during which only Dirac caught a beautiful fish, we ate tender and delicious oysters as fleshy as steaks; I still remember their voluptuous taste. There are also giant clams there, but they aren't really edible. Guessing how I loved fishing, Mrs. Graham invited me to stay on for the tuna fishing in the gulf of Mexico; I was terribly tempted, but the academic year was starting up again, and I had to get home.

Naturally, I was delighted to have met people like Dirac, Bhabha and Zygmund. Dirac had invented a new physics, but also his famous function which doesn't exist and yet which is so useful in theoretical physics. Distributions gave a complete justification for it, and also for the big computations of singular functions for elementary particles which physicists use them for. I had actually made the Dirac function "respectable". He listened to me attentively, curious, as he said, to understand the function he had invented. Dirac was a marvelous man, strange and introverted. After one of his talks, the president encouraged questions, as usual. One participant showed a formula on the blackboard, saying that he didn't understand it. Dirac remained mute. The president asked Dirac if he had heard the question. He answered "I didn't hear any question, I heard a statement: he doesn't understand." He confided to me that I was the only lecturer whose talks he couldn't sleep at. As for me, especially at my age, I'm rarely insomniac during talks. It's most disagreeable. Dirac was also very daring. He terrified me, in Banff, during the trip, when I saw him crossing a precipice on a bridge made of two tree trunks.

Bhabha was a remarkable man, full of intelligence and charm. He directed the Tata Institute for Fundamental Research in Bombay, a high-level research institute for mathematics and theoretical physics, which he had founded in 1945. Thanks to our meeting in Vancouver and my meeting with the mathematician Chandrasekharan, at the International Congress the following year, I was invited to the Tata Institute in 1955. I had the good fortune to meet Dirac again there. Unfortunately, Bhabha died young in a airplane accident over the Alps. As for Zygmund, a world-famous mathematician, he answered every single question one asked him about Fourier series immediately, but his talks were strangely confused. I remember one formula on the blackboard where the letter $n$ occurred three times with three different meanings. I also met the Polish physicist Leopold Infeld, whose interesting autobiography I have since read. He told me this about the USSR: "I always admired Stalin. With the Lyssenko affair, I've lost confidence in him for the first time." For the first time ....

After the conference, I returned to Montreal by train, and then took a different route back to Paris. The museum of Montreal was not very rich, but I discovered paintings by El Greco. It was the first time I had heard his name. Since then, I have become very interested in this extraordinary painter and his inimitable portrayals of Christ; I always thought he should have painted de Gaulle.

My return from Canada was quite exciting. After the airplane trip on the way there, I had thought of going back by boat. My friends advised me to take the *Empress of Canada*. The other empress was the *Empress of France*. Both were flat-bottomed boats which sailed up the Saint-Laurent. Below Montreal, we passed forests of maple trees whose leaves were already astonishingly fiery red and gold, in early September. Most of the trees in Montreal are maples, whose red leaf is nobly enthroned in the middle of a white field on the Canadian flag. I was overjoyed at the idea of seeing them, and not disappointed when I finally did. The rest of the trip was a catastrophe. It's not easy to forget a wild Atlantic crossing on a flat-bottomed boat. Everyone vomited for five solid days, except a few lucky people such as myself, who were equipped with sea legs. However, I didn't close an eye during the entire trip. We kept hearing a gigantic racket of ropes and chains, chairs falling over and so on. I felt my entrails rising and falling within me with the swaying of the boat. During the day, I was usually alone on the deck in a wild wind. We heard the following interesting announcements: *rough sea*, *very rough sea*, *hurricane*. Fortunately, we didn't meet the hurricane. I reached England exhausted, and took a plane to France.

I traveled to other foreign countries in later years, particularly tropical countries. But this was my very first long trip to another continent. It was one of those initiatory voyages which belong to youth. I made a lot of friends, and returned frequently to Canada. In particular, I met a wonderful young woman, cultivated in art and music and full of joy; she radiated contagious enthusiasm. She became a very close friend and remained one until her recent death, which moved me deeply.

Nowadays, everybody travels. But at that time, shortly after the war, it was a rare privilege. When Roubaud came back from the United States in 1946, he ended one of his talks with a slide show, which he presented to an enthusiastic university audience, with the words "This is the lesson of America", which would be unthinkable today. I often found myself to be among the first mathematicians to visit some of the countries I went to. Thousands of tourists have admired the Taj Mahal in Agra, India, but when I went there in 1955, I was almost alone. A few rare and silent people came for the poetic moonlight visit, where you could admire the Taj Mahal much more emotionally than today. There were no postcards to buy, and the slides which I took met with a lot of success among my friends. Nowadays, there are hundreds of tourists milling around, and beer bottles everywhere . . . .

## The Fields Medal

At that time, I had also received an invitation for three weeks to Princeton. I asked for my visa at the American Consulate in Strasbourg, which took care of residents of Nancy. After a lengthy wait, I received a refusal, some time after the trip should have taken place. In the autumn of 1949, I learned that I would never have a visa for the States. It was the period of MacCarthyism and I was catalogued for life as a former Trotskyist.

In the summer of 1949, Marie-Hélène had forwarded to me a letter announcing that I was going to win a Fields Medal at the next International Congress of Mathematicians. I had no idea what it was. A gold medal and fifteen hundred Canadian dollars, awarded to two mathematicians every four years at the International Congress, it was being offered to me for my work on distributions. The letter gave me all necessary explanations and told me that the thing must remain confidential until the day of the congress, in Cambridge, Massachusetts, in the United States. The refusal of 1949 did not augur well for my presence at the congress. The American mathematicians began trying to remedy the situation and get me a visa. The other winner was Atle Selberg, whose proof of the distribution of prime numbers did not use holomorphic functions of a complex variable. A brilliant mathematician of Norwegian origin, he later moved to Princeton. This International Congress was the first one since the war, the previous one having been held in 1936. The International Congress already existed at the end of the 19th century, and the Fields Medal was first given in 1936 to Ahlfors and Douglas. Marie-Hélène added "Here you are, unexpectedly covered with gold and glory." The prize was worth a little less than two months of my monthly salary. It is certainly not the Nobel Prize (worth six million francs nowadays), but since it is given to just two people every four years, it is rarer and harder to obtain than the Nobel. People often say it is the equivalent of a Nobel Prize for mathematics, however at that time no one had ever heard of it. One of the differences between the two awards is that for the Fields Medal, there is an age limit of thirty-five to forty years. I was almost thirty-five in 1950. Fields Medalists are thus much younger than Nobel prize-winners. It was Dieudonné who first established this comparison between the two prizes, in October 1950 on his return from the congress, in an article for the Nancy newspaper *L'Est Républicain*, for which he asked me for a photograph of myself. Nowadays it's a generally recognized thing. Other prizes have been created: the Wolf prize, the Crafoord prize, the Japan prize and the Grand Prize of Kyoto. Thanks to this, the injustices created among mathematicians because of the rarity of the attribution of Fields Medals have been more or less rectified, at least in France where mathematics are particularly brilliant, and where Fields Medals have been awarded several consecutive times. Recently, they have taken to awarding four Fields Medals instead of two, but France has received less of them! Still, together with Jean-Pierre Serre (1954), René Thom (1958), Alexandre Grothendieck (1966), Alain Connes (1983) and finally, two Frenchmen out of the four winners in 1994, Pierre-Louis Lions and Jean-Christophe Yoccoz (these two were in the same year at the ENS and even in the same "taupe"), that makes a total of seven French medalists. I already remarked that thanks to the French university system, research has a way of becoming hereditary here: the last two winners are the sons of the physicist Yoccoz and the mathematician Jacques-Louis Lions, which was not the case of any of the preceding medalists. Wolf prizes were awarded to the mathematicians André Weil, Jean Leray, Henri Cartan, Jacques Tits and Mikhaïl Gromov. Deligne and Grothendieck were awarded the Crafoord prize in the same year, but Grothendieck refused his. Deligne, after several years in France, took American nationality. There's still the Japan prize and the Kyoto prize which was given to Jacques-Louis Lions.

Nobody knows exactly why there is no Nobel Prize in mathematics. There's an anecdote that Nobel's wife was the mistress of the Swedish mathematician Mittag-Leffler (an unbearable character, actually). Nobel, they say, feared that Mittag-Leffler would win the prize. In fact, Nobel was never married, but perhaps he had a mistress seduced by Mittag-Leffler. We'll never know. Anyway, mathematicians like to tell the story.

Obviously, this medal meant worldwide recognition for me. Recognition accompanied me for the rest of my life, but it never left an impression on me as strong as the one I had when I won the Concours Général in high school. I never used the Fields Medal as one of my official distinctions. Still, I'm happy to have it.

Back in France, I needed to think about the possibility of actually going to the International Congress. With almost a year to go, various mathematicial organisations in France and in the US took charge. The American organizers contacted lawyers. Visas to the US are controlled by the State Department. So they laid siege to the State Department. A boycott was considered. Henri Cartan was the president of the French Mathematical Society. After lengthy reflection, a majority of the members decided that they would not go to the congress unless everyone invited received a visa. Quite a few Americans also threatened not to go if any French mathematicians were refused entry into the country. The Americans appointed Jacques Hadamard honorary president for purely scientific reasons. He was then eighty-five years old. I often had occasion to encounter the remarkable integrity and courage of mathematicians, particularly during colonial wars. They knew perfectly well that Hadamard's visa would cause a problem, because Hadamard was a well-known Communist sympathizer. But only scientific considerations counted in their decision. As the date of the congress approached, the requests rose higher and higher in the American political hierarchy. Finally, I received a visa, a few months before the congress, after the personal intervention of Truman himself. The situation of Jacques Hadamard was still uncertain. Out of the twenty-eight Frenchmen appointed to the Harvard Congress by the French mathematical society, sixteen agreed to boycott the meeting if Hadamard was not given a visa. It wasn't a question of the individual, but of conduct to be adopted for all future congresses, regardless of the intensity of the cold war. Henri Cartan gave the list of sixteen people to the direction of the Cunard Lines. Cunard Lines demanded travel to be confirmed at least four days in advance, so we fixed a deadline of July 24, and sent it to the Americans. If Hadamard did not have his visa by July 24, Cunard would cancel all the tickets. The American mathematicians worked like dogs and finally, after once again obtaining the intervention of Truman himself (this was right during the Korean War), Hadamard received his visa on July 23. Cartan telegraphed the fifteen others of us that upon arriving in Paris, we should go to his house to pick up our tickets. The telegram said "Hadamard got visa, phew!" We were in Pelvoux, in the Alps, where a Bourbaki meeting had been held at the end of June. Marie-Hélène and I felt like fooling around, and we responded with a joke telegram: "What's this all about?" It's actually a well-known response attributed to Marshal Foch on one famous occasion. It was really Marie-Hélène's idea, but she never dared own up to it to Cartan until quite recently. When he received our telegram, he was quite horrified – "If Schwartz doesn't know what's going on after all the phone calls, letters and conversations we've had, my telegram must have been completely illegible". He rushed back to the post office, which was a real torment since he was vacationing in Die, in a torrid heat wave, and had no telephone. When he called me to ask what was up, I confessed to the joke. He cursed me and I tried to excuse myself, but it was too late. Since then, I've never called Cartan from anywhere in the world without his immediately asking me "What's this all about?"

The story of the American congress became an example for the future. From then on, whatever country the congress was organized in, we demanded that the organizing society obtain visas for every single participant without exception.

Cartan and Schwartz

Mathematicians gave the example to scientists in other domains, too. In Moscow in 1966, everyone received a visa.

I needed to go back to the American Consulate in Paris, where the consul gave me a visa for exactly two weeks "only for the purpose of attending the international mathematical congress". Thus, I could go to the congress, but not to any other public ceremony, nor to any other university. I left at the end of July, leaving Marie-Hélène and the children at Autouillet with my parents.

The crossing lasted five very agreeable days. From on high, I observed flying fish, which I later saw from much closer up during our family trip to Columbia in 1956. All the French mathematicians were on board. Hadamard, once again, almost missed the boat, because at the critical moment he found he had no ticket. At first he though that the customs officer had kept it. They held up the boat for a few minutes while he searched out the fellow. "No no", said the officer, "I remember you perfectly sir, when I gave you back your ticket, you crumpled it up and stuffed it into your waistcoat pocket." He pointed to the pocket and there was the ticket. When Hadamard entered the conference room, he was greeted by a huge ovation. The congress was not disappointing and I took a lot of pleasure in it. As a Fields Medalist, I gave one of the full hour lectures, where I talked about kernels, a complement to my work which turned out to be very important to distribution theory later on. Kernel theory is used a lot nowadays, particularly in the study of pseudo-differential operators and integral Fourier operators. Directly contradicting the terms of my visa, Oppenheimer announced in a poster at the Institute for Advanced Study in Princeton: "Clandestine lecture by Laurent Schwartz." A crowd came, and my lecture went well. Americans, who are really rigorous about entries onto their territory, are very liberal-minded when you actually meet them. In fact, once my close friend Lipman Bers, a Lettonian Jew who, alas, died recently, told me that when he arrived to the US during the war, after a long and difficult journey taking him through France, where he received his visa, they immediately suppressed his passport. He protested violently to the immigration officer: "But how can I live without a passport?! I'm naked, I can't walk!" The officer replied: "You walk with your legs, not with a passport."

After the end of the conference, I returned to Montreal for a few days, to see my friends of the year before. The return trip from Montreal to Paris was rather painful. One of the propellers of the airplane broke down, which happened quite frequently in those days. The airplane was suppose to land at the nearest airport; in our case this was Goose Bay in Labrador. And we were almost half-way to London! I've always slept well in airplanes, even though we have to sit; that night, I was sleeping happily as usual. When I awoke, my neighbors told me that we were on our way back to Canada. We had to stay in Goose Bay for ten hours. Labrador was even more spectacular than the year before. It was September and already quite cold. The trees were still minuscule, and to my delight, there were quantities of blueberries attached to the branches, which were only a few centimeters long.

My visa problems did have consequences for me over the years. My American colleagues gave up inviting me, and for ten years I did not return to the US. In 1960, it took six months of negotiation to get a visa for a two-month visit to Berkeley. The final decision was always taken by the State Department. I also managed to spend a whole sabbatical year with my family in the US, in 1962–63. It took a lot of months for that confounded visa, too. Later on, I was able to get back to the US from time to time, going through heavy formalities each time. McCarthy had disappeared long before, and plenty of authentic Communists (I wasn't one!) such as Jean-Pierre Kahane had no trouble entering the United States. But since my visa had been refused once, I was blacklisted and had to obtain a waiver from the State Department for each and every visit.

Much later, I was called in by the Consul General who excused himself to me, saying: "It's all ridiculous, but there's nothing I can do about it." Then things became easier, thanks to authorizations obtained by telephone from Frankfurt, and finally, in the last few years, I was able to obtain visas for several visits, so that I could go in and out for as much as one or two years. Finally, my last visa, dated February 7, 1991 *for multiple entries*, is valid from now until $+\infty$. Finally, after waiting for forty years, I can enter the US like any normal person. In all, I only went to the US about eight times. It's very little in the life of a mathematician (four times in twenty-six years, from 1950 to 1976). Most of the good French mathematicians have spent a fair amount of time in the US, at least once every three or four years, and often more. My contacts with Americans were thus quite limited. Since my articles were written in French and Americans don't like to read French, they generally never read anything except the Distributions, which they couldn't do without. But they probably never got to know the other articles, which I had no occasion to lecture on.

This visa story from 1949 had two other important consequences. I couldn't go to the US, and I had gotten to enjoy long trips, so I happily accepted invitations to Third World countries: I spent the summer of 1952 in Brazil, the summer of 1953 in Mexico, January and February 1955 in India. These trips changed my life fundamentally.

On the one hand, because of the war I had abandoned my first butterfly collection. But when I arrived in Brazil, I was stunned by the abundance and diversity of the tropical entomological fauna, and I started it up again. Today I have an enormous collection, one of the biggest in Europe, consisting of nineteen thousand butterflies collected over forty-four years. Butterflies are one of the great passions of my life.

On the other hand, I learned to know the Third World countries, their poverty, their economical and social difficulties, their culture, their art, their hospitality. Before even going there, I was on the side of those countries in their struggle for emancipation. After my visits, thi was no longer an abstract feeling. I penetrated deep into the countrysides to hunt for butterflies, as few people have had the opportunity to do. I have suffered with the Third World, talked to its peasants, but I never, except during my Trotskyist period, considered that it could be the motor for a worldwide revolution. However, I have promised myself to help it in every way. Belonging to the Third World is living in misery, and this misery must be eliminated. I was very interested by the neutralism born in the Bandung conference of April 1955, but I've never been a true Third World enthusiast, simply because some of the principal obstacles to the liberation of underdeveloped countries reside in those countries themselves, in their monarchs and dictators, whom I have always radically opposed. The dictator of an underdeveloped country is an agent of the underdevelopment. Obviously, those tyrants are only too often supported, or even installed, by developed imperialist countries. During my trips to underdeveloped countries, I always participated in activities pertaining to the improvement of education and research; I have contacts with many Third World mathematicians.

## The Soviet visa

In June 1964, I received from the French ambassador in Moscow an invitation to the University of Moscow for two weeks, at any date which would suit me. It was the first invitation of its kind. The USSR was no longer under Stalin but under Khrushchev. But it was certainly not a democracy – and I was a former Trotskyist, after all! I thanked the ambassador by telegram and wrote to Gel'fand that I hoped to visit from October 4–19. Much later, I learned that it took my telegram nineteen days to arrive! Gel'fand told me that everything would be made ready for my visit. For a long time, I had no response to my request for a visa. I waited and waited. At the end of September, still nothing. Finally, on October 3, the day before I was due to leave, I went to the Consulate where a certain Madame Antonova confirmed that there had been no answer. "Your request for a visa is circulating somewhere in Siberia or in the Urals. You should probably give up." Without really thinking, I answered her: "If I give up, I should at least send telegrams to the three members of the Soviet Academy who will be waiting for me tomorrow at the airport." In fact, I had no idea who would be waiting for me. "In that case, the situation is complicated; I'll call the Director of Cultural Relations." By telephone, this person said: "Couldn't you wait a few more days?" This was impossible – even returning on October 19 made me late for the beginning of exams, which I had postponed. And I didn't want to go to Moscow just for a few days. He tried in vain to convince me, but I was stubborn. I had asked for my visa months ago and had no reason to wait any longer. I remembered that in the case of the US visa, firmness had paid. When I said that I was going to telegraph the Academicians waiting at the airport, he said: "Oh no, please, do anything but don't send a telegram!" (I had no idea at that point that a telegram could take nineteen days to arrive.) I insisted sternly. I didn't want to be rude and to let them come to the airport if I knew that I wasn't going to be there. "Anything you want, but not a telegram!" he repeated. Obviously, he didn't want to accept his responsibility. When I decided to leave, Madame Antonova whispered to me "Serves him right." It was eleven forty-five. I reached home at twelve-thirty. That October 3 was a Saturday. When I reached home, Marie-Hélène told me that the Soviet Consulate had called to say that my visa was ready! I called back immediately, but it was closed. I decided that it must be a maneuver to put the blame on me. If I didn't have a visa, it was because I hadn't gone to get it. At the French Direction of Cultural Relations, which had followed all the developments, they advised me to go back to the Consulate as soon as possible, so as to avoid any possibility of blame. I arrived back there at one o'clock. It was closed, but Madame Antonova had been waiting for me at the door for a whole hour, with my visa ready to stamp into my passport. The USSR ambassador to France had been informed of the situation by the Director of Cultural Relations, and he had taken advantage of his right to grant the visa personally without going through Moscow. The USSR ambassador has this right, but the US ambassador doesn't, and the approval of the State Department is mandatory. Equipped with my special visa, I took off on Sunday, October 4. The flight was four hours late because of an improvised stopover in Warsaw. On top of this, I was upset by the whole story, and getting off the plane, I accidentally took the wrong coat. I only noticed it when I saw a pipe sticking out of the pocket of "my" coat ....

Gel'fand, Shilov and Olga Oleinik were waiting for me at the airport. They hadn't been informed of anything, and they'd been there for five hours wondering what was happening.

After the usual greetings, we began an animated conversation. "You know," said Gel'fand, "it was amazingly difficult to get you here." "Well, of course I know!" "How could you know?" "How could I not know?" When I told him about my tribulations, he told me about the Russian side of the story. "When you let me know, via a telegram which took nineteen days to get here, that you had received an invitation from our university, we were astounded. We have been inviting you for years without ever receiving an answer. We got discouraged and stopped inviting you some time ago. And suddenly, you answered an invitation that we didn't even know about! We went to see the rector of the university, Professor Petrovsky (a well-known mathematician). He asked the Minister of Higher Education, who knew nothing about it, and refused to confirm the invitation without knowing more about where it emanated from.

In vain we tried to find out. It was unthinkable for the Minister, or the Rector

Petrovsky, or Professor Gel'fand to contact the French Embassy in Moscow who had transmitted the invitation to me. From delay to delay, nothing happened. Finally, on October 1, they asked Petrovsky to make a decision since I had announced that I was arriving on October 4. The Rector Petrovsky returned to see the Minister. "Frankly," said the Minister, "I can't arrange for an invitation in less than two months from now." "But Laurent Schwartz is arriving in three days!" "What can I do about it?" So Petrovsky pulled out his trump card: "All right, Mr. Minister, if you can't do anything, we'll send a telegram to Laurent Schwartz informing him that his visit is put off till later." "No no," said the Minister, "anything you want, but not a telegram!" "Well, but," said the Rector, "if we don't send a telegram, he will arrive." The Minister saw it: either they sent a telegram or I arrived on October 4. "All right, I'll hurry it up," he said. And miraculously, the next day, he had obtained the invitation and the grant and found a room in the dormitory. I was delighted by this solution, for which I would even be obliged to learn a few Russian words. "Now we know," said Gel'fand, "how to get you to Russia any time we want. Just write to thank us for the invitation! You do what you just did, we'll do what we just did, and there you are. Simple." I'm pretty sure that all this had nothing to do with the fact that I was a former Trotskyist; it was just usual dealings with the administrational mess which reigned in the USSR at that time and still does. Pierre Aigrain, a reputed physicist later to become a member of the Cabinet, had to visit the USSR around the same time as me. Like me, he still had no visa the day before his departure. Unfortunately, the idea of telegraphing didn't occur to him, and he had to cancel his visit. Too bad! Telegrams were apparently most embarrassing to the administration, and thus quite fruitful.

My stay in Moscow went very well. I gave interesting talks and had enriching discussions with mathematicians; I visited the Hermitage museum in Leningrad. Everywhere I went, I was accompanied by a young Soviet mathematician called Leonid Frank. He was Jewish and had not been very well treated, but he had succeeded in obtaining a position as assistant in the university, under the protection of the Rector Petrovsky. He was my guide. He was very nice, sentimental, and he adored French songs, which he sang accompanying himself on the guitar. He translated the talks I gave in French at the university. He later became a dissident and succeeded in fleeing under Khrushchev. He gave an interview on the French radio, talking about what was going on in the USSR. When he was expelled from the USSR, he came to France via Israel. He tried to persuade me to take the day train from Moscow to Leningrad, in order to admire the Russian countryside, but in spite of all his efforts, he was unable to obtain permission for me; I had to choose between a daytime plane or a night train.

It was annoying and I protested, but accepted the night train. He was surprised at my protestations. “Well, but," he said, “if someone wants to go from Paris to Marseille, I suppose he has to get some kind of authorization, doesn't he?" I proceeded to disillusion him. “No, really? I can't believe it! you don't need an authorization?" “No," I insisted. He answered with these pathetic words: “What a mess it must be over there!" We've been excellent friends ever since, and he laughed wholeheartedly when I recently reminded him of that conversation. Obviously, his mindset has changed since those days. I kept very good relations with him. When he left the USSR, he spent a short time in Paris and then settled in Nimègues with his mother for several years. Then he returned to live in Paris, with a position as professor at the University of Reims. A year ago he became ill with cancer, and he died on December 29, 1997. I was very sad.

Since I mentioned Petrovsky, I'd like to tell an anecdote concerning him, but which took place long before my visit. When Petrovsky was still a young man, Jacques Hadamard came to Moscow. Petrovsky was his guide, and had the duty of introducing him, in French, at his first talk, which he did scrupulously. Hadamard thanked him politely for his introduction with these words: “I am sure your introduction was kind and courteous, though naturally didn't understand a single word since it was in Russian." Petrovsky told me this story, adding that he had never spoken a single word of French since. I returned to the USSR in 1966, for the International Congress of Mathematicians, but after that I became highly undesirable there, not as a former Trotskyist but because I was a founder, with Cartan and Broué, of the Committee of Mathematicians[50]. That made it clear that any visa request would be instantly rejected.

## My move to Paris

The year 1952 turned a page in my life. In spite of the excellent young researchers who came occasionally to Nancy, but who never stayed, I decided to leave Nancy. As a source of students, the ENS seemed to have dried up, and I felt that it would always be a struggle to have students. Denjoy, who was in charge of mathematics at the Sorbonne, was trying to recruit analysts, and had already recruited Choquet in 1950. He was eager for me to come as well. I resisted for a whole year, but in 1951, I gave in, applied for the job, and was appointed Maître de Conférences (Associate Professor) at the Sorbonne, although I was a full professor in Nancy; my salary descended back to the beginning stage. While applying, I had a dream which revealed all the difficulty I felt in

50) See chapter 12

leaving Nancy: I had been appointed grocer at the Sorbonne, and I kept telling Marie-Hélène that Roubaud had threatened not to be my client any more if I accepted. And how could we make a living then? I obtained the authorization to spend one last year, 1951–52, in Nancy.

Then came the problem of finding a place to live in Paris, which was suffering from a serious housing crisis. For social reasons, the various governments since the war had prevented rents from rising, and they finally succeeded in provoking exactly the opposite of the desired effect: people with apartments to rent simply did not rent them out, as it was not worth it. No building was going on, yet the population was increasing, and it was impossible to find housing, especially for young people just leaving their families. The way to hell, as they say, is paved with good intentions. We knew it wasn't even thinkable to find a nice free apartment in Paris. However, people renting apartments could exchange without asking permission from the proprietors, who could sue the renters, but usually lost. We succeeded in organizing a triple exchange! Three families made a cyclic permutation. The whole operation had to be done in a single day, because if some apartment was unoccupied even for a single hour, the proprietor could show up and declare the illegality of the whole procedure. My parents had done the same thing when they moved into an apartment on the avenue de Versailles which was actually smaller than their former apartment. With his usual scrupulousness, my father felt like a cheater and a squatter, almost a thief ....

For one day, the wife of the mathematician Myrkil who was working in Nancy, occupied the apartment at 225, avenue Daumesnil in Paris into which we were about to move. The Myrkils were young, generous and cultivated. Myrkil had discovered an interesting but practically unpublishable mathematical result: Dytkine had proved an important theorem with a long, complicated and astute proof; Myrkil had found a simple proof in just a few lines, which rendered Dytkine's result completely obvious. How could one publish such a thing? But their life was smashed by a terrible tragedy – Myrkil had a sudden nervous breakdown and left his house, going out alone and half-dressed in the winter snow; he died very soon after. We were terribly sad; I'm glad to take this occasion to recall his memory.

After numerous difficulties, the day of the triple move finally arrived. It was like a ballet in which all the protagonists had to move simultaneously. It must have been in January 1953 (I had started teaching in the fall of 1952, commuting from Nancy to Paris). A new life was beginning for each of us. For Marc-André, this meant that for a short time he had no lessons and no homework, which caused him to leap and dance with delight. Goodbye, Nancy! We turned our backs on eight years of happiness after the hard years of the war.

Obviously, I did not stop working after we left Nancy. I worked on distributions until 1962, and became mired down in vector-valued distributions; later I will recount how the Algerian war nearly put a stop to my research. I continued studying pseudo-differential operators for some years, and directed two theses on the subject, written by a married couple, the Unterbergers. Then I radically changed subjects, working on geometry and probability in Banach spaces around 1965. Towards 1968–1970, I flung myself body and soul into the grand probability movement, which represented a return to the love of my youth and my memories of Paul Lévy from 1935–36. It was like a new breath of fresh air. But a period begun at the age of fifty-five cannot be like the period of youth. First, I extended Radon measures to arbitrary, not necessarily locally compact topological spaces, for instance Banach spaces, and I worked on the radonification of cylindrical probability. Next, my most important results were on regular disintegration and Markov processes, Paul-André Meyer's semi-martingales and their extension to differential manifolds, differentials of semi-martingales on manifolds, conformal martingales on complex manifolds, stochastic differential equations, and more general infinitesimal stochastic geometry, which is second order. After one collaboration with Meyer and Michel Emery of Strasbourg, I found myself isolated in this domain. It seems to me that there are two reasons for this. The first has to do with myself. As in the case of vector-valued distributions, I consistently tried to deal with the most general cases, which apparently made my work more difficult to understand. I guess I'll always be that way. The other reason has to do with probabilists: they're very ignorant about tensor products and manifolds. If $T_2(V;v)$ is the tangent space of order 2 to a manifold $V$ at a point $v$ and $T_1(V;v)$ is its tangent subspace of order 1, then the quotient $T_2(V;v)/T_1(V;v)$ is isomorphic to the symmetric tensor product $T_1(V;v) \odot T_1(V;v)$. This property is quite familiar to differential geometers and to most analysts, but it seems to be incomprehensible to the majority of probabilists. But it's the key to stochastic differential geometry!

In Paris, I taught in the first, second and third university cycles. My most innovative teaching took place in the second cycle, where I taught mathematical methods for physicists from 1952 to 1962. A large number of avid students took my class each year (four hundred and fifty enrolled in the final year of my teaching the course), and brought me great pleasure. My course gave solid mathematical background to young physics students preparing a Licence: they learned Lebesgue integration, distributions, convolution, Fourier and Laplace integrals and the wave equation. But students preparing the Licence in mathematics also took the course, because these subjects were not treated in the extremely theoretical courses offered for the mathematical Licence.

Starting in 1969, I left the university to become a full professor at the Ecole Polytechnique. I retired, at the age of sixty-eight, in 1983. I published a great deal after my retirement, but the source seems to be drying up; my last publications date back to 1989 and 1994, and sometimes I worry that there won't be any more. It makes me sad. My mathematical memory is deserting me in giant steps. This book shows that my memories of the events of my life have not left me, but that isn't sufficient to create mathematics. I remember the moments and circumstances of my past creations, but much less of the creations themselves .... My beautiful inner castle is deteriorating, the connections are disappearing and sometimes I get lost.

Marie-Hélène, having obtained her doctorate, had sufficient titles to apply for an associate professorship in a provincial university. But we knew that even though she had already done good research, it was highly specialized, and the circumstances of her life had deprived her of the general culture she needed to expand her work. And she didn't have the energy or the memory to skip that stage or to handle it alone. So she was quite pleased to accept a job as assistant in differential and integral calculus, and she did this for two years part-time and then for four years full-time, assisting various professors among which was Choquet. With his typical daring, Choquet taught general topology using general topological spaces – rather than metric spaces, as was usual. Many of his colleagues, including Bourbaki, thought he must have fallen on his head. But it worked very well; Marie-Hélène taught it herself later on. Then she obtained a position as Associate Professor in Reims, and then, in 1963, in Lille. One might ask why she moved from Reims, which is near Paris, to Lille, which is farther off. But she commuted from Paris in any case, and Lille had a bigger faculty and offered her to teach not only the usual Licence courses but also one of the four graduate courses. Marie-Hélène had chosen to learn the mathematical tools she needed by teaching them (mathematicians learn new subjects quite frequently in this way!) In particular, she taught analytic functions of several complex variables, with sheaf theory, differential manifolds and Thom transversality, etc. Her course notes were written up and published in booklets by the Lille faculty, and distributed quite widely. She completed them by teaching a course called *Stratifications of Complex Analytic Sets* at the Tata Institute of Fundamental Research in Bombay in 1965 (we had both been invited there for the summer). Course notes were taken and written up by R. Narasimhan and S. Rajwade, and the book was then published.

Her courses at Lille made agreeable reading for her public. But as a researcher, she was isolated, and as for her articles, she was never sure which public she should be addressing, which theorems she should admit as known, which symbols to use. Her articles were written in a heavy style at first. Later, she became less isolated. In Lille, she immediately encountered graduate students who became colleagues and friends: Bernard Callenaere in 1963, Jean-Paul Brasselet in 1964, and others. Later, she was offered a job in Paris, but she was happy in Lille and chose to remain there.

Marie-Hélène's thesis started her off in the direction she was to follow in all of her research. Even before her thesis, we had received in Nancy, in 1964, a letter from André Weil from Chicago, who talked to us enthusiastically about a young Chinese mathematician named Chern who appeared to be completely renewing the subject of differential geometry. He introduced certain cohomology classes on Hermitian varieties which are now known as Chern classes. I've used them myself, but it was Marie-Hélène who immediately understood how Chern classes could help her generalize one of her main results. Ten years later, she defined the obstructive Chern classes associated to a complex analytic stratification. She almost abandoned the whole thing because of insurmountable difficulties, which forced her to make some hypotheses she found absurd. But then Hassler Whitney introduced his a and b conditions which were exactly adapted to her needs, and she was able to finish developing her ideas, encouraged even more recently by the famous article by MacPherson on Chern homology classes on algebraic varieties. In a word, her mathematical career has been entirely based on a single idée fixe – a very good idée fixe.

Research has always been a large part of both our lives. But shortly after our arrival in Paris, I flung myself into a new adventure: the reform of the Ecole Polytechnique.

# Chapter IX
# The Algerian Involvement

The Hindu mathematician Chandrasekharan, who devoted years of his life to the construction of the Mathematics Department of the Tata Institute in Bombay, making it into one of the best Asian research institutes, literally let himself be devoured by administrative tasks which prevented him from devoting himself to research in his domain. He often publicly expressed regret about this. "But you chose it," people would say. "I only chose five percent of the difficulties," he would invariably reply. His case is exactly the same as mine, except that on top of administrative duties I was also a political militant. Indeed, during the first years of my teaching at the Ecole Polytechnique, my mind was occupied with many things besides the improvement of the Grandes Ecoles. Those who believe that I took pleasure in political activities are mistaken, even though I was passionately involved in them. My eternal temptation to throw myself, in spite of myself, into tasks which I consider as imperious necessities, is what the all too famous Soviet psychiatrists of Brezhnev's time called "torpid schizophrenia", referring to dissidents. In fact, I don't even understand myself! I feel driven by an irresistible force, which is crazy and yet inevitable, and whose price is that I must abandon research.

When the Algerian war broke out, on November 1, 1954, internationalism and anti-colonialism were deeply anchored in my reflexes. I was convinced that the French colonies should be independent. Moreover, I was not an inexperienced left-winger and I knew how to remain prudent in dialogues. I used persuasion with my interlocutors without shocking them by taking positions which they would find extreme. I denounced the crimes committed in the colonies and asserted the necessity of negotiations, but I only actually spoke out in favor of independence when the moment appeared opportune. The events of May 8, 1945 in Setif foreshadowed all that was to follow. A local revolt had ended in the assassination of about a hundred French colonialists, men, women and children, all atrociously mutilated. Such atrocities could only be condemned. But the French government responded with an even more unacceptable massacre. The army sent its forces against the revolting peasants and slaughtered them in such a way that hatred only became worse. The Algerians counted forty-five thousand dead, the French claimed it was one to two thousand; realistically, the number killed was probably around ten or fifteen thousand people.

The colonization of Algeria did not only have evil effects. After a disastrous slide, public health and education had considerably improved. Life had become more modern and the urbanization of the country brought a certain middle class into existence. Railways and roads were constructed, and in some towns telephones worked as well as in France. Most of the French colonialists in Algeria did not possess land; they constituted an essentially urban population. Politically, democracy was reserved to them, whereas the Muslim Algerians, who had no real form of self-expression and almost no place in the Constitution, suffered from a state of dependence and were despised. Those French or Algerians who tried to work for the progress of democracy were more or less gagged. Algeria was part of France, even though the Constitution was not the same in Algeria as in France proper; Oran, Algiers and Constantine were French departments. Furthermore, Algeria was a settled colony; one million French people lived among the nine million Muslims, Arabs and Kabyls. The attachment of Algeria to France was based on this fact.

My sympathies lay with the MNA (Mouvement National Algérien)[51)], led by Messali Hadj. He had been a Communist at first (within the North-African Star), and then in 1937 he had founded the PPA (Parti du Peuple Algérien)[52)], which demanded national independence. When the PPA was dissolved, he continued to run it clandestinely, and then founded the MTLD (Mouvement pour le triomphe des libertés démocratiques)[53)] in 1946. Messali Hadj was a grand nationalist leader, a cultivated, gentle man, ready to dialogue with the French left wing. But he encouraged a personality cult around himself and his activities benefited his movement and his person more than the independence of his country. This had disappointed many Algerians. The FLN (Front de Libération National)[54)] was formed before November 1, 1954 by dissidents from the MTLD who had enough of Messali's constant hesitations. As a reponse to the creation of the FLN, Messali founded the MNA. The nine historic founders of the FLN became known through the coup of November 1, but in reality, the whole movement was based on a well-organized group of just 1500 people. Because it was a small-scale operation, the Aurès uprising was the most important episode of the period. The Algerian war had begun.

51) MNA: National Algeria Movement

52) PPA: Party of the Algerian People

53) MTLD: Movement for the Triumph of Democratic Liberties

54) FLN: National Liberation Front

The French leaders unanimously decided on sending in the artillery. François Mitterrand, who was Minister of the Interior, decided to repress the rebellion. Messali Hadj's MTLD was dissolved on November 5 even though it actually had played no role in the rebellion. In France, left-wing leaders, starting with Pierre Mendès France, began evolving towards more democratic solutions, such as negotiations. However, the serious idea of the independence of the Algerian people penetrated the consciousness of the French left wing only slowly, starting a little before 1961, whereas the FLN had adopted this goal from the beginning. The fissure between Algeria and France was a deep one.

The intellectuals used every possible legal method in their fight against the war; committees, demonstrations, petitions, publication of newspaper articles, reviews, books. The publishing houses Editions de Minuit (directed by Jérôme Lindon) and Editions Maspero were created during the Algerian war in order to support these anti-war publications; many of their books were confiscated, although they continued to distribute them clandestinely. The publishing house Seuil also contributed to the struggle, although less energetically. Four newspapers were extraordinarily active in resisting the war: *Le Monde*, *L'Express* (directed by Jean-Jacques Servan-Schreiber), *France-Observateur* (directed by Claude Bourdet and Gilles Martinet) and *Témoin Chrétien* (*Christian Witness*). Jacques Soustelle called them "the four Giants for French counter-propaganda". In spite of frequent prohibitions, they succeeded in distributing information, sometimes at the price of severe financial loss. The media was thus an indispensable ally of the resistance of the intellectuals. I remember that towards 1960, I tried to explain restricted relativity to Marc-André, who was about 17. I used the same example as Einstein: a train is moving at a speed close to the speed of light, a man standing near the railway measures times and lengths relative to objects or events in the train; a round ball appears to the observer outside the train as an oblong because of the relativistic effects on lengths, and the observer also finds that the clocks in the train are late, because of relativistic dilation of time. But what always seems amazing at first is that an observer inside the train perceives outside objects and events with the same deformation! Marc-André took a little time to comprehend this, like anybody else, and then concluded: "That's the theorem of *L'Express* and *L'Observateur*." After that, I never taught relativity without citing this remark, and it always met with great success.

## The first committees

Very soon, I joined the "Action committee of intellectuals against the war in North Africa", a movement founded in 1955 by Robert Antelme, Dyonis Mascolo, Louis-René des Forêts and Edgar Morin. I met all kinds of intellectuals there, and indeed the fight against the Algerian war welded men and women who might never have otherwise met together in long friendships. Claude Bourdet, with whom I was to sign innumerable petitions, was an indefatigable fighter. Jean Dresch was a Communist, always on the front line. Jean Guéhenno and Daniel Guérin had been among my teachers in 1936. Bianca Lamblin, a high school teacher, played a very dynamic role in the non-Communist left. The Nobel Prize winning physicist Frédéric Joliot-Curie, a Communist, immediately took up our position; Jacques Madaule, a militant Christian, was already at work at the effort of his life, bringing Jews and Christians closer together. From the beginning, the writer François Mauriac stood forth as a Christian against the Algerian war and the use of torture, with his weekly "Notepad" articles published in *L'Express* and his active presence in all the demonstrations. He gave moral support to Marc-André when he took his first steps as a novelist. Naturally, I had the opportunity to collaborate with the biologist Jacques Monod. The Abbé Pierre was working for the poor. I had known Jean Rostand for a long time; as a child, I had read his books on biology with enthusiasm. My road crossed Sartre's many times, in protests against the Algerian war and later against the American invasion of Viet-Nam. Jean-Pierre Vernant progressively distanced himself from Communism from the time I met him, and we became close, partly because of our shared passion for ancient Greece. The diversity of the members of the committee shows how the Algerian war touched off passions in intellectuals of every horizon and every discipline. When the committee was formed, one year after the events of All-Saints-Day 1954, the word independence was hardly pronounced. It was Raymond Aron, a right-winger who did not belong to the committee, who first formulated the arguments in favor of independence, expressing them with striking lucidity already in 1957. Many intellectuals became political militants for the first time. I had been a Trotskyist, but during the Second World War there had been no large mass of people whose combat I could share. Now, however, my faithfulness to the Trotskyist tenets of internationalism and anti-colonialism made me a full-fledged member of the struggle, along with a very large part of French intelligentsia, and ready to play a role on the scale of my deep involvement.

The committee organized many important meetings. One of them took place in the Wagram hall, on January 27 1956; thousands of people came, of which many hundred Algerians. The committee suffered because of the conflict between the Algerian movements FLN and MNA. During the Wagram meeting, a partisan of the FLN wanted to prevent Daniel Guérin from speaking in favor of the MNA. "Supporting Messali during a meeting against the Algerian war," he said, "is like supporting Mussolini in an anti-fascist meeting in Italy!" But the committee met its final demise as a consequence of much more serious events: the Hungarian uprising and Khrushchev's response with the Soviet invasion of Hungary on November 2, 1956. I won't recount the terrible wound, the despair I felt at this event. Many of us in the committee felt that we must inevitably discuss it, in the committee and also publicly. The principle of the rights of peoples to independence was universal and untouchable. The vocation of our committee was to fight colonialism in Algeria, but not to speak out against the events taking place in Budapest would have been a display of revolting partiality. I never thought that we should change the name of our committee or its fundamental goals, which were exclusively related to the Algerian situation, but it appeared necessary to me and to many others that we must at least publish a communiqué expressing our fundamental disapproval of the invasion of Hungary by Soviet tanks. The discussion within the committee was quite violent. The Communists absolutely refused to entertain the idea of mentioning Hungary, and others who were not Communists or even close to Communists agreed with them. The two opposing groups made some efforts at reconciliation. In particular a private effort was made by a group containing the following five people: Gilbert Mury, a Communist who later became Maoist, François Châtelet, a Communist who soon after left the Party, Jean-Pierre Vernant, a member of the Party who was already very close to the non-Communist left and who left the Party in 1968, Jacques Monod and myself. Monod and I were solidly on the side of declaring our feelings; Jean-Pierre Vernant and François Châtelet were more or less sympathetic to us. Mury opposed us squarely. We were unable to find any common ground whatsoever, and the committee died out very soon after the events of November 1956.

Much later, I belonged to the "Colloquium committee for the study of solutions to the Algerian problem". The goals of the committee were much more advanced than mere condemnation of the war. A strong minority consisted of partisans of independence such as Raymond Aron. A large group of militants held regular meetings. I particularly remember André Hauriou, an intelligent and agreeable gentleman who died very young; we often met in his house. I also remember Maurice Duverger, a professor of law in Paris, Régis Blachère who gave me a copy of the Koran with his own commentaries, and Jacques Berque, an Arab specialist at the Collège de France, who particularly surprised me by telling me that he had to look through roughly twenty-thousand pages to prepare each one of his lectures at the Collège. He was more or less Communist. I met Duverger often in the most diverse places; at political meetings and during trips abroad, in Rio de Janeiro and Bombay. I also met Germaine Tillion, an ethnologue who had been deported, and who wrote a remarkable book on Ravensbrück; I admire her enormously. Robert Barrat and his wife Denise, as well as Francis Jeanson, were Christians in close contact with the FLN. Barrat was a kind of intermediary in the conversations between the FLN and Michelet, one of de Gaulle's Justice Ministers; he was sent to prison for two weeks. There were frequent contacts between the opponents via intermediaries; this is often the case during wars. I also often saw Daniel Mayer, the president of the League for Human Rights and a member of the Socialist Party; he was a tenacious and combative man well known for his devastating sense of repartee. Throughout the year 1959, the committee met to discuss a project for the future status of Algeria and for peace negotiations proposed by Maurice Duverger. I believe his project was altogether reasonable; I remember defending it. But it had several opponents on the left, including Pierre Naville, a former Trotskyist. Also, Pierre Stibbe, a lawyer for the historic founders of the FLN (Ben Bella, Boudiaf, Aït Ahmed and Khider, who were imprisoned in Paris after their airplane was hijacked by the French army, but who remained the spiritual leaders of the FLN) confided to me that they also supported our project. Pierre Stibbe, who unfortunately died after the Algerian war, was a very attractive personality; honest, intelligent and generous. The committee continued for a while; I don't know how it eventually ended. The president was Albert Châtelet, a mathematician and an honorary dean of the Scientific University of Paris, who had signed my nomination to the university in 1952. He played a very important role in the Algerian war.

Let me return to the events of 1956. Because of his individual stature, it would have been logical to name Mendès France the president of the Council. He was quite famous because he had signed the Geneva agreements ending the Indochina war, but also for his audacious trip to Tunis during which he had proclaimed the internal independence of the country. However, in the end the presidency was given to Guy Mollet on February 1, 1956. Guy Mollet was universally considered to be to the left of the Socialist Party. A priori, this was not a problem; we were sure that he would seek for a solution by negotiation. But our hopes were soon dashed. His famous trip to Algiers on February 6, where he was received with rotten tomatoes by the French colonialists, made him change his opinion radically; in a single day, he became a staunch defender of French Algeria. On February 9, he appointed Robert Lacoste Resident Minister in Algeria, replacing Catroux who had resigned, and had special powers voted to him by the Chamber of Deputies. What a volte-face in just twelve days! For us, it was a terrible defeat. Mendès France signaled his opposition by resigning from the government. From that time, he kept his own, independent position, but he was unable to effectively work against Guy Mollet. The situation degenerated rapidly. Robert Lacoste, another Socialist, also took the side of the French colonialists. Robert Lacoste, Max Lejeune and Maurice Bourgès-Maunoury should be considered as the Ministers responsible for the use of torture in Algeria. Guy Mollet quickly took several catastrophic initiatives.

On November 1, 1956, together with Israel and England, Guy Mollet organized the Suez expedition against Nasser. His ideas on the subject of the Algerian revolution were extremely primitive. Certainly, the FLN received significant aid from Tunisia, Morocco and other Arab countries, but weapons were furnished above all by Nasser. Mollet thought naively that the influence of Nasser, who had only just nationalized the Suez canal, was the main force behind the Algerian revolution. England was trying to cause Nasser's fall, and naturally, as he was constantly threatening Israel by using and exploiting the Palestinian question, the Israelis wanted to overthrow him as well. The three invading countries obtained rapid victories. However, they were soon confronted with the Soviet and American reactions. The Suez affair occurred at the same time as the Soviet invasion of Hungary, and Khrushchev was being threatened on all sides. He thundered and brandished the atomic threat ("rockets will fly") if the expedition was not stopped immediately. For their part, the United States had been hostile to French politics in Algeria from the beginning (one of the rare positive aspects of their international politics after the war), so that the USSR and the USA coalesced to force France, England and Israel to cease their conquests and return to their initial positions. They did so, with the result that Nasser scored an immense triumph in the Arab world and on the international scene – without striking a blow! I approved the nationalization of the Suez canal, a typical anti-colonialist action. Nasser remained one of the grand figures of the liberation of the Third World.

Shortly after taking power, Guy Mollet decided that bestowing full powers on Robert Lacoste in Algeria was not sufficient, and took the additional step of mobilizing the army reserves. The country's youth had no desire whatsoever to go fight, and there were many reactions and protests. In June-July 1956, unwilling draftees blocked the departing trains and opposed a relatively effective resistance to their mobilization. Around the same time, solid opposition to the war began to crystallize in a group around the newspaper *Témoin Chrétien*. The use of torture had already begun in Algeria, and the Christian group "Spiritual Resistance" had published a pamphlet called *Des rappelés témoignent* (*Draftees bear witness*), in which they recounted a large number of unacceptable atrocities.

Jacques Soustelle, the governor general of Algeria, left France with very liberal intentions. He was a historian and sociologist specializing in Mexican civilization, and well known in university circles. I read his books on Mexico with pleasure. However, when he met the French in Algiers, so passionate about their colony, and saw the corpses of those assassinated by the FLN, so horribly mutilated that he had to rush off and vomit, he changed completely, and used the full measure of his power until the very end of the war against any kind of progressive solution whatsoever. Naturally, I followed this evolution with deep anguish. Our family was to pass the summer of 1956 in Columbia – it was a marvelous summer, but I already felt the war as a heavy weight on my life. I even wondered if I should leave France at all, or simply give up the trip. In the end, we went, and nothing extraordinary happened over the summer. Later, I always made the same choice, although unfortunately it did occasionally happen that during my absences, events occurred for which I probably should have remained in France. But if I had tried to foresee them every time, I would never have left the country – and nothing proves that the course of events would have been altered.

At the time of the Suez expedition, I met certain Israeli leaders, during a meeting organized by a group of French anti-colonialists. The Israelis supported the cause of French Algeria and were scandalized by my attitude. On many other major questions, we did not disagree; I was a firm believer in the development of Israel. I was not exactly a Zionist, but I was certainly not an anti-Zionist and understood perfectly well that after the Shoah, Zionism was a lifesaver for Jews from the entire world. I felt deep connections with Israel, even though I also absolutely defend the rights of Palestinians. My position was close to that of Jean-Paul Sartre or Mendès France. I vainly attempted to explain the significance of our position to the Israelis, and to show them where they were mistaken. Certainly, they had severe – even irreconcilable – conflicts with the Arab countries. There was not a single Arab country which did not ardently desire the death of Israel. This put the Israelis in a difficult situation, leaving them with very little leeway. I thought that if only the Israeli position on certain remote international questions could be brought in line with that of certain Arab countries, friendlier relations could slowly be installed. For example, the destiny of Israel was not directly involved in the Algerian question, and when the FLN called for international support, Israel could have allied itself with the USA to call for Algerian independence, even while preserving relations with France. Or at least, it could have remained neutral, instead of showing itself violently hostile to the FLN. Well before this meeting, I had given some rather strange advice to the Israelis. Several years earlier, Mossadegh had seized power from the Shah in Iran and established a regime of personal power with the support of the Bazar, directed against imperialist powers and England above all. At that time, the United States was less directly concerned with Iranian oil than England. But Mossadegh found himself in dire straits because he could not sell his oil; no Arab country could buy it from him, and he was boycotted by all the great Western powers. He ended up falling, under direct blows from the United States who reinstalled the Shah. The sequel is known to everyone. Mossadegh's experience was very interesting. There was no relation at all between his semi-democratic attitude in favor of the laboring masses and Khomeini's religious fanatism. Perhaps, if we had supported Mossadegh instead of the Shah, we could have spared ourselves Khomeini. My proposition to the Israelis was this: let them build a pipeline and become major buyers of Iranian oil. Mossadegh would then have been obliged, in words if not in acts, to adopt a favorable attitude towards Israel, and the general balance of power would have been modified. The Israelis laughed in my face. Maybe my solution was absolutely crazy (as "you should just do this ..." recommendations often are); I don't claim it was simple, but great political changes in the world have always taken place when States have adopted new and a priori extremely unlikely policies.

## Special powers. Generalized torture

Guy Mollet obtained special powers in Algeria from the Chamber of Deputies, with a vote of 455 in favor, 76 against. Even the Communist Party, which did not want to separate itself from the Socialists, voted in favor. It was a monumental (although perhaps explicable) error, which the Communist Party rectified only much later. Throughout the Algerian war, their attitude remained equivocal, which explains why the Party never played what should have been its true role: leader of the resistance. The Party criticized Mollet's government, but went no further. It systematically opposed the actions of intellectuals, who thus ended up farther to the left than the Communists. The fight against the Algerian war gave rise to a completely new independence on the part of the French intelligentsia with respect to the Communists, which led to a total divorce in 1968. In this, I was well in advance of the others. The evolution of the USSR, from the Khrushchev Report and the invasion of Hungary to the invasion of Czechoslovakia, all helped the wheel to turn.

A terrible period began with the actions of General Massu, and other generals and colonels such as Bigeard, Godard, Trinquier, Challe, etc. Torture developed with incredible intensity. The torture of the bath or electric shocks were used on any "suspicious" Algerian. We have conserved the names of certain specialized centers: the villa Sesini, the El Biar center, the Ameziane farm, but torture was practiced everywhere. Many soldiers of the contingent attended torture sessions or heard the screams of tortured Muslims. Some even lent a hand, without ever admitting it. It is known that the Algerian war left fifty thousand French dead and five hundred thousand Algerians, but no one counts the million or so Algerian torture victims. The banalization of torture became an obsession among the dissidents. Numerous universities protested. A motion against torture "from anywhere" was proposed and voted at almost every faculty meeting of the Scientific University of Paris. Henri Cartan, Roger Godement, Jean-Paul Mathieu, Raphaël Salem and I proposed most of the motions. Little by little, the university faculty came to be in favor of independence.

The atmosphere in Algeria became unbreathable. The partisans of French Algeria organized large-scale agitation, while the Muslims hid, hunted by the French military. The small number of regular troops of the FLN were soon defeated by the French army, and had to resort to indiscriminate acts of individual terrorism, playing out the usual cycle of repression/terrorism. Every day, we heard of new prisoners, new death sentences, new tortures, often with people we knew as victims. Torture, always torture – it was a true nightmare, a horrible tragedy which left us sleepless. We publicly accused Lacoste, Lejeune and Bourgès-Maunoury, and our accusations made slow progress within the public spirit, but did not alter the course of events. During one of the large meetings of the Colloquium Committee, in Lille, I read the report written up mainly by Pierre Vidal-Naquet and published in Mendès France's *Cahiers de la République* (*Notebooks of the Republic*). I still remember a campaign for the engineer Vincent, a former student of the Ecole des Ponts[55)], who was arrested and tortured. I was told by one of his former classmates who informed the entire class of his arrest, but received no support at all from them. One day, we heard that he had been imprisoned in a well, kept there for two weeks guarded by a parachutist, and finally had his throat cut. The whole period was horrible, and yet the decree of January 7, 1957 giving full powers to General Massu marked the beginning of an even worse one. Christian circles grouped around the weekly paper *Témoin Chrétien* were at the center of protest activities; everyone talked about the four M's: Mandouze, Marrou, Massignon, Mauriac. Pierre-Henri Simon published a remarkable book called *Contre la Torture* (*Against Torture*) that same year. Paul Teitgen, the secretary general of the police in Algiers, resigned on September 12, 1957, after having tried in vain to limit torture by ordering the army to keep suspected Algerians under house arrest, so as to be able to pinpoint their whereabouts. He surely succeeded in saving many lives this way; he was a Christian for whom I felt great esteem. He died recently. I

55) The Engineering School of Bridge Construction

had the pleasure of knowing his daughter Geneviève, who collects butterflies; she recently suffered a terrible loss when her husband died after only a few months of marriage. Vidal-Naquet sketched out a topology dividing opponents of torture and war into three political categories: Dreyfusards, Third-Worldists and Bolsheviks. "In some sense, this division happened within each one of us," he wrote recently in the introduction to his book *Face à la raison d'état: un historien dans la guerre d'Algérie* (*Confronted with the Interest of the State: a Historian in the Algerian War*. I was still penetrated with my Trotskyist training, and my intransigent internationalism was a powerful impulse behind my fight against the war; in this, I suppose I was a Bolshevik. But quite soon, I became a Third-Worldist in a sense I will explain. Because of my education, moral considerations always played an important role in my life. My childhood dreams in Autouillet, in Labouise's field, all about knights in armor, "protecting the widow and the orphan", Joan of Arc and the Dreyfus affair were all deeply moral; my joining Trotskyism and my disgust with the Moscow trials were further manifestations of the same feeling. I revolted against the use of torture for the same reasons, like the Christians, although as an atheist. The government mocked intellectuals who raised their voices against torture. Robert Lacoste referred to our action as the "conscience operation" run by "exhibitionists of the heart and the intelligence". Bourgès-Maunoury, after having had Professor Henri Marrou's apartment searched by the police, who went through all his papers, referred to the event ironically, calling him "that dear Professor".

French torture was not the only horror going on in Algeria – we must also take the blind, savage terrorism practiced by the FLN into account. In May 1957, the FLN exterminated a resistance group connected with the MNA, near Melouza, in the village of Meshta-Casba. For the reasons I explained above, I sympathized more with the FLN than with the MNA, but this act deeply horrified me and alas, only half surprised me. That day, I became unable to ignore the fact that whatever happened, a dark future was in store for Algeria. After the Moscow trials, I decided that never, but never would I have any connection whatsoever with Stalinist Communism. But this situation was different. I could not simply cross the FLN off the map. It was the only group really fighting for the decolonization and independence of Algeria, and I had close relations with many of its members and sympathizers. I was among the many people who squarely condemned their act, while still supporting its objectives. For similar reasons, I never acted as a "porter" in the Jeanson group; I esteemed Francis Jeanson and his friends, but I didn't want to join them. I freely defended the FLN without obliging myself to obey it or serve it directly. In Algiers, where information was very confused, a rumor (started by the FLN, no doubt) circulated saying that the Melouza massacre had been perpetrated by the French army. The FLN of Algiers then proceeded to revenge itself by organizing the "corniche d'Algiers" attack against the French. A sinister imbroglio!

## A very long story: the Audin affair

Guy Mollet was fired by the Chamber of Deputies and replaced by Bourgès-Maunoury on May 21, 1957; the right wing did not appreciate his betrayal. But like the SFIO, the French Section of the Workers' International, he continued to believe in his policy, and Bourgès-Maunoury simply went farther in the same direction. The Audin affair began in June 1957. The mathematician René de Possel, whom I have already mentioned meeting during my stay in Clermont-Ferrand in 1940–1942 and who taught me a great deal of mathematics, left for Algiers at the beginning of the war. He taught at the university there. One of his students, Maurice Audin, who was finishing his thesis, followed his advice and came to see me in Paris. He wanted to defend his thesis at the University of Paris and asked me to be in his jury. I found him rather timid and fragile. I believe he was really fragile, but he turned out not to be so timid. I found that his thesis was not quite ready, and that it would take him a few more months to finish it. But he was in a hurry; he was a Communist, and General Massu was persecuting opponents of French Algeria to the point that he feared for his life. The sequel showed how right he was. I told him that he should try as hard as he could to complete the work on his thesis as fast as possible. He left, and I never saw him again. Of all the people who worked for his liberation, René de Possel and I were the only people to have actually met him. I believe I saw him at the end of spring. On June 11, he was arrested. His apartment was invaded by parachutists who took him away, and remained to sequester his wife Josette Audin and their three small children. As Josette had to feed the children, one of the parachutists consented to go shopping. Henri Alleg, a close friend of Audin, had a rendez-vous with him that day. He went to his apartment and knocked; the door was opened by a parachutist and Alleg was caught in the mousetrap. He immediately understood what was happening and tried to squirm out of things by explaining that he was an insurance agent. It didn't work; he was arrested and horribly tortured. He resisted and survived. He wrote a book about his experiences which made a strong impression in France: *La Question*, published by the Editions de Minuit in February 1958. Madame Audin remained for several days with no news of her husband. The whole story has been recounted in detail by Pierre Vidal-Naquet, in a book which he devoted to the affair in 1958 (which was reedited in 1989 with additions). Josette Audin was allowed to leave her house freely on June 15, and she immediately took to sending letters and telegrams to French and Algerian authorities, insisting that she be allowed to know the whereabouts of her husband. Until about the 1st of July, she was simply told that he was in a secure place. But Maurice Audin was already dead. He died on June 21; he was ignobly tortured by the parachutist Lieutenant Charbonnier, and as he resisted, Charbonnier strangled him in a burst of rage. Probably Charbonnier did not mean to kill him, but Audin was exhausted by torture and died immediately. The parachutists were frightened and tried to cover up the assassination. They invented a whole story about how Audin escaped during a momentary stop of their Jeep on the road, and added all kinds of details to make it sound more realistic. This legend was made public after July 1. But no crime is perfect. The parachutists' version was full of contradictions and was soon disproved. Everyone was soon convinced that Audin had been killed on June 21 by Lieutenant Charbonnier. Such things happened all the time. In some cases, Muslims were so seriously tortured that they were no longer "presentable" and they were killed, their bodies thrown into the sea. One often heard talk of Bigeard's "shrimps". Others were thrown to the ground or into the sea from helicopters. The French explored an infinite scale of deaths and tortures. Many villagers were also deported and kept in camps which were certainly nothing like Nazi death camps, but where nevertheless they were uprooted, forced to lead unnatural lives and seriously undernourished.

One of the telegrams sent by Josette Audin reached me, following me all the way to Bombay where I was spending the summer. I received it in July, and Audin had already been dead for weeks. I sent numerous telegrams to all the French authorities from Bombay. Obviously, it was wasted trouble: I never received the slightest response. At the beginning of the academic year, in the fall of 1957, I was feeling very sceptical about the type of battle we could undertake in Audin's favor, when two original proposals modified the situation. The first came from Possel, who proposed to have Audin defend his thesis *in absentia* at the University of Paris. I didn't know such a procedure existed, but it seemed possible. Dean Pérès courageously accepted our initiative. We organized the thesis defense in December 1957, with Favard as president of the jury, myself as referee, and Dixmier in the jury. The thesis was defended before a public which completely filled the room; it consisted of numerous mathematicians but also many non-mathematicians, particularly journalists. François Mauriac came and reported the event in his "Notepad" in *L'Express*. The session was deeply moving. Favard was not really a man of the left, but he was an honest man who felt unable to accept Audin's death. As president of the jury, he asked in a loud voice: "Is Maurice Audin present?" Obviously, no response was forthcoming, and as we had decided, he asked René de Possel to undertake the defense. Possel spoke for about three quarters of an hour. After the deliberation, we declared Maurice Audin a doctor of science, with the grade "très honorable" (summa cum laude). Actually, the thesis contained several errors, which was normal since Audin had not had time to finish it. When we had it printed, Dixmier and I added four pages of corrections. I didn't try to hide them, quite the contrary; when the jury came forth to give the results of its deliberations to the public, I clearly stated that we had found errors, which was quite normal since the thesis was unfinished, and that we intended to reproduce them faithfully and simply signal their existence at the beginning of the text. And indeed, we were able to rectify and complete the proofs of most of the theorems Audin stated; clearly many of the errors were really due to lack of time. If we had actually introduced our corrections into the text, rumors would have circulated and no one would ever have been able to distinguish what really came from Audin and what from us. Our choice made everything absolutely clear.

Audin's thesis *in absentia* served as a catalyst in the combat against torture. When Dean Pérès was interrogated by the press, he told them that given the circumstances, the defense *in absentia* was normal university procedure. This was not completely true. I felt particularly concerned by the whole affair and published an article in *L'Express* called "The revolt of the university", an expression which had been coined by the barrister Thorp. Even more than before, the university began to play a truly avant-garde role, with mathematicians in the forefront. In journalistic circles, which are always about two centuries late, it is still the custom to use the word "intellectuals" only for people in literature and the arts. When they talk about "the intellectuals", they refer only to them. It is true that literature and social sciences are essential to today's society, but scientific intelligence also has a fundamental function. It is not merely a series of automatic operations, but a grand discipline of the spirit, a culture and a form of thought which constantly transforms knowledge and society. It demands a high degree of rigor. Throughout the whole of the dark history of the Algerian war, scientists unquestionably joined a list of intellectuals in literature and the arts which would be far too long to cite here, to form the avant-garde of the struggle. And it is not a mere question of chance.

My photo was on the cover of *L'Express*. At the next course I taught, in the Institut Henri Poincaré, the students stuck that cover page to the clock. When I finally glanced up at the time and noticed it, the students, who had been waiting for that moment, burst out laughing and clapping; they were expressing their solidarity with my article. In general, student circles were strongly opposed to the war. The UNEF (Union Nationale d'Etudiants Français)[56], directed by their president Gaudez, played a first-class role in the resistance.

The number of students passed from 140,000 in 1954 to 250,000 in 1962, between the beginning and the end of the Algerian war, and unlike what happened later, this augmentation occurred without a corresponding lowering of the average cultural level. Although students had possibilities of postponing the draft, the mass of students was still strongly affected by the war. The news they received of conditions on the front, and of the practice of torture, suppressed all their enthusiasm for the war. During that period, a majority of students joined the union, because at that time, the union represented their real interest, whereas a few years later, it came to represent only a few tiny political groups. Apart from the UNEF, after the 1961 schism, the FNEF (Fédération Nationale des Etudiants Français)[57] was founded, and its tendency was towards Algeria remaining French. After the war, student unions grouped only a few uninteresting lobbies, some right-wing groups, Christians, Communists and Maoists, with a few Trotskyists amongst them – and involved only a small minority of students. Later, I abandoned all contact with the unions, but I had been very close to the UNEF under Gaudez, and also to the UGEMA (Union Générale des Etudiants Musulmans Algériens)[58], which was supported by the UNEF. The UNEF thus took on the role of a large union undertaking numerous initiatives, among more or less decomposed political parties, unsupported by the other French unions, such as the SNI (Syndicat National des Instituteurs)[59] and the FEN.

After the emotion aroused in the whole country by what had been revealed about the tortures, Guy Mollet had created in 1957 a Commission for the Safe-keeping of Individual Rights and Liberties. Apart from the already existing committees, the barrister Thorp had set up a Commission for the Preservation of Judiciary Institutions. Guy Mollet's commission had twelve members, of whom the most eminent were the magistrate Pierre Béteille, president of the Commission, the rector Pierre Daure, a physicist, the reputed lawyer Maurice Garçon, the ambassador André François-Poncet, president of the French Red Cross, the doctor Charles Richet, discoverer of anaphylaxy and president of the Order of Doctors, Robert de Vernejoul, who later entered the Academy of Sciences, where I met him – he lived to the age of 103 – and Robert Delavignette, a former governor general of the colonies, a man of great rectitude who appeared

56) UNEF: National Union of French Students
57) FNEF: National Federation of French students
58) UGEMA: General Union of Algerian Muslim Students
59) SNI: National Syndicate of Elementary School Teachers

to me to play a dominant role. The barrister René Thorp and the Vice-President of the Council of State, René Cassin, refused to join the commission, as a protestation against its lack of means. The commission traveled to Algeria where it was quite simply fooled by the parachutists. They were informed of the visit beforehand and guided the commission through their visit to the El Biar center, a well-known torture center (where Audin was tortured and killed), naturally after having evacuated the prisoners and cleaned the whole place thoroughly. The commission was obliged to observe (some of them quite innocently) that everything was in perfect order. In spite of all, they were able to note such a number of atrocities and irregularities perpetrated in Algeria that Guy Mollet refused to allow the publication of their devastating report which came out on September 14, 1957; this caused Delavignette, Emile-Pierre Gérard and Maurice Garçon to resign. In my article in *L'Express*, I asked: "Must the mathematicians organize a commando to steal the report of the safekeeping commission?" We did not have to, in the end, because *Le Monde* procured the report and published the entire thing, which once again had a tremendous effect on the country.

## The Audin Committee

In the autumn of 1957, the protest movement of the intellectuals was strengthened by the creation of the Audin Committee. Maurice Audin's assassination mobilized tens or even hundreds of thousands of consciences. The "initiators of the committee" were a group of four people: Michel Crouzet, a very efficient Communist (who later turned extreme right-wing and campaigned for Le Pen), Luc Montagnier, another Communist at the time, who lately became famous for discovering the HIV virus, Jacques Panijel, an excellent scientist working at the Pasteur Institute, and Pierre Vidal-Naquet, a specialist in ancient Greece who turned a new page in the history of the Algerian war. At that time, I did not know any of the four of them. I believe it was André Kahane, the brother of Jean-Pierre Kahane, who came to propose the creation of the committee to me. At first I was quite reticent. There were already so many committees (there always are in circumstances of this kind) that I wondered if it was really useful to create yet another one. Finally, I let myself be convinced. It was a new type of committee, and the Audin affair didn't resemble anything else. The meetings of the fifteen permanent members were held in the home of Geneviève and Pierre Vidal-Naquet or at the Lalandes'.

Many very interesting people participated in the committee. Among those who regularly attended meetings was Madeleine Rebérioux, an active member of the Communist Party, although she fully understood the deficiency of their attitude in the struggle against the Algerian war. Vidal-Naquet and I shared her positions. She became one of France's best historians, a professor at the University of Paris 8, a specialist of the history of Jean Jaurès. Elizabeth Labrousse and her son, a mathematician, were among the faithful, as were Louis and Hélène Lalande. Hélène was actually a relation of mine, the niece of my grandmother, and her real name was Lehmann, but she had chosen to keep the name she had used on her false papers during the war, as I had used Sélimartin. Josette Audin had sued X for the murder of her husband, choosing a team of lawyers consisting of Jules Borker, Robert Braun, Roland Rappaport and the barrister René William Thorp. The Audin Committee also took lawyers; Robert Braun, the barrister Thorp and later Robert Badinter who was just a beginner at the time and came to play an essential role in the activities of the Audin Committee. These lawyers devoted a great deal of time and energy to the affair, and contributed largely to the defense of our cause. The Committee also sued. I won't describe the trials in detail; they've been described and analyzed in Vidal-Naquet's books. Josette Audin gave us all the documents we needed, and stayed in close contact with the committee. Her extraordinary energy was one of the essential aspects of the affair. Thanks to the intervention of ministerial circles in Paris, she obtained a pass to spend a week in Paris and attend the thesis defense *in absentia* of her husband. She also held a press conference in which Borker explained the legal aspects of the affair to the press. Borker was and remained a Communist; he was extremely sectarian and opposed many of our initiatives. In particular he wanted to reserve for himself a monopoly on the publication of the findings of the committee, which we refused. We kept our independence by publishing our own communications, in *Témoins et Documents* (*Witnesses and Documents*), a bulletin created to denounce torture, organized and directed by Maurice Pagat. He was in contact with resistants from every group, particularly Christians.

With hindsight, it is certain that the Audin Committee was the first model of what is now commonly called a non-governmental organization. Later, committees were spontaneously created to denounce all kinds of injustices or threats to civil liberties. Their importance has continued to grow, and now there are probably more than a thousand such committees in France. I attended the meetings very assiduously; progressively they turned into one of my main activities. The Audin affair and the Committee probably changed my life. During the Occupation, I had searched in vain for some form of political action conforming to my personality; I found it in this form of combat against the Algerian war. There was no political ambition in my actions, even less any idea of obtaining power, although the committee of course needed a president and a treasurer. The first president was Albert Châtelet, and I was the second. But the position did not correspond to any real hierarchy; we were all equal.

Our first task was to discover the true circumstances of Audin's death, and in particular to neutralize the faked story about his escape from a Jeep. The book which Vidal-Naquet wrote about the affair, which was worthy of a Sherlock Holmes investigation, brought us the decisive elements. Anyone who has seriously studied the affair accepts Vidal-Naquet's conclusions. We now know more or less exactly what happened. However, we never won our cases, one of which took place in Rennes and reminded us of the second Dreyfus trial of 1899; Audin's murder was never officially admitted. At first, Josette Audin's suit was to be tried in Algiers. The judge was called Bavoilot. He was strongly in favor of French Algeria, and justified the use of torture to obtain confessions, as long as it was practiced directly in the judge's office! Fortunately, a liberal Justice Minister, Edmond Michelet, came to the aid of the resistants, although for a long time he was careful to respect his duty of official neutrality, particularly under de Gaulle. He obtained the transfer of Josette's trial to France; it took place in Rennes on April 11, 1959. But he could not or would not go farther, and no elucidation at all resulted from the trials. The final conclusions were made by civil parties, the Audin Committee in particular. The only result was that Lieutenant Charbonnier, the murderer of Audin, was removed from direct police operations for some time. He ended his career by retiring in 1981 with the rank of Colonel and Commander of the Legion of Honor.

## Why Audin?

Why did Audin become an emblem for all the victims of the Algerian war? He was not a typical case. He lived in Algeria, but he was French and not Muslim. However, he was not an anonymous figure; he was an intellectual at the university and a mathematician, circumstances which awakened sympathy in the corresponding circles in France. He was tortured and assassinated and the crime was covered up, so his story became a legal question. Defending Audin did not necessarily mean that one had to hold any given views on the future of Algeria; it was enough to reflect on the destiny of one man, without necessarily having to deal with the defense of the million other torture victims. Were we conscious of all these reasons? Perhaps partially, because we were the same kind of "atypical case" as Audin. In our later actions, in the light of the Audin experience, we always preferred to deal with cases which would have greater "effects". Was that a sign of Machiavellian cynicism? I don't believe so. There are innumerable victims of torture in the world. It is impossible to defend each and every one. It is far more striking to evoke a single person than a thousand. Thus, the defense of a single, symbolic victim simultaneously defends all victims, which is an essential and honorable motivation.

The Audin Committee did not deal uniquely with the case of Audin. It became a center of struggle against the use of torture in general, and its action was very efficient. Eventually, a serious disagreement arose between the committee and Maurice Pagat. He continued publishing *Témoins et Documents*. After the Algerian war, he lost his job at the general electric company because of his political activities. He became unemployed and suffered all kinds of difficulties for many years. He finally obtained a pension from the electric company and devoted himself to helping the unemployed, and organizing them into a very effective union which published an excellent bulletin called *Partage* (*Sharing*). As for us, we took to publishing a bulletin called *Vérité-Liberté* (*Truth-Liberty*). Little by little, we increased the printing; once, we followed up an idea of Ida Bourget and sent out free copies to the thirty thousand mayors of French towns. Unfortunately, reactions were few and far between. But many preachers and elementary school teachers joined or supported the committee. I know that one always tends to exaggerate the result of one's own actions, but I really do believe that our cause enjoyed a certain popularity.

We published every specific case of torture that we heard about. When the OAS (Organisation de l'Armée Secrète)[60)] began to wreak havoc, de Gaulle's regime mobilized its own secret agents against it. Soldiers faithful to General de Gaulle did not hesitate to use the same barbaric methods against the OAS and more generally against French people in Algeria that had already been used against Muslims. The French community in Algeria soon became extremely worried about the spreading tales of torture, showing that they had certainly not been ignorant of the use of torture before. We publicly denounced these tortures. We also had to deal with the extremely complicated situation of the harkis. The harkis were native Algerians who were also regular troops of the French army and fought for French Algeria. Tens of thousands of them were guilty of nothing, but an important group of several thousand became torturers, turning cellars of Parisian buildings into centers of torture for Muslims living in France. They were a minority, but they must not be forgotten. After France made peace with the FLN, France denied the existence of the harkis and forgot about them. Some of them returned to Algeria, but they had left such terrible memories there that villagers frequently took revenge on them, or on their wives

60) OAS: Organization of the Secret Army, a French group based in Algeria, in favor of French Algeria and violent methods

and children. The French government remained deaf and dumb, ignoring their martyrdom in Algeria. Alone, the Audin Committee protested loudly to the new Algerian government. In general, we did not only denounce the "crimes of the French army" (the title of a book by Vidal-Naquet), but also condemned the crimes of the FLN. For instance, when the FLN tried and condemned two French soldiers (Le Gall and Castera) to death after a mock trial, our energetic reaction probably prevented them from repeating the experience. Naturally, our committee was in close relation with the lawyers of the FLN, both in Algeria and France. The action for the truth of the Audin affair remained popular in Algeria, and there is even a public square named for Maurice Audin, in the center of Algiers.

The Audin Committee did not immediately dissolve after peace was made. Torture continued in Algeria at the beginning of Ben Bella's regime. The French army had given the example by making torture into a banal practice. (To be fair, Algerians were already quite familiar with violent procedures.) The methods were known, it sufficed to use them. One of the former French members of the Audin Committee, Arnaud Spire, was himself tortured by the police of Boumediene while on a trip to Algeria, for ridiculous reasons. He gave a press conference when he returned to Paris. When Jacques Panijel traveled to Algeria, we contacted Ben Bella through him; he received him well and said some "kind words" on the subject.

In 1958, the Audin Committee and others published a brochure called *Nous accusons* (*We accuse*), which listed all cases of atrocities we had heard of from lawyers. Our brochure was confiscated, but many people read it. The committee had many adversaries of different types. Some of them were right-wing, and even Gaullists, because although it is true that de Gaulle tried and finally succeeded in making peace in Algeria (after four years, as long as the war had lasted before he took power), he was nonetheless strongly opposed to a committee denouncing torture. Malraux, the Minister of Information, frequently declared both that torture would stop and that it had already stopped; in reality, it never stopped. Of course, we had adversaries from the extreme right starting in 1961, one of which was the OAS. On the left, the Communist Party was tenaciously hostile to the Audin Committee because they felt we were meddling in their business and they disliked our total independence with respect to any Party or organization whatsoever (one of the most interesting aspects of the committee). We never aligned ourselves with any other group. Against all odds, the committee survived from 1957 until 1962. It was never reduced to silence, and all the documents it obtained were published, even though *Vérité-Liberté* itself was confiscated several times. I have met Josette Audin and two of her children, Pierre and Michèle, a few times in recent years; we are on excellent terms. We know who they were, they know who we were. Michèle is a brilliant mathematician at the University of Strasbourg. A group of mathematicians have organized a Maurice Audin prize, by subscription, given to one or two mathematicians every year. Michel Lazard received it the first year and then Jean-Pierre Kahane, André Néron, Pierre Cartier and Paul-André Meyer.

Under de Gaulle, only a single trial against torture took place. The French elementary school teacher Sanchez, who was mobilized in Algeria, had joined together with two companions to torture to death a young Muslim woman, supposedly because she had not voted in de Gaulle's referendum in 1961. The trial was carefully organized. But to everyone's surprise, Sanchez was acquitted on January 16, 1962. Sanchez returned to France and was given a job as teacher in a school. The Minister of Education, Lucien Paye, possessed a moral conscience and immediately suspended him. "Give us back Sanchez!" was the cry of the parents of the schoolchildren, who demonstrated in his favor, apparently seeing no problem with the idea of entrusting their children to a man who had tortured a young woman to death for no reason. Paye refused to give in – today, one would probably have to give in, as no Minister dares resist massive demonstrations nowadays. In spite of all, Sanchez surely continues to live a peaceful and agreeable life. His story inspired a very good movie, called *Muriel*, by Alain Resnais.

Another case which became extremely famous was that of Djamila Boupacha. She was horribly tortured by parachutists for days, and raped with a bottle. She was defended by Gisèle Halimi who sued her torturers. Gisèle Halimi was a well-known lawyer; she was amazingly courageous and dynamic and her actions during the Algerian war were as remarkable as her combat for the rights of women. She created a Djamila Boupacha committee, in which Simone de Beauvoir and Anise Postel-Vinay played an important role, and which of course I joined.

## Self-determination

I will not describe the events of May 13, 1958. The majority of the left was against de Gaulle, who came to power by the events in Algiers on May 13, the Committee of Public Health, Massu's call to de Gaulle on May 14 and the pseudo-fraternization of May 16. We all remember de Gaulle arriving in Algiers on June 4, throwing his arms into the air and shouting: "I have understood you!" He had understood the agitators, partisans of French Algeria, and everything was to begin over again. It was infuriating. People often have said that "understood" did not mean "approved". I don't think de Gaulle was actually trying to fool anyone; I think he was imagining a federalist solution. He came only slowly and ambiguously to the idea of a solution by negotiation. He put out the idea of the "heros' peace" in October 1958. I think this idea was unacceptable to the FLN. Shortly after, in September 1959, he gave his famous speech on auto-determination, which was a fundamental step towards agreement with the FLN. Unfortunately, in March 1960, during a visit to the military in Algeria, he declared that he "would never lower the French flag in Algeria". For years, he spent his time changing his mind and saying contradictory things to different people. What was he thinking exactly?

Ivan Craipeau convinced me to join the New Left, although I felt a certain reticence. But it turned out to represent a fruitful initiative. I was not enthusiastic at first; it was a minuscule organization which reminded me of other, analogous experiences. But it merged on December 7, 1957, with the MLP (Mouvement de Libération du Peuple)[61], a Christian workers' group. Together they formed the UGS (Union de la Gauche Socialiste), a party which soon numbered several thousand members and took an active part in the struggle against the Algerian war. The Audin affair had already begun. Gilles Martinet soon came to lead the UGS, with Claude Bourdet. As I said earlier, I had known Martinet during the war; he was a former Communist who had left the Communist Party because of the Moscow trials. Bourdet was a militant Christian and a very early resistant. We all remained close friends. Bourdet died recently, after a whole life of militating. What divided us was not so much our view on the Algerian war as our support or lack of support for de Gaulle's "coup". At the Socialist Congress in Issy-les-Moulineaux, on September 11, 1958, the defeated anti-Mollet minority seceded and on the same day they founded the PSA (Parti Socialiste Autonome)[62]; their temporary secretary was Edouard Depreux. Thus, there were two left-wing parties which were neither SFIO nor Communist. Yet a third sprang from the ranks of the Communist Party. On July 2, 1958, forty-nine former members of the Communist Party formed the Tribune group of Communism, publishing a bulletin of the same name. One should also mention two isolated individuals of enormous political stature, Pierre Mendès France, a former radical, and François Mitterrand, a former member of the UDSR (Union Démocratique Socialiste de la Resistance)[63]. These two men did not belong to any party; the non-Communist left was in a state of ferment.

61) MLP: People's Liberation Movement

62) PSA: Autonomous Socialist Party

63) UDSR: Democratic Socialist Union of the Resistance

Even before these last events, Daniel Mayer conceived a grand project of unification, and on the morning of July 7, 1958, he summoned about forty people to the office of the League for Human Rights, rue Jean-Dolent. Important people were present, representatives of political groups and unions, united against the Algerian war and the government which had emerged from the events of May 13. The UFD (Union des Forces Démocratiques)[64] was created that day. A temporary bureau of ten members was chosen: in alphabetical order, Albert Châtelet, Edouard Depreux, Jean Hyppolite (the director of the ENS), Maurice Lacroix (of the Jeune République), Gilles Martinet, Pierre Mendès France, François Mitterrand, Francis Perrin (a well-known physicist), Laurent Schwartz and Robert Verdier, a member of the minority of the SFIO. We were filled with hopes, but our goals were ambiguous. Over the next two years, we held numerous meetings, and several times I wrote up the minutes or a text of our agreements.

Our discussions were sometimes quite tortuous. Should we be a league? a coalition of organizations? should individuals or organizations join us? should we be a new political party and accept only individuals as members? Sometimes I had enough of the interminable disagreements which reminded me of the Trotskyists (although in reality they occur at every meeting of this type). Mendès France of course wanted a party. The UGS did not want another party at any price, because it would necessarily be a rather undefined party, whereas the UGS itself was very structured. There was also the problem of the two individuals, Mendès and Mitterrand, and their status in the party. This is not the place to recall all the details of that period. The PSU was not really born from the UFD, but the UFD was unquestionably a kind of antichamber for it. The books which have appeared in the last few years tend to present the attitudes of Mendès France and Mitterrand within the UFD as extremely contrasting, with Mendès convinced that Gaullism would not last and that de Gaulle would not succeed in making peace in Algeria, and Mitterrand more moderate and perspicacious, believing in long-term Gaullism.

But my own memory is completely clear and quite different; I remember that Mendès and Mitterrand held identical opinions, both asserting aggressively that de Gaulle would never make peace and inciting all of us to act in consequence. Although I managed to feel a little hope sometimes, in spite of de Gaulle's frequent flip-flops and contradicting promises, on the whole I felt as they did, incredulous and pessimistic about the likelihood of his making peace. And can

64) UFD: Union of Democratic Forces

it even be called an error to have been pessimistic on the day before Melun[65], since Melun failed? However, it must be said that Mendès, Mitterrand and myself completely lacked perspicacity at the beginning of 1962. Martinet was already talking about the soon-to-arrive peace; he was more lucid and better-informed than we were. It is true that as a journalist, he had access to certain sources directly from the ministry.

In November 1958, the first legislative elections after May 13 routed the left. The candidates from the UFD obtained only 5% of the votes, about eight hundred thousand. Mendès was defeated with 8.5% of the votes in his department; the Communists descended from one hundred forty-five deputies to just eighteen. The classical left-wing parties were discredited, the Gaullist regime was triumphant. I was incredulous about our poor showing and asked Mendès if it could be because of the way the voting was counted by region. He lucidly replied that no, I should look directly at the evidence: the votes spoke for themselves and we had to accept a large-scale defeat. During the presidential elections on December 21 – at that time the president was elected by eighty thousand notables – Albert Châtelet of the UFD ran alongside de Gaulle and the Communist candidate Marrane; Châtelet received six thousand seven hundred votes, which was 8.5% of the total, and Marrane 13%.

The existential crisis which took place within the UFD over the question of whether it should be a coalition of organizations or a bona fide party and what the personal roles of Mendès and Mitterrand should be ended up being fatal. Something else was needed. That is when the negotiations between the UGS and the PSA were begun; both parties called themselves Socialist. The negotiations began in 1959, and they were long and tortuous. They came to a head in 1960, at a time when the situation in Algeria was becoming extremely serious. The problem posed by Mendès and Mitterrand remained crucial on both sides. Depreux really desired that Mendès France personally join the PSA. But for that, he would have to accept the word "Socialism". An absurd quarrel over words, which continues today. The French always adore fighting over words – I stopped being interested a long time ago – and we are also the country of religious wars. Algeria was in fire and flame, we had to fight against the Gaullist regime and alas, also against the regime of the military in Algiers; we had to modernize our own country. The working class and the unions certainly had their place in the whole process, but they were not the only ones to play a role, indeed their role was not even the main one; I had been convinced of this

65) The Melun negotiations of June 25–29, 1960 between de Gaulle and the FLN ended in deadlock.

since my discussions with Paul Ruff at the ENS. Depreux ended by obtaining from Mendès France a sentence in which he pronounced the word Socialism: "... economic and social progress, via planning, full employment, reconquering the rights of workers, in view of realizing the Socialist idea". Thus, he passed the degrading entrance examination to the Socialist Party (without anyone ever making it clear exactly what "the Socialist idea" was supposed to mean). Mendès joined the PSA on September 22, 1959.

Most of our divergences were political, not personal. Mitterrand's situation was the contrary of Mendès France's; he wanted to join the PSA, but they didn't accept him for personal reasons. He had worked for too long in Mollet's ministry; Mendès had very quickly resigned, as had Alain Savary, and Paul Teitgen had resigned from his job at the police headquarters of Algiers. Mitterrand remained until the departure of Guy Mollet in 1957, and thus, even if it was not his desire, he sanctioned the use of torture in Algeria while he was Minister of Justice. Nor did people forget that he had worn the francisque[66)] under Pétain. The revelations published in a recent book by Péan, with the agreement of Mitterrand, were not known at the time, but the affair of the emblem certainly was. At the time, Mitterrand asserted that he had received the francisque when he was already in London, and that he had never worn it. We know today that he received it in France, at the beginning of 1943, and wore it. But even in 1959, there were suspicions. Other questions about his presence in Vichy during the war persisted. Everything which has been recently revealed about Mitterrand's Vichyist past leaves me with a most disagreeable impression, but at the time I did not participate in the many discussions about his entering the party, whereas the affair of Mendès France was essential for me. There was also the story of Mitterrand's "dirty affair", in which I was entirely on his side. An extreme right-wing henchman called Robert Pesquet informed him that an attack was being prepared against him or his son. The same day, October 16, a mysterious shot was fired at his car, and he quickly stepped out of it and hid in the gardens of the Observatory. The whole scenario had been set up by Pesquet; he had predicted exactly what would happen, and before the even he wrote a letter describing the event and deposited it with a bailiff. On October 21, he took it out and made it public, claiming that Mitterrand had actually organized the attack together with him. I feel that Mitterrand had simply panicked at the attack, which was normal considering the atmosphere, and fell into a trap. The machination was successful and the lie triumphed since the Senate actually voted that his parliamentary immunity be lifted. I was scandalized by the credulity of the public and

66) emblem of the Vichy government

the unjustified accusations against Mitterrand, and said so to everyone around me. The Audin Committee defended him. At that time, I had no idea that my son would be kidnapped in 1962 and that after a well-orchestrated publicity campaign, public opinion would accuse him of having run away and invented the story of the kidnapping. Even later, Gorbachev was captured by putschists and publicly accused of having organized the kidnapping himself. Even if the victim emerges unharmed from this kind of operation, public opinion does not always forgive him. However, I avoided talking about the affair with Mitterrand, who felt wounded by it, as I had the impression he preferred not to talk about it. Mitterrand had made the mistake of trusting that dirty fellow Pesquet and not confiding in anyone. If he had spoken to any of his friends, we would have advised him to consult the Minister of the Interior, the police would have been warned and later accusations would have fallen flat. Mendès held a position similar to mine and I fully approved of it. But everything had happened in a single day. Mitterrand's error was not comparable with the enormous crime he was accused of, having organized a fake attempt against himself. Nonetheless, he was not admitted to the PSA, nor, later, to the PSU, for personal reasons whether justified or unjustified.

Mendès took a central place in French political life. He was deeply attentive to social problems even if he did not call himself a Socialist; he would have modernized the country and its industry, he would have promoted high-level research, but circumstances played against him and it was de Gaulle who ended the Algerian war and raised the levels of industry and scientific research. There was not enough room for the two of them. I always observed with a feeling of bitterness how left-wingers verbally shot down Mendès, calling him a "neocapitalist democrat". In November 1967, during the second session of the Russell Tribunal on the American aggression in Viet-Nam, held in Roskilde in Denmark, the television suddenly announced his death. In the midst of a movement of general consternation, everyone began praising him, describing his exceptional valor as a statesman. Sartre won laurels by comparing him to Trotsky: like Trotsky, he claimed, his destiny had been stolen by men and by society. One hour later, the television rectified its error: it was Mendès France's first wife who had died. Of course we regretted her death, but nevertheless we felt relieved, and everyone immediately began severely criticizing him again. I told the anecdote to Mendès, who found it very funny, especially the comparison with Trotsky.

The discussions about unity finally ended with the politically indispensable fusion of the UGS, the PSA and the Tribune of Communism. The fusion took place officially at a congress held in Issy-les-Moulineaux, exactly where the schism between the PS and the PSA had taken place on April 3, 1960. I was the president of the founding session of the PSU, thanks to my reputation as a scholar and the obvious honesty of my political activities, but also because I had not been involved in any of the conflicts preceding the fusion; I appeared neutral to everyone. There were many foreign delegates, such as Pietro Nenni; representatives of the Italian Socialist Party, of Tito, of the Spanish POUM, the Israeli MAPAM and even El Mehdi Ben Barka himself, the Moroccan leader who was later assassinated in France (I had met him during a dinner at the home of Claude Bourdet). We had to take certain precautions around the latter two, because it was unthinkable to have them speak directly one after the other. Even though I referred in my speech to "a marriage of reason", I felt extremely enthusiastic about the creation of this new force, which was to become the "big" party of a "true" non-Communist left, with more than fifteen thousand members, and which could weigh heavily on the events of the Algerian war. The national secretary was Depreux, the two assistant secretaries were Martinet and Longeot. The party grew rapidly because it answered an obvious need. However, I was unable to feel the personal involvement I had known in my youth. I was not a member of the PSU who belonged to the Audin Committee, but a member of the Audin Committee who was joining the PSU. In the end, I did not even attend many of the section meetings. In fact, I forgot all about my participation in the fusion congress until Lionel Jospin reminded me of it in 1983. I don't want to disappoint my former comrades; I shared the successes and failures of the party with them. It was the major tool in the political training of left-wing youth for at least a decade. I met many important people there, such as Robert Verdier, Jean Verlhac, Jean-Marie Vincent, Marc Heurgon, Michel Rocard, Serge Mallet, Jean Arthuis, Claude Estier, Pierre Hespel, Jacques Nantet, André Philip, Philippe Viannay and many others.

Sometimes I took certain important decisions without even informing the PSU, which never was upset with me about it. I don't remember for what reason Verlhac and Martinet had asked to see Pierre Vidal-Naquet and me about a position we had taken in the Audin Committee which displeased them. Our discussion took place in the home of Panijel; the tone was amiable and it ended without conflict. Mendès was an even more independent member than Vidal-Naquet or myself. His membership was superficial; he didn't even attend the founding congress.

## The Jeanson network and the manifesto of the 121

The year 1960 was rich in events. At the end of January and the beginning of February, the Algiers barricades (fortunately) fell. Lagaillarde, their organizer, was imprisoned, and his trial took place in November. All the accused in the Barricade trial were acquitted on March 2, 1961. The Jeanson network of "porters" was discovered on February 23, Jeanson was arrested and his trial opened on September 5. Neither Vidal-Naquet nor I had joined his network. I felt esteem and friendship for him, but even though I supported the cause of Algerian independence, I did not want to help the FLN to organize terrorist attacks, with money or arms, or to lose my liberty and depend on their decisions. The Melun talks were held on June 25–29, but ended in failure. De Gaulle demanded a ceasefire before negotiating, and the FLN insisted on negotiations before any ceasefire, and was not yet ready to make concessions. The manifesto of the 121 appeared on September 5, 1960, the day Jeanson's trial began, which was not a coincidence. The left was so exasperated by the continuation of the war, and in particular by the failure of Melun, that some spectacular action seemed necessary. Hostility to the war was increasing among mobilized youth, and cases of desertion and insubordination were becoming frequent. We had several links to student groups. The manifesto was written by a small group of intellectuals and signed by one hundred twenty-one people who solemnly proclaimed the right of young people to insubordination in the Algerian war. The terms of the manifesto did not appear completely appropriate to me, but at least it existed. Sartre and Simone de Beauvoir of course signed it, as did Pierre Vidal-Naquet and several others I have already mentioned. I was in Berkeley at the time and did not hear about it, so that the manifesto appeared without my signature. I canceled the rest of my stay in Berkeley and urgently returned to Paris, read the text and signed it. Many more than one hundred and twenty-one people later associated their names with the text. If I had been a young man drafted into the army, I would probably have deserted, in spite of the inevitable consequences this would have had for my career. Marc-André, who was seventeen, would have done the same. The text provoked strong reactions throughout France; the ultra-extremists became wild, the government and its Ministers were enraged. In Algeria, the Muslims who learned about it welcomed it with sympathy. Those who signed it were soon sanctioned. Vidal-Naquet was suspended from his university job for almost a year. Towards the spring of 1961, I wrote a private letter to the Minister of National Education, Lucien Paye, explaining to him that if Vidal-Naquet, a member of the Audin Committee, was not immediately restored to his position, then I, as the president of the committee and a signer of the manifesto, would feel compelled to resign from the university. The letter was written in a courteous and confident tone which reflected the esteem I felt for Paye, but it was nonetheless intransigent. He immediately answered me that it was his intention to restore Vidal-Naquet to his position in Lille soon, and he did so.

As for myself, my sanction consisted in the fact that the Minister of the Defense, Pierre Messmer, removed me from my position at the Ecole Polytechnique. In a personal letter, Messmer wrote to me: "As I have been told, the commanding general of the Ecole Polytechnique invited you to inform him whether or not you signed the declaration on the right to insubordination in the Algerian war. You confirmed to him that you had signed the manifesto. The position you have taken is incompatible with the exercise of a professorship in a military establishment. It would go against honor and common sense for me to allow you to occupy such a position any longer. Attached to this letter please find a declaration that you cease to be employed by the Ecole Polytechnique."

I immediately answered by a letter sent to the Audin Committee, which published it in *Le Monde*. Here is the text of my letter: "I regret the necessity of abandoning my teaching activity and my very interesting students. But that is not the question. You found it necessary to accompany your decision with a letter claiming that it would "go against honor and common sense" to allow me to occupy my position any longer. If I signed the declaration of the 121, it was partly because for years I have seen torture go unpunished and torturers rewarded. My student Maurice Audin was tortured and assassinated in June 1957, and you, Mr. Minister, signed the exceptional promotion of Captain Charbonnier to the grade of officer of the Legion of Honor, and the promotion of Captain Faulques to the grade of commander of the Legion of Honor. I repeat: "Honor". Coming from a Minister who has taken these responsibilities, considerations on the subject of "honor" cannot but leave me cold."

I wouldn't deny a word of this letter today; it still seems to me that the word "honor" in the mouth of the Minister is tantamount to usurpation. Several professors from the Ecole Polytechnique expressed their solidarity with me by publishing letters in *Le Monde*. Two hundred former students signed a petition (one of them, John Nicoletis, was particularly active). It wasn't that they adopted my views on the Algerian war, but they protested against a decision of a political character, in a school which traditionally did not take political opinions into account. After my answer, I received eighty letters of warm approval. Independently of the text itself, which was not without influence in the development of French resistance to the war, my letter had diverse consequences. For instance, it was necessary to find a successor for my position in the year 1961–62. But out of feelings of solidarity, not a single university professor applied for the job. At the last moment, Gilles Legrand, a professor at the Ecole Polytechnique who had taken over from Gaston Julia teaching geometry, accepted to replace me for a single year, but absolutely not for longer. That left it necessary to find someone for the following years, and gave them some time to think.

In the meantime, I had gone to law to the administrative tribunal. I was defended by André Mayer, a famous lawyer who defended me gratis as many lawyers defending political causes did, and who pleaded independently of the political opinions of his defendants, which is deontologically perfectly natural. He did not plead for my ideas, but against the loss of my position. Similarly, he defended General Salan, who he considered had also been unjustly fired, even though he certainly had more sympathy for my ideas than for those of General Salan. I remained on excellent terms with him, but unfortunately he died only a few years later. After his wonderful defense, the administrative tribunal came to the following judgment: my firing had indeed been illegal – but only because of a certain technical detail, namely that I had not been shown my dossier! Messmer appealed, and this time I was shown my dossier and fired. I sued again, this time before the State Council, but things ended differently. First of all, I should say that my dossier was shown to me by General Cazelles, the successor of General de Guillebon, and it was absolutely empty. I signed on the fact that I had consulted my empty dossier. If the State Council chose once again to go against the Minister, it would need to find other reasons. General Cazelles was very amiable with me and explained his views clearly: they were the same as those of the former students at the Ecole Polytechnique and all my colleagues there, namely that the Ecole Polytechnique should remain outside of all political ideology. He assured me that the Ecole Polytechnique itself played no role in my firing. I thanked him for his support.

At the Ecole Polytechnique, everyone saw the approaching deadlock. No new candidate would apply for my position, and the internal situation was bound to degenerate. For example, my students mentioned me abundantly in their annual review. In spite of his sympathy for me, the General could not allow the review to appear; it would have caused too many difficulties with the authorities in the ministry. He forbade its publication, a rare event in the history of the school. The effect outside was deplorable.

At that time, the Ecole Polytechnique was regarded with contempt by a large part of the University. My firing only aggravated the situation. If I had not been re-employed, I believe I can say without false modesty that the future of the Ecole Polytechnique would have been deeply modified. It became urgent to begin indirect negotiations between Messmer and myself, with mutual concessions. I have never believed in extremist politics. Several professors from the school acted as mediators, in particular Ullmo, a general examiner in mathematics whom I have already mentioned, and Bernard Grégory, a young and very valuable physicist who belonged to the physics department and worked in the team of Leprince-Ringuet; both were former students of the Ecole Polytechnique. Also Louis Armand, the president of the Administrative Council of the school, announced that he hoped, one way or the other, to see me return to my job. The only possible compromise appeared to be the following: my job would be open to applications for the year 1963–64, and I would have to apply for it, which was of course a concession on my part since I refused to admit that my firing was legal, but also on the part of the ministry since they would be obliged to accept me again. I spent the year 1962–63 in the USA. Grégory, who came for a week, asked me to accept this solution, and my brother Bertrand transmitted to me a letter from Louis Armand encouraging me to do so. Compromises are necessary, and this one was not disastrous either for Messmer or for me. Through Bertrand, I answered Armand's letter, saying that the solution was all right with me, and I guessed that Messmer would manage to present it in some way that made it all right for him. I ended my letter with two Latin words: "*Messmer demerdetur*"; *demerdetur* being the third person singular subjunctive of the reflexive verb *demerdari*, which of course does not exist but was the Latinization of "se démerder"[67]. But when I dictated this letter to my American secretary, she separated the words differently and typed "Mess merde merde tur" which neither Bertrand nor Louis Armand were able to understand. Bertrand asked me to explain, and transmitted my explanation to Louis Armand, who found it very funny. And that was the end of the story: I returned to the Ecole Polytechnique in 1963, and of all the classes I taught there, that was the one which gave me the most satisfaction.

Thus, it turned out that my conflict with Messmer ended well. In our correspondence at the time of my firing, he had hit below the belt, and so had I. Later, in 1964, following a report by myself and Leprince-Ringuet, discussions on reforms at the Ecole Polytechnique began, and Messmer came to interrogate the directors of the school on their opinions of our project. Messmer himself fundamentally agreed with us, but de Gaulle was against us. I was sure that Messmer would summon me, and also sure that he would make no mention of our former difference. He listened to me very politely, reflected and agreed with me. Then he added: "Now, I would like to talk to you about the past; it's

---

67) an untranslatable (and quite improper) French word used to mean "manage somehow"

just as well for us to know where we stand." He told me that he had fired me because he found it unacceptable for a professor in a military school to encourage insubordination among draftees. What was I thinking of to sign such a text, given my situation? Firstly, I told him, we had not called upon anyone to desert, but simply declared that insubordination was morally acceptable. Secondly, if I had been in a real military academy like Saint-Cyr, it would have been legitimate to fire me, although even then it would have been contrary to the statutes of government functions. But the military aspect of the Ecole Polytechnique was folkloric; I was training scientists, not soldiers. He appeared to understand my point of view, and we separated on good terms, after a conversation which remained entirely courteous. Another man would have discussed the proposed reforms with me without alluding to disagreeable past events.

## Sit-ins and demonstrations

Following the declarations of the government and the multiplication of torture in Algeria, many public demonstrations took place. Intellectuals felt mobilized and they came to the fort of Vincennes, where they sat on the ground in a silent protestation against the war. Naturally, such demonstrations were forbidden. Policemen arrested the participants and threw them into buses. Thus, on April 30, 1960, a large number of us, among which I remember Jacques Dixmier and his wife Suzanne, Alfred Kastler, Pierre Vidal-Naquet, Henri Marrou, Paul Ricœur and Germaine Tillion, spent part of the day sitting in front of the fort of Vincennes and part of the night in a cellar where I remember having a long conversation with Germaine Tillion. I participated in many such sit-ins. Once, I remember that I couldn't participate for some reason, and Marie-Hélène and Marc-André went separately to demonstrate on the Champs Elysées, with about a thousand people. They had met up while being arrested, then were separated somewhere in the suburbs in the middle of the night. I was not particularly worried, because hard-line methods were not yet in use. However, around one o'clock in the morning they still had not come home, and since I was at home, I telephoned the police station to find out where they were. I came upon an officer in a somewhat drunken state, who said: "Monsieur, if your wife is not at home at one o'clock in the morning, then she is not where she should be!" I couldn't obtain anything else from him. Both Marie-Hélène and Marc-André ended up returning home later in the night, but not all arrests ended so easily. In 1961, Paul Thibaud was violently beaten in a police station. Whenever we participated in demonstrations against the Algerian war, they were always forbidden, and the police would charge us and hit us over the head with large clubs. I was never personally struck, but Vidal-Naquet was once, when I was marching next to him, and in my emotion I tripped over a milestone and collapsed. I must frankly admit that I was afraid during these demonstrations. Later, we created some useful strategies. We would loudly call for a demonstration in some given place, which was immediately forbidden, and meanwhile we would communicate with hundreds of people to let them know that it would really take place in some completely different place. When we arrived, the place would be empty – and suddenly, someone would shout "Fascism won't pass!"[68] and demonstrators would appear all over the place by spontaneous generation; sometimes up to a thousand people would start marching, shouting out slogans. Onlookers standing on balconies would applaud and the police would take so long to get to us that the demonstration would be over before they got there. Sometimes, when they got there too late, the police blindly struck any passers-by who happened to be around. One Swedish woman tourist who was violently molested while coming out of the metro was much talked about.

The manifesto of the 121, partially motivated by the Jeanson trial, set off a whole chain of political consequences. The UNEF organized a gigantic demonstration which aimed both to assert itself as a political and union force and to put a brake on the temptation to illegality expressed in the manifesto. Indeed, although the draftees engaging in insubordination were a small minority, the idea was spreading "dangerously" and risked throwing many young people into illegal situations which would not be meaningful unless it happened massively. The manifesto proclaimed their right to insubordination and committed itself to defending them. The UNEF obtained the agreement of the main unions and left-wing parties. It had been obliged to modify the place and date of the demonstration several times, but finally, after various hesitations and refusals, it was fixed for October 27 at the Mutuality, together with the FEN, the CFTC (Confédération Française des Travailleurs Chrétiens)[69], and the FO, and in spite of the belligerent refusal of the CGT which was itself holding a meeting calling for short strikes, in another place but at the same time. The PSU fully supported the UNEF, the Socialist Party and the Communist Party refused. I myself was involved in these carpet-seller negotiations. The government had forbidden all demonstrations in the streets. The demonstration in the Mutuality was a resounding success; thousands of people crammed into the hall and around it. Gaudez and Denis Forestier, the secretary general of the SNI, gave excellent speeches. In spite of the prohibition, the PSU organized a long parade of several

68) This slogan is just one of the many adaptations of the anti-Franco cry "No pasarán!!"

69) French Confederation of Christian Workers

thousand people which the police charged savagely for over an hour. Many of our friends were struck or beaten: Roland Dumas, Gisèle Halimi, Madeleine Rebérioux, Depreux, Verdier, Mitterrand. The UNEF had proved itself, so had the PSU. Analogous demonstrations took place in many provincial towns.

This demonstration had a considerable effect in Paris and in Algiers. In Paris, it led de Gaulle to pronounce himself in favor of an Algerian Republic (on November 4) and to organize a new trip to Algiers from December 9–13. Once there, he had to face a violent demonstration by the French Algerians, but he remained even more impassive than Guy Mollet in 1956. However, the demonstration was followed by an unexpected Algerian response, a massive demonstration organized by the FLN, led by green and white flags, under the astounded eyes of the soldiers who supposedly controlled the Algerians perfectly. To the stupefaction of the Algerian French and the soldiers, the Algerians loudly acclaimed de Gaulle. At that point, everyone realized that the FLN had morally won the war, even though it took another fifteen months to end the story. In the flush of these events, de Gaulle organized a referendum on his Algerian politics (on January 8, 1961) which seriously fractured the left. As on other occasions, his referendum asked two opposite questions, which had to be answered simply by yes or no. The vote was in favor of or against auto-determination and the organization of public powers in Algeria during its establishment. A vote in favor reflected the desire for peace and the confidence that de Gaulle was able to establish it. It was clear that the majority of the population would vote in favor, and indeed seventy-five percent of the population did so. The right-wing voted in favor, although the extreme-right voted against. The PS voted in favor, the PC against, and the PSU, after long hesitations between voting against and boycotting the vote, finally voted against.

This does not mean that everyone had understood exactly what was meant by auto-determination. The people who took care of our garden in Autouillet asked us what it meant; they didn't know the word and had not been able to find it in the dictionary. They thought it meant determination that moved as quickly as an auto(mobile).

The referendum was massively victorious, but the OAS had just been formed, in February 1961. Shortly afterwards (April 22–25), French putschist generals seized power in Algeria, intruded into Corsica, and threatened to land in Marseille and to march on Paris.[70]. Perhaps they did possess the military means to do this. The situation was complicated for all the protagonists. Pessimism

70) At that time, many young French soldiers had been drafted into the army and sent to fight in Algeria; they formed what is referred to below as the "French Algerian army".

and panic reigned. Panic is really the right word. Michel Debré gave a speech, on the television, exhorting the French people to go to the airports, on foot or by car (we used to paraphrase him: on foot, on horseback or by car). Military measures were taken to protect the airports, but they might not have been sufficient, because the French Algerian parachutists were extremely experienced. Furthermore, the army was deeply divided and nothing proves that the French army would have accepted to resist an attack by the French Algerian army. Apparently many if not most observers believed that the whole thing was an attempt at intimidation. I believe they were mistaken. For us, who had lived through all the former events, the situation was extremely obscure. My family was in Paris when I was alone in Brazil in March and April 1961. Claudine was only thirteen and not yet very politically conscious, but she made astonishing progress during those two months, and described the April putsch to me in detail. I've always regretted not keeping her letter.

Just at that moment, my heart was doubly rejoiced. I had gone for a few days to the national park of Itataya, where I was allowed to hunt butterflies, and brought back an ample harvest of ultra-rare specimens. I learned from an enormous headline in a Spanish paper that the American landing in Cuba, in the Bay of Pigs, had failed. "*La expedición fracasó*" was the headline in Spanish (which I knew well; fracasó is pronounced "fracasso" with the accent on the "o"). This verb "fracasar", which means "to fail" always enchanted me, and particularly in this case. About the Algerian situation, I saw several possibilities. First, I thought that if the French Algerian army landed in France, they would find the French population solidly against them. But that was not so sure, since after all, shortly before, Daniel Mayer and I had tried to organize a unified meeting between the CGT, the Force Ouvrière, the Socialist Party and the Communist Party, to try to set up a joint action against the ultras. It was an absolute fiasco: Socialists and Communists answered jointly that "the situation was not yet serious enough" for them to be able to overcome their mutual disagreements. The whole situation was intolerable and reminded me incredibly of the grotesque situation in Germany just before Hitler came to power. If the situation was not yet serious enough, it would certainly become so when it was too late to do anything about it. At any rate, I felt certain that the whole situation bristled with danger. The most sordid possibility I could conceive of was the following: proposing a political compromise to the agitators in the form of some ministries in a new government, which would be a complete defeat of de Gaulle's ideas, which were not yet completely concrete. In 1961, the Algerian war was in full swing. In 1959, General Challe had received the green light for a final attempt to eradicate the FLN, called "operation Jumelles", and he had succeeded in pacifying the main regions of Algeria. But he had not destroyed the FLN which subsisted in several pockets, and which upheld the morale of the population, as shown by the Algerian demonstrations of 1960. This compromise would have been a slow slide towards a seizure of power in France by the fascist generals. General de Gaulle, who probably had the whole of the French population behind him, was nonetheless quite isolated in the political arena, and all of his attempts at peace had failed. He showed what he was capable of with his magnificent speech in which he referred to the putschist generals as "a retired foursome", and his call to the draftees in the French Algerian army to remain faithful to the mainland and not to defect to the OAS; thus he turned the whole situation around within a few days. The army mobilized against the agitators, who soon understood that the large mass of French Algerian soldiers would not join them. One by one, they capitulated and gave themselves up to justice. *Le Canard Enchaîné*[71)] headlined: "Fortunately we have a General in power, otherwise we risked having a General in power." This chain of events appeared to be linked by a relationship of cause-and-effect: the Jeanson trial, the manifesto of the 121, the demonstration of the UNEF, de Gaulle's trip to Algiers, the demonstration of the Muslim masses, de Gaulle's referendum and the formation of the OAS. I thought so at the time, and I still do. Some historians believe that the intervention of the intellectuals (and students) and the masses was negligible, and that in the end, de Gaulle made peace when and how he wanted it. This is absolutely false, as false as the idea that he vanquished the ultras when and how he wanted – it took him another four years.

## Plastic bombs

Starting in 1961, members of the OAS in Paris took to setting plastic bombs in the apartments of people involved in the combat against torture. I believe I was one of the first victims. It was a rather underdeveloped bomb, which exploded in the garden in front of my apartment building (I lived on the fifth floor). It blew in the windowpanes of the ground-floor apartment, belonging to the guardian of the building, and shards of glass flew into the apartment and landed on a cradle containing a baby, which fortunately was not hurt. Several of my colleagues had similar experiences. Roger Godement and Louis Lalande suffered more serious attacks; powerful explosives placed in front of their homes burst both of their front doors, although by a miracle no one was hurt. Public reactions to these incidents was generally quite negative, and a

71) *The Chained Duck*; this satirical political newspaper takes its name from "canard", a slang word for newspaper, to refer to the freedom of the press or lack thereof.

few demonstrations were organized. A long list of my friends and relatives were attacked: Claude Bourdet, Jean Daniel, Maurice Duverger, Clauder Estier, André Hauriou, Alfred Kastler, Serge Mallet, Gilles Martinet, Emmanuel Le Roy-Ladurie, Jean-Paul Sartre, Pierre Stibbe, and others I've surely forgotten. In my case, my neighbors in the building showed exemplary solidarity; if they had been ill-disposed towards me, they could very well have accused me of creating problems. Instead, they discreetly pretended not to know who the attack could have been directed against. Quite the contrary, they always spoke as if we were all equally concerned, and we were very moved.

At that time, Marc-André had already received quite extensive political training, but Claudine had not. She endured my endless conversations with Marie-Hélène and Marc-André about the Algerian war without ever really feeling the situation as a concrete reality. She thought that the First World War was the 1939–1940 war and the Second World War was the Algerian war. She did feel that we were all threatened. We had had to take a series of protective measures against further plastic bombs. I had even brought several dozen boxes containing my rarest butterfly specimens to a friend's house. But we took many other precautions. It was almost as difficult to live normally as it had been during the Occupation! Several plastic bombs were placed in boxes which were attached to people's front doors in the night, set to explode as soon as the door was opened, so that whoever opened it risked serious injury. Marc-André was nineteen, and already at the university. Claudine was in high school and left the house every morning at seven thirty to go to the Claude-Monet school. Thus, she was the first to open the door every morning. We attached a long string to the door-handle which went all the way to the farthest room in the house, and managed it so that she could open the door from there in safety.

For months, members of the UEFJ (Union des étudiants juifs de France)[72] guarded our house. I knew its members well; we had many interesting conversations. Their ideas were very advanced, and their views on the war were close to mine. I was very grateful when they proposed to guard my apartment. All day, two of their members armed with cudgels stayed on the staircase or on the landing between our floor and the floor below. If someone got off the elevator on the fifth floor, they could easily see if he was carrying a suspicious package. We gave them food and had long conversations with them. They were really nice. I always felt very grateful to those young people for all they did for us. There were also a couple of policemen guarding 37, rue Pierre Nicole, stationed in the garden in front of the house, but they never interrogated anybody and I believe

72) Union of French Jewish Students

they didn't even know our name nor why they were there. Our Jewish guards would have immediately noticed any danger and reacted instantly. We always went up to the sixth floor in the elevator and walked down to the fifth floor to enter our apartment. Alfred Kastler had guards at his apartment as well; he lived just a few doors from us, in the rue Val-de-Grâce. He had very categorical views on the war and was extremely threatened. He was always accompanied by a colleague when he went to his laboratory, often by the well-known physicist Brossard, his closest collaborator. We often went around together, as we attended the same political meetings, and he would drive me in his car. I was one of the rare people to have had a Nobel Prize winner as chauffeur! Later, I often ran into my former guardian angels, for instance Lucile Rosenberg and Bernard Kouchner.

## October in Paris

Starting in the autumn of 1961, a mandatory curfew was installed for all Algerians living in Paris. Every Algerian on the streets after eight thirty in the evening was arrested. On October 17, the FLN organized an enormous pacific demonstration against the curfew. Algerians – the majority of them women – massively took to the streets after eight o'clock and walked towards the place de la Concorde. Of course, they were not armed. The police chief Maurice Papon, whose nefarious role during the Occupation has recently come to public attention, had more than ten thousand male demonstrators arrested. They were transported to special police locales where they were savagely beaten and many of them killed in a full-fledged massacre. Over the next few days, more than a hundred bodies floated down the Seine. The massacre took place during a meeting of the Audin Committee. We learned about it from one of our members, who arrived late. Should we rush to the scene? I believe I remember that it was already too late; furthermore there were only about ten of us and many more would have been needed. Instead, we hastily gathered a large quantity of documents and published it in our next issue of *Vérité-Liberté*. The press discussed the affair for a few days, after which total silence reigned. Only a single deputy, Claudius-Petit, spoke out against it in the Chamber of Deputies. And then everything was forgotten for decades. It was one of the least spoken-of events of the Algerian war. Jacques Panijel decided to make a film of the event to keep its memory alive. Alone, he made several trips to the slum in Nanterre where many Algerians lived. During the filming, all exits were guarded by members of the FLN. He filmed many conversations with Algerians, and took photos of their serious wounds. The film is deeply anti-racist, very lively, but difficult to watch. He projected it privately a few times, but no firm accepted to distribute it publicly, ever. Recently, the book *Octobre à Paris* (*October in Paris*) by Jean-Luc Einaudi recalls and recounts those painful days. I always avoided bringing politics into my teaching. But exceptionally, Jean Dresch, Alfred Kastler, Robert Ricatte and I decided to say a few words to our students at the beginning of our first class following the massacre. It was very sober, and the students applauded. Most French people today know absolutely nothing about October 17, 1961.

There were many demonstrations against the war in the last weeks of 1961. But the left never succeeded in unifying itself. They were always agitatedly forming new leagues and committees against the OAS. For instance, they formed the League of Action for an Anti-fascist Rally. During interminable discussions, they tried to unite the Communist and non-Communist left wings. But under the firm leadership of Guy Besse, a Stalinist of the old school, the Communists succeeded in making sure that the members of the League were uniquely intellectuals and university faculty members. It was like shutting us into a ghetto. The Communist Party did everything to prevent people from other classes of society from joining. I refused to allow myself to be thus limited. Obviously, in practice, probably only intellectuals would have wanted to join anyway, but I refused to accept that as a condition imposed from outside. I was sad to see friends like Jean Dresch and Madeleine Rebérioux against me; they were somewhat independent of the Communist Party, but not enough to fight its weight in such matters. (They have evolved since then.) After difficult negotiations with the Communist Party, the League joined a rather undefined structure called the FACUIRA (Front d'action et de coordination des universitaires et intellectuels pour un rassemblement antifasciste!)[73] FACUIRA – a boring and absurd name which probably had the effect of driving people away! I left the meeting in disgust. The whole idea was still-born, which was Guy Besse's goal in the first place. Marc-André actually joined the thing for a short time. The philosopher Meyerson coined a strange formula: "We proclaim: fascism will not pass! It is doubly wrong: Gaullism is not fascism, and it has already passed." Marc-André was thrilled. I pointed out to him that the fascism we were supposed to be incriminating was not de Gaulle's regime, but the OAS, and that we were supposed to be preventing it from passing. His interest in verbal thrusts and parries soon disappeared.

73) FACUIRA: Action and Coordination Front for University Faculty and Intellectuals, for an Anti-Fascist Rally

The FUA (Front Universitaire Antifasciste)[74], was a much more interesting organization. It consisted mostly of young people, Trotskyists or Communist students who had ruptured with the Communist Party. Marc-André was an active militant. It was a very aggressive group; its members actually engaged in physical fighting with members of the OAS, and occasionally even civilly arrested members of the OAS and brought them to the police, who was pleased with the unexpected collaboration. The police had antennae in all left-wing circles, but not in fascist circles. Thus, it had difficulties in the struggle against the OAS, whereas students from all sides met in their university classes and knew each other.

After the plastic bombing of Malraux's house, in which Delphine Renard, the little daughter of the guardian, was wounded in the face and blinded, a public (although naturally forbidden) demonstration was organized in Paris on February 8, for peace in Algeria. I believe a good thirty thousand people took part in it. For the first time, different left-wing groups acted together. At the end of the demonstration, police crushed hundreds of people into the entrance of the Charonne metro station and other stations, so severely that there were five deaths in Charonne and three in Voltaire. The public was more disgusted than ever and on February 13, 1962, a gigantic and this time authorized demonstration gathered five hundred thousand people. On that day, de Gaulle finally understood that he could no longer put things off, and he accelerated the Evian talks. The Evian agreements were signed on March 18, and a ceasefire was proclaimed on the 19th. On the 18th, the UNEF organized a celebration in the courtyard of the Sorbonne. A militant from the UGEMA gave a talk and waved a tiny green and white flag (the colors of the FLN) which he took out of his pocket; he was warmly applauded. Shortly before, the UNEF had established an organic liaison with the UGEMA; it had addressed a letter to the president of the Republic, asking if French students should consider themselves at war with Algerian students. A few days later, I was summoned by the rector Roche, who transmitted to me a formal "blame" from the Minister of National Education, for having attended a speech by a student representative of the FLN, with whom we were at war. I pointed out to him that we were no longer at war, and he replied that we were still at war until the next day, adding: "I can't tell you how well I understand you." Then: "The Minister told me to tell you not to reply to him on any account." The Minister, Paye, remembered my reply to Messmer after I was revoked from the Ecole Polytechnique.

74) FUA: Anti-Fascist University Front

## Marc-André's kidnapping

In February 1962, we were struck by a terrible event. In the middle of the worst period of the OAS, Marc-André was kidnapped.

On Wednesday, February 7, around ten in the evening, Marie-Hélène received a phone call from an unknown young person, telling her that there was a project to kidnap Marc-André. We took the information very seriously, but unfortunately we did not judge how urgent it was; we simply asked the advice of our political friends. We agreed that the first precaution was for Marc-André to go out as little as possible, and only in our black Fregate car. In fact, that was the most absurd solution. We had the impression that the phone call was not a threat, but a warning emanating from some young man who had heard something and thought things were going too far.

The very next day, Thursday, February 8, Marc-André went shopping in the car. He had barely entered the car when he saw someone on the back seat holding a revolver. Someone else immediately entered the car by the drivers' door and pushed Marc-André onto the back seat. A third person, a slender, blong young man called the "sosia" (double) by his comrades, stuck his head into the passenger's window, exchanged a few words with the others and then drove off in a black Fregate identical to ours. Marc-André was made to wear dark glasses which prevented him from seeing anything, and told that they had nothing against him, that they were simply obeying orders. As for the double, he drove very quickly to Coignières, near Rambouillet, and parked his Fregate outside the Hotel of the Gourmand Capucin (we learned all this later). The wife of the hotel keeper noticed the car, and saw a young blond man leave it with a bag over his shoulder. She and her husband were surprised to see a car which did not belong to a client of the hotel, but they didn't think of observing the license plate, which would have been extremely helpful for the subsequent inquiry. The young man walked off in the direction of the village Maison-Blanche, towards the forest; four people saw him take the path to the forest, carrying his bag. The four descriptions were actually rather different from each other, as is often the case with witnesses. The witnesses believed that the young man was Marc-André and that the black Fregate was ours.

As for our Fregate, it drove for three hours without actually going very far from Paris, but Marc-André didn't realize this. When night fell, his dark glasses were removed. During the night, the Fregate joined the other Fregate which the double had left parked; they took it away and left our Fregate in its place. Marc-André was shifted from our car into the other one. The exchange took place right in front of the hotel, but it was night, the lights were off; in fact

Marc-André had no idea there was a hotel there, and the hotel keepers had no idea that the cars had been exchanged. They must have been in a room which did not give onto the road. At any rate, our car was found in front of the hotel the very next day.

Marc-André was taken away in the other Fregate, and they drove to the edge of the forest. When he stepped out of the car, he was wearing the dark glasses again, but he was able to see the double, who came up to the car and said to the others: "It won't work." What wouldn't work? I can only guess: perhaps their original plan was to take Marc-André somewhere else? Perhaps someone had changed his mind because of the risk? In that case, perhaps the forest was only a secondary plan in case the first one failed. In any case, the double had not remained inactive during the three hours. They put scotch tape around Marc-André's dark glasses (a similar technique was used when Mrs. Dassault was kidnapped) and took him deep into the forest. The group remained there until Friday evening, changing camps on the Friday morning. One of the kidnappers went shopping on the Friday morning and Marc-André was given pâté sandwiches, oranges and milk.

Marc-André spent most of the time immobile, lying on the ground, freezing (he was wearing his usual warm snowjacket, but ordinary trousers and nylon socks, and the weather was extremely cold). His kidnappers were no better dressed than he was. Most of the time, they stayed at some distance from him, listening to a transistor and talking amongst themselves. Marc-André heard some fragments of conversation: women, cinema, not much politics. He remembered hearing them say "That's bad for us", after hearing on the radio about the deaths in the Charonne metro station. He couldn't figure out what social class they came from, but they repeated to him several times that they were only executing orders. On the Friday, he heard on their radio the speeches from a meeting in the Sorbonne courtyard, and to his amazement, he heard the voice of one of his friends from the FUA telling the other students the license plate number of our Fregate! Later, he told us: "It was the only moment of joy in the whole kidnapping. I remember feeling my anguish lessen for a while. I wanted to ask my kidnappers a question, but seeing that they weren't looking at me, I restrained myself." Later, police verified that what Marc-André had heard really had been on the radio, but they said he was mistaken about the time at which he had heard it. Anguish, freezing cold and lack of sleep had reduced him to a miserable state, but he never gave way to panic. He tried to avoid provoking dangerous reactions in his guards, who were armed (this occurred during a period when the Algerian war was threatening to turn into a full-fledged civil war in France). In the afternoon, he noticed that they were showing signs of worry.

The police search for Marc-André began during that afternoon. On Thursday evening, not seeing Marc-André at home, I thought that although he had said he wasn't going to, he must have decided to join a (forbidden) demonstration against the OAS plastic bombs, in particular the one which had blinded little Delphine Renard; this was the demonstration which ended with the deaths in the Charonne metro. I spent the evening telephoning police stations and hospitals, hoping to receive reassuring news of him. After hours of this, we finally began to believe that the kidnapping we had been warned of had actually occurred. In the middle of the night, I called Marie-Hélène, who was in Reims, and told her to come home. Jacques Panijel spent the night with me. I telephoned Matignon, the residence of the Prime Minister, and police headquarters, who immediately sent an officer. They began the search by broadcasting our license plate number on the radio and asking for news from anyone who had seen it. Their call was heard in Coignières, and in the middle of the day, the hotel keeper of the Gourmand Capucin telephoned the police to say that our car was in front of his hotel. In the afternoon, Marie-Hélène went to Coignières with the police. They brought a police dog; we were very hopeful because the weather was dry and only 24 hours had passed. He snuffled and sniffed and searched, and with the help of his master he obviously tried hard to obtain results, but in vain. We had him smell the car, and we had him smell a package of Marc-André's clothes, but he did not find the slightest trace of him, just as though he had never left the car. In the end, this turned out to be normal, since Marc-André and the kidnappers had gotten out of our car and immediately into another one, so that they had left no trace which a dog could follow. A witness had seen someone get out of a car and walk away on foot, but it was the double and not Marc-André, and he was getting out of his own Fregate and not ours. A dog can make a mistake, but his sense of smell will not be fooled by two cars or two young men who look alike. In fact, the behavior of the dog should have made us realize that Marc-André had left in a second car. But nobody mentioned the possibility, and that evening we were in a state of deep dismay. The police contined to search; towards nine-thirty they began to explore the edge of the forest with their electric lamps which they shined among the trees. Then they stopped for the night.

The night was clear, and Marc-André and his kidnappers saw the lights from their camp. He became hopeful, and they probably became fearful, because soon after they picked up their things and moved off without saying a word. Seeing this, Marc-André turned away from them and began quietly walking towards the place where he had seen the lights, fearful of seeing his kidnappers reappear. They did not return, but he could not find the edge of the forest and was obliged to spend a whole second night there. He was so exhausted that he kept walking and walking in a state of half-sleep. Early on Saturday morning, after some more walking, he found his way out of the woods and entered a hospitable farm. He immediately telephoned us from there to reassure us. We called Matignon, the police and some friends who lived very near Coignières. They immediately went to get him, and found a young man looking haggard from his adventure, but visibly perking up, drinking a generous cup of café au lait in the reassuring atmosphere of the farmhouse. They understood nothing of the confused story he began to tell them – two cars, meeting heard on the radio, a young man called "sosia" etc. .... Soon my brother-in-law Jean-Claude came from Versailles to get Marc-André and bring him back to Paris. Claudine had spent the whole of Friday with a friend of hers, exhausted and anguished.

## A machination

The government, the police, the radio, the press which quickly alerted the public (not to speak of our friends and students) – everyone collaborated in finding Marc-André. On the Saturday morning, the dramatic kidnapping ended in the happiest possible way; anything might have happened in that difficult period near the end of the Algerian war. But immediately afterwards, complications occurred whose result was that after everything had been done to rescue Marc-André, everything was done to shut his mouth about what had happened.

As soon as he was found, without "losing" a single hour in verification, a campaign was begun to spread the idea that Marc-André had run away, and that the whole kidnapping story was an invention. Adolescents run away frequently and kidnappings are rare. Perhaps it is natural that some of our friends mentioned the possibility to us. We had even considered it ourselves, but when we saw our son on his return we knew it was absurd. Nonetheless, we kept the idea in our minds when we listened to his story and observed his behavior. But a group of people went so far as to explain that they were well-informed and spread the story as a certitude backed up by ample proof; they addressed themselves to left-wing figures and explained their view with the confident, discreet air of those who know much more than they choose to say. Thus, they succeeded in spreading a little doubt; in the name of charity, indulgence and friendship they spread an atmosphere of suspicion. We were even charitably warned that if we went to law, Marc-André could find himself accused of contempt of court for lying to the investigators. Of course, we stuck to our guns. We sued X for kidnapping and sequestration. Our lawyer was our old Trotskyist friend Yves Dechézelles, a deeply sensitive man, and the judge was M. Delmas-Goyon, a judge for minors who always displayed tact and intelligence. "It's just too much," Marc-André told us already on that first morning, on the telephone from Coignières, "I just spent two horrible days in the hands of kidnappers and they're saying I ran away." He was right to feel indignant; we were indignant too.

He was interrogated for over twelve hours, in four sessions, at our home, by the most qualified investigators of the criminal brigade. They were always courteous, but even though Marc-André was supposedly being interrogated as a witness, it was impossible for him not to feel that he was actually a suspect. In particular, during the last session, on the evening of Wednesday February 13, he was told that all of his declarations had been taped. This session was more solemn than the others, and apparently had been prepared with a view to obtaining collapse and confession. Without his knowing it, my wife and I were installed in the next room, with orders not to intervene "whatever happened". During the interrogation, every possible objection was opposed to his statements: his story was insufficient, imprecise (they could not point out any actual contradictions) as to the way he was dressed, the fact that he had not tried to escape (perhaps he could have, but the kidnappers were armed. Since then, we have heard of hundreds of cases of kidnappings and the victims almost never try to resist or escape.) Of course he was nervous and tired, but he remained perfectly calm, answered all the questions tranquilly and reasonably, saying what he knew and what he didn't know, so that in the end the session petered out rather miserably. An eminent doctor told us that in cases of runaways, the runaway inevitably ends up contradicting himself, because he is necessarily in a state of mental imbalance, and incapable of constructing a flawless but false version of events. However, in spite of the exhaustion due to spending two nights without sleep in a glacial forest, Marc-André simply did not contradict himself at all. As a scientist, I was shocked by the methods used in the investigation. When a mathematician looks for the solution to a problem, he doesn't know what he is going to find; he feels around simultaneously in several directions, ready at any moment to change his views if necessary. It seems to me that a police investigation should use this method even more. But instead, they spent far too much time trying to decide whether Marc-André had been kidnapped or had run away, and they did not sufficiently explore the various possibilities of collaboration with kidnappers in the direction of Coignières. Afterwards, it was too late. Marc-André did not "collapse"; his story only seemed complicated because the reality was complicated, but it held no contradictions either with facts learned *during his absence* or with facts obtained *after the event*. He remained calm and tranquil the whole time, and all his friends and relations who saw him afterwards were scandalized by the calumnies directed against him.

But the "proof" which appeared to make the strongest impression was the sheaf of witnesses from Coignières. At the very moment when Marc-André said he was in our car with his kidnappers, "he" had been seen leaving "his" car, alone, by the wife of the hotel keeper and by a group of four people in Maison-Blanche, walking towards the forest. But the car was not ours, and "he" was the sosia. And in any case, statements by witnesses should always be treated carefully, and their ability to identify someone should be tested. Those who recognized Marc-André first were shown a photo of him, at a moment when of course everyone thought he must be the person they had seen; then they saw his picture in newspapers and on television, and finally they identified him face to face.

In the first few days, the press constantly cited one particular witness, a schoolteacher from Coignières, who claimed to have seen a young man telephoning from the town hall on Thursday, February 8, and believed it was Marc-André. On February 12, he was brought face to face with Marc-André in Coignières and formally identified him. The very next day, he hastily telephoned us and the police to rectify his error – he had just met the person he had seen telephoning, who confirmed the fact. This person was neither Marc-André nor the sosia, but a complete stranger who had telephoned from the town hall. The schoolteacher was extremely sorry to have caused us trouble, but the press never bothered to straighten out the story.

In fact, many elements of the inquiry showed that there was no question of a juvenile flight. For example, Marc-André did not possess a transistor nor money to buy one, yet he was able to say what he had heard on the radio. If one were to believe that Marc-André had invented the whole story, then he must have planned it carefully several weeks ahead of time, with help from accomplices. It comes out to be more complicated than the kidnapping which already appeared too complicated. It doesn't fit in well with the known facts. But most importantly, it simply wasn't in Marc-André's nature to behave that way. He was not at all sportive, had never gone camping, and he was rather clumsy about the practical aspects of daily life; furthermore he was fundamentally honest. As a child, he was naive, never sly or even mocking; he was utterly frank and we never had the slightest problem of honesty with him. Disagreements, arguments, generation gap, yes: lies, no. He was not very aggressive; not aggressive enough.

In our eyes, the whole campaign was really just a rather succesful political operation. The day of the kidnapping, February 8, was also the day of the Charonne metro massacre. The next Tuesday, February 13, half a million people attended the funeral of the victims, demonstrating their disapproval of the government's leniency with the OAS. Once can consider that on that day, the reigning powers truly adopted the idea of peace in Algeria. In spite of the passionate opposition of the OAS and all the ultras to de Gaulle's policy, the government was linked to the right-wing via a continuous chain of intermediaries, who were respected by all parties. The government which had sprung from the events of May 13 was sentimentally much closer to the ultras than to the intellectuals or left-wing students. Its collaboration with the latter only lasted strictly as long as was necessary for Marc-André to be found. Immediately afterwards, it returned to its ordinary business of not allowing the left to mark a single point against the ultras. From Thursday to Saturday, we received the unanimous sympathy of the public, which was simultaneously expressing its reprobation of the OAS. But the government wanted to prevent this sympathy from turning into any advancement of the ideas we represented. We had no unhealthy intentions of using the event to obtain publicity. But it is a fact that independently of everything we did, Marc-André's return after his kidnapping, which was closely followed by public opinion, the police and the government, marked a point against fascism and in favor of peace in Algeria. The political reasons for the police campaign against Marc-André are so obvious that it still amazes us today that so many people were taken in by it.

The investigation ordered by judge Delmas-Goyon began very soon, and gave a large number of interesting results on the practical organization of the kidnapping, the camping place in the Coignières forest, and the group of people the kidnappers belonged to. However, the kidnappers were not found. We don't know if they were linked to the OAS; there were only too many violent extreme right-wing groups at the time. On January 24, 1962 (fifteen days before the kidnapping), Salan gave the following order: "I order that spectular, indiscriminate kidnappings be organized immediately by all sections and subsections, aiming at Gaullist families and people in power. The victims should be treated properly but kept prisoner as long as necessary."

A few months later, the judge was obliged to declare that the case should be withdrawn: "The information gathered by the investigation has made it possible to direct a search, but the search has not yet succeeded in identifying the authors of the kidnapping of Schwartz Marc-André. Thus we now close this procedure with a withdrawal of the case, and order that the dossier be deposited in the Clerk's Office, to be opened again in the case that new information arises." Alfred Kastler expressed the conclusions to be drawn from the whole sinister story in an article in *Le Monde*, calling it: "[...] a new method for attacking the opponent, not physically, but morally. A typical example was the machination directed against François Mitterrand. Direct attacks frighten, but they also provoke reactions [...] and a wave of indignation which seals the unity of the opponent. Turning a tragic affair into a burlesque comedy is a much more advantageous method. This method was also used on the occasion of the kidnapping of young Marc-André Schwartz [...]. The strangeness of an unexpected story, misunderstandings and tendentious rumors were subtly exploited. Malevolent insinuations were complacently amplified by the press, which took a perfectly straightforward story and made it "troubling" and "mysterious" [...] The lesson we should learn from this is that we, also, should accept all the risks, both physical and moral. The goal is worth it."

Marc-André had been a relaxed, enterprising and active boy. The kidnapping had an effect on him like an enormous blow to the head. He lost his audacity and became timid. Once, in Autouillet, the small lateral lens of his telescope was stolen, but *he didn't dare to complain* because he "was afraid of finding himself *accused*". In the end, my mother dealt with the event and discovered the thief, a relatively inoffensive young man. Until then, Marc-André was scientifically minded, interested in all sciences. Several years earlier, he had been enthusiastic about astronomy. Once, in 1959, when he was sixteen and we were on vacation near Die, he gave us a terrific lecture on astronomy. Soon after, the whole family went to visit the observatory in Saint-Michel-de-Provence, where he made extremely interesting observations with the large telescope. Paul Couderc gave us a wonderful talk, really directed at Marc-André. We bought him a thirty-five centimeter (parabolic) mirror telescope which he set up in Autouillet and learned to manipulate very well; to our great joy, he showed us the rings of Saturn.

After the kidnapping, however, he decided to become a writer. The Mercure de France published some short, modern poems by him which were appreciated by connoisseurs. He met the poet Yves Bonnefoy, who received him warmly and advised him to keep his independence by having a job outside of literature. But very rapidly, his psychological state began to deteriorate; it became worse and worse and we saw it happening. Apart from writing novels, he was incapable of doing anything. In 1965, he asked for help from a good psychiatrist, and in September 1966, he wanted to enter a psychiatric clinic chosen by his psychiatrist. In 1970, he changed clinics. Towards 1966, he absolutely ceased speaking aloud; he only spoke in a very low voice, although what he said was perfectly coherent. It was very impressive. When we asked him why he spoke so softly, he answered that speaking out loud gave him the impression of re-opening an old wound, but he didn't know what. His neurosis progressively turned into psychosis, one of the worst kinds, suicidal psychosis. He was always perfectly lucid and intelligent and he worked all day at his novel, one of several which he had begun. He finished it in 1970, to the amazement of the director of his clinic, who had never yet had a patient write a complete novel. The novel is called *Autumn* and was published by Berger. It is beautifully written, well-thought, original, and full of sweetness and delicacy, but also tragic. It was favorably received by François Mauriac. But even this victory did not distract Marc-André from the idea of suicide. He talked to me about it and we spoke of it often. He became completely obsessed by it; he tried to forget about it, but it kept coming back, more and more strongly. Well before finishing his novel, already in 1967, he tried to commit suicide. On March 21, 1971, he succeeded; he went to a fair and shot himself in the temple. It was a terrible tragedy for all of us, even though we partly expected it; it was his sixth attempt. He had fully assumed the responsibility for his act, and it was a courageous act, destined to end a suffering which had become intolerable, and from which he could no longer see any possible escape. We often regretted many of his attitudes, but his problem was far beyond us.

We will never forget the child and the young man that he was, always gay, very high-minded, eager to learn, sweet and nice; a friend. For me, he was a true friend as well as a son. I talked about science with him, and I even discussed my ideas about the Algerian war with him before taking certain decisions. He painted and played the piano beautifully. He adored butterflies like I did, captured them and identified them perfectly; starting in 1963, our collection became a joint collection. The double wound of his kidnapping (two freezing nights of anguish in the forest) and the unjustified accusations against his honesty were certainly among the major causes of his illness.

## Death threat

Shortly after Marc-André's kidnapping, a student at the Ecole Polytechnique told me that a group of ultras had decided to assassinate me. The threat was obviously true and precise. It is very difficult to protect oneself if one is specifically aimed at. I could have worn a bullet-proof vest, but I was not advised to. Instead, my friends and police headquarters recommended me to never take the same routes when going anywhere, to avoid routine and to go to the university or to meetings, if possible, at a different time every day. Again, I found myself examining everything around me as I had during the Occupation. All my precautions could have been totally useless. I told a member of the ministry of National Education about it; he was a biologist from the university whom I knew quite well, and

I hoped he would help me obtain some kind of protection. He found the threat worrisome and decided that I did indeed need protection. Of course, such threats always contain an element of psychological warfare, but there were still real reasons to take serious precautions. After all, Marc-André's kidnapping threat had turned out to be more than just psychological warfare!

My brother Daniel went to see Michel Debré to persuade him to provide my children Marc-André and Claudine with bodyguards during their journeys to high school and university. One armed plain-clothes policeman, Mr. Sorba, was given the job of protecting all of us. He was a very nice man and I kept up a correspondence with him after it was all over. He knew every detail of my schedule and accompanied me closely. One possibility was a motorbike or even bicycle assassination, a simple and well-known method. However, it is certainly more difficult to kill someone who is permanently accompanied, especially if his guardian angel happens to be an armed policeman.

## A catastrophic result

We thought that it would be possible to create a multi-national State in Algeria, something like South Africa today, with religious issues dealt with separately, but in which French Algerians would have their place. In particular, there were many Jews among the French Algerians, and it seemed as though it would be possible for these Jews to cohabitate with Muslims. I admit that we did not display much perspicacity on these questions; we had not thought about them enough. We had not thought for an instant that victory in the Algerian war would produce a monumental, practically unsolvable problem. It does not reflect well on our political judgment, and almost constitutes some excuse for the hesitations of the French regimes during the war, because the French in France attached great importance to the destiny of their compatriots in Algeria. It wasn't that we were indifferent to the problem, but we oversimplified it. Like others, we protested against the armed attacks of the FLN and against their particularly cruel individual attacks, often accompanied by mutilation. On the other hand, Algerian independence was an unquestionable necessity and we were right to fight for it. Nobody, today, seriously believes that Algeria could have remained French, forming a pocket of three departments. Algeria has twenty-five million inhabitants and France has fifty-six million; democratically integrating Algeria into France would have given the Algerian people enormous power within metropolitan France. Economically and politically, it would not have been viable.

We were also too timid in our criticism of the single party, even though as a Trotskyist, I was well aware of the dangers inherent in a one-party system. The single party already contained the germ of all the disasters which followed the end of the war. Ben Bella was imprisoned by Boumediene for many years, and completely isolated. He was allowed only two things: books and permission to marry. I met him many times in Paris, with his wife, after his liberation. A Ben Bella political party with a reasonable internal structure and a newspaper still exists today. The horrible terrorism exercised by the FIS (Front Islamique du Salut)[75] has become an inextricable problem. It is useless to lay emphasis on a problem which by now everyone knows about, and on which my views correspond to those of the majority of the French people. I never had any further relations with the government of Algeria, and still do not. The victory of the FIS in any form would be a disaster for the entire world, and in particular for Algeria, North Africa and France. Muslim fundamentalism is an international danger, close to the danger of fascism during the Second World War. It should be battled with fierce energy. It is not that I am in favor of the manner in which the Algerian military is now suppressing democracy in order to perpetuate a deeply corrupt regime[76]. Like many democratic organizations in France, I support all Algerian democrats. But of all possible outcomes of the current struggle, the worst would certainly be a complete victory of the fundamentalist party FIS, which would turn Algeria into another Iran with all its dramatic consequences. One cannot deny that today's terrorism is a consequence of the terrorist methods used by the FLN during the Algerian war. It has been said, with some justification, that the FLN lost militarily but won politically. At the end of the Algerian war, there were almost no more armed FLN troops; only a few terrorist kernels remained, but these were unconditionally supported by the masses. On the contrary, the French army possessed absolute military domination, but were politically disavowed. The situation was the opposite of the American war in Viet-Nam.

Understanding does not mean excusing. One can defend the point of view that if the Algerians were unable to establish democracy without having recourse to the use of terrorism, then it would have been better for them to renounce independence and declare that the country was not yet ready for it. But I believe that we cannot rewrite history. The obstacles that successive French governments put in the way of even the most microscopic developments of democracy in Algeria made revolt inevitable. I doubt that Algerians who were aware of the

75) FIS: The Islamic Salvation Front is a powerful fundamentalist political party in Algeria today.

76) The Algerian government annuled recent elections won by the FIS.

situation really had the choice of waiting five, ten or twenty years for democracy to establish itself, little by little but inevitably. Revolts are always premature (except perhaps the French revolution of 1789), especially when they burst forth in countries which have remained politically underdeveloped and primitive by the will of colonizing forces. The moment when a people is ready to revolt in order to seize independence necessarily occurs earlier than the moment in which it is ripe for a democratic government. It is in the nature of things (Trotsky's theory of permanent revolution lucidly took it into account). That is why countries which have emancipated themselves almost always became dictatorships dominated by a single party. The only exceptions today are India and Portugal after 1974. In any case, the situation in Algeria today is a catastrophe.

## Militating or Research

Except for the Occupation, the last part of the Algerian war was the most painful period of my existence. Battling against the war and protecting my family was not my sole activity, but it devoured an enormous portion of my time and energy, and obsessed my thoughts. During the years that the war lasted, until 1962, I did not stop thinking about it for a single day. I could not give my seminar at the University of Paris. My student Malgrange, who was teaching in the provinces at that time, organized a remarkable one-year seminar in Paris, which was much appreciated on an international scale. He absolutely wanted to call it "the Schwartz seminar", "in order to get people to come," so he said, but also in order to help me through this difficult moment. He had full responsibility for the seminar, and in all justice, it should have been called the "Malgrange seminar". I wrote the preface for the published version. At the time, I was a professor at the University of Paris and also at the Ecole Polytechnique. My three accumulated tasks constituted an insane enterprise, whose consequence was that I almost completely abandoned research, even though I cared about it more than anything else. Research activity is not troubled by changes or momentary interruptions, but it must remain a permanent activity of the mind. But my mind was completely occupied by the Algerian war. I still remember days which started out well but which continued badly, as I received yet another telephone call about an urgent case. I cannot complain; I "chose" it myself, and a free man always chooses what he does. And I am deeply convinced that as a single individual overflowing with relentless activity, I was a non-negligible factor in the peace in Algeria, while working on modernizing the Ecole Polytechnique at the same time. A factor, only a factor, but more than a drop of water in the ocean.

However, what I did not choose was the immense size of the task I set myself. After the Evian agreements in 1962, there naturally remained many reasons to stay interested in Algeria. But I was frustrated about not doing research, and decided to recycle myself. I was invited by New York University, and with my family (except Marc-André, who chose to remain in Paris) I spent a marvelous year at the Courant Institute for Mathematical Sciences, where I led a seminar, did a lot of research and relieved myself of the burden of Algeria.

# Chapter X
# For an Independent Viet-Nam

## The war in Indochina, 1946–1954

Back in France, in the autumn of 1963, I immediately flung myself into a new struggle, against the Viet-Nam war. It is impossible to understand this struggle without knowing the origins of the situation.

The Japanese, who had occupied Indochina during the war, were expelled after their surrender in August 1945. Starting on the 2nd of September, after a rapid conquest, Ho Chi Minh, the leader of the Viet-Minh front against the Japanese and French occupations since 1941, proclaimed the independence of the RDVN (République Démocratique du Viet-Nam)[77] in Hanoi. But France had other plans in mind. General Leclerc was appointed the superior commander of the ground forces in Indochina, under the orders of Admiral Thierry d'Argenlieu, the high commissioner of France in the Far East. Leclerc went immediately to war and reconquered every large city in the South, and the war of Indochina began. Even at that stage, peace still would have been possible. Already from the middle of the Second World War, groups in London were planning for the future. De Gaulle associated the greatness of France with the greatness and the modernization of its empire. He wanted to put an end to old-fashioned colonialism, and envisioned first and foremost economic modernization via large-scale industrialization. But political changes turned out to be minimal; great changes occurred within the French Union, but they came too late with respect to the actual events, in the cases of Algeria and Indochina.

The grand idea for Indochina consisted in the Federation of the five provinces of Annam, Tonkin, Cochinchina, Laos and Cambodia, under French direction. French was to remain the official language, and the language of all higher education. The different States would retain only internal sovereignty: their exterior representation was to be entirely handled by France, possibly with local agents in French consulates and embassies. The five countries would be

77) RDVN: Democratic Republic of Viet-Nam

allowed to maintain national armies, but the French Union would have troops based locally as well, all of which would be placed under French command if necessary. No universally elected parliament of any kind would be recognized, only a legislative power given to an Assembly of the States consisting of five national delegations and a French delegation. Executive power was to be given to the French high commissioner, with the assistance of federal commissioners appointed by him! This impressive edifice reeked of the purest colonialism, albeit in a new style. Compared with Ho Chi Minh's declaration of independence of the three States of Viet-Nam, and his creation of the RDVN, the anachronism of the French plans was frightful, and they were completely unacceptable to the Vietnamese. The Fontainebleau conference in July 1946 left open a possibility of pacific collaboration between France and the RDVN: the latter would be directed by Ho Chi Minh, and would be free but not independent, remaining within the French Union. Ho Chi Minh was ready to make great concessions to avoid war, confiding at that time to several people that he was not even sure if he would not be repudiated by his party upon his return to Hanoi. But the conference did not end with an agreement. What is now called the Fontainebleau agreement was merely the *modus vivendi* signed in extremis on September 14th, the day before Ho Chi Minh's departure. All these efforts were completely nullified by several bloody military skirmishes, followed by the bombing of the civilian population of Haiphong on November 23, 1946, which killed thousands, on the orders of the sinister Admiral Thierry d'Argenlieu (who ended his days in a monastery).

Lerclerc, who had agreed with the idea of a compromise, was replaced by a whole sequence of generals, ending with Navarre (1953–54) who lost the battle of Dien Bien Phu. De Gaulle wanted to continue the war. France had capital in Indochina; we knew the "Hevea plantations", the "Hanoi tramways" and the "Tonkin coal mines" by heart, and their stocks were highly valued. The Indochinese Bank was prosperous. In spite of the paucity of food in Indochina, France actually imported rice from there! This economic set-up was important, but we still wonder today how it could have been important enough to justify the long and abominable war with Indochina. Starting in 1947–48, the Cold War became the underlying influence in the war; it was the fundamental factor in determining the US intervention.

It was not a "normal" war, declared between powers recognized by the UN. France considered it as an insurrection against its legal power, and both sides refused to apply the Geneva conventions. The army corps was made up of rag-tags and bobtails. Probably the majority of them were regular soldiers of Leclerc's army, fighting out of patriotism or simply continuing their professional military activity. There were French soldiers, but also Africans from French colonies, anti-Communist Indochinese, legionaries, former militiamen and even German SS officers come to retrieve their virginity .... Former collaborators and militiamen were taken from prison and sent to Viet-Nam, where massacre and torture were rampant. I found myself confronted with something which obsessed me for the rest of my life: the use of torture as a weapon in a colonial war. Ordinary soldiers tortured, and so did officers, to obtain information or simply for fun. Several French newspapers (but not enough) denounced the practice: above all the *Témoin Chrétien*, *Libération*, *Franc-Tireur*, and the Communist newspapers. Paul Mus, a professor at the Collège de France, a former counselor to the French high commissioner, who had played an important role in the resistance in Indochina in 1944–46, wrote a front-page article in the *Témoin Chrétien*, headlined "No, not that!" He repeated the tranquil chat of a French officer, telling how "We captured a woman .... I hung her naked from the ceiling by her wrists, and we worked on her for three days. She never said a thing. The most amazing thing is that on the third day, she managed to get down and fled into the bushes, in the state she was in ...." Another young officer showed a journalist the skull of a Vietnamese soldier, telling how he had cut off the head of his horribly screaming victim, cleaned it by boiling it for four hours, and now used it on his desk as a paperweight; it was perfectly normal, everyone did it. I still own a sheaf of newspapers of the time, recounting stories of this kind. Later, high-level French officers from the war in Indochina went to Algeria: Godard, Massu, Trinquier and so on. And doubtless the methods of torture used in the Algerian war had been well practiced in Indochina. It is certain that our adversaries also practiced torture, which we formally condemned, but they certainly never did it anywhere near as much as we did. The minister René Pleven wrote, giving precise examples, that we were systematically exaggerating the torturing practices of our adversaries: his courageous article only appeared in a small provincial newspaper (the *Petit Bleu des Côtes-du-Nord* of March 20, 1949) and was not printed by the Parisian press. The number of losses in each of the two camps is eloquent: one hundred and seventy thousand dead in the French expedition to the Far East (of which only one fourth were actually French) as against more than half a million soldiers and more than two hundred and fifty thousand civilians on the Vietnamese side (these numbers were given by Colonel Bonnafous, a historian involved in the war, who believed in colonialism and who wrote a doctoral thesis on Indochinese prisoners which was judged very objective). A famous letter from General Beaufort to the high commander of the Indochinese army (March 11, 1955) recommended the non-publication of the number of French prisoners who died in Viet-Minh prison camps, since in that case it would become necessary to publish also the number of Viet-Minh prisoners dead (or killed) in French camps, whose total was probably more than nine thousand. I was horrified by these bloody revelations, and over a period of more than thirty years of colonial wars, the accumulation of horrors of this kind never ceased to torment me. Most people did not speak of them; Indochina was a long way away. Delsarte and Dieudonné were disgusted by the war, but Delsarte kept repeating that the problem was "insoluble", and Dieudonné thought that it was impossible to bring back the expeditionary corps because of the lack of boats, and that the only possibility was to continue the war. Furthermore, various shady financial deals were making certain doubtful politicians quite rich – I still remember the "piastre" scandal, discussed at length by the newspapers.

I had lectured to groups of intellectuals and workers on the position of the IVth International on the subject of the Indochinese war (in favor of independence), while I was still a Trotskyist. Even when I broke with Trotskyism in 1947, my position didn't change. But this was very little. In Paris, I would have been able to act in concert with others, but Nancy was a small provincial town, and in spite of my unlimited moral involvement, my possibilities for actual activity were very limited. The defeat at Dien Bien Phu (May 17, 1954) which led Pierre Mendès France to agree to a peace treaty, was a relief which I remember to this day. Unfortunately, the terms of the treaty were not respected, and the general elections planned in the South for the year 1956 and neglected by France never took place. As for the Americans, they desired instability in the region, and ended up by bringing the war to the North; the game was not over, but merely postponed.

The Viet-Minh fighters took a considerable number of prisoners, without really having planned or organized it; this happened in particular during the battle of Cao Bang in 1950 and the battle of Dien Bien Phu. Taken to prison camps in horrible conditions, after endless, exhausting forced marches barefoot through the jungle with almost no food, the prisoners suffered horribly. After the last exodus, the losses were more than sixty or seventy percent. The life of the prisoners in the Viet-Minh prison camps was a torture in itself: insufficient food (partly because of general difficulties due to French blocking of supplies), a climate unbearable to Europeans and worse than anything, the plague of all expeditionary forces, tropical illnesses – tuberculosis, beri-beri, parasitic diseases such as amoebiasis and malaria. With no medicine, thousands of prisoners died. Torture also took place in certain camps known as re-education camps. In the Viet-Minh camps of French prisoners, people died of illness, of exhaustion and of executions; the mortality rate was higher than fifty percent. In France, a complete and total silence reigned as to the fate of these French prisoners: they were simply abandoned. The French government no longer wanted to talk about Indochina, and at that time I knew nothing about what was happening (otherwise I would have protested). The thesis of Colonel Bonnafous was the most important document on the subject. The prisoners were made to submit to "re-education" and propaganda by the Vietnamese. Georges Boudarel, a French teacher in Saigon and a profoundly idealistic and anti-colonialist Communist, passed to the side of the Viet-Minh, and found himself promoted to "re-educator" in prisoner camp number 113. Personally, I think he should never have accepted the position, and I always told him so. In 1991, at a public lecture in the Senate, a former prisoner in an officers' camp publicly called him a torturer, and a national campaign of calumny was started against him (which is still going on). He was called an executioner, torturer, assassin, and held responsible for every death, even those from illness. I've read books about the Boudarel affair, and articles in the press; I've heard radio and television programs on the subject, especially at the time when Boudarel sued for slander. I saw from this example how he was made into the "French torturer of the Indochinese war", a pariah never to be mentioned again, a leper – all through a story created by inventions, deformations of reality, omissions, "organized" witnesses whose memories were surely modified after forty years. This reassured me and many of my friends that the horrors he was accused of were merely inventions, and I publicly took a stand against this national slander, one of whose goals was quite simply to burnish the image of the colonial expedition and to discredit the anti-colonialist effort. On this point, I regret that the political left gave in so easily. Such an affair should have been impossible in France. These lies were similar to those which René Pleven had denounced at the time. The affair has calmed down somewhat, but Boudarel was still wronged. He knows Viet-Nam exceptionally well, and he wrote a remarkable book on it, called *Les Cent Fleurs écloses dans la nuit du Viet-Nam* (*The Hundred Flowers Blooming in the Night of Viet-Nam*). He recently played a non-negligible role in the slow but certain democratization of the country.

## Viet-Nam: North and South

The Indochinese war opposing France and the RDVN (the Democratic Republic of Viet-Nam, called Viet-Minh by the French) ended in 1954 with the Geneva conventions. Independently of Laos and Cambodia, which had particular destinies of their own, Viet-Nam was divided into two parts by the seventeenth parallel. The North was left to the Communists who ran the country under President Ho Chi Minh. The South became an autonomous country under the monarchy of Bao-Daï; the president of the Council was Ngo Dinh Diem. But it was planned in the Geneva conventions that general elections should take place in 1956 in both Viet-Nams, with the goal of unifying them under a single president. The agreements were signed by the RDVN, China, Laos, Cambodia, France, England, the URSS and the USA. The USA did make a special declaration, however, which insisted on some restrictions, while agreeing with the general tenor of the treaty.

In the North, a Stalinist style Communist regime settled in solidly. Ho Chi Minh was a traditional Marxist, but strongly influenced by the Chinese. We witnessed an extraordinary exodus of Catholics to the South; later, for a long time, they constituted an anti-Communist force. Successive reforms were imposed from the top, popular participation was reduced to absolute obedience. Agrarian reform on the Chinese model (1953–56) was accomplished in a horrific way. Teams of apparatchiks arrived in the villages, confiscated even the tiniest parcels of land – in any case, there were no grand landowners (the huge majority of people possessed between 0.5 and 2 hectares) and a number of modest landowners were condemned to death by "popular tribunals" where votes were counted by hand-raising. Soon afterwards, a campaign of "rectifying errors" was begun, and the apparatchiks of the first reform were condemned to death. This period has left most Vietnamese, including the leaders, with extremely painful memories. The dissident novelist Duong Thu Huong has written an excellent description of it. Later on, I had several opportunities of talking with the Prime Minister, Pham Van Dong, who told me that he had actually wept over the situation, as he witnessed the multiple executions of small landowners who had played the role of heros and patriots during the struggle against French colonialism. On the other hand, it is true that extremely progressive large-scale reforms were introduced, in particular in health-care and education. Although cases of malaria remained in the provinces, the disease was almost completely eradicated from the townships, through the influence of the famous doctor Pham Ngoc Thach, a man of extraordinary devotion and competence (at the end of his studies in Paris, in 1934, he was given the position of assistant in consumptive diseases at the Faculty of Sciences in Paris). His daughter Colette Thach, whom I also knew well, lived in Paris during the Viet-Nam war, and was included in all our efforts, particularly the Russell Tribunal. In the South, however, malaria remained a large-scale problem. With incredible bad luck, Doctor Thach, who made only a single trip to the South, caught a pernicious form of malaria there and died very quickly. The Vietnamese jungle possesses one of the most unhealthy tropical climates in the world, and the impressive decrease in the occurrence of tropical fevers and the major tropical diseases was no mean accomplishment.

At the beginning of the war with the US, in 1963, the North had become to some extent a modern state, still underdeveloped but already quite well-organized. Famines, which had been frequent during the Japanese and then the Chinese occupations (by Chiang Kai-Shek's troops) and after 1945, disappeared completely. It is true that rice production was only just sufficient, but rice grows very efficiently, and production was not hindered.

Progress in education was also considerable. When the French left, Indochina had one of the world's highest percentages of analphabetism – almost 90%! The French colonialists were filled with contempt for the Indochinese masses and even for the elites. Several Vietnamese friends who spent time in France during the colonial period told me this. In France, these men were treated like free men among other free men, but at home, they immediately encountered the arrogance of the colonialists. This disdain left deep traces in the Vietnamese spirit. For instance, I knew Ta Quang Buu, the future Minister of Higher and Technical Education in independent Viet-Nam. He had been taught the famous "our ancestors the Gauls"[78)] in his Saigon high school. He told me how struck he was by the free atmosphere which reigned in Paris during his studies of higher mathematics. The French students considered him as an equal. But upon his return to Saigon, after his Licence, he immediately encountered the usual contempt of the French and was made to feel that he was less than nothing. He was deeply attached to the "France of France", and contributed a great deal to cultural relations between France and Viet-Nam. His case is an excellent illustration of the ambiguous nature of cultural relations between colonizers and colonies.

The Vietnamese mathematician Le Van Thiem, who studied in France as a foreign student at the ENS (class of 1941) had left France, during the war, in 1942, and gone to study in Switzerland with Rolph Nevanlinna, where he worked on holomorphic functions of one complex variable. After defending his thesis in Paris in 1948 under Georges Valiron (as I did), he returned to Viet-Nam right in the middle of the war there; he reached the country only after an endless trek on foot through the jungles of Thailand and Laos. In the free zone, he founded the High School of Fundamental Sciences, of which he became the director, at the same time as directing the High School of Pedagogy. There were practically no books in these schools; everything had to be built from nothing. Le Van Thiem had brought only one book with him: *Die Methoden der Mathematischen Physik* by Courant and Hilbert, and Ta Quang Buu, the future Minister of Higher and Technical Education, possessed only Bourbaki's

78) The typical history curriculum taught to French children was exported unchanged to the colonies . . ..

*General Topology*! These two schools played fundamental roles in the education of the country's educators, scientists and technicians: thanks to these people, education in the country reached a reasonable level even though it took place in conditions of total isolation. Many of the scientists and educators of Viet-Nam today received their first schooling in those institutions. They also served as a basis for the immediate re-opening of Hanoi University, after the Geneva conventions in 1954, with a completely Vietnamese teaching staff; this was a great exploit. Le Van Thiem was appointed dean of the Faculty of Sciences of Hanoi, and was also the founder and president of the Vietnamese Mathematical Society, the founder and main editor of the mathematical journal *Revue de mathématiques*, as well as of the journal *Acta mathematica vietnamica*, a journal edited in foreign languages and distributed around the world. While always prodigiously active, Le Van Thiem continued doing mathematical research and gave a talk at the International Congress of Mathematicians in Vancouver in 1978. Almost all Vietnamese mathematicians of today were his students or students of his students. He always worked closely with Ta Quang Buu, who founded and was the first director of the Scientific and Technical Institute of Hanoi, the biggest research center, often called the "CNRS of Vietnam". In a word, at the beginning of the 60's, North Viet-Nam was a much more developed state than any other state of south-west Asia, although one of the poorest. The population at the time was about twenty-five million inhabitants in the North: in unified Viet-Nam today there are about seventy-two million people.

The South on the other hand lived under a ferocious reactionary dictatorship. The French installed a puppet king, "His Majesty Bao-Dai", who represented only the rich of Indochina; he was an habitué of spas and vacation centers. In fact, after Dien Bien Phu, France became completely uninterested in Indochina. Nearby Cambodia, a country of 8 million people, was directed by prince Sihanouk who had himself called Samdech Norodom Sihanouk, Samdech meaning something like "Your Grace". This colorful personality was a "personal friend" of the *Canard Enchaîné* newpaper and a grand client of French spas. He loved his country and was doubtless a "patriot", if a rather fantastic, even facetious one. Laos, an underdeveloped and sparsely peopled country with just two million inhabitants, was led by Souvanna Phouma, who represented American interests, but whose brother the prince Souphanouvong was close to the Communists. Prince Souphanouvong directed the Pathet Lao, a group close to the RDVN.

The regime of Ngo Dinh Diem exercised a dark dictatorship. Dissidents were arrested, jailed and tortured. In the prison of Poulo Condor, which was functioning at high capacity, refined tortures were the order of the day. A certain number of Buddhist monks burned themselves to death there – and Madame Ngo Dinh

Nuu, the sister-in-law of Ngo Dinh Diem, actually mentioned “barbecues” at a conference in the USA .... The Americans were directing Ngo Dinh Diem’s regime. American imperialism of the time was incomparably stronger than it is today, virulently conquering on a world scale. The Vietnam War with the US is essentially due to Eisenhower and then to Kennedy. This is not a unique example, in historical terms, of a truly democratic regime which behaves like a brutal imperialist in other countries. In fact, Athens behaved like this in antiquity, and so did France more recently; French behavior in its colonies could not compare to its internal regime. In the States, first Kennedy and then Lyndon Johnson ran their country in a deeply democratic manner, even helping the progress of the cause of Afro-Americans. But it is impossible to deny the imperialist nature of their foreign policy, especially Johnson’s. The same, alas, cannot be said of Richard Nixon, who was not only rapacious outside the country, but whose internal policy evolved into catastrophe as well. American aid to Diem’s dictatorship, beginning in 1956, reached the sum of several million dollars. Under his regime, there were forty thousand political prisoners at the end of 1958 and one hundred and twenty thousand in 1961. All these numbers have been confirmed by the CISC (Commission Internationale de Surveillance et Contrôle)[79)], whose mission was to verify the application of the Geneva conventions, and which was made up of trained representatives from three countries: Canada, Poland and India. The dictatorship naturally provoked a whole resistance movement in the South, which became known as the NLF[80)], and gained strength over time. Diem was finally assassinated by order of the Americans, on November 1, 1963, and replaced by Minh, “fat Minh” as he was called, on October 30 1963, then by Khanh on January 30, 1964, then by Nguyen Cao Ky, and after 1965, by Nguyen Van Thieu, who remained in power until the end of the war (1975). In Nancy, where I had personally and actively participated in the struggle against the war in Indochina, I followed these events with very little means of protestation and a relative sense of impotence. A few petitions, a few meetings, but nothing sensational. At that time, I didn’t know any of the important Vietnamese leaders, and the NLF developed mostly after 1960. The arsenal held by the North and the NLF of the South was quite rudimentary (and partly obtained from remains of French armaments). A group of Vietnamese later told me that they spent four years before figuring out how to use a 75 mm cannon taken from the French.

---

79) CISC: International Commission for Surveillance and Control

80) NLF: the National Liberation Front of South Viet-Nam

## The war

This funereal situation was generated above all by the non-application of the Geneva conventions. In spite of what had been expressly stipulated, no elections were prepared in 1956. So Indochina remained indefinitely divided into two parts by the 17th parallel. The CISC and some of the world powers tried to persuade the country to prepare elections, but the USA arranged for Ngo Dinh Diem to refuse categorically. Under these conditions, war was inevitable. The official beginning of the war was the Gulf of Tonkin "incident", on August 4, 1964. The American fleet had been floating off the Vietnamese coast for some time already. The Americans claimed that a small North-Vietnamese boat had fired a torpedo at a boat from the VIIth fleet. This is perfectly inimaginable; it's difficult to understand why the Vietnamese would have risked such an attack against an extraordinarily powerful fleet. But the Americans took this as a reason to begin bombing Viet-Nam, first from their fleet and then from airplanes based in South Viet-Nam, Thailand and the island of Guam. The bombings became more and more intense each day. Americans sent in more and more troops; first fifty thousand, then one hundred thousand, then two hundred thousand men. During the worst of the war, there were half a million Americans fighting in South Viet-Nam against the NLF. This army perpetrated atrocities and massacre, but at the same time they were desperately wading in the mud and being decimated by the Front.

Truman's doctrine, which was taken up by his successors, was the domino strategy: the Communists were going to seize one country after another and conquer all of South-East Asia; Viet-Nam was just one piece, which would inevitably be followed by Laos, Cambodia and Thailand. The opposition to this mechanism, called *containment*, was also used to counter USSR expansionism in Europe via Communist popular democracies. This was the most intense period of the world-scale cold war, with the dominating figure of Foster Dulles (the USSR was not innocent, either!) The Korean war of 1950 was just one move in this grand geopolitical game. The country had been divided in two along the thirty-eigth parallel during its peace treaty with Japan, with a Communist North which quickly fell under Chinese influence, and a right-wing dictatorship in the South, which did however remain independent. Stalin decided that North Korea should attack South Korea, in a move calculated to trap the Chinese. At that point, I must say, I disagreed with most of my left-wing friends. In spite of the terrible dictatorship of Syngman Rhee in the South, I thought that the UN should intervene under American influence, because the invasion of the South by the North appeared inacceptable and dangerous for the future. The USA, particularly under Johnson, had put forth the idea of escalation. The total ignorance of Western strategists about what was happening in Asia is quite amusing. They really thought that Viet-Nam was a Chinese pawn, and that it was supported by the whole of China's power. And indeed, in 1964 it was certainly to be feared – I feared it myself – that the Viet-Nam war would be followed by an American attack on China. One of the most important American statesman encouraging "escalation" was McNamara. His evolution is rather curious to observe. After having been a hawk during the Viet-Nam war, he became director of the World Bank. While the World Bank was playing its role of smothering the Third World, however, he unexpectedly played the role of moderator. We had cursed McNamara during the war, but afterwards we were tempted to support him. Recently, he wrote a confession – late, but nevertheless sincere – in which he admits having been wrong for years, and takes responsibility for many innocent victims. There were already a fair number of neutralists in the US, particularly the senators Mansfield and Fulbright, but the resistance of the students and the pacifists had not yet begun.

On February 7, 1965, the US began bombing North Viet-Nam, in revenge for an attack of the NLF. The close relations between the North and the South cannot be denied; the North actually furnished limited quantities of logistics, arms and perhaps even troops to the South. All the prisoners captured by the Americans or by the Southern dictators were Southerners, and the very primitive arms they captured were all of Southern provenance. The Tet offensive[81)] was a military disaster which annihilated the Southern troops, and the combat with the North then became decisive.

The Americans and the dictators never succeeded in establishing themselves outside of the large towns. The rest of the country was under the domination of the NLF. The aerial bombings were mostly aimed at northern Viet-Nam, although they did not spare the free zone in the South.

At that time, American imperialism dominated the entire surface of the planet. One by one, the countries of South America were taken over by dictatorships. A dictatorship of torturer-generals seized power in Brazil in 1964 (that day, the headline of the *New York Times* was "A new day rises in Brazil"); they would be excellent candidates for the doubtful prize of having used the most refined methods of torture. American torture schools were opened. After the death of the

---

81) During the Tet offensive of January 31, 1968, VietCong troops launched a sudden and unexpected attack on American and Southern Vietnamese fighters; the resulting slaughter represented a turning point in the opinion of Americans, who came to consider that it would be impossible to win the Viet-Nam war.

Perons, Argentina became a dictatorship just as bad as Brazil's. Chili was taken over by Pinochet in 1973; he was largely supported by the North-American TTC (Telephone and Telegraph Company) against Salvador Allende. The colonel Jacobo Arbenz Guzmàn, who had established a progressive regime in Guatemala, had already been overthrown in 1954 by a coup d'état organized by the US, directed by the American Peurifoy, ambassador to Guatemala City. In 1965, General Wessin Y. Wessin led American military intervention in the Dominican Republic; he bombed Santo-Domingo and quelled the uprising of the population. The Americans replaced the social-democrat Juan Bosch, who had been elected president in a perfectly democratic manner, with the industrialist Balaguer, who is still there. In Asia, the US supported the dictatorship of Chiang Kai-Shek in Taiwan, that of Syngman Rhee in South Korea and that of general Suharto in Indonesia. In 1966, Suharto eliminated the progressist regime of Sukarno which was allied to a pro-Chinese Communist Party, and its five hundred thousand members were pitilessly persecuted and even killed. In the Philippines, General Marcos took power. Johnson constantly complained on radio and television about the number of duties overwhelming him. And indeed, during his presidency, the world became covered with dictatorships. That period remains a sinister one in my memory.

It is of course necessary to mention the Cold War and the division of the entire world in two, when discussing the Viet-Nam war. And yet, the outcome of this war was eventually determined by the peculiarities of the Vietnamese people. The partition of Viet-Nam into two parts was entirely artificial; families were naturally spread over the two halves, and many parents and children were brutally separated for over twenty years. A relatively high level of political tolerance prevailed within the country, since every family contained Communist and anti-Communist members.

## First steps

When I returned from the US, I began my teaching at the Ecole Polytechnique and the university, as well as a seminar at the ENS which was really difficult to prepare: the Cartan-Schwartz seminar on the Atiyah-Singer index theorem. That year, my political involvement remained modest. But political activity arises from many circumstances. In Viet-Nam, the situation became worse and worse. Modestly at first, I set to work. The Communist Party and the Peace Movement directed by André Souquières organized an ambiguous form of support for Viet-Nam, under the slogan "Peace in Viet-Nam". The notion of peace implies that the belligerents are of equal strength. But obviously, this was not the case. It was a war of independence quite analogous to the Algerian war, except not against the former colonizer, France, but against the aggression of American imperialism. Independently of the dictatorship exercised in the South, this aggression took more and more serious forms. Massacres, tortures, assassinations, bombing of civil populations, construction of strategic villages around the towns, in which hundreds of thousands, then millions of Vietnamese were interned in camps surrounded with barbed wire. These camps had nothing to do with concentration camps. They mainly contained peasants, who were regrouped around the towns in order to remove them from the free zone occupied by the NLF, but they were removed from their usual activities and forced to depend on food brought in from the outside. The Vietnamese fighters of the Front were like fish in water – the Americans wanted to suppress the water. The true slogan of the Peace Movement should have been "Victory for the NLF" rather than "Peace in Viet-Nam". But words are not so important.

I had had plenty of time to observe that independent Algeria had not evolved in the direction we had hoped. I was not altogether surprised by this. I also suspected that a unified and liberated Viet-Nam would not necessarily function as a Western-style democracy. Hanoi's Communist regime was typically Stalinist even though Stalin was long dead, and it remained so; Viet-Nam was probably the last of the Stalinist countries. Even in 1976, Marie-Hélène found portraits of Marx, Engels, Lenin, Stalin and Mao Zedong in our hotel room, in the consecrated order. Marie-Hélène protested and informed them that it was very disagreeable to us to see a portrait of Stalin. The ministerial director of cultural activity in Viet-Nam innocently responded: "But Stalin is unanimously considered in the whole world as a great revolutionary." We tried to tell him that he was twenty years late and that in the rest of the world, everyone talked only of Stalin's crimes, of which he appeared to know nothing, but the portrait was removed from our room. The point of my combat was not so much to establish the basis for a future Vietnamese society, which I hoped would be democratic, but to help the Vietnamese people liberate themselves from the pitiless aggression of the most powerful imperialist force in the world, which employed the most barbarian methods. I detested Johnson much more than I had detested Guy Mollet. During the war against the USA, the Vietnamese Communist Party (VCP) represented the Vietnamese masses reasonably well.

During the years 1965–1966, I contacted various unions and the Communist Party, in the hope of undertaking some concerted action. My undertakings were more or less individual. The group which was fighting most actively (and even that was not very active) was the Interunion University Collective for the Fight for Peace in Viet-Nam. This Collective grouped various teachers' unions. I already mentioned Madeleine Rebérioux, who was the main player. She was still a member of the Communist Party, in spite of the Khrushchev Report and the invasion of Hungary, but her doubts were beginning to deepen. During the Algerian war, I had admired the fact that she did not adopt the habits of the party, and worked with the Audin Committee without imposing her ideology on us; in any case she was beginning to have her doubts about this ideology. While sharing our ideas about the necessity of helping the Vietnamese people, she wanted to do it under the direction of the Communist Party, and that is what she did within the Collective. I always considered Madeleine Rebérioux a most remarkable person, and we worked closely together, but I have certain reservations, which I did not keep from her during the struggle against the Viet-Nam war. Her difficult relations with the Communist Party and her eventual rupture with it were very wounding for her. She detached herself only very painfully from the Party. People like Jean-Pierre Vernant left the party of their own accord, whereas it might be said that Madeleine actually waited until she was excluded from it, in 1968. Altogether, I was not particularly pleased by the passive manner adopted by the Collective in the struggle for Viet-Nam.

A new movement soon appeared. Jean Schalit, a dissident member of the Union of Communist Students, came to my home with a most fantastic proposition: he wanted to organize "Six Hours for Viet-Nam" in the Mutuality Hall. The idea of a six-hour meeting was wonderful; I immediately discussed it with the Interunion, the Peace Movement and even the Communist Party. During a meeting between Communist and union teachers and a little group consisting of Alfred Kastler, Pierre Vidal-Naquet and myself, the Communist Party loudly opposed our idea. It ended up by accepting it only on the condition of directing the meeting by itself. I have to admit that I let myself be completely dominated by the Communist Party during that preparatory meeting. Even worse, Schalit, who had had the idea in the first place, was excluded from the organization, and they even tried hard to prevent him from sitting on the tribune. He was not a man to let himself be overridden and installed himself on the tribune with authority in spite of everything. Vietnamese citizens in France played important roles in such meetings. Two particularly important groups were the Union of Vietnamese Students in France, led by Giao (who was a moderate Communist at the time, although he ceased to be one long ago) and the general delegation of the RDVN in France. The only existing embassy was Thieu's South Vietnamese Embassy, as France had not formally recognized North Viet-Nam, which was a paradox since the existence of North Viet-Nam was a result of the Geneva conventions. Nonetheless, the RDVN had a delegation in Paris whose leader was Mai Van Bo. The "Six Hours" also had an artistic aspect: Vietnamese students came and danced dances of their country. The event was unexpectedly successful, and was much spoken of abroad, but it seemed clear that it had no future.

Starting in September 1966, I began discussions with the other organizers of the "Six Hours", in the hopes of organizing another, similar but even bigger meeting. I encountered considerable reticence. One of the last preparatory meetings brought together the group I was working with and the organizers of the previous meeting. Around me were my friends Vidal-Naquet, Kastler, Schalit and maybe a few others. In the Communist group were Jean Chesneaux, a deeply convinced Communist (who evolved later on) and a historian of Viet-Nam who traveled there frequently, and the most hard-line Communist of all, the mathematician Jean-Pierre Kahane. He is really an excellent mathematician with a powerful mind, and I've known him for a long time. His scientific judgments are always absolutely objective, and his activities in various university organizations is very efficient. We've had many interesting discussions, but I can never forget that he is not just a Communist but a true "follower". He still is one today, he has been a member of the central committee of the French Communist Party for years, with all that that implies. We appreciate our mutual frankness, but it is clear that we cannot agree! I never understood how he could keep on and on approving the Communist Party politics, which has undergone so many radical changes over time. At any rate, our heated discussion continued till nearly midnight, at which point I was really exasperated and finally said: "It's useless to attempt to come to any agreement, because we won't succeed, so we will just go ahead and organize a meeting by ourselves." And off we went to undertake this new adventure. I felt as though I was taking quite a risk for Viet-Nam.

## The National Viet-Nam Committee

Quite soon, the group working on the new meeting shrank to exactly five members. Fortunately, some of them were very prestigious, which was necessary in order for them to be able to preserve their total independence. Bartoni belonged to the Christian movement; he was a professor of union rights at the university, which gave a certain aura of originality to his otherwise very impartial contacts with the leaders of all unions. The others were Alfred Kastler, Jean-Paul Sartre, Pierre Vidal-Naquet and myself. That day, we undertook a double initiative: the creation of a national committee called CVN (Comité Viet-Nam National)[82)]

82) CVN: National Viet-Nam Committee

and the organization of an extension of the "Six Hours" by a "Six Hours of the World for Viet-Nam", which we planned to hold at the Mutuality Hall in early October. We blindly flung ourselves into the vacuum, but not without some feelings of apprehension. How could we, five people helped by a few friends, organize "Six Hours of the World for Viet-Nam" in the Mutuality, without any support and even with the opposition of the Communist Party? Jean-Pierre Kahane did not restrain himself from writing a protestation against our initiative in *Le Monde*; he called us a little group of dissidents who stole the title "Six Hours" in order to work for our own advantage. Apparently, his article did not cause much reaction. The new meeting was also declared in the columns of the same journal – by a declaration of the Five!

The rupture between the Communist Party and the intellectuals, which broadened considerably during the Algerian war, widened even further during the Viet-Nam war and became complete in May 1968. The Communist Party was reduced to a marginal party. The five of us began to distribute propaganda in our respective circles, and rapidly received expressions of interest and even support. Of course, we substituted "The NLF will be victorious" for the Philistine slogan "Peace in Viet-Nam". I asked for a meeting with Mai Van Bo, the delegate of the RDVN in France, and he received me warmly. He was a man of sharp intelligence and a subtle diplomat. He was worried about the separation between us and the Communist Party. He met with us, so he was able to form a clear estimation of the personalities of the leaders of the new movement, but it still appeared dangerous to him to go ahead without the Communist Party. His reaction was not sectarian, it was the reaction of a man preoccupied by the problems of his country. "I regret that things are turning out this way," he told me, "but since it is decided, we must make sure that the new Six Hours meeting is a success." Thus we obtained the support of the Vietnamese.

In the end, the meeting in the Mutuality was an unbelievable success. The hall was completely full, and it became clear that the majority of French anticolonialist militants no longer were interested in what the Communist Party had to say. The preceding "Six Hours" meeting had displayed only a semblance of unity. If it had been possible, I would have preferred to collaborate with the Communists, but the rupture was inevitable.

After this reinvigorating adventure, several new movements were initiated. One group put forth the idea of subscribing a billion francs for Viet-Nam, and they succeeded brilliantly. Many provincial meetings analogous to the Paris meeting were organized. Communication between French and American students and teachers increased, because numerous Americans and people from all over the world had participated in our "Six Hours of the World". The most moving moment was when Steve Smale, the American mathematician from Berkeley, shook hands with Mai Van Bo, in spite of the war between their two countries. This was one of the things which made the Viet-Nam war very different from the Algerian war: the North Vietnamese government was fully internationalist in the traditionalist Marxist sense of the word, and they kept in contact with Americans who protested against the war, which reinforced the concept of internationalism. University students in the USA had started to protest against the American aggression much earlier than in France. The University of California in Berkeley was the spearhead of the protest, and Steve Smale was one of the leaders of the Peace Committee in Berkeley. Smale was one of the world's most reputed mathematicians; he had received the Fields Medal at the International Congress of Mathematicians of Moscow in 1966.

After our almost unexpected success, the CVN became extremely active. It had thousands of members. Protests against the American army became very frequent in France, in particular in the neighborhood (though not too near) of the American Embassy. The police dispersed the demonstrations, but not with the same violence as during the Algerian war; they simply gave a sharp order to disperse which was rapidly obeyed. As the elected president of the bureau of the CVN, I usually had the role of negotiating with the police; it wasn't always easy. The CVN took a very hard-line political stance with respect to Americans. One of the most important demonstrations took place during the first bombing of Hanoi. I learned of it from a phone call from Mai Van Bo, one weekend at eleven thirty in the evening: "I need your help. I've tried telephoning the Communist Party office to ask if they could organize a demonstration against the bombings, but in vain; I didn't reach anybody. So I'm calling you: can you organize a demonstration of the CVN as quickly as possible?" The bridges linking us to the Vietnamese authorities were firmly established. After that, every Vietnamese official visiting France wanted to meet with us.

When Kissinger and Nixon ordered the bombing of Hanoi by B 52's, at the end of December 1972, we organized an enormous demonstration. Another more "solemn" demonstration was ordered for the day of Nixon's second inauguration, on January 20, 1973. International tension was extremely high during the bombings. However, they did not last long and were quickly followed by peace agreements between Viet-Nam and the United States. The Thieu government remained in the South, but the NLF obtained a legally recognized existence. The deputies of the third force were also recognized, but large numbers of them were imprisoned by Thieu because they didn't have the power to resist him, whereas the NLF represented considerable power in the South. The carpet bombings of Hanoi by the B 52's were particularly barbarous. I remember that the American singer Joan Baez, who militated for years for Viet-Nam, happened to be there during the bombings, and that like everyone else, she had to bury herself in one of the holes dug all over the city and covered with metal covers. The fact that the bombings did not last long is symptomatic in many ways. For one thing, they were murderous for the civilian population; the Vietnamese government could not have resisted long under the American firepower. It was absolutely necessary to immediately conclude even a precarious peace. As for the Americans, they could not keep the war up much longer either, and not only because of international disapproval. Indeed, until then, no B 52's had been shot down by the Vietnamese anti-aircraft rockets because they emitted waves which disturbed the radar frequencies of the rockets. But an unexpected event occurred during the Hanoi attack: the B 52's found themselves unable to disturb the Vietnamese rocket radar frequencies and in just a few days, thirty-four giant airplanes were shot down. The Americans possessed only two or three hundred of these airplanes in all. Thus each of the belligerents had excellent and urgent reasons to sign a peace treaty as soon as possible. How did the Vietnamese succeed in changing their rocket radar? The secret was revealed to us by the representatives of the Vietnamese delegation in France. The anti-aircraft rockets had been delivered to Viet-Nam by the USSR, which did not know how to effect rapid changes in the frequencies. But the Vietnamese engineers were very good with machines and they studied the radar themselves and modified their frequencies. They laughed heartily about the whole story.

Along with the CVN, another important group was founded; it was called the CVB (Comités Viet-Nam de Base)[83], and its tendency was Maoist. The cruelties of the Viet-Nam war had augmented the mass of people revolted by the United States, the Maoists among them. They had even seized the direction of the UNEF. In the story of the Algerian war I recounted the degeneration of the UNEF and how Gaudez had transformed the group into several UNEFs all run by different people and no longer representing the mass of students. The leftist UNEF was at that moment led by hard-line Maoists. There was a certain preponderance of Trotskyists in the CVN (for instance the JCR (Jeunesse Communiste Révolutionnaire)[84], led by Alain Krivine. The tension between the CVN and CVB was sometimes very high, but personally I was always in favor of united action and sometimes we succeeded in overcoming the obstacles and acting together. The CVN had a central organization with a central committee

83) CVB: Basic Viet-Nam Committees

84) JCR: Revolutionary Communist Youth

and a political bureau of which I was the president, whereas the CVB was theoretically founded on a basic democracy and consisted of a large number of entirely independent committees. They were active during the Viet-Nam war, and they had good relations with the Vietnamese delegation in Paris where Mai Van Bo had been replaced by Vo Van Sung, a warm and gentle man who had no reason to choose between the various organizations defending Viet-Nam.

One day, towards Christmas of 1966 I believe, I received a telegram from Ho Chi Minh in my apartment in the rue Pierre Nicole, thanking us and congratulating the CVN for our energetic defense of Viet-Nam in its struggle. At first, I did not attach any particular importance to it and kept it to myself; I just mentioned it to Alain Krivine on the telephone while discussing something else. He was amazed: "You're crazy, it's very important, you have to publish it immediately in our bulletin of the CVN!" And indeed, the telegram represented official recognition of the role of the CVN in the struggle against the war. For everything concerning logistics, official relations, administrative and financial problems, the French Communist Party was and remained the unconditional ally of Viet-Nam. No unexpected harm could come from it. But for political and moral support for Viet-Nam, the only true ally was the CVN, even knowing that sometimes the CVN took unexpected and complicated initiatives. Shortly after, we invited a Cuban representative close to Castro, Melba Hernandez, to a meeting of the CVN; she had distinguished herself during the war against Batista by attacking and taking the Moncada barracks. She read a message from Castro filled with greatness and even grandiloquence, carrying his full signature: Fidel Castro Ruiz. A new international consecration.

There were some conflicts within the committee. The average tendency was rather violent and Trotskyist, but there was an even more extreme Sartrian tendency. Sartre even went as far as asserting in front of the whole Mutuality: "Those who are here protesting against the Viet-Nam war uniquely for humanitarian reasons, because children are being killed, have no place among us." I violently protested against his idea. The role of morality in politics has inspired my entire life. Sartre rejected this for a long time, shamelessly proclaiming "Science – arseskin, moral – arsehole." He was hostile to the introduction of a humanitarian dimension in politics. Because of this, the truly political causes he defended were sometimes just, for instance in the case of the Algerian war, the Viet-Nam war, Israel and Palestine, but sometimes completely false, for instance when he joined up with Stalinism and Maoism. However, towards the end of his life, his position evolved under various influences. For one thing, he became passionately interested in the book he was writing about Flaubert and at that point modified some of his ideas on purely intellectual work; before, he considered that attaching any value to a literary work was a sign of a bourgeois or petit-bourgeois mind. On the other hand, the trial of Bruay-en-Artois also led him to adopt some new positions. A notary was accused of having assassinated a young girl according to a logic of public opinion analogous to the logic of the class struggle: the accused was the notary public, the most reactionary citizen of the village. Sartre protested against this amalgamation, bringing the debate back to the simple question of truth and proof. At that time he went through a sort of crisis, and a sense of morality finally came to play a role in his combats. During his crisis, he became isolated from the group of his usual friends and admirers, including Simone de Beauvoir who remained loyal to Sartre's former maxim on science and moral. Both of them were deeply affected by his crisis. I learned about Sartre's psychological evolution near the end of his life, which brought our positions closer together, from Sartre's adopted daughter Arlette el Kaïm. She asked me to write him a letter in which I confirmed this new proximity, which I did willingly. Sartre continued to militate indefatigably against war.

Alfred Kastler expressed a moderate tendency to the right of the CVN when he stated, at the grand meeting: "If we want peace in Viet-Nam, the Vietnamese will have to make a certain number of concessions." He was booed by the over-excited public. I didn't agree with him, because the Vietnamese were fighting for their independence; North Viet-Nam was a savagely bombed country and I did not see what possible concession there was for them to make. They could not accept to abandon the NLF of South Viet-Nam, which was under terrible repression by a ferocious dictatorship. The conflict worsened during the election of the first central committee of the CVN. Although I didn't really agree with his point of view, I energetically protested the fact that Kastler was not elected. He was an eminent scholar and well-known for his morality and his militantism; we needed his help, and the central committee was not to be monolithic. He was also supported by Pablo Raptis, representative of the "Pablist" tendency of Trotskyism, who always attached importance to intellect and to the views of intellectuals. The predominance of Trotskyists within the CVN created some difficult situations. Personally, I never believed that any party should take over an organization. However, I also found that the PSU was too weakly involved in the CVN. I was still a member of the PSU and I would have wanted to see it playing its part in the orchestra. I tried to meet with the main leaders. Our discussion made it clear that they were afraid to become too deeply involved because of the Trotskyist domination, and I tried to convince them that Trotskyists were dominating precisely because the PSU was not sufficiently present. Anyway, all these questions were important only among a small minority; the mass of our members were students, teachers and intellectuals who did not belong to a particular political movement and if anything, sympathized more with the PSU. It is true that the most active members, who were usually members of political parties, all belonged to the central committee. My discussion with the PSU did not really give any result. There was yet another movement which became more and more important, and which without being altogether Maoist was still to the left of Trotskyism. The CVN was becoming more and more extremist.

## The mystique of violence

The Viet-Nam war and the escalation of the United States provoked such repugnance in young people that they became attracted to every form of violent struggle. I was very afraid of this tendency and did not stint my efforts to neutralize this "at any price" attitude. I was reproached for this by some members of the central committee. According to them, simple demonstrations taking place at some distance from the American Embassy were just not interesting, not important, inefficient and sorry-looking. I knew it was true but I didn't see how to go farther. After all, we weren't about to set bombs in front of the American Embassy! This tendency towards violence expressed a feeling of powerlessness faced with a horribly brutal war. In a way, I lived May 1968 several months before May 1968. Some young people said to me: "Schwartz, the horrible situation created by this war and its frightful consequences in the world make violence necessary. Armed struggle must develop, now." This movement was strongly influenced by the arrival in Paris of Stokely Carmichael, who was famous in the USA for his struggle for the rights of blacks (he ended up allowing himself to heed the siren call of capitalism). Carmichael declared that American blacks were brothers in arms of the Vietnamese fighters. He was not hostile to the idea of armed struggle against white power. Most of the groups of young people struggling against the Viet-Nam war were tainted by this mystique of violence. In Germany, the movement gave rise to the Red Army Faction, founded by Andras Baader, Gudrun Ennslin, Ulrike Meinhof ... "Baader's band", which turned towards violence. In order to attack American interests, they attacked American installations in Germany, or people representing the development of military power in Germany. They were responsible for many acts of terrorism. They were almost all arrested and imprisoned, in inhuman conditions of solitary confinement. Supposedly all of the prisoners committed suicide after the failure of the Mogadishu episode in Somalia: air pirates seized a German airplane in the Mogadishu airport and demanded the liberation of "Baader's band", but the anti-terrorist German group called to the scene of action succeeded in freeing the airplane by an extraordinarily sophisticated maneuver, ending all hopes for liberating the prisoners by such methods. That was the end of the Red Army Faction in Germany. In fact, the members communicated with each other even while in solitary confinement, in particular through their lawyer Heinrich Klaus, who must have transmitted the arms they used to kill themselves. Young people everywhere in the world sympathized with them; the son of one of my friends said to his parents on the day of the suicide, which some people suspected of being an assassination: "If they were killed, *we* will never forgive *you*." An extreme left-wing terrorist movement sprang up in Italy, the Red Brigades. There is a Red Army Faction in Japan. Japan collaborated with America against Viet-Nam. An extremely violent Marxist-Leninist sect, which did not hesitate to use torture on its dissident members, sprang up in Japan. The American extreme left behaved similarly: the so-called *weathermen* slid towards a dangerous insanity which recalls the insanity of today's fundamentalist Islam. They took their name from a song of Bob Dylan: *You don't need the weathermen, you tell from where the wind blows.* It was the armed branch of a political association, the SDS, Students for a Democratic Society. One of their groups actually decided to proceed to the slaughter of white babies in order to protect blacks. One Sunday, they filled a car with explosives, meaning to park it in front of a church. The car exploded too early, with the driver still in it, provoking the self-criticism of the group. One day at a sidewalk café in Paris, near the Pantheon, a brilliant American mathematician told me that he found that blowing up electric pylons was perfectly justified (although he never did it himself). The son of a Japanese mathematician tried to convince me that one should blow up university computer systems. The only common feature of all these movements was a paroxystic style in the struggle against the Viet-Nam war, an acceptance of individual terrorism, unconditional support for Palestinian terrorists and a generally more or less concealed anti-Semitism, called "anti-Zionism". Already at that time, many international terrorists learned their trade in Palestinian camps in various Arab countries. International terrorism, which strikes everywhere in the world today, was born exactly at that time, before May 1968. Traditionally, France is not a violent country. Quite the opposite; in the face of the Nazi threat, alas, we saw only too much pacifism and collaborationism. Nevertheless, I was horrified by a certain verbal violence which began to be heard within the CVN; I no longer recognized my own child. Benni Levy shared my experience; he was one of the founders of Maoism in France, and he also watched the movement he founded slide towards a state of exacerbated violence. The Cause of the People took revenge for the murder of the worker Pierre Overney at the Renault factory by kidnapping a Renault engineer. After that, Benni Levy took it on himself to dissolve the organization which he had created. A French Maoist movement still exists, a prisoner of some of the most scandalous opinions of Mao and Deng Xiaoping, such as the defense of Pol Pot and of the massacres in Bangladesh perpetrated by western Pakistan.

The majority of the members of the CVN were pacific and level-headed people. We were all furious and passionate, but only a minority appeared ready to engage in extreme solutions. Some months later, the movement of May 1968 was born. May 1968 was first and foremost an outburst of joy, even of joyous insanity. Some violent tendencies existed, although they were not the most striking ones. In the heat of their desire to change society, the sixty-eighters had conceived the destruction of certain social fortresses, first and foremost the University, but also technological and industrial production. Some groups in 1968 really tried to destroy the computers of the scientific universities, which were judged to be responsible for the consumer society and for exploiting the workers. Sometimes the Trotskyists had to be called in to protect the computers. But the radical current, the same one which I had seen develop within the CVN some months earlier, remained a minority. The young people who were attracted by "armed struggle" consulted Jean-Paul Sartre. He received them warmly and assured them that he understood them very well. I believe he moderated them more than he encouraged them. In his inimitable manner, Sartre was often virulent with words, but rarely, if ever, in action.

Activity in favor of Viet-Nam slowed down considerably after May 1968. The leftists oriented their combat around the liberation of the Third World, partly because there was no absolutely pressing political cause within the country. The events of May 1968 gave them the means to express their aspiration to overthrow society. Anti-imperialist activity slowed down. From the time when Laos and Cambodia were invaded or bombed by the USA, the journal of the Viet-Nam committee changed its name to *Front Solidarité Indochine* (*Indochina Solidarity Front*), whose slogan was "Viet-Nam, Laos, Cambodia – Indochina will be victorious!" But after the supposed peace treaty of 1973 between the USA and Viet-Nam, for which the Nobel Peace Prize was attributed jointly to Henry Kissinger and Le Duc Tho (neither of whom deserved it in the least, but it must be said that Le Duc Tho refused it), it stopped. Without officially being dissolved, the movement slowly disappeared of itself, and ceased to exist when the North definitively won the war, by taking Saigon on April 30, 1975.

## The Russell Tribunal

The Russell tribunal was created about the same time as the national Viet-Nam committee. Lord Bertrand Russell, a British citizen, was famous both as a mathematician and as a philosopher. In 1910–1913, together with Alfred N. Whitehead, he had published *Principia Mathematica*, in which they showed that "all of mathematics" could be written without a single word of any language, but simply by a succession of mathematical signs and symbols such as $\forall$, $\exists$, $=$, $\Rightarrow$, the minus sign and letters, following certain precise rules. In this way, Russell founded axiomatic theory in mathematics. His axiomatics are not used today, but that is not important; the principles were his. In every serious logic, there is an "empty set" $\emptyset$, which contains no element. This is written $\exists\emptyset, \forall x : x \notin \emptyset$, i.e. "There exists $\emptyset$ such that for all $x$, $x$ is not in $\emptyset$." One could imagine a "full set" containing all mathematical objects as elements; it would be written $\exists\Omega, \forall x : x \in \Omega$, or "there exists $\Omega$ such that for all $x$, $x$ is an element of $\Omega$". The impossibility of the existence of such an $\Omega$ is proved in a theorem of Russell.

He was given the Nobel Prize in literature in 1950. He was also famous and unanimously respected in Great-Britain for his untiring combat in favor of peace and liberty, for his humanitarianism and his political morale. He was imprisoned in 1918 for rejecting the First World War. Together with Albert Einstein, he founded the Pugwash movement, which worked throughout the decades of the Cold War to gather international scientists, particularly Russians and Americans, for discussions on nuclear arms. This movement received the Nobel Peace prize in 1995, in the person of its president Joseph Rotblat. It is said that Russell once tried to put up a poster on 10 Downing Street, but he wasn't tall enough. A policeman of free England came up and offered kindly to put the poster up for him, but added: "I must tell you, Professor, that it is illegal, and that I will be obliged to take this poster down immediately." Russell created Bertrand Russell's Foundation to help his work, and supported it with his own money. It was run by several militants, most of whom were Trotskyists.

The idea of a tribunal was put forth by Bertrand Russell himself. It was received enthusiastically by the Vietnamese and by Ho Chi Minh. There was already a Commission in Viet-Nam, presided by Pham Van Bach, which listed American crimes and had collected a huge number of documents. Ho Chi Minh proposed that the Tribunal send a mission to Viet-Nam to take the entire set of documents from the Vietnamese Commission. The Tribunal would judge. Russell and Schönmann quickly protested: that would make it equivalent to a "mock tribunal". To be truly impartial, the Tribunal would have to run its investigations by itself, and send its own agents to Viet-Nam to interview people and verify the documents. In particular, he insisted that in spite of the difficulties created by the war, the agents of the Tribunal should enjoy complete freedom of movement. Very wisely, Ho Chi Minh acquiesced to all these demands. The Tribunal consisted of: Bertrand Russell, Honorary President, Jean-Paul Sartre, Executive

President, Vladimir Dedijer, President of the Sessions, and then in alphabetical order, Günther Anders (a German writer), Mehmet Ali Aybar (a professor of law at the University of Istanbul and a member of the Turkish parliament), Lelio Basso (a professor of sociology at the University of Rome and a Socialist member of the Italian parliament), Simone de Beauvoir, Lázaro Cárdenas (son of the president of Mexico, who did not attend the sessions of the tribunal), Lawrence Daly (a British unionist), Dave Dellinger (a famous American pacifist), Isaac Deutscher (an English historian, author of a very interesting life of Trotsky), Melba Hernandez, Mahmud Ali Kasuri (a lawyer at the Pakistanese Supreme Court), Kinju Morikawa (a Japanese jurist), Carl Oglesby (an American writer), Soichi Sakata (a professor of physics at the University of Nagoya), Laurent Schwartz, Peter Weiss (a Swedish writer). The German writer Abendroth was replaced by Sara Lidmann, and Stokely Carmichael by Courtland Cox.

I conscientiously took up my full duties before the tribunal. It held two sessions each of which lasted a week, and the members learned to know each other well and became friends. I became particularly close to two of the members: Vladimir Dedijer was a Yugoslav historian; he was strongly Stalinist and had been a personal friend of Tito and spent the whole war alongside him. He told us a thousand episodes from that period of his life. He became a dissident in 1950 and was harassed in all kinds of ways. Finally, he was forced into exile, but he preserved a great admiration for Tito, although he was angry with him for many reasons. His son had also been persecuted, and finally committed suicide. Vladimir was a diabetic with fragile health, but he was vivacious and always ready to talk about his rich past and his future full of projects. His wife was even more fragile than he was, and never really recovered from the death of their son. One evening, he went out with friends to relax after a difficult session, and his wife remained at the hotel, too tired to accompany us. He mentioned this to me and asked me to dine with her. I talked a great deal to distract her, but I could feel that she was penetrated by infinite sadness. We often saw Dedijer again, in many different places, and he was always the same. He died several years ago. Lelio Basso played an important political role in Italy. He was an independent leftist Socialist, very close to the Italian Communist Party, which unfortunately led him to criticize the foundation of a united Europe (which seems anachronistic today). At the same time, I don't know why, he had close relations with a great many Italian bishops and archbishops, so that all in all he was connected with many different political movements and enjoyed tremendous prestige, while always displaying moderation and tolerance. He had a broad knowledge of law and politics, and a prodigious library. After the sessions of the Tribunal, he created a Lelio Basso Foundation which was essentially a documentation center.

Other tribunals came together in the following years to judge various important issues, and many of them were presided by Lelio Basso. Everyone liked his extremely dynamic and kind secretary Linda Bimbi.

The Tribunal was supported by dozens of other people who participated directly in its work. I wouldn't be able to list all of their names. Jurists and doctors helped us enormously, writing up reports. One of the most important ones was Leo Matarasso, who wrote up a report on genocide and continued to work for various tribunals after Viet-Nam. He went through a period of depression around May 1968: he used to say that he could associate neither with the old fogeys nor with the young beatniks. I felt rather close to this sentiment myself, and to the person who expressed it. I also want to mention Yves Jouffa and Gisèle Halimi, whom I've already mentioned in the chapter on Algeria; she traveled as a witness to Viet-Nam. Joë Nordmann was an orthodox Communist at the time (he has changed since), and this created a problem with the Russell Foundation. Mrs. Russell sent Ken Coates to protest against his speech at the tribunal. I had to plead Nordmann's defense, arguing that Nordmann's objective role would surely conform to the objective tendency of the Tribunal, and that we could hardly exercise political censorship. It is true, however, that the role of the Communists was somewhat contradictory. They boycotted the CVN and kept their distance even when they joined the Russell Tribunal.

Many doctors examined the medical consequences of wounds caused by anti-personnel, phosphorus and napalm on the health of the Vietnamese people; the most active among them was Marcel Francis Kahn, a high-level rheumatologist. Our mutual close relations never ceased; he was and still is a Trotskyist. I must also mention Alexandre Minkowski, who acquired an international reputation for his treatment of premature babies. He came before the Tribunal to report on the consequences on children of the chemical products and exfoliants poured out by American airplanes. I often run across him in the street, he always stops on the sidewalk of the boulevard de Port-Royal to ask me: "How can we get the French to understand that they're idiots?" I'm afraid he never succeeded. He worked at promoting good neonatal services in many Third World countries, where he is a much-loved figure, particularly in Cambodia and Bangladesh. Jean-Michel Krivine, another close friend, also played an important medical role in the Tribunal.

Several historians took part in the sessions. Let me cite Jean Chesneaux and Gabriel Kolko. Kolko's lectures were absolutely remarkable. He was a professor of history at the University of Pennsylvania, and I learned a great deal from him. We saw each other many times afterwards, as he spent long periods in France. I should cite other collaborators: Madeleine Riffaud, a very active and agreeable Communist journalist, Denis Berger, Jean Bertolino, a free-lance journalist, etc.

We sent twenty-six witnesses to Viet-Nam. All of them obtained a visa without difficulty. They traveled all over North Viet-Nam, in spite of the difficulties of the war. They saw the churches, the hospitals, the schools, the dikes and everything which had been destroyed by the bombings. It is important to recall that the American troops had fought uniquely in the South, attempting to diminish the free zone. They never penetrated into the North, but they bombed it massively. The witness verified the minutious precision of the notes taken by the Vietnamese commission on American crimes, and they made checks and cross-references. The Vietnamese Commission indicated the dates of the bombings and the numbers of victims with precision. Our witnesses interrogated the inhabitants of bombed villages; of course they needed a Vietnamese interpreter, but that did not make it possible to cheat. And in any case, many of them were Vietnamese or knew the language. Roger Pic, a photographer, and the Dutch film director Joris Ivens took many photos and made some films. The accuracy of the Vietnamese documents was beyond discussion. I made the trip myself in 1968 and visited many of the bombed sites. The Tribunal finally concluded, objectively and impartially, that there had been massive, deliberate and systematic bombings of civilian targets. The leper's colony of Quinh Lap had been bombed. The immense colony was next to a small town of about two thousand inhabitants, where former lepers went to live when they were cured. The Americans certainly knew the purpose of the buildings, yet they were bombed. Lepers are naturally particularly vulnerable in difficult circumstances, and generally cannot be transported. Many churches were bombed, always on Sunday mornings during mass, or in the evening at vespers – times deliberately chosen to make sure the churches were filled with the faithful. Buddhist pagodas were not spared either. The incessant conflicts between Vietnamese Catholics and the Communist government ended with the massive exodus of the Northern Catholics to the South. The Vietnamese Catholics I spoke with were particularly shocked by the bombings of the churches.

The Americans also bombed dikes and dams, causing floods and destroying the irrigation system of the rice paddies, which was primitive but efficient. Except during a particularly dry season, rice paddies are inundated because rice needs water in the early stages of its growth, and must be replanted in water. The sight of innumerable Vietnamese women, bent over for entire days planting the rice in the water, became a familiar one to me during my visit. Fortunately, unlike in many other countries on the African, Asian and American continents, bilharziosis does not exist in Viet-Nam, so that there is no inconvenience for the women to work with their feet in the water. Millions of Egyptian agricultural workers near tributaries of the Nile suffer from this terrible illness. However, malaria is much more common in Viet-Nam than it is in South America or Africa. Fortunately, the methods of Dr. Thach and the sanitary methods of the government of the North had practically eliminated it except in deep forest regions. Each rice paddy is surrounded by a protective dike, simply built out of mud and earth, less than a meter high, and reinforced with stones. If the water ran out, even in the plains of Viet-Nam (around Hanoi), it would travel to lower regions and there would be no irrigation in even slightly higher regions. The water level must be maintained in spite of the irregular rainfall, and every day the water levels in neighboring rice paddies have to be equalized. Big poles are placed along the dikes every two or three meters, and a pail is swung along from one rice paddy to the next. Two women spend part of the day equalizing the water level: they toss the pail, suspended from ropes, back and forth to each other with a harmonious ancestral gesture. The parts of Viet-Nam located in the deltas have actually replaced this primitive manual method with a hydraulic system. The rice paddies are full of minuscule fish. Rice is the staple food of the Vietnamese, who eat about a pound per day per person. Its price is extraordinarily low in the towns, where everybody manages to eat enough, more or less, even on very low salaries. All the villages I visited had at least one or two little restaurants serving only rice soup (*com pho*) with maybe a few little fish in it. One also needs protein, after all, and part of the protein is furnished by the little fauna of the rice paddies; they also eat a little pork and sometimes buffalo. I ate buffalo several times. The little children of the village carry big nets and go fishing, wading in the rice paddies, carefully so as to catch the fish without destroying the rice shoots. Some much bigger nets are used by groups of peasants. I've talked a lot about the work of women and children; it's because the men were mostly away fighting at the front. The Vietnamese plains around Hanoi form an immense landscape of rice paddies, full of people busy with their activities of replanting, irrigating and fishing. A significant surplus is obtained by the importation of Chinese rice. The bombings did not succeed in starving Viet-Nam!

From one end of the country to the other, almost every wood and stone house in every village had been destroyed by the bombings. The schools and hospitals were relocated in bamboo huts, identical to the residential huts, so that the airplanes could no longer distinguish between a house and a school or a hospital. Most of the villages had no electricity. The doctors and surgeons used little bicycle lamps for light during operations. Bicycles represent the universal form of transportation; there are almost no cars on the country roads or in the towns. The bicycles are sometimes ridden by several Vietnamese together. It's one of the gay and attractive aspects of the town of Hanoi: the streets are filled with many-headed, two-wheeled creatures. I often saw a car having the greatest difficulty circulating amongst the bicycles. Every time a car appeared in a village, groups of children greeted it with laughter and applause, calling out "Soviet uncles!"

The bombings of the villages were observed by the Tribunal. The bombings of dikes corresponded to the information contained in the Vietnamese documents. Dikes, schools, hospitals and industries were the main targets of the Americans. At the end of the war, there were almost no "solid" buildings in the countryside. Fortunately, fearing excessive international disapproval, the Americans did not take the risk of bombing the towns, except near the end. I saw a very well disguised factory in Haiphong, whose chimney was built almost horizontally so as to send the smoke out as far as possible from the building, which remained almost invisible.

The Tribunal proved clearly that the American objective was to kill as much of the civilian population as possible. They made massive use of anti-personnel bombs. An anti-personnel bomb is made of a large bomb which, when it explodes, frees a multitude of tiny ball bearings, some of them equipped with a metallic point. Their impact on a concrete wall is basically inoffensive. But on a human being, they cause tremendous harm. The Tribunal medically examined many people of every age wounded by these bombs. The ball bearings lodge inside the body, and are practically impossible to extract except by radioscopy, but the Vietnamese do not possess millions of radioscopy machines. Later, the metallic ball bearings were replaced by plastic ball bearings which cannot even be located by radioscopy. The ball bearings remain in the body of the victim body for the rest of his life, causing him to suffer from internal hemorrhages. The Tribunal examined many wounded children. The French physicist Jean-Pierre Vigier was our "military expert" for examining the various armies.

Village life went on more or less as usual as long as the village was not bombed. Whenever the airplanes came, the peasants rushed to hide in shelters, all crowded together with their livestock. Each school had its shelter nearby, underground, which teachers and students could reach very quickly. Even the organization of the baccalauréat took the risks into account. In the villages and groups of villages, the centrally prepared examinations were brought by bicycle on the day of the test. In case bombing prevented one of them from arriving, a second bicyclist left soon after the first one, then a third. The livestock was generally looked after by children; even very small children lay on the back of a buffalo, all ready to guide it to the nearest shelter if necessary. Buffalos endure the climate perfectly well, but they do not possess the strength of an ox. However, they are extremely docile. The bombs spread random destruction, sometimes setting huts on fire, but it was rare that a village was completely destroyed. As soon as the airplanes left, the huts were reconstructed. Finally, only huts made of straw and dried earth remained, some of which were used collectively, although they appeared identical to the others. Similarly, the bridges made of tree trunks were rapidly reconstructed. It seems clear that extremely barbarous methods were used expressly to maximize the sufferings of the Vietnamese people of the North and the South, with a determination to destroy which was incomparable to anything employed in the Algerian war. The American General Curtis Le May declared: "We'll send Viet-Nam back to the Stone Age." The population hung on to its survival, with continual attention to protection and reconstruction. I remember a picture postcard which was distributed everywhere in the world, showing two smiling little girls on their way to the village school, wearing huge straw hats covered with leaves as camouflage.

Several of our witnesses traveled beyond the seventeenth parallel, to the liberated regions of the South. They were able to judge the terrible pressure that the American air war exercised on these zones. Marcel-Francis Kahn reported that he had had to plunge to the neck into a watery marsh, his head covered in leaves, in order to escape the surveillance of a helicopter, and that the least movement of a pig or a few hens in a village gave rise to machine-gun fire. There were Oradours[85)] in the South. The best known was the one in My Lai, reported by the international press. We watched an unbearable film about torture victims. We were overwhelmed by the declarations of a young American of about twenty-five, Peter Martinsen. He had been sent to Viet-Nam very young, and had no particular prejudices. But he was demoralized by the operations he was obliged to perform in Viet-Nam. He beat Vietnamese during interrogations and saw torture sessions every day. Once he had obliged men and women who had taken refuge in a large underground pipe to emerge by sending in smoke, so that they staggered out half-asphyxiated. One young girl of sixteen had caught pulmonary congestion or pneumonia. The boy was touched and tried to watch over her to see if she could be cured, but she died in front of him. For a young man, to watch the death of a young girl with the feeling that he is guilty of her death is obviously traumatic. From that day, he saw the war differently. When he learned about the existence of the Tribunal, he wondered if it was his duty to report what he had witnessed. His professor of psychology, whom he esteemed

85) Oradour was a village in France where a division of German SS officers perpetrated a violent massacre for essentially no reason (revenge for a minor incident), in the summer of 1944.

highly, advised him to act as a witness. The last question of the Tribunal was: "Does it disturb you to testify against your own country?" he answered: "It does not disturb me. For the first time in a long time, I feel that my hands are clean." We were deeply moved. His declarations are of course written up in the minutes of the Tribunal.

The first session of the Tribunal was to take place in Paris. But de Gaulle refused to allow Sartre to hold it there, in very haughty terms, and the Tribunal moved to Stockholm for the sessio of May 2–9, 1967. At the opening of the session in Stockholm, we were welcomed by the Minister of Foreign Affairs. He was against the type of manifestation we represented, but the freedom of Sweden made it possible for him to allow us to do our work without restrictions of any kind. Our deliberations occupied the entire day, and continued after dinner. The deliberations of the final day even went on throughout the whole of the final night. We had a functioning system of simultaneous translation. For the sessions which took place in big halls, the translators were located in little cabins. It was amusing to observe their gesturing: each translator looked at his orator and imitated him while speaking, even though in principle he was alone and seen by no one. The discussions continued during our common meals. We had no leisure time at all. My mathematician friends Lars Gårding and Lars Hörmander, who lived in Lund, would have liked me to come and give a mathematical lecture, but the schedule left no time for it and it was also too tiring to add a single activity.

The Tribunal had to answer five questions asked by Bertrand Russell. The first three questions were answered during the Stockholm Tribunal. After that first session, we sent the results of our judgment to many countries. In France, the CVN organized public meetings where the results were announced; they were highly attended, mostly by young people. The Tribunal was a very particular, singular initiative. It was not an ordinary tribunal, and did not formally pronounce a judgment, or rather, it did pronounce a judgment but had no power to follow it up with any punishment. There was a jury, but no judge. Several times, we allowed people from the audience to speak, including Americans who defended Dean Rusk's policy.

The second session of the Tribunal was held in Roskilde, Denmark, near Copenhagen, from November 20 to December 1 1967, and ran similarly to the first session. In the second session, Bertrand Russell's final two questions were answered, and a final judgment was rendered.

The crucial question we had to answer was this: had a genocide taken place? We unanimously answered that it had. Today, I am still disturbed by this answer, and I no longer know exactly what to think. Pierre Vidal-Naquet had expressly telephoned me from Paris, pleading with me to answer "no". Was he right? We, the members of the Tribunal, were the only ones to have seen and heard everything, and we were all overwhelmed. Leo Matarasso and Lelio Basso furnished remarkable juridic arguments. Sartre developed an admirably coherent political analysis. But how could one speak of genocide, when thanks to their remarkable system of self-protection, relatively few Vietnamese actually died? Were Americans really killing them *uniquely* because they were Vietnamese? Certainly not in the towns, but the anti-personnel bombs and the exfoliants in the countrysides did lend support to this hypothesis. The Nazis were reproached with genocide of the Jews and the Gypsies, but the massacres of the populations of occupied countries (particularly the USSR) were not considered as genocides. I think we should avoid banalizing the terms "genocide", "torture", "crime against humanity". So? The war against Viet-Nam was anything but banal. Could we not speak of a genocide in the course of being performed and which would have been completed if the war had continued, which did not depend only on Nixon but also on the national and international consequences of continuing the war? I now incline to the belief that we were mistaken, but I am not absolutely sure. Merely reading the notes of the Tribunal is a horrifying experience.

There is, however, one point on which we undoubtedly acted badly, and I feel ashamed of it. Sartre's text on genocide mentioned the genocide of the Armenians in 1915. Immediately, the left-wing Turkish juror Mehmet Ali Aybar lost his temper. According to him, it was the Armenians who committed a genocide against the Turks by allying themselves with the Russians during the First World War. He stuck to his guns against the general furor provoked by his words and threatened to resign if we mentioned the Armenian genocide. Obviously, his resignation would have been very embarrassing for the tribunal. After a discussion following dinner which lasted for several hours, the Tribunal gave way and the Armenians were not mentioned. If I had tackled Aybar persuasively enough, saying that I would resign myself if the Armenian genocide was suppressed, and that no member of the Tribunal had the right to use blackmail, I could have convinced him. However, it remains that Sartre received the approval of the Tribunal for his statements on genocide, and the juridic lecture of Matarasso, who did not belong to the Tribunal, did mention the Armenians.

The second and final session was closed by a solemn declaration by Lelio Basso, a remarkable speech which will not be forgotten. The details of every session are noted in two small books, one with a blue cover and one with a red, published by Gallimard. They are excellent sources of historical information. Moreover, we enjoyed ample press coverage during and after the sessions.

After the second meeting of the Tribunal, there was a regrettable conflict with

Bertrand Russell. The Russell Foundation required the Tribunal to reimburse the cost of the sessions! But we were absolutely incapable of such a thing. I was given the task of negotiating with Russell. I found the great man to be formidably stubborn. I argued that the Foundation had provided all the money without ever discussing with us. The Foundation had selected the twenty-six witnesses, more than we had requested, and paid for their trips to Viet-Nam, and for the trips of certain Vietnamese and Cambodian people to France, and many other things. And there had been no bills of any kind, at any time, just this enormous, unexpected global bill at the end. We had thought that the Russell Foundation was very rich and it had not occurred to us that it would ask for any reimbursement at all, and here it was asking for everything. I discussed at length with Ken Coates, but the true mediator was Ralph Schönmann, the secretary of the Foundation. He was a deeply devoted professional militant, but desperately sectarian. It was arranged that I should meet with Russell in England. But the day before the meeting I found myself glued to my bed with a terrible flu and high fever. Russell (Schönmann, really) wrote me an insulting letter, claiming that under these conditions he refused to see me. There were no further discussions, and the problem remained unsolved. Shortly after, the members of the Foundation and Mrs. Russell in particular decided they had enough of Schönmann, and Russell agreed to exclude him from the Foundation. My good relations with Ken Coates and the Foundation were later restored.

When the Tribunal was dissolved, we decided to continue our activities in the form of a Tribunal of Peoples, apt to judge in situations where certain peoples were gravely oppressed. Lelio Basso was the president and Rome was the center. There were a great many activities in which I did not participate even though they were very important; I participated only in a tribunal on torture in Brazil (1974) and two sessions on the war in Afghanistan (1981–1982), each of which lasted a week. I only have one life.

The question of whether an initiative such as the creation of the Russell Tribunal is opportune or not can give rise to an infinitely long discussion. It had a considerable impact in Viet-Nam. It gave courage to the Vietnamese. They even edited a *Russell Tribunal* stamp to keep it in memory. It also renewed reflection on the Viet-Nam war in France and in the USA. I always believed in the formation of a real international tribunal, having the right to *judge* anyone, in any country, anywhere, with an international jury (the accused being allowed to choose his own lawyers) and to condemn him. All the dictators, the Pinochets and other South American, African or Asian generals, find some country to take refuge in, or receive amnesty. And thus, they escape judgment, while petty delinquents do not; thus it is the greatest international criminals who are spared.

The international tribunal could require extradition. The torturers of the Algerian war were amnestied; France had the right to amnesty them, and maybe could not have done otherwise (personally, I think they should not have been amnestied, and doing it was a catastrophe which left deep traces both in France and in Algeria).

But another country or a simple individual would have had the right to demand an international warrant of arrest. France would have refused to extradite them, but at least they would have been confined to France. It's not much of a punishment, but it's better than nothing. Anyway, the existence of such a tribunal would be a sword of Damocles suspended over the heads of dictators, and would perhaps cool some of their ardor. They could also be judged in absentia. I am perfectly aware of the many obstacles to the formation of such a tribunal. Practically, it would only be possible the day there were no more dictators in the world! I know I'm just a dreamer of beautiful dreams. I did, however, discuss it at length with the prosecutor André Boissarie, who undertook the judgment of French collaborators after the war. He was a remarkable, rigid juror, and had many thousand Frenchmen condemned to death after 1945. Yet he was a generous man, but immense crimes had been committed. He considered my idea to be totally idealistic, but like me he was revolted by the enormous number of great international criminals who were free thanks to amnesty, and finally agreed with my project. Today, it is possible to have recourse to the International Penal Tribunal in the Hague, where the criminals of Yugoslavia and Rwanda are to be judged. And Amnesty International is working officially for the creation of a permanent international criminal court.

Activities in support of Viet-Nam did not end in 1968, although there was less militantism. Bombings continued, and so did demonstrations in France against them. Nixon's presidency was a terrible period. The group of the Tribunal met regularly from 1968 to 1975 in the home of Madeleine Garaudet, a devoted militant Communist. Alas, she died prematurely of cancer, and we were all very sad. Every two weeks, I met Jean-Paul Sartre, Leo Matarasso, Marcel-Francis Kahn and his wife Réna, Mireille Gansel, Roger Pic and often Maria Jolas. An official Vietnamese delegate who was our liaison with Viet-Nam also came often. Sometimes we had quite serious arguments. For instance, Viet-Nam refused to make public the names of American pilots taken prisoner, which was against the Geneva conventions. Our discussions were not in vain since names later came to be communicated somewhat more freely. He kept us informed on the development of the war, so that we were able to modify our communications. Our meetings often ended late in the evening, and Réna Kahn brought me home in her car; sometimes we stopped at the door of my house and went right on with the discussion. It was a traumatizing event when she died very suddenly of an acute attack of leukemia, less than two weeks after one of these meetings.

Maria Jolas was a truly extraordinary person. She was American, but lived in France for decades. One of her daughters, Betsy Jolas, is a well-known musician. Maria was linked to the American anti-war movement, and was our link with it. Every American dissident who came to France necessarily visited her, including Joan Baez and Jane Fonda. She was a great animator, a true symbol, and a woman of incredible courage. Many meetings were held in her house, as she was very hospitable. I believe she lived until the age of ninety-five, but the end of her life was difficult. She was also well-known in Viet-Nam, and kept up the contacts between Americans and the official Vietnamese circles in Paris, particularly with Mrs. Binh, the head of the national delegation of the NLF of South Viet-Nam in Paris (and at the Paris Conference in 1973).

## A trip to Viet-Nam in 1968

After the closure of the Tribunal, the Vietnamese government invited me to spend three weeks in Viet-Nam. It was a few months after the Tet offensive (January 31, 1968) which had been a murderous military defeat but an undoubted psychological victory. First I had to spend two weeks in Phnom Penh, in Cambodia, because I was to be transported to Viet-Nam via Cambodia, by the weekly airplane of the CISC. But often that airplane did not make its trip, or did not stop in Phnom Penh. During my two weeks in Cambodia, I had met a Frenchman appointed Charles Meyer, and his Chinese wife. Charles Meyer was the personal counselor of prince Sihanouk and introduced me to Cambodia. I also met a young man, Alain Daniel, who was doing his civil service there, and who took me butterfly hunting in the countryside. I had not forgotten to bring my butterfly net. I did not capture anything very interesting, but I kept contact with Alain Daniel who was always very involved with Cambodia. I also had time to admire the beautiful Cambodian dances.

A Chinese traveler who was a friend of the Meyers had just returned from China. The couple asked him about the Cultural Revolution, which was in full force. "It's very simple," he explained: "there is the bourgeoisie and the proletariat, and you have to choose (it's the *proletarian* Cultural Revolution). Mao chose the proletariat." Bravo, it was indeed very simple. The Meyers did not seem to be convinced. I only suggested that the peasants also played a certain role in China. "Negligible," he answered.

I couldn't travel, because the airplane of the CISC was liable to arrive at any moment (in spite of its fixed schedule). Instead of visiting Angkor, I did mathematics and proved a theorem on the $C$-spaces introduced by Erik Thomas. A $C$-space is a topological vector space in which every series which converges after multiplication by any numerical sequence converging to zero is itself convergent. I showed that the spaces $L^p$ $(0 \leq p < 1)$ are $C$-spaces; Erik Thomas had showed it for the spaces $1 \leq p < +\infty$. It was my father-in-law who had asked me the question, with $L^0$; if a random series converges in probability after multiplication by any numerical sequence converging to zero, does it converge in probability? I wrote out by hand a note to present to the *Comptes Rendus de l'Académie des Sciences* and sent it to Paul Lévy. It inaugurated several articles of mine and of my students and inspired developments which were the object of my first seminar at the Ecole Polytechnique, called "The Big Red Book" (1969–1970). Finally, after two weeks, the airplane of the CISC arrived. Several of the other passengers were going to Hanoi in order to leave immediately for Saigon. The French air hostess told me: "You should go to Saigon, Hanoi's not very amusing." Our arrival was quite impressive because everything was dark on account of the bombs. I was lodged in a nice little hotel in which I was the only guest and where I took all my meals, served by a young man who introduced each dish to me "meat, fish, bird" (chicken) before giving it to me. From my window, the view gave onto a lake and part of Hanoi. Like gracious shadows, lovely young Vietnamese girls passed on bicycles in their black silk dresses.

I had a busy program, daily visiting bombardment sites. Not a single stone house remained in the countryside, and shells had left huge craters everywhere, but rice paddies filled the landscape. The Vietnamese particularly wanted to show me certain places like their "university in the countryside", which was hidden in a straw hut identical to all the others; it was the classroom and laboratory of a mechanics department. A blackboard was covered with Euler's hydrodynamic equations. The war had not sent Viet-Nam back to the Stone Age. Later, the professor of this class taught at the University of Hanoi, and I saw him again. I attended several primary and high school classes in huts. The example chosen to explain the parabolic flight of a projectile was invariably a bomb falling from a rapid airplane. In a little model weaving industry which was proudly displayed to visitors, we were expected to admire the materials, but even more the secondary education of the workers, who were all women (the men were fighting at the front). Certainly, there was some propaganda in all this, but I myself chose the questions I asked them: the area and the volume of a sphere, the approximate value of $\pi$, second degree equations, a system of two linear equations in two unknowns, and to my surprise, they were even familiar with Ohm's law of electricity – even though the whole community was deprived of electricity and lived with candles! Naturally, this performance is not to be exaggerated. Of course, secondary education existed in Viet-Nam, including in the villages, and Ohm's law was taught, with or without real examples of electricity. But this particular example was a successful experiment of continuing education in a small weaving factory. I also heard some musical concerts (in particular the famous one-stringed *dan bau*, played by a virtuoso), and singing. I briefly saw an anti-aircraft battery, though I wasn't able to locate where I saw it. I learned to absorb liters of tea and give a speech after every visit, a primordial rite in Viet-Nam. Above all, I learned to love the country whose liberty I had been fighting for for more than twenty years.

Between the many workshops we held on the future of the struggle led by the CVN or the repercussions of the Russell Tribunal, I met the Prime Minister Pham Van Dong twice for three hours; he was a very open-minded man. He took me to see Ho Chi Minh, and we chatted for an hour. One felt that Ho Chi Minh was entirely concentrated on a Vietnamese victory in the war. He lived very simply. For example, he thanked the USSR for giving him the Lenin Prize, and excused himself for not leaving his country at war to go and receive it. The personality cult around him was incomparable with what was being developed in other Communist countries, even though one military distinction was "Worthy Nephew of Uncle Ho" and every peasant's hut was decorated with a small portrait of Uncle Ho. My interview with Pham Van Dong and Ho Chi Minh was on the following subject. On March 31, 1968, Johnson publicly announced that he would not run for re-election as president. At the same time, he promised that the bombings north of the nineteenth parallel would soon stop. Peace negotiations in Paris, avenue Kléber, had begun in May 1968 under rather poor auspices, with endless arguments about the shape of the table, etc. During the whole of the following period, the bombings between the seventeenth and nineteenth parallel were terrible. The region was devastated. The North underwent two thousand five hundred bombings a month at the beginning of 1968; the fourth zone (between the seventeenth and nineteenth parallel) was bombed six thousand five hundred times in August alone. Johnson's goal was simply to render the region uninhabitable, so as to cut Viet-Nam in two. He failed, but the inhabitants spent a large part of their life underground. This was still the case when I arrived in September 1968. Pham Van Bach had asked me if I wanted to visit the fourth zone, and I accepted, knowing that they would take care to protect me as much as possible. But in the end, they gave up the project because of the risks and the complications. This was the situation during our three-way conversation. Ho Chi Minh explained to me that Viet-Nam would never give up and that real negotiations would begin only when the bombings stopped completely everywhere. This is what he told me to tell my friends in Paris. When I returned to Paris and told them, in early October, they didn't believe me. Heeding various other clues, the press concluded that the conference would begin on a serious basis; yet the bombings were continuing north of the nineteenth parallel. I didn't understand at all. I couldn't see why Ho Chi Minh and Pham Van Dong would have misled me on such an important point just a few days before. Why? But Lyndon Johnson announced a complete end to the bombing on October 31 and the Vietnamese had not given way. The American elections took place on November 4, so that Johnson left his successor with a clear situation. The negotiations did not actually seriously begin until January 1969. We know that Nixon started bombing and even carpet bombing Hanoi with B 52's again at the end of 1972, after the beginning of his second term, and that the "limping peace" was only finally reached in 1973.

To understand these 1968 attempts and hesitations, it is necessary to recall the background of the situation. The Americans had occupied Khe San, a site on a hill at the frontier of the North. But they were surrounded by the Vietnamese and could only leave via helicopter, which would have been the admission of a humiliating defeat. The Vietnamese approached nearer each day by digging tunnels under the hill; the American troops in Khe San heard the incessant sound of their hammers. It would have been another Dien Bien Phu for the Americans. This is the situation which drove Johnson to accept Ho Chi Minh's conditions. After his speech on October 31, the Vietnamese abandoned their pressure on Khe San.

In 1968, there was a certain coldness between Viet-Nam and the USSR under Brezhnev. The USSR did not really desire a Vietnamese victory (neither did China, as later events showed). Khrushchev had proposed that the two Viet-Nams be represented separately at the UN, but nobody had accepted this idea. The USSR furnished Viet-Nam with the heavy weapons necessary for its defense, but absolutely nothing more. It sent many more heavy arms to Arab countries than it did to Viet-Nam. Viet-Nam could not attack the American VIIth fleet floating off its coast. When I mentioned the possibility, I met with complete silence. I should observe that during this period of nearly hot cold war, a Vietnamese bombing of the VIIth fleet by Soviet rockets would have been a sure way of setting fire to the powder. Viet-Nam itself oscillated at length between pro-USSR and pro-Chinese positions. Shortly after the Russian-Chinese schism of 1960, Viet-Nam was pro-Chinese. But when Richard Nixon met Mao Zedong in 1972, all the portraits of Mao disappeared from Viet-Nam. Then there were the Cambodian frontier attacks against Viet-Nam, the genocide in Cambodia supported by China, the Vietnamese invasion of Cambodia in 1979, the consecutive Chinese attack on Viet-Nam, which had become frankly pro-Soviet, with the reservations we already pointed out. Even today, the leaders of the Vietnamese Communists are still divided between the Maoist tendency, the Khrushchev-type "revisionist" tendency and the neutral tendency which dominated in 1968. There are all kinds of different movements within the power center. Viet-Nam simultaneously needed Soviet aid for heavy armaments and Chinese aid for light armaments. Ho Chi Minh had succeeded in performing the exploit of negotiating with two enemies to obtain permission for the heavy arms sent by the USSR to be delivered over Chinese land; permission was granted on condition that the arms be contained in trains driven by Vietnamese. Le Duc Tho (now dead) was a resolute defender of the Maoist line, and the author of the worst abominations committed in Viet-Nam; he was the damned soul of the country. His terrible sanctions were accurately described by the well-known novelist Duong Thu Huong; she was imprisoned for her writings but is now tolerated thanks to strong international pressure. Le Duc Tho remained powerful until his death, and his faction is still powerful.

Pham Van Dong asked me to deliver a lecture to the political and union leaders of Hanoi, to explain to them the conditions for forming a national Viet-Nam committee and directing its activities. He knew about our difficulties with the Communist Party, and to some extent so did my audience, so that I could not dissimulate them. He didn't even ask me to do so, since within certain limits, he really wanted them to be well-informed. So I gave an objective but moderate one-hour lecture which interested them. The Vietnamese officials knew perfectly well that I had been a Trotskyist, but they let it pass. As for me, I was not unaware that they were Stalinists and thought I had no illusions about the political regime which they would set up as soon as the war was over. Alas, even what I expected would have been better than what actually happened.

I was allowed to see an American pilot taken prisoner. During the hour we spoke together in the presence of one of his jailors, he impressed me with his intelligence and frankness. I introduced myself to him as a member of the Russell Tribunal and explained the objectives of the Tribunal, but promised him that our interview would remain confidential. His airplane had been shot down. He had come down in a parachute, and as soon as he had reached the ground, he had been surrounded by Vietnamese soldiers who protected him from a group of peasants who flung themselves at him with obviously aggressive intentions, and took him away. In prison, he was well-treated and appeared in good health. He had many books, and had become deeply interested in the history of Viet-Nam. "I know it better than the history of my own country," he told me. The Vietnamese people deserved liberty, and now he found himself unable to justify the bombings he had participated in with the best intentions. I knew that brainwashing of prisoners existed, but this one appeared to me perfectly sincere, tranquil and open. "Did you have at least some of these ideas before coming here?" I asked him. "Not really, but American opposition to the war and particularly Dr. Spock's pacifist propaganda had started to trouble me." Dr. Spock was very popular for his teachings on the rearing of children, judging that educational methods were usually too repressive. He was criticized for having gone too far and created a "Spock generation" of lax and lazy adults, so that in the end he backpedaled and even gave vent to serious self-criticism. He also fought against the Viet-Nam war, and never expressed any regret about it. "I think I don't know enough to have a clear idea about the political choices we should make," the prisoner told me, "but I'm convinced that we should stop the bombing, not only north of the nineteenth parallel, but everywhere." Shortly after, he succeeded in having a letter to his wife transmitted to me. He did not mention our meeting but described his life and sent his best wishes to his family, who had no news of him at all and didn't even know whether he was still alive! I pointed out to the Vietnamese that this was totally in contradiction with the Geneva conventions and against human rights. I transmitted the letter to his wife, telling her who I was and relating our conversation to her. She thanked me with a postcard of flowers, telling me that I had "done well". In spite of all, when he returned home after the war, he did not try to contact me, although he had my address from his wife. Was he really frank when he spoke to me? Or was he reconverted by the American military when he returned home (he was a commander and wanted a career). I also knew his address, but I never tried to find out what became of him; it wasn't up to me to do so.

My departure from Paris to Viet-Nam took place in the second half of August 1968, a few days after August 21, the day of the invasion of Czechoslovakia by the troops of the Warsaw Pact. I was very anxious and troubled. I was going to have to remain without news for weeks. My trip was supposed to last three weeks, but more than two weeks went by in Cambodia. I was all the more unhappy as the Vietnamese government had just publicly stated its position of support for the Soviet invasion. Fortunately, the French Embassy in Hanoi helped me, and had the telexes of the French newspapers and press agencies hand-carried to my hotel. Unfortunately, telephone calls to the Embassy never got through. As I had been told on my arrival that I would be received by Pham Van Dong, I let him know that I wanted to talk about Czechoslovakia. The day before our meeting, Pham Van Bach, the president of the Commission of Vietnamese Inquiry into American Crimes, whose objectivity I had appreciated during the sessions of the Tribunal, met with me for two hours to explain the invasion of

Czechoslovakia to me. I had to hear every single one of Brezhnev's Stalin-style slogans: Dubcek's fascist, pro-American attitude, the capitalist development of the country, etc. etc. I thought I was hearing the insults poured over Vychinsky at the Moscow trials. I counterattacked on every point, but with no success. It did not augur well for the next day's meeting. But contrarily to my expectations, the meeting was perfect; it's better to address God directly than his saints. We only spoke about Czechoslovakia after two hours of serious work, and then it was he who asked me to explain my point of view, which I did for half an hour. He answered very soberly: "We think that Czechoslovakia was evolving towards capitalism. But we are very unsure about the whole question. We can see that you do not agree with us, nor does Sartre; we are blamed by our best friends in France and in the United States, in particular the members of the Tribunal. Even the French Communist Party is not sure! So perhaps we are mistaken. I understand your position perfectly and do not feel able to refute it. In any case, you understand the difficulties of our situation. I promise you that the Vietnamese government will no longer state a position on this question, and I ask you and all of our friends to forget that we did so and to continue to bring us your support, as you have in the past, and of which we are greatly in need." I promised and we spoke of other things. I was very moved. Never in my life did I hear a head of state express himself so humbly. Usually, heads of state "are never wrong". Pham Van Dong and I remained good friends. I saw him during each of my later trips to Viet-Nam and he always sends me New Years' wishes. When he visited Paris under Giscard d'Estaing, he invited Marie-Hélène and me to a small lunch. He personally welcomed all the members of the Tribunal in Hanoi, and many other French people who traveled to Hanoi later on.

During my visit to Haiphong, in 1968, I saw a hurricane. I had already seen a hurricane named Diana in Princeton, but it was just an inoffensive tail, although it had uprooted the century-old willow in the middle of the green lawn. In Haiphong, it was a gigantic cataclysm. I was shut into a solid room and ordered not to move at any price. The calm in the eye of the hurricane was terrifying, but then everything started up again worse than before. Many houses lost their rooves and all lost their window panes. The streets were filled with fallen trees, and with Vietnamese people sawing them up and gathering the logs for firewood. France is a marvelous country; rivers occasionally overflow and flood villages, but there are no hurricanes. Underdeveloped countries often seem to lie near tropics, and they have to endure hurricanes on top of everything else. The Haiphong hurricane claimed more victims than some of the American bombardments. I also met Vietnamese mathematicians in 1968.

I am coming to the end of my stay in 1968. It was October and I needed to return to France quickly; I didn't want to spend another two weeks stranded in Cambodia, so I needed to find another mode of transport than the airplane of the CISC. I asked to return via China: a plane from Hanoi to Canton, a train from Canton to Hong-Kong, an airplane from Hong-Kong to Paris. In spite of the moral and military support which China gave to Viet-Nam, I soon realized that relations between the two countries were anything but amicable. The Chinese visa was long and complicated to obtain. The trip from my hotel to the airport was already complicated. We had to cross the Red River by the Paul Doumer bridge (he had been the governor general of Indochina from 1897 to 1902, before becoming President of the Republic and then being assassinated by a lunatic in 1932). It was a magnificent bridge (Long Bien in Vietnamese) but it had had been bombed and repaired many times. It was a mass of patches, but it still held, and it was a fundamental communication route for Hanoi. However, it had a central lane reserved for trains which ran alternately in one direction and the other, and on each side of this central lane was a narrow lane for cars. Every time there was even the tiniest accident, an enormous traffic jam inevitably ensued. There was a tiny sidewalk for pedestrians and bicycles on each side. With the usual Vietnamese prudence, my suitcases had been transported to the other side of the bridge on the previous day. Now, I found myself in another car, with no suitcases – and the inevitable accident occurred. The whole bridge was immobilized, and I had to get out and walk over the whole bridge to the other side, where I was able to join the other car containing my suitcases and get to the airport after all.

The building of the airport in Canton contained at least a hundred portraits of Mao Zedong, of different sizes. My neighbor, a French consul in Hanoi who was returning to France by the same route as I was, remarked: "Wonderful, I was wondering what the face of the head of China looked like!" In Canton, I was picked up by the Vietnamese consul with two assistants. A Chinese soldier asked me to empty my pockets. There was a Vietnamese coin in one of them, a dong, worth less than a franc. He reproached me: "You know it is forbidden to enter here with undeclared foreign currency!" All this for just one dong! I authorized him to confiscate it. "I don't have the right, Sir, it is you who are guilty for having brought it." Deadlock. The Vietnamese consul remained impassive; obviously he had nothing to say. The soldier went to report to customs; he stayed away for three quarters of an hour. He returned at midnight to inform me: "Sir, the director of customs has decided to render you the exceptional favor of returning your dong." There were no further incidents. Good night to Canton, train for Hong Kong. A portrait of Mao was enthroned in the aisle of the train. The train hostess distributed a copy of the *Little Red Book* to every passenger and chanted a verse which everyone droned together with her. In China, like in Viet-Nam, the countryside was made up of an infinite series of rice paddies, and we saw men and women at work, in little teams of about fifteen people, directed by a man whose job consisted in brandishing an immense red flag with the effigy of the great leader. When I described them to Grothendieck, who had spent weeks in Viet-Nam teaching mathematics in a country university, he remarked ironically: "Maybe it would do them some good to be bombed so they could measure the uselessness of their flag-bearers." When I returned to Paris at the beginning of October, part of the French press and some right-wing deputies attacked me because of the nature of my trip to Viet-Nam. I remembered suddenly that when I left, I had been too late in requesting the official authorization of the Minister of National Education, which should always be done well in advance of any trip to a foreign country, and had never received an answer. But my relations with the French Embassy in Hanoi had been perfect. Seeing me under attack, Edgar Faure sent me of his own accord a pre-dated authorization for my trip, and intervened personally at the tribunal of the Chamber of Deputies. He was an elegant Minister.

Because of its diversity and its intensity, this trip left me with a powerful impression which has never faded. I felt myself linked once and for all to Viet-Nam. I often returned later on, but more for scientific reasons. Marie-Hélène and I were both invited to the Hanoi Polytechnic in 1976, and I returned alone in 1979, then again in 1984 with Marie-Hélène and alone in 1990, invited by the Minister of Higher Education and Training.

## Scientific cooperation

During my stay in 1968, I also made contacts with Vietnamese teachers and researchers. At the International Congress of Mathematicians in Moscow in 1966 there were many conversations about the Viet-Nam war, and such meetings became a habit in later Congresses. I actually regret this a bit; the goal is to have people come together who would normally never meet, which is fine, and also to show that mathematicians are human beings sensitive to the problems of their time. But in spite of improvements, the organization of the meetings takes time and steals from the mathematical activities which are the essential reason for the existence of the Congress. In Moscow, I was one of the organizers, together with the Americans Moise, Martin Davis, Steve Smale and Chandler Davis. It was a semi-failure, but it did give me a chance to get to know the Vietnamese who were present at the Congress, particularly Nguyen Dinh Tri and Hoang Tuy. One anecdote is that Steve Smale, a Fields Medalist at the Congress and a defender of Viet-Nam, was interviewed by the Soviet Press. The interview took place in the street, after a session. Smale, who had plenty of presence of mind, added: "But I also want to say this, and I beg you to take note of it: I was scandalized by the Soviet invasion of Hungary in 1956." They did indeed note it; you could tell because Smale was no longer able make a single move without being followed and accompanied, and found himself being promenaded firmly through all the tourist traps, until he finally categorically refused to go.

During my trip to Hanoi in 1968, I saw the mathematicians whom I had gotten to know during the Congress; we were still friends, and I gave a lecture on partial differential equations. We discussed mathematical life in Viet-Nam. The universities and the research centers had all taken refuge in the countryside, as had the Hanoi Polytechnic, of which Nguyen Dinh Tri is today the vice-rector. In spite of the extreme difficulties of the war, Viet-Nam had time to think about its scientific future and sent its best students to write their theses in the USSR or the other Communist popular democracies. Thus, mathematical training was actually better in wartime Viet-Nam than in many of the neighboring countries of East Asia, and it remained one of the best for years. Tri and Tuy had spent four years in the USSR working on their theses. During that time, they were separated from their families, but were spared the dangers of the front. Germany had followed a similar policy of inequality in drafting soldiers during the First World War, which had given it a tremendous advantage in scientific status after the war, because France had lost a great part of its intellectual elite at the front. I found that the scientific community in Hanoi was very impressive.

I also met Ta Quang Buu, an extraordinary person, well-known in France. During the war between France and Indochina, he had been the Vice-Minister of Defense, and had signed the Geneva conventions in 1954. He was Minister of Higher and Technical Education for a long time, a truly remarkable Minister who contributed much to the progress of science in his country. He had studied in Paris, where we had managed to have the same professors, more or less at the same time, without ever meeting each other. Then he returned to Viet-Nam, taking only one book with him, Bourbaki's *General Topology.* He continued to do a little mathematics for his own pleasure, for one hour early each morning; "don't say a word of this to Pham Van Dong," he told me with a smile. He knew enough to solve many problems: his mathematical culture was very wide, though of course superficial. He was present at some of the mathematical lectures which took place in Viet-Nam, and asked pertinent questions. Once, he even asked the difference between distributions, ultradistributions and hyperfunctions. After 1975, he invited many foreign mathematicians to Viet-Nam, particularly from France. Constant exchanges, reviews and seminars went on between the

Hanoi Polytechnic and the Ecole Polytechnique in Paris. It is important to note here the open-mindedness of the directors of the Ecole Polytechnique, who supported these exchanges in spite of the obvious inequality of scientific level. Tri and Tuy traveled all over the world and acquired the spontaneous reflexes of democracy. Tri's first contact with an edition of *Le Monde* was a fascinating experience for him and for us. He was particularly intrigued by the presidential election in France which occurred during his visit. For years, we have been able to openly discuss political subjects with Tri and Tuy with no embarrassment on either side. They said that the professors of mathematics in the USSR acted like Chinese mandarins[86)]; they did so somewhat less in Viet-Nam, not very much in France and not at all in the United States; I think they were right. The Vietnamese mathematician Lê Dung Trang, a naturalized French citizen, played an important role in the invitation of many foreigners to Viet-Nam, as did Frédéric Pham, Alain Chenciner, Pierre Cartier and myself. I already mentioned Grothendieck's sojourn in Viet-Nam during the war. André Martineau also spent one month in Hanoi to give some lectures, and returned enthusiastic. At the International Congress of Mathematicians of Nice, in 1970, Grothendieck and Martineau recounted their trips to Viet-Nam. There were a dozen over-excited left-wing mathematicians, who vociferated that Viet-Nam needed rifles and cannons, not books and mathematicians. Martineau had spoken about the training of future Vietnamese mathematicians in the USSR, to prepare the scientific future of the country. First you have to win, retorted the others, then you can think about the future. We defended the right for a country at war to make social choices, but our ideas left them totally cold. They were only a minority but they spoiled the whole session with their heckling.

Today, slowly, very slowly, Viet-Nam is becoming democratic. The newspapers still write in a solid language of Communist formulas, but the people now discuss freely, at least in intellectual circles. My colleagues invite me freely to their homes, which was not the case during my first trips. Among those who have not traveled, shyness and ignorance of the wider world are dominating. Ta Quang Buu was very intimate with us, and was once again a candidate in the elections. We asked him if he traveled for his election campaign; he answered us simply that since he was the only candidate of the party, he had no reason to do so. He had certainly never seen a presidential election in France!

It is impossible for me not to mention our mathematician friend Hoang Xuan Sinh. We have known each other for decades. She studied in France where she passed the agrégation, and then took the momentous decision to write her thesis

86) deeply respected intellectual leaders of respectful, close-knit communities

under Grothendieck. He received her in his several-acre domain; he was in the full force of his ecologist period, cultivating the land and raising livestock. As one could imagine, faced with such an enormous task all alone and knowing nothing of the job, he failed completely. Mrs. Sinh, who is very fond of him, says that he couldn't distinguish a cabbage from a linden tree. At any rate, he gave her an excellent thesis subject which she brought back to Viet-Nam. She never wrote him a single time to ask for help, but returned to France with the solution all ready, to the immense surprise of Grothendieck himself. She defended her thesis in Paris, before a jury consisting of Cartan, Grothendieck, Schwartz, Verdier and Zisman. Grothendieck wrote the report on her written thesis, Verdier on her minor (oral) thesis, and with this she made a superb entry into French mathematics. She later returned to France several times. Ta Quang Buu, who was always eager to promote scientific values, watched over her in a country where feminine talents tend to pass unrecognized (Vietnamese women mathematicians, he told me, can be counted on the fingers of one hand). He did not quite succeed. She was appointed director of the Pedagogic Institute, which is of considerable importance, training future elementary and high school teachers, but she devoted herself to it and had no more time for research. She belonged to the highest instances of the Union of Vietnamese Women, and often represented Viet-Nam abroad. She untiringly rendered great services to her country. In these last years, she and her colleague Bui Trong Luu have founded the Thang Long University Center, the first free University of Viet-Nam, which functions thanks to private support from Vietnamese and foreign individuals and firms.

From top to bottom, the *deus ex machina* of this new university has been represented by Bui Trong Lieu and his wife Colette, who live in France. They are absolutely discreet and fearfully efficient; he is a professor at the University of Reims and we know him well. He is the brother of Bui Trong Luu. The sensational idea of the university received the support of the party and the state, thanks to their confidence in Sinh. Candidates are admitted to the university only according to their scientific value, with no consideration of political criteria, which is not the case in other establishments. At first, the university was a spectacular success, but it has lessened lately. Mrs. Sinh gave herself unstintingly to the task, sometimes going down to the street in the morning to usher in students who might be tempted to remain hanging around outside. At that university, like everywhere else, ardor at work requires control and stimulation. In general, the students feel happy in the university. Sinh taught many courses in Thang Long herself and recruited high level teachers and researchers, offering them a slightly higher salary than elsewhere. The laboratories and library are in her apartment and in Buu's. She often meets the country's leaders, in whose presence she can speak frankly; she is full of humor and presence of mind. One winter, Marie-Hélène mailed her a heavy sweater. At the post office, Sinh was told that she had no right to receive packages from abroad, except from family. She immediately replied that Marie-Hélène was her cousin. "With a name like that?" "Uh, sure, she married a Frenchman!"

Viet-Nam remained very poor. It makes the scientific life extremely complicated. They don't have enough money to buy books and journals for their libraries. Scientists from many countries send them books and articles at their own expense. China sends them photocopies of many journals. When a Vietnamese mathematician is invited to a conference abroad, there is always a subscription to finance his trip. I learned that the Vietnamese consulate used to lend these travelers a suit and a pair of shoes for the duration of their trip. This, at least, is no longer the case. Obviously, a mathematician who spends time in France earns much more than he would at home, spends less out of habit and must give a good part of it to the country (this is not only true in Viet-Nam, but also in Sweden, a Social-Democratic country). He still remains richer than his colleagues, so that trips abroad are very desirable. Plenty of French mathematicians travel abroad for the same reasons. The most important of the French organizations which massively organized the scientific development of Viet-Nam was the Center of Scientific Cooperation with Viet-Nam, directed by Henri Regemorter; he does not just preside over exchanges of books and people, but he also goes to Viet-Nam to discuss questions, projects and plans relative to the scientific development of the country. One of his best collaborators, Didier Dacunha-Castelle, is a professor of statistics at the University of Orsay. I should add that the French Embassy in Hanoi and the Asia directors in our Foreign Ministry have always been very cooperative.

When we gave lectures in French in Viet-Nam, they were translated sentence by sentence into Vietnamese. We had to organize examinations of this type at the agrégation. Many scientists have the detestable habit of occasionally mumbling sentences into their beards. The translator is a mathematician, but even so he is not supposed to have to guess what we are saying, so everything must necessarily be stated clearly and precisely. It is a significant pedagogical exercise. Lectures last twice as long as usual, which is no small feat, given the climate. We generally had about fifty very attentive listeners. The lectures were of course organized by Ta Quang Buu. In 1976, just after the liberation of Saigon which became Ho Chi Minh City, the students welcomed us in a crowd. The people accompanying us tried to separate us from the students, who were avid to speak to French mathematicians. We protested, and the young Saigonese, who are quite rebellious, refused to move off. They asked us very interesting questions, and we conversed in French. Ta Quang Buu then accompanied us to the marvelous site of the Along Bay, comparable to the Rio Bay, and then to Cao Bang. The Viet-Minh had won a historic victory there against France in 1950, and it is also the location of the famous, venerated cave where Ho Chi Minh lived while he was being hunted by the colonial French police. The Chinese destroyed it expressly during their "punitive" invasion in 1979. We spent the night in a little hotel in Cao Bang, where I set up my mercury lamp against the window, in front of a little terrace, to attract moths. Ta Quang Buu came to watch my nocturnal hunt. It was really a pity, he told me, that I hadn't been with them on the Ho Chi Minh trail in the depths of the jungle, because there was an abundance of beautiful butterflies there (and shell holes, also). The bombings had been inexpressibly violent there, but the forest is so dense that the American pilots never located the paths used by the jungle fighters. We talked about politics and about Asian production modes, about Trotskyism and anti-Semitism and bureaucracy; about everything. The Vietnamese didn't really understand what anti-Semitism was (apart from Nazism). I explained it to Ta Quang Buu, citing examples of the ways in which the powerful official anti-Semitism of the USSR manifested itself. According to him, this was impossible since Lenin had condemned anti-Semitism. I tried to tell him that Stalin and Brezhnev were not Lenin, but I don't think I convinced him. He gave the following convincing analysis of bureaucracy in Viet-Nam: "It has three sources: mandarin-style bureaucracy imported from China, the French colonialist bureaucracy and the Socialist bureaucracy." He always tried to keep science independent of politics. The political opinions of their parents and their own position with regard to the party tended to count much more than the actual scientific value of students trying to obtain scientific promotions, prizes or other advantages, even enrollment in the elite universities. It was obviously an immense privilege to be sent to write a doctoral thesis in the USSR instead of being sent to the front, without even counting the future prospects it opened up to them. But the students who were chosen uniquely for political reasons sometimes obtained purely decorative theses in the USSR and then overloaded the scientific community with incompetent teachers and researchers. This practice of the intelligentsia was called *ly lich.* Ta Quang Buu was firmly opposed to it and himself controlled the attribution of scholarships to the USSR, with the support of Pham Van Dong, who was always closely interested in the scientific development of the country. Apart from these individual meetings with scientists whom he knew personally, he summoned a few dozen of the best ones regularly one or several times a year for an informal discussion. But the divergences and tendencies at the summit of the State were powerful, and Pham Van Dong's support was insufficient. The nomenklatura obtained a compromise: Ta Quang Buu was fired, but his policy was applied. The *ly lich* did not completely disappear but at least it weakened considerably.

When I turned seventy in 1985, Nguyen Dinh Tri and Ta Quang Buu organized a one-day workshop in Hanoi (in my absence). After the mathematical lectures, there was a summary of my life, and a lecture by a biologist on the development of colors in butterfly wings. Ta Quang Buu spoke of our hospitality, adding with humor (but some truth) that in seating oneself on our antique living room chairs one risked breaking one's bones.

When Ta Quang Buu passed away in 1986, I lost a friend. For his country, he was more than a great Minister. He was good and generous; you could see it in his face. His death was felt painfully, especially in the cities, because he was very popular as Minister of Technical Education. He had lived a very full life.

Vietnamese students are generally very studious. Once, in a large deserted street of Hanoi, towards midnight, I was passing in a car and I saw a student installed in the middle of the sidewalk, sitting on a chair in front of a table covered with papers, under the light of a big street lamp; he had chosen this unusual spot because he had no electricity at home. The street was also the source of other, more unpleasant discoveries. In 1976, in Saigon, I saw a Vietnamese women with a black child. I was told that these children of American fathers did not have the right to go to school. An analogous fact shocked me deeply in Cuba in 1968: the schoolchildren all wore identical uniforms except for the children of political prisoners who had to wear a different uniform.

My 1990 trip was a trip of scientific cooperation. When I became president of the CNE (Comité National d'Evaluation)[87] from 1985 to 1989, I sent all of our publications to Nguyen Dinh Tri, who transmitted them to higher places. The Minister of Higher Education and Training asked me to come in October 1990 to evaluate higher education in Viet-Nam. It was a considerable task for such a short time. I wrote the evaluation report after visiting the universities and high schools of Ho Chi Minh City and Hanoi; it is about forty pages long and is still rather confidential. Fundamentally, at the end of the war in 1975, thanks to an excellent scientific policy, Viet-Nam was one of the best of the South Asian countries, but this is no longer the case. As in other domains, it seems that Viet-Nam won the war but lost the peace. After its victory against American imperialism following a pitiless war, it enjoyed considerable prestige in international left-wing circles and in Third World countries. It could have

87) CNE: National Committee of Evaluation of the Universities

become a model in some sense, although not one of the "dragons" of Southeast Asia, since it was exhausted by decades of war and almost destroyed, and the United States continued, unacceptably, to obstruct its development as much as possible. The South would have deserved and wished to be given its proper political place, instead of which, it felt invaded. Mrs. Binh had been appointed to the key post of Minister of National Education in unified Viet-Nam, but most of the other leaders of the NLF were ignored. And above all, the country was confined in the stifling atmosphere of bureaucratic Communism which was so ill-suited to the South, where brutal collectivization was imposed. Democracy was not at the rendez-vous.

The proportion of Ph.D.'s among all university teachers is about the same as in neighboring countries. The level of the best researchers is perhaps better, but the average level is worse. The number of students with respect to the number of inhabitants, and their value, and scientific research in the universities, have all dropped drastically in Viet-Nam, which is now far behind the other countries in Southeast Asia. Taking the different types of students in early stages of training into account, Viet-Nam has 31 for 10,000 inhabitants, Thailand 127, Indonesia 37, Malaysia 42, the Philippines 250, France (in 1990) 240. It is a defeat for Viet-Nam. The fundamental cause lies in the desire for egalitarianism and the anti-intellectualism inspired by the ideology of the regime: teachers' salaries are much too low, \$5 to \$16 per month in 1990, no more or barely more than workers' salaries. Teachers in higher education (all teachers, in fact) are obliged to have a second job, for example in commerce, and cannot do research, even after a good thesis. The profession is no longer attractive and except for the elite institutions, the students are no longer motivated for work which will bring them no reward, as the budget for national education is very low (10% of the national budget in 1990, compared to 22% in Thailand, 20–25% in France). The Hanoi university has only one ancient photocopying machine, which doesn't work half the time, and four micro-computers. But in the town, there are photocopying shops all over the place. The students are forced to make long trips, sometimes to get to their workplace. Moreover, individual establishments are isolated from each other. When the physicist Louis Michel arrived in Hanoi with his wife in 1984 at the same time as Marie-Hélène and I, he noticed that his lecture was posted up at the Polytechnic but not at the CNRS, which was very far away in any case. There is no monthly bulletin containing the various activities. Faxes were inexistent in 1990, the telephone worked badly and was very expensive, two researchers never telephoned each other, letters took three weeks to reach Europe. I had a telephone in my hotel room, but I could never reach the Embassy (which was probably tapped anyway) and I was awakened all night by calls which were not for me. Every time I met Pham Van Dong, I complained to him about the postal and telecommunications problems, so that in the end he forestalled me humorously. A scientist cannot work in isolation. The telephone works perfectly well in Taiwan, with a tax per pulsation as in France, which is very economical. I devoted four pages of my evaluation report, 10% of the total, to the problem of communication.

The Vietnamese were aware of these problems. The directors and presidents of the various establishments and the Minister "played the game" and frankly spoke to me about their problems, with an obvious desire to seek for improvements. In spite of my knowledge of the country, I would not have seen these problems myself. The Minister then traveled to Paris to rediscuss everything with me. I believe that many things changed after my report. In France, I can't imagine a Minister of Education – too often they think they know everything – who would actually take the evaluation reports of the CNE into account. However, the universities did so. The Vietnamese knew my deep attachment to the country and were grateful for my severity. The atmosphere of the trip was one of sincere close collaboration and friendship. The goal was not so much to "judge" but to find solutions to the problems, together with them. I proposed many solutions and they reflected on them with me. Principally, I noted that when they established an inegalitarian system during the war, privileging intellectuals, their educational system and research had taken a giant step forward, and that it should be done again.

According to my usual habit, I took a three-day butterfly trip between Hanoi and Ho Chi Minh City, in the big forest of Cuc-Phuong. The splendid forest overjoyed me, but it was a poor season, and the hunt was mediocre.

## Cambodia

Cambodia remained outside of the war for a long time under the reign of Prince Sihanouk. He kept it in a delicate state of balance between the United States and Viet-Nam, keeping up relations with both countries. Part of the Ho Chi Minh trail which was used by convoys of Vietnamese soldiers, crossed a region of Cambodia called the Duck's Beak because of the form of the border at that point. When Richard Nixon decided to set off a blitzkrieg against the Duck's Beak, he encountered energetic protests from Americans of all circles, obliging him to abridge the invasion. In 1970, Sihanouk, who was vacationing in France as he often did, was overthrown by a coup d'état led by marshal Lon Nol, who established a pro-American military dictatorship. From that day, the Viet-Nam war extended to Cambodia and intensified in Laos. The three countries were systematically carpet-bombed by American B 52's. The guerrilla war against Lon Nol was led by the Cambodian Communist Party of Pol Pot, called the UNFK (United National Front of Kampuchea); the soldiers were called soldiers of the Khmer Rouge. My first, vague suspicions date back to 1975. Foreign journalists who went to the Khmer Rouge disappeared; no news came from them and no one knew whether they were alive or dead. I had good relations with Ok Sakun, of the Khmer Rouge's delegation in France (Lon Nol also had an embassy in France), and I met with him several times. When I asked him about the foreign journalists, he answered that everything would be fine. It was not fine. During a conversation in the avenue Henri-Martin, Decornoy described the Khmer Rouge soldiers to me, saying they were very narrow-minded and very tough. Sihanouk had taken refuge in China, where he was well-received. Together with the government of Mao Zedong, later Deng Xiaoping, he fully supported the Khmer Rouge. Yet, a rumor started that he had been horrified by what he had seen during a visit to Pol Pot, in the "free zone" of Cambodia. But it was only a rumor. So I had only doubts and suspicions to go on, and meanwhile we supported the Khmer Rouge.

Things changed radically when the Khmer Rouge won the war (before the Vietnamese) on April 17, 1975, and emptied Phnom Penh of its entire population in a single day, including the hospitals. It should have opened our eyes, but we didn't understand anything of what was going on. I interrogated the Vietnamese leaders in Hanoi in 1976, but they didn't understand what was going on either. The attitude of the Khmer Rouge was so opposed to traditional Marxism that it was flabbergasting. I believe I remember that Ok Sakun left Paris without my seeing him again. I learned later that he was a firm supporter of Pol Pot. The situation was clearly serious.

I believe that it was not until the year 1976, thanks to the book *Cambodge Année Zéro* (*Cambodia Year Zero*) by R.P. Ponchaud, that we became certain that the Khmer Rouge was organizing a genocide in its own country. They divided the entire population into "ancient people", those of the countrysides, and "new people", city dwellers, and they undertook to eliminate the new people entirely, because they were the source of all corruption. It resembled Nazism: a group of people were to be eliminated simply because they belonged to the "new people". The executioners employed all kinds of horrific methods: insufficient food, exhausting work, forced marches, elimination of the ill, pitiless punishments (every error was punished by suffocation with a plastic bag placed over the head), daily terror of the decisions from the Angkar, a kind of secret and mysterious army, tortures (torture centers have been located). The peasants, the "ancient people" were treated similarly in case of disobedience, but not with the goal of systematic extermination. The warriors of the Angkar were young, almost children, sometimes less than fifteen years old, who had been made into fanatics and who accomplished their tasks with sadism. I often cite them as an example to those who claim that the young are always right. Pol Pot's troops were mostly adolescents; the Hitler Youth was also savage. Pol Pot broke with every country. During a solemn ceremony, his supporters threw thousands of dollars into a river. Father Ponchaud was our principal source of information; he had lived in Cambodia for many years. A Cambodian, Pin Yathai, described how he emerged from hell, having lost his entire family, in an extraordinary autobiography called *The Murderous Utopia*. He was the only survivor of the eighteen members of his family. After a hallucinatory flight lasting four weeks, he arrived in Thailand in June 1977. He found himself alone with his wife in the jungle; he left her for a quarter of an hour's exploration, and never saw her again. She was the last remaining member of his family. I bought several copies of his books which I gave to friends to read. It was soon out of print and Laffont didn't want to re-edit it. I invited Pin Yathai to my home several times. He was very pessimistic about the future of Cambodia. He now lives in an Asian country.

At the beginning of 1977, together with a few friends – Father Ponchaud, Jean Lacouture, Vidal-Naquet, Maria Jolas – we decided to establish the truth, and edited a little bulletin called *BISC: Bulletin d'information sur le Cambodge* (Information Bulletin on Cambodia). We felt ourselves to be desperately alone. Certainly, we knew things that others did not know, but why did they not know when we knew? There was not a single Communist with us, not even former members of the CVN or the Tribunal. Jean Lacouture published a remarkable book about Cambodia which enlightened many people, but it was insulted by many others, particularly Communists who accused him of primitive anti-Communism. For them, it was absolutely necessary to empty Phnom Penh because the population had been corrupted by Lon Nol, which was partly true. At that time, I broke with Gabriel Kolko, whom I had become close to during the sessions of the Tribunal, because he wrote a very insulting letter to Maria Jolas. The same went for Noam Chomsky, who is a remarkably intelligent man, whom we were quite fond of previously, but who continued to support Pol Pot far too long.

The invasion of Cambodia by the Vietnamese army on December 25, 1978 was a truly unexpected event. The army was motorized and they took Phnom Penh on January 7, 1979, defeating Pol Pot's army which fled to the western part of the country, near the border with Thailand. Viet-Nam established a new regime in Cambodia, led by Heng Samrin, a former Pol Pot supporter who had become afraid for his life and had fled to Hanoi. This created an incredible international imbroglio. The Western countries, led by the United States, which had never digested Hanoi's victory in 1975, accused Viet-Nam of every possible crime.

Certainly, the goal of the Vietnamese invasion was not to end the genocide, but Cambodia constantly attacked Viet-Nam at the borders, in various barbarous and primitive ways (it is said that the Cambodian soldiers ate the livers of their prisoners, but nothing is less certain!), and Viet-Nam had enough of this enemy who continued to harass them, who was the bridgehead of China, and who was obviously less well-armed than they were. Of course, refugees like Heng Samrin certainly made them aware of the genocide. Some of my best friends told me that they would never forgive Viet-Nam for the invasion. It is true that they were not aware of the genocide. But the reality of the facts was beginning to be widely known, although the press disseminated an incredible mixture of information (on purpose, I believe): yes, there had been a genocide, but it had been executed by the Vietnamese invaders against the Cambodians! Today, the facts are well-established, but those who read the press articles from 1979 can verify what I said. The Communists, who had called Lacouture all kinds of names, changed their attitude from the day that Viet-Nam invaded Cambodia, and organized aid to the new Cambodia under Heng Samrin, which practiced a more liberal form of Communism. About the same time, in 1979, they signed a long text in *Le Monde*, congratulating the USSR for its civilizing influence in Afghanistan.

In Hanoi, in 1979, I spoke at length about Cambodia to Pham Van Dong. The Vietnamese invasion had saved Cambodia from genocide, I told him, but now they had to let it re-establish democracy. I compared the situation to the Indian invasion of Bangladesh in 1971. Pakistan was a country made up of two disjoint components, which is obviously an unhealthy situation. Today's Pakistan is to the northwest of India, in the Indus plains region; its capital is Islamabad. To the east lies what used to be oriental Bengal, a Muslim country (occidental Bengal remained part of India) now called Bangladesh, whose capital is Dacca. On March 25, 1971, the regime of Islamabad led by marshal Yayâ Khan organized an incredible massacre of the Bangladeshi population, killing hundreds of thousands of people, and installed a regime of military dictatorship. A leaden cover lay over the country. I belonged to a committee for the liberation of Bangladesh, which from the very start contained a harmonious and open-minded mixture of people with both right-wing and left-wing views, who were interested mainly in moral issues. Although Bangladeshis are Muslims, and not particularly fond of Hindus, more than eighteen million of them took refuge in India, which welcomed them, lodged them and fed them. But soon the task became too heavy for India. Moreover, of course, India felt that it was the perfect opportunity to inflict a devastating defeat on Pakistan, which could certainly not resist an Indian attack on Bangladesh for very long. So, armed with military and logistical help from Moscow, the Indian army finally invaded Bangladesh in a spectacular and massive parachute landing over Dacca on December 4. Pakistan surrendered on December 12 and our committee was dissolved. The world emitted a protest on principle against the invasion, but in fact everyone approved of it (except for China, which always defended Pakistan against India for geopolitical reasons, just as it had armed Pol Pot's Cambodia against Viet-Nam). International public opinion exercised strong pressure for India to abandon Bangladesh, which it did easily on March 12, 1972. In any case, it would have been unable to lead a Muslim country of seventy million inhabitants. This was the example I cited for Pham Van Dong. I was pleased by the Indian invasion, but on condition that India leave the country after having freed it, which it did. I was pleased by the Vietnamese invasion of Cambodia and the fall of Phnom Penh, but now Viet-Nam should leave the country. I used every ounce of persuasion that I possessed and left a letter for him behind me when I left. Pham Van Dong was always very cooperative and promised me that the Vietnamese army would leave Cambodia as soon as possible. I believe that he really meant it. In the end, though, it took ten years (until 1988). There were good reasons for this, because under American pressure, the environment in Viet-Nam was hostile to him, and he had to defend his security above all, in difficult conditions. China led a little invasive war against the northern border of Viet-Nam "to punish it". Viet-Nam resisted perfectly well, and China backed off, but it was still a war. The Vietnamese had not completely vanquished the Khmer Rouge, of which a residue continued to wage guerrilla warfare in the western part of the country, planting mines more or less everywhere, under the protection of China and Thailand, so one can understand that the Vietnamese were not in a hurry to leave. But the situation improved drastically when they left, and I believe that in spite of all, they could have left several years earlier. During my discussion with Pham Van Dong, we also talked about the future of Cambodia. I suggested calling for the return of Sihanouk. It's not that my opinion of him is particularly high, because the man is extremely versatile. He had been sequestrated from 1975 to 1978 by the Khmer Rouge who had killed a large part of his family. However, he detested the Vietnamese even more than the Khmer Rouge, and this visceral and ancestral hatred was shared by many of his people; of course the Khmer Rouge also made every effort to encourage it. Pham Van Dong had a certain esteem for Sihanouk because of his past attitude.

He was ready to accept this solution and pull his army out of Cambodia. But at that time, Sihanouk was no more than a puppet in the hands of the Chinese leaders, and was spouting rhetoric against Viet-Nam. After the Vietnamese left, there were democratic elections in Cambodia where Sihanouk reigns without being king. Power is shared between many factions, but the weakened Khmer Rouge remain dangerous and destructive. We formed a Cambodia committee a few years ago, under the leadership of Pierre Max, president of the association Aid for the Third World. He knows a great deal about the problems of Cambodia, and the only goal of our committee was to get rid of the Khmer Rouge, without otherwise involving ourselves in the internal politics of the country. In spite of our good will, we did not succeed, because the re-establishment of peace in Cambodia depends on geopolitical forces which are beyond us. There are several other non-governmental organizations of aid to Cambodia, and French university professors such as the doctor Minkowski, the mathematician Pierre Schapira, and the physicist Colette Galleron visit the country. But Cambodia was martyred by the Khmer Rouge, and everything remains to be done. Misery is widespread, educational and medical establishments lack absolutely everything. Many Cambodian intellectuals speak French. Cooperation with France could be enormous ... and in fact it is almost non-existent.

## The dissidents

As Pham Van Dong was one of the most liberal of the Vietnamese politicians and he was Prime Minister, I always took advantage of my trips to Hanoi to talk to him about human rights and to give him a list of people I wanted to see liberated. Some of these people were prisoners of opinion supported by Amnesty International, others had been pointed out to me by friends – it depended more or less on chance. There were not very many each time. About one third of them were eventually freed; eight cases out of twenty-two. I later learned that my intervention had really counted in their liberation; the prisoners themselves had been told this. Pham Van Dong told me several times that he considered it a good thing if one third of his orders were actually followed! I was particularly touched by the case of a Buddhist, Ly Dai Nguyen. I felt a great deal of sympathy for these generous and courageous people, persecuted more or less everywhere in Asia. His sister Chân Không had spoken to me about him. She was a serene and pacific person who had spent her life inserting herself with incredible audacity between opposing parties and saving lives.

In April 1979, I persuaded Pham Van Dong to invite Amnesty International to come to Viet-Nam. He knew nothing of the apolitical and neutral character of the organization. I explained it to him at length and told him that even if an official visit from Amnesty might cause some trouble, it would also clearly illuminate the most important points to be changed in the judiciary and prison policies of the country, and would make fruitful exchanges of information possible. Until then, Viet-Nam had never answered any of Amnesty's demands. He first asked me to personally transmit the invitation, and to take part in the delegation. I told him that it was up to him to make the move, and that I should not be part of it, and he promised me to think about it. As I did after each trip, I left him a letter on the subject. And I simply let the direction of Amnesty in London know what I had done, via the militant activist Lise Weill, responsible in France for relations with Viet-Nam. A few months later, in summer, I was in Autouillet when I received a phone call from the Viet-Nam Embassy in Paris, informing me that the Vietnamese government had invited Amnesty and expressly requested the Embassy to inform me. The direction of Amnesty in London learned about it before I did, but did not let Lise Weill know, so in the end she learned it from me, which is strange. The mission took place in December 1979. I believe it was a partial failure, because neither of the two parties were sufficiently well-prepared, and there was no further visit to Viet-Nam, but researchers from Amnesty traveled there openly from time to time.

Very recently, I called on Pham Van Dong about the liberation of a particular prisoner. There are numerous dissidents of varying tendencies in Viet-Nam. One of the best known among them, Nguyen Kac Vien, is one of the intellectuals of the party; he visited France several times and I met him often. He is brilliant and persuasive and always has an answer to everything. Until recently, he was an inflexible defender of the regime in its most Stalinist form. But over the last ten years, he has multiplied written and oral protests against the bureaucracy and in favor of a complete separation of the Communist Party from the center of power, which of course would render the Vietnamese Communist Party useless. His criticisms are biting, and he is listened to in every circle. Thanks to this, he protects other dissidents, since he has contacts with the highest officials. He has never been bothered. The famous novelist Duong Thu Huong is more radical. Her books on the daily life of the people during the war and the different stages of the revolution are widely read. Her attitude towards the country's power center is extremely severe; she consideres that even if the party truly represented the interests of the people and the country during the war against France and against the United States, between the two wars and after the second one, it took advantage of its many privileges to set itself up above the people, all of which is alas true. She appears to deeply respect Ho Chi Minh. Towards 1991, after a public meeting in Ho Chi Minh City, in front of five hundred auditors in the hall and as many more outside, she was arrested. It took an international petition of intellectuals who had defended Viet-Nam during the difficult period to free her, but even then she was kept under strict surveillance. When she asked to visit France, where she was invited by the Society of People of Letters, she produced a battle of tendencies within the Vietnamese Communist Party. She remained very demanding and uncompromising, but she kept her contacts with the leaders of the Communist Party, who finally acceded to her request. This complicated kind of game is quite typical of today's political atmosphere. Her trip to France, from August 1994 to January 1995, was very successful, punctuated with public meetings and personal visits. Marie-Hélène and I were deeply touched that she insisted on meeting us, in our house, wanting to tell us that she knew and esteemed our "Vietnamese past".

Certain dissidents held even more radical positions. Hoang Minh Chinh is a perfect example. A militant from the age of fourteen, he joined the Vietnamese Communist Party from prison. His first five years of prison were spent during the World War. In 1954, general Giap decorated him for extraordinary valor. He was a well-known intellectual. After having been the secretary general of the Youth Union, he was appointed Rector of the Philosophical Institute until his arrest in 1967, in the middle of the Viet-Nam war, for having opposed Maoism. Starting in 1981, he accused (as Zola accused) Le Duc Tho for his attacks on democracy, which got him imprisoned once again. Nobody abroad knew of his arrest. If I had known about it, I would certainly have tried to intervene with Pham Van Dong in 1968. Chinh spent eleven years of his life in prison and nine under house arrest, and was often tortured. His case was described in detail in the book *Les Cent Fleurs écloses dans la nuit du Viet-Nam*, by Georges Boudarel, a connoisseur of Viet-Nam. Nguyen Trung Thanh, the collaborator of Le Duc Tho, who had had his orders executed in the past, obtained access to the files in 1993 and discovered that they were either empty or filled with trumpery documents, and he tried to have the victims rehabilitated. He went to see the secretary general of the Communist Party, Do Muoi, who showed the understanding of an open-minded man but who could do nothing against his opponents. Hoang Minh Chinh was arrested again on June 14, 1995, at the same time as his friend Do Trung Hieu, for unknown reasons; perhaps because of the approaching 1996 meeting of the central committee of the Vietnamese Communist Party, which exacerbated passions. Hoang Minh Chinh's wife was received by the leaders of the party, who proposed the liberation of her husband in return for "tiny, very tiny confessions", but they refused to give way, upon which she was also threatened with arrest. A petition carrying the signatures of more than fifty important people who had actively participated in the defense of

Viet-Nam in the past was circulated. I signed it, and I wrote personally to Do Muoi, to tell him that the repression exercised against these intellectuals was not only a characteristic abuse of justice and an attack against legitimate human rights, but also a monumental error which Viet-Nam, which was attempting to re-enter the concert of nations, would pay for dearly. I also sent a letter to Pham Van Dong. Even though he is already very old, very tired and almost blind, and lives in retirement, he is still the most prestigious of former leaders and an important counselor to the central committee of the Vietnamese Communist Party, and he is still heeded. It was the last letter I wrote to him demanding a liberation. The trial of Hoang Minh Chinh and Do Trung Hieu took place on November 8, 1995. Hoang was condemned to a year of prison and Do to fifteen months; they had no lawyer. Only six members of their families were allowed to attend the trial; the court was full of policemen – but the trial was shown on television, another sign of the ambiguity of Vietnamese internal politics. Thus every Vietnamese was able to hear the self-defense of Hoang and Do. Even their condemnation was a compromise; many Vietnamese at the summit of the hierarchy were against it.

Let us return to my 1979 trip, during which a certain coincidence had important repercussions on my butterfly collection. The flight to Hanoi made an unplanned stopover in Vientiane, the capital of Laos. The passengers were invited to the best hotel in town, the Hotel of a Thousand Elephants. Near Autouillet, in Montfort-l'Amaury, I had met a collector of Coleopterae named Rondon; he owned more than one hundred thousand specimens and had discovered many new species. He had left France to hunt Coleopterae in Laos, but I didn't know his address. He was almost the only person to have gone hunting in Laos. I had practically no chance of finding him during the single night I was spending in Vientiane. Like the other tourists, I was having dinner at the hotel of a Thousand Elephants, sharing my little table with an English lady who lived in Vientiane. During the conversation I couldn't resist asking her if she knew anything of Rondon. "But I do know a Mr. Rondon who collectes Coleopterae!" she exclaimed. I was astounded. He lived two hundred meters from the hotel. Immediately after dinner, I left the hotel to go to his house. The street was a surrealistic spectacle. The grand bourgeoisie of Vientiane had the habit of wasting electricity without paying their bills, so the General Electricity company had cut off the electricity in the area. It was late at night and the street was lit only by a few candles on the edge of the sidewalk. People stood in the eerie light, chatting together in front of their darkened houses. In front of the house located at Rondon's address, I stopped and called out towards the window: "Monsieur Rondon!" The window opened and there he was. We talked for two hours in the light of the candles. When he hunted his Coleopterae at night, with his mercury vapor lamp, a large number of moths also collected around it. He generously proposed to send me all the moths he captured for an entire year. Thus, I received postal packages from Laos containing all kinds of Sphingidae, Attacidae and other families of moths. I shared this manna amongst my collector friends, some of the Sphingidae going to Jean-Marie Cadiou and the Attacidae to Claude Lemaire. It was an excellent year, full of extremely rare species. In particular, there was a species of Sphingidae called *Clanis acuta* and described by the naturalist Mell of which only a single, male specimen existed, in the Berlin Museum. But its description contained errors and needed to be rewritten, modifying its name which had already been used. Rondon sent 10 specimens, both male and female, which we shared. Cadiou redescribed it, baptizing it *Clanis schwartzi* (the Latin version of my name is "*schwartzus-schwartzi*"). Actually, he should have called it "*rondoni*". Rondon died the following year. In 1984, Marie-Hélène and I passed through Vientiane on the way back from Viet-Nam. We met Mrs. Khamvene, an energetic and friendly lady who used to help Rondon with his hunting. She detested the regime of her country, but she remained in Laos until the death of her mother, then left for the United States to join her daughter. She had had to endure numerous "re-education sessions", and her son-in-law had been arrested and never heard of again.

## The boat people

The story of the boat people is the story of a particularly tragic episode in the history of Viet-Nam. The Vietnamese political regime has been Communist for decades. Democratic methods are not used. The NLF of the South fought remarkably throughout the war, and the population of the South paid heavily for their freedom. But when the North invaded the South, it spread its power everywhere, relegating the leaders and militants of the NLF to second place. As for the "third force", that of the non-Communist middle classes which had fought Thieu's regime and which had been savagely repressed by it, they were quite simply excluded from any kind of political participation in the new, unified Viet-Nam. Many intellectuals in the South were very resentful of the power of the North. Lacouture wrote about this in a book written in 1976, entitled *Voyage à travers une victoire* (*Voyage Through a Victory*), coining the neologism "northmalization", very ill-received by the Northern leaders. The South felt conquered. Moreover, near Saigon, in Cho Lon, there was a solid Chinese community of good merchants, who monopolized the rice commerce. The Communist regime could hardly avoid confronting this community more or less severely, and they did so pitilessly. It is often said that the Southerners voted with their feet; on the arrival of the Northern fighters, they massively abandoned the towns. It is also said, and there is truth to it, that the Algerian FLN was militarily defeated by France, but won the war politically, and that Viet-Nam defeated the United States militarily, but suffered a political loss. In the South, under Thieu, especially in Saigon, miserably poor people and prostitutes lived alongside a rich bourgeoisie, and the stores were filled with goods. A population used to a market economy abruptly confronted the restrictions of a Communist economy, with extremely low, almost unbearably low salaries. As one can imagine, there were murmurs of revolt, which took the form of a flight abroad, anywhere in the world. Hundreds of thousands of South Vietnamese fled. In November 1978, a boat containing nearly two thousand refugees was turned away from Malaysia.

This event had the effect of bringing the fugitives into the limelight; they were given the name of boat people. Their departures were clandestine, but it was very officially known that candidates for exile had to pay enormous sums in gold to intermediaries, who were well-known to the leaders and dignitaries of the regime, because corruption was generalized. The boat people traveled in disastrous conditions. The boats were usually old wrecks which were barely seaworthy, and they were the prey of pirates. Thailandese fishermen attacked them, raped the women and young girls, killed the men and carried off everything they could. Some of them managed to find refuge in Western countries, where they were welcomed, and moved in and became integrated, But many died at sea or were taken to refugee camps in the Far East; others lived hidden in horrible conditions. I signed onto a French operation called "A boat for Viet-Nam", and I was able to help a few refugees. But at the same time, a political campaign filled with hatred against the Vietnamese government began in France, particularly among left-wing intellectuals who now hated the same Viet-Nam which they had adored and idealized before. I categorically refused to participate in this movement, and considered it as no more than blind propaganda. Certainly, the Vietnamese government carried heavy responsibility for the fate of the boat people, but the rest of the world carried even more. Who had plunged Viet-Nam into war, if not France from 1945 to 1954, then the United States immediately afterwards, supporting the Southern dictators and attacking the North in 1964, in a terrible and ferocious war specially directed against the civil population, until the limping peace of 1973? Viet-Nam, already miserable under French colonization, emerged economically exhausted from its decades of war. It was the fifth poorest country in the world, just before Bangladesh. The big rich countries, the United States first and foremost (the famous Viet-Nam syndrome) practically directed the World Bank and they all boycotted Viet-Nam. The war was not fully over until 1975, and now it was 1979. I continued to support Viet-Nam just as before. The ambiguous and complicated attitude I adopted reminds me of a conversation I had had long before with Mai Van Bo. A mathematician we had invited to France for a short stay under a program of French-Vietnamese cooperation, had "chosen liberty" and wanted to remain in France. I tried to show him that he was putting the cooperation in general and me in particular into a delicate situation. But he had reflected carefully and conscientiously before making his choice. I gave up my protestations and helped him find a job in France. When I met Mai Van Bo, the Vietnamese organizer of the cooperation, to inform him of the situation, he asked me: "When a Vietnamese comes to France via the cooperation, do you exhort him to stay?" I answered him that of course I did nothing of the kind, that he could trust me on that, and that on the contrary I encouraged the Vietnamese to return home to work for the future of their country, but that if one persisted in wanting to remain in France, then it was my duty to help him. He understood that I could not act any differently. Naturally, the situation of the boat people was very different. They were refugees who reached France or other countries by risking their lives, and they needed help even more. However, the uninitiated cannot imagine to what point the Vietnamese (not all of them, obviously) are attached to their country, and often they try to keep contact with it even when they live abroad. Mai Van Bo truly approved of me. He hoped that the mathematician I had spoken to him about, and who meant to become French, would keep contact with his family which remained in Viet-Nam, and return as often as possible to visit. Pham Van Dong always wanted Vietnamese expatriates to consider themselves not as exiles but as friends from afar. Today, after decades of violent internal conflict, we are finally seeing a kind of reconciliation between Vietnamese refugees deeply hostile to the regime and others close to it; this reconciliation seems to be based on a middle road of critical adhesion to the country itself.

I expressed my opinion on Viet-Nam and the boat people in an article in *Le Monde*, dated December 18, 1978, cosigned with Madeleine Rebérioux. I would sign it again today if I needed to. I understand, however, that others felt differently. When I arrived in Viet-Nam in 1979, the French Embassy explained to me the deep influence my article had had in Viet-Nam. It was the first article they had seen outside of *L'Humanité* which was not a cry of hatred against them. After a long discussion at the summit, they had decided to publish it in their daily newspaper *Nhan Dan* (The People) – suppressing all the critical parts! That was one of my numerous subjects of discussion with Pham Van Dong in the year 1979.

## Vietnamese in my heart

Viet-Nam changed a lot after the disappearance of the Soviet Union, which obviously caused a considerable shock. Little by little, the country installed a market economy with all its good and bad consequences, but a very orthodox form of Marxism-Leninism is still taught in the universities, although the students couldn't care less about it. The Minister to whom I sent my evaluation in 1990, and to whom I openly expressed my reservations on this point, reacted by smiling humorously. The stores, which used to be empty, are now filled with all sorts of goods even in the smallest villages, but few people can actually procure what they need. However, thanks to the fact that they often work at many "little jobs", the people are able to buy more than one would think a priori. There are still a lot of bicycles in Hanoi, but there are also motorbikes and even cars.

On the other hand, the political habits at the top have not changed at all. In order to please the politicians, the newspapers continue to write in a language of formulas all too familiar to the depressed connoisseur. The example, naturally, comes from the top. For instance, a few years ago, a central committee meeting seriously concluded that "Viet-Nam would pursue the march towards Socialism with ever increasing success, with the help of the glorious Soviet Union" (which no longer existed). Fortunately, the public speaks out much more freely, even in the street and in public places, and is more and more interested in what goes on in foreign countries. In a word, underneath the mantle of official declarations, Viet-Nam is changing. The generation of the old political leaders is still prestigious, but no longer really governs. The young generation which will change Viet-Nam and make it into a democracy does not govern yet. Power belongs to an intermediate generation which lived exclusively during the war, and which lacks both the political references of the older generation and the modern ideas of democracy. The young people can acquire some idea of democracy via contacts with foreign countries. There are many different movements within the government; some leaders are locked into their orthodoxy, others are open to modernization. The case of the dissident Duong Thu Huong is quite typical of the situation: one can have broken with the regime and yet, directly or via intermediaries, dialogue with the leaders at the highest level. That would have been unthinkable in Brezhnev's USSR.

Viet-Nam marked my life. I learned about colonized Indochina by the book written by Andrée Viollis in 1931, which I read *sixty years ago* in 1935. I was twenty. My fight for the liberty of Viet-Nam was the longest fight of my existence. I loved and will always love Viet-Nam, its landscapes, its extraordinary people and its bicycles. I am a little bit Vietnamese. Meeting a Vietnamese person or hearing Vietnamese spoken in the bus (even though I don't know the language) makes me inexplicably happy. My sentimental fiber vibrates for the country. Many left-wing people share this feeling. The Vietnamese do not forget me and many students write to me, calling me "the godfather of all Vietnamese." Sartre also had the right to this title, but I don't believe he remained as attached to Viet-Nam as I am.

# Chapter XI
# The Distant War in Afghanistan

## The Soviet invasion

Afghanistan always escaped being colonized. Both England and Russia, then the USSR, considered it as a buffer state separating the British empire from the Russian State. It was governed by a dynasty whose last king, Zāhir chāh, abdicated in 1973, overthrown by his cousin Daoud who proclaimed a republic. The American scholar Michael Barry, who knows the country well, described it superbly in his very interesting 1984 book *The Kingdom of Insolence* as a country whose central power had no authority whatsoever outside of the capital. The rest of the country was in the hands of tribes from various ethnic groups (Pachtouns, Tadjiks, Uzbeks and so on) or religious groups (always Muslim, whether moderate or fundamentalists, shiite or sunni), and was governed by independent local chieftains. This confusing situation was actually more or less stable. With the support of the USSR, the Communist DPAP (Democratic Party of the Afghan People) prepared a coup d'état on April 27, 1978; the USSR had always wished to achieve an opening to the "warm seas", but it broke with Anglo-Russian tradition in doing so. Daoud was deposed and assassinated. The Afghan Communist Party was divided into two rival sections, the Khalq and the Partcham. Before 1979, the Khalq held power under the direction of Taraki, and then of Hafizullah Amin, an incredibly brutal leader who exercised an inhuman repression. He was so unpopular that in 1979 he was on the point of being overthrown in favor of non-Communists.

To avoid that, the Red Army (105-th parachutist division with motorized troops) forcibly invaded the country on December 27, executed Amin and installed Babrak Karmal of the Partcham in his place. The Soviet government claimed to have been called in by Babrak Karmal, but we know that in fact he was simply brought along in the Red Army trucks. The Red Army quickly occupied the large towns and the main roads, but it was incapable of establishing control over the countryside. Babrak's regime was much more open, but it was dominated by the Partcham Communist Party. For their part, the Afghan people quickly organized a resistance. They held the countryside and in particular the mountains, and even some main roads which were not paved. The rest of the world was not fooled: this was a Soviet invasion. President Carter announced a series of economic sanctions against the USSR, and organized a boycott of the Moscow Olympic Games in 1980. The European countries were more "prudent", but some of them did recall their ambassadors from Kabul. The Islamic Conference decided to give military support to the Afghan resistance, and Pakistan, which neighbors Afghanistan, actually became deeply involved throughout the entire time of the invasion. Most of the resistance parties organized their headquarters in Peshawar. At the general assemblies of the UN in January and November 1980, a large majority demanded the immediate and unconditional departure of the "foreign troops". A new "Viet-Nam war" was beginning. The Soviets, who had obviously believed in a swift invasion encountering no resistance, showed a complete ignorance of the Afghan situation and forgot the lesson learned from Viet-Nam, of the impossibility of subduing a hostile countryside. It was the first time that the USSR, which usually supported third world countries against the West, actually engaged in an imperialist and colonialist war against a third world country. On the international level, this created a completely new situation for left-wing public opinion. In France, the non-Communist left chose to support the Afghan resistance – personally I remember speaking openly of "Soviet imperialism" – while the French Communist Party chose to support the invasion. A text explaining that "the Red Army was bringing Socialist civilization to a backward country still living in the Middle Ages" appeared in *Le Monde*, signed by a thousand French Communists. It wasn't completely false either: Babrak Karmal made some efforts to modernize a very backward country, but he imposed them using brutal methods intolerable to a people proud of its independence.

Many young people in France were quite embarrassed by the whole thing, although there had been precedents such as the invasions of Hungary and Czechoslovakia, or the events concerning Pliushch and Shcharansky. However, the invasion of Afghanistan rapidly had major consequences. Personally I considered it as one of the major events of the last quarter of the century and believed that it would last for years, and end with the fall of the Soviet regime. Support groups for the Afghan resistance were formed in all the Western countries and in Japan. In France, a support movement was started up by members of the PCR (Parti Communiste Révolutionnaire)[88], which was a tiny marginal Maoist Party. Three young people in their thirties, Jean Freyss, Jean-Paul Gay

88) PCR: Revolutionary Communist Party

and Jean-Pierre Turpin, all intelligent politicians, set to organizing the movement with energy and sagacity. Freyss had been a classmate and a friend of Marc-André; he contacted me (later, he became less involved with this movement, and turned towards the defense of the Kanak people of New Caledonia). His approach squared with my own feelings and I joined his group immediately. Several others joined us as well: Mike Barry of course, and Bernard Delpuech, a journalist who traveled extensively to Islamabad and Peshawar and Afghanistan, Camille Perret, Jean Elleinstein who left the French Communist Party at this point, Hervé Barré, Emmanuel and Isabelle Delloye, etc. During all those years of war, we remained close. Our "collective" organized a demonstration on June 6, 1980, at the Mutuality: "Six hours for Afghanistan". The hall was full and Freyss presided the meeting: many intellectuals such as Arthur and Lise London, Hélène Parmelin and Antoine Spire were present. On December 2, 1980, our newspaper the *Afghan Struggle* began a campaign called "a million for Afghanistan". Mike Barry and Hervé Barré went to Afghanistan. Later (January 24, 1981), the "collective" transformed itself into the SMAR, the Support Movement for Afghan Resistance, which became a real association according to the 1901 law, of which I was elected president. The movement became well-known thanks to the *Afghan Struggle* and to the events of the war itself, and we came to know a variety of Afghan groups; we met with representatives of the parties in Peshawar and representatives of the local chieftains of the resistance (these were not always the same people!) as well as with support groups in other countries. Michel Verron joined us in May 1981. He had worked for UNESCO in Kabul on a literacy program, and he had a deep understanding of the Afghan reality; he left the Communist Party to become a pillar of the SMAR, and later joined the BIA (Bureau International d'Afghanistan)[89] and the MADERA (see page 430) as well.

The Soviet invasion had high and low points over the years. The governmental army suffered from constant desertions. In September 1981, the government extended military service by two or three years, but the young people took to massively supporting the resistance. After important military operations in 1982, the year 1983 was fairly calm and there were even several efforts at diplomacy. A delegation of resistants was received by Margaret Thatcher and President Reagan. However, the year 1984 saw the greatest military offensive yet by the Soviet Union, with high-altitude bombings, troops transported by helicopter to the mountain crests, and threats by the USSR to bomb Pakistan. These offensives continued in 1985, but Gorbatchev came to power on March

89) BIA: International Afghanistan Bureau

11th of that year with a desire to relieve himself of the difficult situation. He tried desperately to put an end to it once and for all, but the resistants became stronger and stronger, supported by military material from foreign countries. So Gorbatchev began some prudent negotiations with the United States. It was high time. In Erivan, capital of the Republic of Armenia, two hundred parents demonstrated, on May 20th, against sending their sons to the Afghan front. The war was more and more unpopular, it was becoming impossible to conceal the number of dead, and the Soviets hardly understood what they were actually doing in Afghanistan. Kabul government aviators destroyed planes on the Shindad base, and two Mig helicopters spontaneously left for Pakistan. The first Soviet-American exchanges took place in June. In 1986, admitting that the Afghan war was a "bleeding wound", Gorbatchev dismissed Babrak Karmal and replaced him by Najibullah, in the hopes of a "national reconciliation" – without however shedding doubt on the Communist regime. It's always stupefying when a government engaged in calamitous politics sinks endlessly deeper and deeper into the mire.

The Afghan resistance had become an offensive on all fronts. In 1987, the government of Kabul (Najibullah) ordered the retreat of the Red Army into its security perimeter and sent delegations to sixty-six capitals of non-aligned countries to obtain some support, but in vain. Afghanistan was also a non-aligned country .... After years of asking, the Afghan resistants obtained the delivery of some Stinger ground-air missiles which could be manipulated by a single man, and the Soviets, using low-altitude helicopters, finally lost control of the air. Internal protests increased in the USSR; the newspaper *Ogoniok* published an article by a veteran declaring that the Afghan war was a "dirty war". Sakharov, freed by Gorbatchev in 1987, immediately demanded the cessation of the war. On May 15, 1987, Gorbatchev announced the complete evacuation of all Soviet troops from Afghanistan, within 10 months and without any compensation. However, the war actually ended only in 1988. The conditions of the evacuation were fixed in an agreement between Najibullah and Pakistan, and guaranteed by the USSR and the USA. The Soviet intervention had lasted over nine years, from December 27, 1979 to February 15, 1989, about as long as French wars in Indochina and Algeria, and US war in Viet-Nam. It seems that that is the period of time necessary for a government to "understand". But in the preceding cases, once the issue was understood, it was understood, whereas the USSR immediately began a new war in Chechnya.

## The BIA, International Afghanistan Bureau

Our support of the Afghan people was always very active. There were continual meetings, colloquia, international forums, and trips; I won't enumerate them. In November 1981 we decided to found the BIA (see page 423), because of the needs of the Afghan resistance and also because our group had become so large, absorbing many foreign delegations. I was elected president, Jean-Paul Gay and Michel Verron were permanent paid staff, and Camille Perret was the secretary (all French). I remember receiving the Italian delegation (Ripa di Meana) as well as Belgian, German and Norwegian delegations. into the BIA, which had a large office. what with trips to Afghanistan and Pakistan, lectures and the publication of the *Afghan Challenge*, that our finances were always in difficulty. But we received support from all over Europe, for instance from the French Ministry of Foreign Affairs and from the European Parliament for colloquia.

Islam is the religion of all Afghans. The alliances between the various tribes were always unstable. In 1981, there were two alliances. On the one hand, the fundamentalists, uniting seven groups: the Hezb-I-Islami of Gulbuddin Hekmatiâr (extreme fundamentalists), the Hezb-I-Islami of Khales, the Jamiat-I-Islami of B. Rabbani (a hinge group), and four small parties all with Islam in their name. On the other hand, the moderate alliance consisted of three groups: the Harakat-e-enqelab of Nabi Mohammedi, the Mahaz-e-Islami of Gaylani and the Jabha-yemelli of Mojaddedi. These alliances functioned until 1985. In 1984, the Afghan resistance sent a unique representative, Rabbani, to the Arab Conference of Casablanca from January 16–18 (he was later received by the secretary general of the French Ministry of Foreign Affairs). But their unanimity did not last. Ahmed Shah Massoud, the "Panshir lion", a prestigious military leader who led significant victories against the Soviets, was among the more moderate Islamists of Rabbani's party. However, none of them was really secular; Islam was their cement. But they were not Arab, which explains why Arab support was always modest although constant. What were we doing in the middle of this situation? We were defending the Afghan people against Soviet aggression, but we had no illusions about the future of Afghanistan after the end of the war. We never took sides between the different religious or tribal groups, and we kept up direct relations with representatives from all the groups. But we needed to remain prudent about certain doubtful aspects of the religious conflict. For instance, the parties were unanimously opposed to the Soviets when they allowed girls to attend public schools. Many cultured women from Kabul and other cities had excellent reasons to be in favor of the Soviet invasion.

For decades, people from western countries visiting the USSR could easily perceive the mess which reigned everywhere. The economy was delapidated and society was decomposing. According to the moment, long lines formed in front of food stores or clothing stores. During my stay in Moscow in 1964, I sometimes had to wait as much as an hour and half before being served (this happened in Poland as well). There was no telephone book in Moscow, so if you didn't know someone's phone number, you couldn't phone him. Socialist economy used no checks, and payment at a distance was possible only by postal order. University professors received their salaries every fifteen days, and lined up in long queues in front of the treasury to get it in cash. But a solid rumor said that the army, which had an enormous budget and a formidable organization, was the strongest in the world and functioned perfectly. The Afghan war showed that this was untrue: the army was just like the rest of the society. Relations between officers and privates were execrable, the fighting troops were inefficient, many soldiers deserted or suffered the ravages of alcoholism. The army battling the Afghans was in a state of decomposition. Every sector of Soviet society was disintegrating, even the nuclear plants (Tchernobyl). Probably space exploration is the only domain in which the Russians had a truly advanced and well-supported position. The sciences, except for biology which was destroyed by Lyssenko, never ceased developing in the USSR, even under Stalin, but the disastrous economic situation is now causing many scientists to flee to the United States. But enough of this digression. The Afghan mujaheddins usually treated Soviet deserters and prisoners reasonably well. The prisoners were taken to Geneva under the control of the Red Cross. However, whenever they conquered a town, the mujaheddins immediately set to massacring the inhabitants. We had to inform them of the Geneva conventions and teach them good behavior. The lesson was learned – more or less. But the Soviet acts of barbarism and torture were innumerable. Gorbatchev had called the war "a bleeding wound"; it accelerated the disintegration of the army, while bonding the resistants.

After the meeting of the Russell Tribunal for Viet-Nam, a permanent organism was created for the purposes of organising similar meetings in the future. This organism, the Permanent Tribunal of the People, was located in Rome at the Lelio Basso Foundation. The Permanent Tribunal of the People met many times, but I was almost never able to attend their meetings. At our request, however, the Permanent Tribunal organized two sessions on Afghanistan, one in Stockholm in May 1981 and one at the Sorbonne in December 1982. The president of honor was Vladimir Dedijer, the president François Rigaux. There was a very interesting discussion at the first meeting, with Fred Halliday. The meetings, like all the preceding ones, were highly dramatic. Witnesses spoke of numerous massacres of the civil population, revenge and ferocious torture on the part of the Red Army. Entire villages had been erased from the map. The conclusions were severe against the USSR, which was unanimously condemned. As it should be, the procedure was entirely rigourous; thus for instance, we came to no conclusions on the subject of the purported chemical warfare. Certainly, there were strong reasons to believe it was taking place, which we pointed out, but there were no actual proofs. This was a great disappointment for the Afghan fighters. Several of them said to us: "But we saw the rotting corpses of our comrades, whereas the Koran says that if a Muslim soldier dies in combat, his body will never rot." I succeeded in persuading them that our objectivity was the guarantee of the influence our judgement would have on international public opinion. The unbearable statements of a young girl from Kabul, who was interviewed afterwards by journalists, spoke of torture. Locked in a local torture center, she saw human arms and eyes lying on the tables. I congratulated her publicly for her courage, the precision of her statement and her answers to questions.

I remember one difficult moment when one Indian Communist juror became irritated about one of the accusations against the USSR: "I don't understand," he said, "how the USSR can be accused like this! In every international conflict, the USSR has been on the side of peace. Here, it is attacking, and that is why I have accepted to be a member of this tribunal, but it is merely a tiny particular case; we must not be too severe."

The Afghan people suffered terribly. Out of a population of twenty million, more than one million died. Nearly six million people became refugees, two million in Pakistan, one million in Iran and three million as "internal refugees," having fled their village or town because of the bombings and invasions. This created a tremendous problem. The idea of an international conference on the subject was initiated by the BIA, and the conference was held in Geneva in November 1983. The collaboration of the general organizations representing refugees was essential. We prepared our participation in the conference with an information-seeking mission to Pakistan, to Iran and to the interior of Afghanistan: Michel Verron, Jean-Paul Gay, Michel Foucher (a geographer), René Garrigues (a doctor) and Charles Lilin (an agronomist) returned with abundant documentation.

Afghanistan was ruined by the Soviet invasion. Even when the Russian troops left, in February 1989, power remained in the hands of Najibullah in Kabul. The war was far from over. The refugees did not return. Combat never ceased between the power in Kabul and the resistance. I won't describe the results of these combats, although they were murderous. Najibullah held Kabul and some other towns; the resistants held all the rest. No international organization intervened in this stage of the war. But the United States generously armed the resistance, and the USSR transported food, basic products and arms to Kabul by air. The situation was burlesque. The resistance slowly but surely surrounded and laid siege to the capital. The fall of Najibullah became inevitable.

During this long period, we were not inactive. Apart from the actual trips, such as those taken by Delpuech, our correspondent in Peshawar, which were the source of a wealth of information, we organized a number of small meetings in Paris, inviting roughly fifty people to each: members of the BIA, delegates representing the various Afghan political tendencies (whose trips to Paris we financed) and important French politicians and diplomats. I met the greatest number of Afghan delegates at these meetings: Homayoun Assefi, Modjaddedi, Rabbani (who was only recently the president of Afghanistan). We discussed the possibilities for negotiating with the different Afghan factions, and the conditions for a general peace. Everyone who attended these meetings gained something from the exchange of views. At the last meeting, we invited the First Secretaries of the embassies of the US and the USSR in France. They did not say a word, but listened attentively to the proceedings. Several Afghan delegates expounded on precise peace plans, giving a role to each actor and counting systematically on the departure of Najibullah, the last symbol of the Soviet occupation. The end was approaching. The last episode in the history of the BIA was rather comical. As president of the BIA, I received a letter from the ambassador from Kabul in Paris, congratulating (!) the BIA on its interest in the destiny of the Afghan people, and asking it to act as intermediary between Kabul and the Afghan resistance! He had confidence in us because he knew that the Afghan leaders did. He claimed to be sent by Najibullah (we were never really certain of this; perhaps he was acting independently). I must say I was amazed. After having informed my colleagues at the BIA, I spoke to Louis Joinet, a counselor to the Prime Minister, and a close friend, whom I had met when he was national secretary to the union of magistrates. He advised me to go ahead with it, using all necessary prudence. I became a diplomat!

The rule in such cases is that we placed ourselves on the same level as Najibullah: the ambassador wrote to me in the name of Najibullah, and a member of the bureau of the BIA, Jean-Paul Gay actually, wrote to him in my name. We proposed a meeting in a neutral location: a room in the Ecole Normale. He accepted the principle. We did however warn him that: 1) we needed a letter from Najibullah confirming his request, 2) we would not be able to act as intermediaries in the negotiation because we were not neutral, but friendly to the resistance, but we could discuss with the leaders of the resistance and persuade them to meet with him; we could even attend the meeting, but without taking part in the discussion; 3) before anything, we needed to discuss with the leaders of the Afghan resistance. A necessary condition to the success of such a meeting was that the Kabul party accept the departure of Najibullah, though some of his Ministers could remain in a post-negotiation government. He accepted everything. Najibullah would leave, the negotiation would deal with the conditions of his departure. We decided to do nothing before receiving a letter from Najibullah himself. It never came. In fact, it seems that this ambassador had simultaneously made other contacts. He was depressed and couldn't take any more; probably he was not sent by Najibullah, unless Najibullah couldn't take any more either. A few weeks later, the resistance captured Kabul and Najibullah left the capital as a prisoner on April 16, 1992.

The different factions struggled for power for several years. The war continued. In 1994, a new force appeared, that of the "religious students", the Taliban. These men came directly from Koranic schools and their only culture was the Koran, and in fact uniquely the most reactionary elements of the Koran. They quickly won a series of victories, arriving near Kabul almost without fighting, profiting from the war-weariness of the population, who had had to ceaselessly endure the looting and corruption of the tribal leaders. They were even quite well received in the southern countryside and towns, where their form of Islam remained relatively moderate. At first, they were defeated before Kabul by General Massoud. But they returned and took Kabul on September 27, 1996, at which time they assassinated Najibullah and hung him from a street lamp. In fact, the Taliban was completely supported and armed by Pakistan, which thus gained influence in Afghanistan, but also by the United States. They are Sunnite Muslims (the Iranians are Shiites), and deeply fundamentalist. The tribes of northern Afghanistan were Tadjiks and Uzbeks; the Taliban are Pachtouns. As soon as they arrived in Kabul, they applied the Islamic law Sharia: chador for women, beards for the men, hands cut off for stealing and stoning of adulterous women, suppression of all schools for girls: these appear to be their only projects. The future of Afghanistan appears dim, especially for women. But perhaps this regime is less stable than that of the Iranian Ayatollahs: Afghanistan has always been the kingdom of insolence. The truth is, Afghanistan was destroyed by the war; the elite disappeared and was replaced by a political vacuum. The American government wanted above all to combat Iran's possible influence. Pakistan always feared a "Teheran-Kabul-India" axis. The situation is still quite fluid. Massoud and his allies occasionally attack. Pakistan doesn't want the Taliban to achieve a true supremacy; it wants control itself. Washington is concerned by the opium culture, which is contrary to the principles of Islam but a source of riches which is difficult for the Taliban to reject; Washington also fears a true Pakistani domination. I wouldn't hasard a prediction of even the near future.

One might claim that we shouldn't have expended so much energy on the expulsion of the Soviets and the Communist powers; their social program was much better than the Taliban's! But with or without a social program, the Red Army was a foreign conquering army, bringing bombings, massacres, torture and mass executions. The Afghans unanimously revolted against it; it was universally hated, it was impossible not to support them. The final result is execrable (and it was actually foreseeable, not in the form of Taliban domination, but at least in the form of civil war). But it was the Communist and Soviet repression which gave rise to civil war and the Taliban. Similarly, the French left fought against the Shah of Iran, who had a program of industrialization, medicalization and education for his country, but who was tyrannical, brutal and a torturer (his political police, the DINA, burned prisoners on a grill); the regime of the Ayatollahs is obscurantist and uses torture even more. If we had known, we would all have preferred the Shah to the Ayatollahs, but he was not endurable because of his excesses, and it is precisely his excesses which allowed the rise of the Ayatollahs. If we also consider what happened in Algeria and Cambodia, we might actually become discouraged from helping third world peoples in their legitimate defenses against imperialist attackers. But it is the imperialist attacks themselves which produce the regimes which follow them, by their pitiless repression. Thus we must support third world revolts – but it makes one feel like tearing one's hair out!

After the departure of Najibullah, the BIA disappeared, having no further role to play. We considered contributing to civil peace by using our past prestige to unite the representatives of the different factions in discussions, but we soon saw that it was hopeless.

Apart from the BIA, the MADERA (Mission d'Aide pour le Développement des Economies Rurales d'Afghanistan)[90], a non-governmental organization founded in 1988, should also be mentioned. This organization worked alongside the BIA, but was administratively independent. It sent working teams to Afghanistan, where salaried (French and Afghan) members taught Afghan peasants about modern agriculture and agronomy, cattle raising, forestry, economy, road construction, education, medical and veterinary problems etc. This enormous organization employs more than four hundred salaried members in field work, of which about ten are French expatriates. Thanks to its success, it has a considerable budget; it receives about seven million francs a year from

90) MADERA: Aid Mission for the Development of Rural Economies in Afghanistan

the European Union, and controls the use of these funds; its total annual budget is about fifteen million francs. For many years, I was its president, and together with Michel Verron or Jean-Paul Gay, I went to request funding from the ministry of Foreign Relations. After a certain point, I asked to be replaced. The work involved in running such an enormous organization was too much for me, and a more "industrial" direction was becoming necessary. I was replaced by Michel Verron, who worked almost full time at it (he asked to retire in 1997). Monique Otchakovski-Laurens is the general delegate. MADERA also helped the Afghans to learn about the exercise of democracy, justice and law, and to allow women to speak. When a local chief opposed the presence of women in our working team, MADERA left, saying that it would return only when women would be accepted as the equals of men. The local population put pressure on the chief until he recanted his order. The culture of poppies, by far the most lucrative for the peasant, has recently expanded, as in many other third world countries. Anyone can measure the consequences on the country. MADERA combats this tendency and has left every poppy-cultivating region. The BIA has disappeared, but MADERA continues to help the Afghan reconstruction. There are other similar non-governmental organizations, in particular very good Scandinavian teams.

## The limits of involvement

In comparison with the French people who became involved in the struggle against the Algerian war or the Viet-Nam war, the number of those interested in the war Afghanistan war was tiny, perhaps just a few hundred. Why? In part, because the aggressor was the USSR. The USSR no longer exists. However, in spite of the Pliushch affair which began in 1974, there was a powerful taboo associated with it back in 1979 when the Afghan invasion took place. Moreover, Afghanistan is very "far away"; less than Viet-Nam in actual kilometers, but much more in the public mind. It is a country with no modern history; an unknown country. I myself knew next to nothing about it when I became involved. Very few people had actually visited the country. Yet an ancient local civilization exists. Since the time of Alexander the Great, Herat is a large town full of Asian and Greek art. We tried to organise an exhibition of the splendid artistic treasures of Herat in the Grand Palais Museum in Paris, under the direction of Dupaigne, the vice-director of the Museum of Man. We wanted it to be cultural and apolitical, we wouldn't have dealt with the war. I submitted my project to the Ministers of National Education and of Foreign Affairs, and obtained their good will and some grants. But then the project ran into an obstacle. We could not obtain the agreement of the Afghan ambassador to Paris; diplomatic relations between the two countries had not been broken off. At any rate, we were almost alone in a combat which, as I said above, dealt with a major event. The press almost never discussed Afghanistan. Almost none of the "big consciences" were troubled by it. All the people who were really active were "new arrivals," young people. The old fighters whose causes I had defended were absent. As far as fame was concerned, I was more or less the only possible president. On the other hand, it was a period of tremendous overwork for me. Let me digress here. I was at the end of my university career, but my activities had doubled. I was involved in the Commission du Bilan, created by the Prime Minister Pierre Mauroy, directed by François Bloch-Lainé, to write the *Bilan de la France* in May 1981. We had about six months to do it, from June to December 1981, and it was a full-time job. For this commission, I wrote the fourth volume together with Renée Ribier and André Staropoli. Called *Teaching and Scientific Development*, it was five hundred and seventy pages long and discussed elementary and secondary school teaching, universities and "grandes écoles," research and technology.

The Savary law on higher education, which was passed on January 15, 1984, was then in preparation. A large number of university professors considered it to be very negative, and I still believe that it carries a heavy responsibility for the terrible decline of the University. Most of it should be annulled. These professors became massively involved in an association for the defense of science, called QSF, Quality of French Science. It was founded on the initiative of Georges Poitou, director of the ENS, and soon had five hundred members. Once again – I found myself president! I devoted a lot of time to writing articles, and published a book, *To Save the University*, which appeared at the same time as the Savary law. We were defeated, but QSF continued the battle, with however a great reduction of membership, and is still continuing. I am now honorary president, and Pierre Merlin, a professor of urbanism at the university of Paris VIII is the president. Finally, the National Committee for Evaluation of Universities, CNE, whose creation I had suggested in my report to the Bilan in 1981, was decided in the Savary law (that was one good thing!), and created by Jean-Pierre Chevènement in May 1985. The seventeen members were appointed by the president of the Republic, and – once again I found myself plumped into the president's chair! Since I had insisted on the creation of this committee, I could hardly refuse. It was a useful and extremely interesting experience, I think, but it caused the worst period of overwork of my existence. From May 1985 to May 1989, I devoted myself to it full time. For the first time in my life, I had an office job: Mondays, Tuesdays and Thursdays I was in the CNE office from

Schwartz teaching

nine-thirty to six or seven in the evening. I ate a cold lunch brought from home, having no time to eat in a restaurant. Fridays, Saturdays and Sundays, I stayed home the entire day to do mathematics. I didn't have a single "distraction" in the week. Naturally, I was helped by the collaboration of the other members of the committee, for instance the secretary general André Staropoli. Mlle. Pouderous, the former secretary of the direction of the Mathematical Center of the Ecole Polytechnique, handled the secretarial work at the CNE and I had several secretaries and delegates with different missions.

On the other hand, several hundred experts chosen by us wrote up remarkable reports on the various disciplines represented in the universities. I alone wrote the complete evaluation of the Louis-Pasteur University of Strasbourg, which was the first university to request an evaluation. The work was very interesting and useful to the universities, who took our recommendations largely into account. However, they were obviously not useful to the Ministry of National Education. I never found any proof that a single person there ever read our reports. It is obvious that someone should have been appointed in the Ministry to read them, to brief the Minister on them, and to discuss them with us, as they do in Great Britain. I promised myself to write a vengeful article one day on the incompetence of the central administration. I have no room to write it here; actually it would take a whole book. Thanks to my Spartan schedule, I did not have to interrupt my mathematical research. I read deep and important articles, for instance the one by Jean-Michel Bismut and Nicole Berline on the Atiyah-Singer index and Watanabe's article on Malliavin calculus (which I should have known before), and I published seven articles during this period.

Given the conditions, it was clear that I was simply unable to be a fully present president of all the different organizations dealing with the situation in Afghanistan. However, I was not a ceremonial president either. I followed everything that went on, participated in every initiative and took care of the most important procedures myself. But I also functioned a great deal via delegation of power, and held regular meetings of the principal actors in my house. Our numerous conversations on every aspect of the war revealed a convergence of views so complete that I knew I could count on them with complete confidence. This constant delegation of power was well-received by the organizations. It gave more responsibility to everyone, so that the various missions were accomplished in an atmosphere of freedom which was personally gratifying and very efficient. I thank all of them for the friendly collaboration which endured for so long, in the defense of a just cause which remained almost unknown in France.

# Chapter XII
# The Committee of Mathematicians

During my long struggle for human rights, I tried to support the rights of certain oppressed peoples, Algerians, Tunisians, Moroccans, Madagascans, Vietnamese and Afghans. But one of my battles remains separate from all of those: the battle to defend the rights of certain individuals, prisoners of opinion and above all, mathematicians. The case of Maurice Audin inspired many of these new battles.

## The Pliushch affair

The Soviet mathematician Yuri Shikhanovich, a logician, was the Russian translator of the first volume of Bourbaki's *Set Theory*. He was arrested in September 1972 and judged in November 1973, without the presence of a lawyer; he never had the slightest contact with a lawyer either before, during or after his trial. After the trial, he was condemned to be sent indefinitely to the psychiatric hospital in Yakhroma. The judgment stated that "His symptoms are not serious. He has a great formal intelligence and thus reasons logically. However, and this appears more serious, he has interests outside of his speciality." This judgment requires no commentary. They didn't find it necessary to prescribe any treatment for him. I was told about his case in 1973 by Tania Mathon, a psychiatrist working for the CNRS. A Frenchwoman of Russian origin, she had traveled to Moscow several times in the preceding years, spoke Russian very well and had many ties in Russia, particularly with dissidents. She remained closely associated with the Committee of Mathematicians which was formed later, and played an important role in it – I call her our *mathematician Honoris Causa*.

Shortly afterwards, I learned of the arrest of Leonid Pliushch, which had actually taken place earlier. He was a young computer scientist who had been arrested in January 1972 and judged in July 1973 for having belonged to an organization for the defense of human rights. He also had no contact with a lawyer at all, and was condemned to be sent to the *special* psychiatric hospital of Dniepropetrovsk, in Ukraine. Without ever telling him what his treatment consisted of, he was subjected to high doses of painful medicines which provoked swellings, exhaustion and fever and rendered him incapable of reading and writing. His wife, Tatiana Pliushch, managed to visit him in January 1974 and found him profoundly changed.

A communication was sent to the newspaper *Le Monde*, alerting the paper to the Shikhanovich case. On February 8, 1974, some mathematicians (Marcel Berger, Michel Broué, Claude Chevalley, Claude François Picard, Laurent Schwartz, Jean-Pierre Serre, Jean-Louis Verdier and Michel Zisman – Cartan couldn't make it) were received, very politely, in the Embassy of the USSR in Paris, by Mr. Valentin Dvinine, the counselor of the Embassy, and by Valery Matissov, second secretary of cultural services, around a table pleasantly laid for tea. We demanded the immediate liberation of Shikhanovich since the diagnosis of his medical condition was not serious. We left them several volumes of Bourbaki in French, dedicated to Shikhanovich, but they were never transmitted to him. He did, however, manage to let us know that he had received what we sent him by post.

We asked about Pliushch also, but they claimed in the Embassy to know nothing about his case. In fact, they hinted that there was no such person. They explained to us that the special psychiatric hospitals were organized to take care of important people, such as Academicians, and the patients there received even more qualified care than those in ordinary hospitals. We only obtained a promise that they would keep us informed. A few days later, we received a petition in favor of Pliushch, signed by Soviet VIPs whom we we were to encounter frequently later on: Andrei Sakharov, his wife Elena Bonner, Tatiana Velikanova, Sergei Kovalev (who was recently appointed by Yeltsin to the position of "delegate of the President for Human Rights" and who, in this capacity, tried to help the Chechen people, but ended up resigning because of Yeltsin's intransigence), Andrei Tverdokhlebov, Tatiana Khorodovich. In USSR, some dissidents were locked up on trumped-up pretexts, but others were able to communicate with foreign countries, so we received their messages and were able to make them known. It took a lot of patience and determination to communicate by telephone; it took hours, and when you had obtained the communication, you kept getting cut off and having to start again. As time passed, the telephones of our partners were cut more and more, and it became impossible for them to telephone us. Tania Mathon spent a large portion of her time making these telephone calls.

At the beginning of 1974, the American mathematician of Lettonian origin Lipman Bers, who was in contact with us about Pliushch, proposed to Henri Cartan to found an international committee to defend Shikhanovich and Pliushch. Cartan and I set to work without delay, and Michel Broué soon joined us. French mathematicians joined us massively; quite soon we had five hundred members. Other countries such as Israel, Switzerland, Sweden, the United States, Canada

Claude Chevalley

(where Israel Halperin was particularly active), and Japan (impelled by the energy of the mathematician Iyanaga), created analogous organizations. In an incredibly short time, committees appeared in dozens of countries. From the very beginning of our activity, we were in close contact with Pliushch's wife, Tatiana Jitnikova. She had been fired from her teaching job in 1972, was constantly harassed and even threatened with the removal of her children from her care.

The increase in numbers and power of our international committee was amazing – we soon had eight hundred members in France and two thousand around the world. Shikhanovich was liberated on July 5, 1974, without having been subjected to treatment of any kind. An international conference on psychiatry took place in the USSR (Tbilisi) on October 15, 1973. In a letter to our committee, the Soviet authorities claimed that two western psychiatrists had confirmed that Pliushch was mentally ill, whereas in fact we were aware that those same two psychiatrists had co-signed a letter protesting the detention of the latter. We wrote this back to the Soviets, who never answered.

At the International Congress of Mathematicians in Vancouver (which I did not attend) in 1974, Michel Broué organized a meeting of two hundred and fifty people, with Lipman Bers as president, for the liberation of Pliushch (this activity made it practically impossible for him to attend any of the talks at the conference!) Lipman Bers was then a professor at the University of Columbia in New York; he was also the president of the university, a member of the American National Academy of Sciences, president of the American Mathematical Society and a personal friend of many of the Soviet mathematicians at the conference. In a remarkable speech, he mentioned Angela Davis, a black American philosopher then in prison in the United States and defended by many democratic people and in particular by the Soviets, which the Americans considered to be an unjustifiable intervention in their internal affairs. The case of Pliushch was quite analogous; his liberation was a simple question of justice and principle. The Soviet mathematicians at the conference did not react. Tatiana was able to visit her husband on July 23, 1974, and found him in a slightly less miserable state; in particular the exhausting insulin injections had been stopped. At that 1974 conference, Michael Atiyah, Lipman Bers, Henri Cartan, Israel Halperin and Shokichi Iyanaga put out a petition which obtained nine hundred forty-eight signatures.

When Tatiana saw her husband again on February 10, 1975, he was hardly recognizable. She immediately alerted the international committee; indeed she played an essential role in saving her husband, at that point and during the whole story. Avital Shcharansky was no less pugnacious, later, in the defense of her husband Natan. Similarly, Josette Audin contributed enormously to the elucidation of the death of her husband Maurice Audin. In fact, wives played a fundamental role in every one of our struggles to free some prisoner, and I ended up thinking that this energy born of love was not just an aid, but an indispensable condition of success. On December 16, 1974, a very courageous letter was published by thirty-eight Soviet mathematicians, among which were Andrei Sakharov and the well-known physicist Yuri Orlov, who had been excluded from the Armenian Academy of Sciences, Alexander Lavout, another mathematician, the wife and daughter of Andrei Grigorenko, the famous general who had spoken in the defense of the Crimean Tatars and who had himself been sent to a special psychiatric hospital, the daughter of general Yakir, Elena Bonner, the wife of Sakharov, Tatiana Khodorovich, who was always very active and who came to see us when she emigrated to France, Tatiana Velikanova, who was later arrested and whose case we defended, and Anatoly Marchenko.

In France, there were three truly active members: Michel Broué, Henri Cartan and myself. We met regularly with Tania Mathon. The "affair" was practically run by the four of us for several years. I already knew Broué, but not very well; our present great friendship dates back to this activity. He's an excellent, internationally recognized mathematician, and at that time he was also a very militant Trotskyist (which I respected a lot – it reminded me of my youth – but I used to tease him, predicting that he wouldn't always be a Trotskyist). He's much younger than I am, and also reminds me of Vidal-Naquet, whom I met as a very young man. Thanks to his extraordinary dynamism, he was the pillar of the committee. He and I almost always agreed; we were guided by the same sense of ethics. With Cartan and Vidal-Naquet, he is one of the few people whom I invariably consult before taking a moral or political decision. We put out a regular bulletin, which was entirely edited by Michel Broué; it was an enormous job. We sent it to all the French members of the Committee, and to certain foreign correspondents who translated it into their language and sent it to their members; we also sent it to major human rights organizations. Some people reproached us for defending only mathematicians. But it seemed to me that although of course we were interested in each and every occurrence of injustice, coming to the aid of our colleagues was in the nature of a primordial duty.

By selecting a single individual, Pliushch, instead of a collectivity, we were applying the lesson we had learned from the creation of the Audin committee: the defense of an individual contributes to the defense of all similar victims and reinforces that defense by supporting opposition to dictatorship or abuse of power. The Committee of Mathematicians continued to defend individual cases. The Bourbaki seminar, which took place several times a year, adopted a particularly welcoming attitude to the Committee, in the Pliushch affair and all the later ones as well. By tradition, politics is entirely absent from the seminar. But a box for donations did circulate amongst the mathematicians during the first talk, which took place on a Saturday. Furthermore, after each Bourbaki seminar, about thirty people met regularly in the Henri Poincaré Institute to keep informed about the situation of the Committee and the people it was defending.

During the meetings of the four of us, we reflected on the past and decided on future projects. We wanted to avoid engaging in politics in the strict sense of the word. We demanded the liberation of Pliushch, but we forbade ourselves to express any views on the Soviet regime. This view made it possible for a few Communist mathematicians to join the Committee. There was no hierarchy. We were three active mathematicians, plain and simple. Broué called himself the secretary, in that he did the secretarial work and wrote up the bulletins, but it didn't correspond to any well-defined power. We had neither president nor vice-president. We wanted to do absolutely nothing which could repel even one single mathematician. If we had taken even the weakest of political stances, or if we had committed any act which did not receive universal approbation, the whole of our activity would have suffered. This ensured a total democracy within the committee. However, from the point of view of actual work, the committee was not democratic, since there was no elected hierarchy and "consequently", as I always said, no bureaucracy. Broué, Cartan and I had no power over anything whatsoever. The power the committee had obtained as a whole was absolutely spectacular; we re-awakened the activism of mathematicians who had already expressed themselves in the Algerian war, but without ever openly expressing any political opinions.

The following year, things began to move seriously. In the *Paese della Sera*, on April 27, 1975, there appeared a letter signed by twelve Italian Communists – the Italian Communist Party was much more open than the French one – in favor of Leonid Pliushch. Among the signataries, Lombardo Radice, a remarkable Communist and a member of the central committee of the PCI, worked closely with the members of the French Committee. At the beginning of 1975, an essay appeared signed by two Soviet prisoners in the camps, Vladimir Boukovsky and Doctor Semion Glouzman, called *Manual of Psychiatry for the Use of Dissidents*, giving advice to dissidents who found themselves arrested or sent to psychiatric hospitals. In France, it was published by *L'Esprit* (*The Spirit*) and distributed by our Committee. It made a lot of noice in a number of Western countries. At the beginning of 1975, Tatiana sent us an extremely dramatic letter: "The mathematician Pliushch no longer exists; there remains only a man at the extreme point of suffering, he has lost his memory and every faculty of reading, writing, thinking, he is infinitely ill and exhausted. I ask only one thing: that my husband be given back to me, that they allow us to leave this country. The right to emigration is the only thing which I absolutely require." This plea for help incited us to take a radical step in order to increase the intensity of our action: we organized an International Day for Pliushch on April 23, 1975, in collaboration with Amnesty International. That day, a press conference was held at the Henri Poincaré Institute, with Amnesty International and Jean-Jacques de Félice. The Day received the support of a large number of well-known figures; in particular we received an admirable letter from the Soviet mathematician Shafarevich, the support of the unions, that of the central committee of the League for Human Rights, of the Committee against special psychiatric hospitals in the USSR, and that of the thirty-eight Soviets I already mentioned. The delegation who organized the demonstration was received at the Embassy of the USSR by Yuri Victorin; it was one of the very last times that the Embassy consented to receive us. "You'll see, if ever Pliushch is freed one day, you'll see for yourselves that he's crazy," the secretary of the Embassy found it proper to inform us. We formally denied this statement, and I added (more from optimism than from reasoning) a statement which turned out to be prophetic: "Be careful, mathematics are very powerful in the world. I warn you that in a short time, Leonid Pliushch will be spoken of in every major world capital." He answered that I must be insane as well as Pliushch. The International Day, orchestrated simultaneously in several countries, had an enormously important effect.

Everything encouraged us to augment our power even further, which we did by organizing a grand meeting at the Mutuality Hall in Paris, on October 23, 1975. Once again, the logistics were completely handled by Broué, Cartan and myself. I had inherited the presidency, the speakers were Michel Broué, Jean-Jacques Marie who represented the "Committee for the immediate liberation of political prisoners in Eastern Europe", the lawyer Jean-Jacques de Félice, a psychiatrist, Dr. Ferdière, representing the "Committee against Special Psychiatric Hospitals in the USSR", Marie-José Protais representing the French section of Amnesty International, the British doctor Barry Lowbeer, a psychiatrist from the "Working Group on the Internment of Dissidents in Mental Hospitals", Louis Astre representing the Federation of National Education, Claude Payen representing the CGT-FO, Dominique Taddei, the national secretary of the Socialist Party, the writer Claude Roy, Denis Sieffert representing the (non-Communist) UNEF of the rue Soufflot, Roger Pannequin, a representant of the "Committee of January 5 for a free Socialist Czechoslovakia", and finally, our collaborator Tania Mathon. There were also Ennio Di Giorgi, a remarkable Italian mathematician (who recently won the Wolf Prize), who represented the Italian section of the Committee and worked closely with the French, the lawyer David Lambert from the League for Human Rights, and Henri Cartan. This meeting was the climax of our campaign for Pliushch, and I must say we found it absolutely extraordinary. The Mutuality Hall was completely full. Given all the people who entered and left constantly, there must have been several thousand people in all. The press was present, and the meeting was described in numerous newspapers. The journalist Christian Jelen wrote in the newspaper *Libération*: "The Mutuality Hall was full to overflowing. Thirty important figures representing various unions and political, humanitarian and scientific organizations were on the tribune. Trotskyists, Socialists, independents, young people, a lot of young people, the same ones who are now protesting in the streets against Pinochet and Franco, and who yesterday were protesting the aggressions in Viet-Nam .... Perhaps, the other evening in the Mutuality, a new conscience was awakened. Perhaps, also, some taboos are about to fall.“ One such taboo certainly fell that day: henceforth, we had the right to denounce any government, including the government of the USSR, if it was found to be attacking human rights; there were to be no more sacred exceptions. Another taboo: we henceforth felt we had gained the right to interfere in the internal affairs of another country, and indeed, this was not merely a right, but a duty, if it is for the defense of human rights.

We received numerous messages from people who did not actually participate in the meeting: from the actors Yves Montand and Simone Signoret, from the Spanish UGT[91)], from Lombardo Radice, from the writer Vercors, the CFDT, Peter Reddaway, the Spanish POUM, Sakharov, Lipman Bers (the differences between countries are interesting: later, Lipman Bers told us that such a meeting could never be held in the US), the PSU, Jiri Pelikan, a group of teachers, a workers' union, high school students, well-known French and international figures, Jean-Louis Lévy, the grandson of Alfred Dreyfus. the Association of Public School Mathematics Teachers. Pliushch was indeed talked about in every major capital of the world. The Committee of Mathematicians had become a force to reckon with. And the whole of the logistics rested on the shoulders of just four people. Michel Broué had produced an enormous multitude of tracts, put out a bulletin calling for the meeting and at the last moment, Henri Cartan had put up posters on the doors of the Mutuality and inside it. The speeches were punctuated with frenetic applause. I rarely saw such political excitement in the Mutuality. Louis Astre, the national secretary of the FEN, brought this atmosphere to a paroxysm by gesturing towards the tribune with these words: “There are two empty chairs up there, which should be welcoming the representatives of the CGT and the Communist Party.” One fact shows how much prestige the Soviet Union still enjoyed at that time. When the bulletin announcing the meeting appeared, Tatiana's words “let us leave this country” had stimulated a strong reaction on the part of the Socialists, who found these words “anti-Soviet and highly unsuitable”. If they had read the bulletin earlier, they would have refused to attend the meeting at all.

The same day, the Communist leader Georges Marchais organized a compet-

91) UGT: Spanish General Union of Workers

ing meeting in the defense of freedom in the Parc des Expositions, not to help Pliushch but to harm our movement. However, his meeting did not meet with much success. Of course, the Communist newspaper *L'Humanité* accused us of organizing our meeting the same day as his! But at the point we had reached, that kind of maneuver had no effect on us.

The consequences of our meeting came rapidly. Two days later, on October 25, an editorial by René Andrieu appeared in *L'Humanité*: "The case of Leonid Pliushch does not leave us indifferent, and we have tried to obtain information on his subject. If it is true – and unfortunately we have not been convinced of the contrary – that this mathematician was interned in a psychiatric hospital uniquely because he took a position against certain aspects of Soviet politics or even against the Soviet regime itself, we can only state our complete disapproval and demand that he be freed as soon as possible . . ." The text could not, however, resist tossing a few insults at the Socialist Party: "Not you, and not that," and recalling the period of Guy Mollet. I had to agree with some of it, but that period dated back twenty-five years, and in between there had been the invasion of Hungary.

Andrieu's sentence was decisive, and represented a turning point in the Pliushch affair; the French Communist Party took a stance which could not be ignored by the supporters of the USSR, who found it intolerable and indefensible. Broué, Cartan and I thought that this sentence could not have been written without the personal green light of Brezhnev, and that it must signal a turnaround in the Soviet position. Andrieu's article also mentioned the pain caused to the Communist Party by the publication of the Khrushchev Report; at the time the movie *The Confession* came out, the Communist Party had denied any knowledge of the report. The Committee asked to meet Pierre Juquin, who received us cordially. All this was recounted in the 7th bulletin of the Committee. The 8th bulletin was able to proudly headline "After Pliushch, Massera, Bukovsky and Gluzman" – because in between the two, Pliushch had been freed! Shortly after Andrieu's article, we learned from Tatiana that he was about to be freed.

On January 8, 1976, he was led out of the hospital, dressed in a new suit, and put in a train for the town of Mokhachevo, forty-five kilometers from the border town of Chop in Czechoslovakia; he was given no further information about his destination. The next day, a car brought him to Chop where he met his wife and his two children and learned from his wife that he was going to be freed! In Czechoslovakia, they still didn't feel completely relaxed. They were under the spotlights of the *Pravda* journalists. On the Saturday morning, they took the train for the border town of Marchew, on the Austrian side, where Tatiana was able to declare with emotion, after crossing the border – "Finally free!" They were met by Michel Broué and Tania Mathon from our Committee, Barry Lowbeer from the British committee and the Austrian section of Amnesty International. It was an immense victory. However, Pliushch had still been receiving injections in Dniepropetrovsk just a few days before, and he was in poor shape. He was content to declare, quite simply: "You know, I'm still a Communist." "Well, that's fine," Lowbeer told him with a pat on the back. "I admire Trotsky a lot," added Pliushch. "Good," responded Lowbeer once again, with a new pat on the back, "so does Broué." Fantastic. But a lot remained to be done, because the only country they had the right to emigrate to was Israel, like all the other emigrants from the USSR at that time, all of whom were Jewish. But they were not Jewish, and they wanted to come to France. They traveled to Vienna by car, and Broué telephoned Astre in Paris. Astre went to see Jacques Chirac, the then Prime Minister, and everything was arranged in a wink. He then telephoned back to Broué: "Go together to the French Embassy in Vienna, say you're Michel Broué and the visas will be given to them immediately." It was done, and twenty-four hours later, they touched down in Orly. A crowd of friends and journalists awaited them, but Pliushch accepted to see only a handful of people, and no journalists. One felt he was still rather stunned by it all. When I introduced myself to him, he introduced himself back, "Leonid Pliushch", as though I might not have known who he was! The international press was present, and once more, Leonid Pliushch was discussed in every capital of the world. Broué and the Pliushches were driven by cars accompanied by a motorcycle brigade to Montereau, where they stayed with Broué for a month.

Cartan and I visited them there. It was a very interesting day. Tatiana knew a little French, the children didn't, but Pliushch, who knew only Russian, Ukrainian, Latin and a little German, had a translator. He can't have been very talented for languages, because years later, he still hardly knew any French. However Tatiana, who spoke better and better, found a teaching job. The children were sent to school and rapidly learned perfect French. The FEN and the SNES (Syndicat National de l'Enseignement du Secondaire)[92] gave generous financial help to the Pliushch family. Pliushch gave a press conference one month after his liberation, in the presence of representatives from every political, social, scientific, artistic and literary group in France, as well as national press and television – it was a real information festival. I sat next to Yves Montand and Simone Signoret. Later, Joan Baez invited Broué and the Pliushches to her hotel room, and sang for them. People considered that Pliushch represented

92) SNES: National Syndicate of Secondary Education

a tendency rather like Dubcek. They asked him if he wasn't embarrassed by Communist doctrine. "The Communism of the Soviet Union is completely degenerate," he replied. "Why should I, just because of this degeneration, abandon the luminous ideas of Communism?" He wasn't actually so far from Trotskyism or the PSU. He avidly read a number of books, and contacted various political groups. Little by little, partly under the influence of Soviet and Vietnamese immigrants in France, he became more and more anti-Communist. He ended up becoming very close to a Vietnamese anti-Communist group. I certainly don't blame him, any more than I thought to blame Solzhenitsyn for his opinions. Everyone is free to react in his own way to what he has seen and lived. At any rate, he represented a fundamental moment of my life and history, and now he was a free man; that was the main thing.

The Committee of Mathematicians continued its activity as before, but it never happened again that any head of state, either of the USSR or another, abandoned to us so completely the monopoly and the glory of a liberation. Under the multiple pressure of public opinion, certain leaders intervened personally in our later cases, in particular the president of the United States, but also the French president and Prime Minister. In Pliushch's case, it is clear that his liberation was obtained essentially by the Committee of Mathematicians, not alone of course, but with the aid of Amnesty International, the League for Human Rights, and other committees and organizations, and above all the massive support of public opinion. On this one particular point, we had succeeded in vanquishing an empire. We never tried to use the question for political purposes; we wanted to liberate Pliushch and nothing else. Of course, Broué, Cartan and myself were not indifferent to the fact that we were actors in the beginning of the end of the Soviet empire. This could be leveled at us as a reproach, but we can only answer that no neutrality can annihilate feelings, and that we took a thousand precautions to make sure that this consciousness had no effect on our actions – and it certainly did not have any essential effect. Obviously, many mathematicians shared our ideas, but this was 1975, and the Soviet union only crumbled under Gorbachev, in 1987. Throughout our entire campaign, we received no support at all from the French Communist Party, even though they must have received our revelations as a frontal shock. After all, the Communist mathematicians were still part of the French mathematical community, and they knew the truth about Pliushch's situation. I've heard – I've never been able to verify this story, nor really tried to – that a few days after Andrieu's declaration, the party summoned a group of Communist mathematicians and asked them if what the Committee was saying about Pliushch was true. Apparently they all responded "Yes". And they were all just as convinced as we were.

## The Committee overwhelmed

The Pliushch affair was over. Saving Massera, Boukovsky and Gluzman was the logical next step.

Galvanized by its victory of October 1975, the Committee of Mathematicians decided to hold a second meeting in the Mutuality on October 21, 1976, for the liberation of six people: Vladimir Bukovsky and Semion Gluzman (USSR), Edgardo Enriquez (Chili), Victor Lopez (Bolivia), Jose Luis Massera (Uruguay) and Jiri Müller (Czechoslovakia). We received a moving message from Bukovsky's mother. Pliushch was present at our meeting. The central committee of the Communist Party, however, didn't even answer our invitations for a long time. But tensions were rising and the Communist Party finally understood that its absence would cause even more talk than at the preceding meeting. On the evening of October 13, I found a letter of Gaston Plissonnier under my door. He insulted me copiously, simultaneously claiming that I had not invited him, and announcing that the Communist Party intended to participate, in the persons of Pierre Juquin and Jean Elleinstein. It was no joke being the president of this very stormy meeting. Juquin could hardly pronounce a word without being hissed. I tried my best to calm the crowd; I kept on repeating that like the others, Juquin had been invited by the Committee of Mathematicians, and requesting silence when he spoke. He, meanwhile, supported the freedom of all the prisoners on our list, without restriction; he pronounced each of their names in a loud and clear voice. It was a turning point in the role of the French Communist Party. However, the silence only lasted a few moments, and then the racket began again, worse than before. I can't deny that it was both inevitable and well-deserved, since Juquin also felt the need to remain faithful to tradition and vilify the Socialist Party for the Tours secession. Six million copies of his speech were then edited by the Communist Party and distributed all over France. Certainly, it was necessary to display the change in attitude, since change in attitude there was. The reaction of the Tass news agency, on October 22, was quite interesting: "The meeting was organized by a group of people known for their anti-Soviet and anti-Communist views .... To make their anti-Soviet activity more plausible, they simultaneously pretended to defend victims of fascist arbitrariness in Uruguay and other countries ....

It is even more incomprehensible for Soviet public opinion that the representatives of the French Communist Party should have found itself involved in a dirty operation of this kind. Whatever the motives and considerations which guided them, their participation at the tribune of the Mutuality helped only those forces which are irreducibly hostile to the ideals of liberty, democracy and Socialism which the French Communist Party has always defended." A change, albeit a modest one, had occurred. This change was perceivable in Czechoslovakia in 1968, in the form of a few swallows who alone could not bring the spring, but who announced the beginning of the end of winter. Spring itself arrived only in 1986 with the advent of Gorbachev.

In January 1977, Boukovsky, Lopez and Müller were liberated. I received a long letter from Lopez. Boukovsky was liberated via an exchange between him and Corvalan, a political prisoner in Chili; he met with quite a success upon his arrival in France. In the airport, on being asked his opinion of this exchange, he replied "I hope they exchange Leonid Brezhnev with General Pinochet." He wrote a long and extremely interesting autobiography. Locked up for three years in a psychiatric hospital, he had published *Dissidence: A new illness in the USSR* in 1971, and together with Glouzman, he had written the famous *Psychiatric manual for the use of dissidents*, dedicated to Pliushch. In 1972, he had been condemned to two years of prison, five years of camp and five years of internal exile; he certainly owed his early liberation to the new international climate. Glouzman, who had been condemned in 1972 to seven years of prison followed by three years of internal exile for having published a counter-expert report on the psychiatric judgment which had sent General Grigorenko to a special psychiatric hospital, remained in prison.

Mathematicians held widely varying opinions on the success of this second large meeting. I myself believe we may have made a mistake. We had promised never to engage in politics; we thought we were not doing so when we defended Boukovsky and Glouzman (and nobody said we were), nor Massera of course, since he was a mathematician. But we were reproached for having included the others. We had good reasons to be interested in Jiri Müller, who was one of the student leaders of the Prague Spring (although not a mathematician) in 1968, and was serving a five year prison sentence starting in 1971 for having distributed tracts. But Enriquez was a Chilian engineer, who had taken refuge in Argentina but was delivered back to Chili where he was imprisoned, and Lopez had been secretary of a miners' union on Bolivia and was arrested during a strike, deported to Chili with another thirty union leaders and imprisoned in the deep south of the country. It was perfectly natural to desire their liberation, and to feel indignant over their fate as a personal matter, or even within another organization. But it was unthinkable to really mobilize the full mathematical community over a long period for these cases. Furthermore, many young people had attended the meeting uniquely for the purpose of heckling Juquin, which the newspaper *Le Monde* did not fail to point out.

Our sacred rule never to engage in politics was transgressed during this meeting. We accepted that fact. The internal cohesion of the Committee of Mathematicians did not suffer from it too much, because we chose to discuss the whole question openly in our bulletin number 9.

Jose Luis Massera was a Uruguayan citizen of Italian origin. He was a mathematician of international standing. He had forty-three publications, including several original results on differential equations and a book written jointly with his student Schäffer. He had taught practically every Uruguayan mathematician, and formed Uruguayan mathematics practically single-handedly. His prestige in his country was gigantic. There was even a Massera Street, actually named for his father, but he was just as famous. During my previous trip to Brazil in 1952, I was received by Massera during a short trip to Montevideo and Buenos Aires. I met several brilliant Uruguayan mathematicians on that trip. Massera was a noted Communist and an influential political personage. We know that Uruguay had been a democracy in Latin America (it was even called the Switzerland of Latin America), which was economically quite prosperous thanks to leather and meat exports. But Colonel Pacheco had established a "moderate" dictatorship in 1971 against President Bordaberry. Various grave injustices gave rise to the birth of the Tupamaro movement (also called the "Robin Hoods"); this movement was efficient in combating corruption and abuse. The Tupamaros behaved reasonably, however they held a trial, which was altogether regular, during which they condemned and executed an American living in Uruguay who had played a sinister role in the government. Nobody in the world considered this to be a barbarous act, yet that day, the dictatorship adopted a series of pitiless totalitarian measures, imprisoned huge numbers of people and took various kinds of revenge on the population. Attempts were made to demoralize the prisoners and break them psychologically; they were locked in sinister prison called "Libertad", where they heard the screams of torture victims throughout the day. They tortured Raoul Sendic, the main leader of the Tupamaros, and crushed the movement forever. Jose Luis Massera and his wife Marta Valentina de Massera were later imprisoned there.

Massera was the adjoint secretary general of the Communist Party of Uruguay. This party was very legal and legalistic, compared to Pacheco's dictator party; it was not a rogue movement like the Tupamaros, but it did have an important arsenal at its disposal. When Massera replaced the secretary general who had fled abroad, he wanted to suppress the arsenal, but he didn't have time. It was discovered, and Massera was arrested on October 21, 1975 (around the time of our meeting for Pliushch). He was imprisoned in the "Libertad" prison and tortured, as was his wife Valentina (who was forced to remain standing on one foot for two days). During one interrogation, he was violently pushed while his feet were tied, so that he fell forward and seriously fractured one of his legs. After that, things went a little better; he was operated and put in a cast. We learned about his arrest only in December, through some Venezuelan mathematicians at a conference in Caracas. Anyway, his arrest was only announced officially in December. The Committee of Mathematicians immediately began to work on his case. Four hundred and fifty mathematicians signed a petition, and a delegation consisting of Yvette Amice, Cartan, Dieudonné and myself was received at the Uruguayan Embassy in France. The ambassador was someone of an older generation, which means that he had not yet been replaced by an agent of the military. He was very kind and seemed to understand us perfectly. We felt that he approved of us; he said he would do what he could, but that it was not much. We gave him a written message, signed also by Jean-Pierre Kahane, a mathematician who was also a member of the central committee of the French Communist Party but who couldn't make it to the Embassy. This ambassador received our delegation several times after that, always with the same kindness, although he could never do anything more than transmit our messages to Uruguay. This went on for several years, and then one day our ambassador was replaced by another one who was particularly ill-mannered. So we re-adjusted and began considering another mode of action. Several articles about Massera appeared in the French press; one of them, written by Cartan, appeared in *Le Monde*. At that time, we also began working on the case of a young Communist teacher named Roberto Markarian, arrested in Uruguay in November 1976. His situation was particularly tragic because his wife suffered from Hodgkin's disease, which could not be treated properly in Uruguay. She survived, however, and was able to complete her treatment after they were freed. I met them in Paris where they came to visit me. In 1975, although this seems incredible, there were more than five hundred thousand political prisoners, out of a population of less than three million. Dieudonné published an article on Massera's mathematical work. Dieudonné was strongly conservative and violently anti-Communist, but he was rigorously intransigent on the subject of the value of human rights. We never stopped campaigning intensely in favor of Massera. One of the most active people in our group, the Canadian mathematician Israel Halperin, wrote to thousands of people, and made a list of more than one hundred Nobel Prize winners who supported Massera. He got innumerable Canadians used to stamping the following message on every single envelope they mailed: "Uruguay is the country where Jose Luis Massera is in prison". Valentina Massera was liberated in 1978. Massera himself was liberated only in March 1984, after nine years of prison.

Dieudonné and I both went to Montevideo to try to visit Massera. Dieudonné went on July 26–27, 1979, and was rudely sent away by the prosecutor of the tribunal, Silva Ledesma. But he was courteously received by the ambassador Giandebruno, secretary general of the Ministry of Foreign Affairs, and was able to meet with Massera's wife and his lawyer.

I went to Brazil in the summer of 1980. I had decided before starting my trip to spend two days in Montevideo to try to see Massera. I obtained a visa from Rio with no difficulties. As soon as I arrived in Uruguay, I met the French ambassador there; he had good relations with the regime of the generals (this is usually the case). I agreed with him that he should let them know about the purpose of my visit; he promised to help me, and over lunch we discussed a plan of attack. Above all, he told me to contact the Italian and US ambassadors. The Italian ambassador, who was deeply interested in the case of Massera because he was Italian, gave me a lot of information, but was unable to help me otherwise. He sent me to three other key members of the Uruguayan regime: the Minister of Foreign Affairs, the president of the tribunal Ledesma, and the commander in chief of the army. The French ambassador, who knew his way around, predicted to me that ambassador Giandebruno, the Uruguayan Minister of Foreign Affairs, would receive me kindly, that Ledesma would throw me out with insults and that the commander in chief of the armed forces, who was really the principal leader of the regime, and who had his doubts about the whole affair, would "do his best to receive me" but finally manage to be unable to do so. And it all happened exactly as he said! Giandebruno received me courteously, even reminding me of the existence of Massera Street in Montevideo, he told me that Massera was universally respected and that the Uruguayan people were proud of him. But he had conspired, his arsenal had been discovered; it was logical that he should be in prison. He promised me nonetheless to intercede in his favor, but did nothing. Then I tried to see Ledesma; I didn't even succeed in getting him on the telephone, however I did get myself threatened with the loss of my passport and to be thrown out of the country. I had also requested a meeting with the head of the armed forces. They said they would call me back to fix a meeting. And when I returned to my hotel from the ministry of Foreign Affairs, I found a message from the commander in chief waiting for me. Most Uruguayan citizens were hostile to the regime, and the people at the reception desk considered me as a worrisome individual. The message, which had been sent during my visit to Giandebruno, as I'm sure they were fully aware of, said that he wished to see me immediately. I called back at once, only to learn that he had left for several days ….

The American ambassador received me immediately. It was during the presidency of Jimmy Carter, who intervened favorably in every part of the Third

World. I spoke to many people in Latin America about Jimmy Carter during that time, and everyone put a lot of hope in him. Alas, he made a lot of mistakes. He eventually fell on account of the fifty-four members of the US Embassy in Iran, whom he tried to have rescued by a mission which failed miserably and made him look ridiculous. The US ambassador, whose door was closely guarded by Uruguayan soldiers, obviously didn't receive many visits and was delighted to see me. He asked me what he could do for me and in particular if I was having trouble getting around Montevideo. He offered me his car and chauffeur, which I actually used for the whole afternoon.

When I telephoned Mrs. Massera from a noisy bar in town (there were no public telephones), the only thing I was able to understand was that she was inviting me to come over immediately. She was waiting for me at the door of her house and was quite surprised to see me get out of the car of the American Embassy. "Oh caramba!" she cried when I told her my name; she hadn't understood from my telephone call who I was. She received me kindly. She was not surprised by the attitude of the American ambassador, but it was the first time that anyone had actually shown such interest in her story. I had dinner with her and her sister, a doctor, and Massera's lawyer. We talked at length about her husband's life. She was deeply anxious: "Muchas veces pienso que mi marido se me muere en prison", which literally means "Often I think that my husband will die on me in prison" ... it's typically Spanish to insert that tender "me" into the sentence, meaning "he'll die, and I'll lose him". In spite of all my efforts, I did not succeed in seeing Massera.

I had to be contented with leaving a modest package with his wife, from the Committee of Mathematicians. She wrote to me that he had been deeply touched by our visit and by the attention of the Committee whose activities he was aware of. But it was very difficult to communicate with her from Paris because it was necessary to wait as much as five hours to obtain a telephone connection, and even then we had to be prudent and not give our names.

The status quo lasted several years. The Communist parties of different countries, who had remained entirely indifferent to the Massera affair, became concerned with it suddenly, several months before the departure of the generals. They were much more powerful than the Committee of Mathematicians. The campaign was organized in France by the mathematician Gabriel Mokobodzki. They told us everything they were undertaking, but didn't really work with us. They did ask Cartan to join their committee, but they didn't ask me or Broué. Cartan accepted, since two committees are better than one. And finally, a foreign Communist was able to meet with Massera. Things were changing in Latin

America. As soon as he came out of prison, after the regime fell, Massera returned to the University with pomp and circumstance, recovered his former job and took on a fundamental role in the reconstruction of the university. Then he took a trip around the world with his wife and came to France. He didn't talk to me about the past, but hoped for a positive evolution in the USSR where Gorbachev had begun his *perestroika*. He thanked the Committee of Mathematicians warmly. I also saw Markarian and his wife, who came to dinner at our house and were very nice. The Masseras also visited Dieudonné and Cartan, though not Broué.

We were also concerned with the situation in Czechoslovakia. The mathematician Vaclav Benda, a young lecturer and the speaker since February 1979 of the famous Charter 77, had been arrested in May with nine of his comrades and accused of subversion, then condemned in October to four years of prison. A French delegation composed of Patrice Chéreau, an actor representing the AIDA (International Association for the Defense of Artists), Jean-Pierre Faye and Alain Challier from the International Committee Against Repression, Jean Dieudonné from our committee and Paul Thibaud from *L'Esprit*, went to his trial. Of course, they were not allowed to attend the proceedings, and in fact all of them except Dieudonné and Chéreau were expelled from the country; those two were able to hold a press conference in front of thirty or so militants for Charter 77 and for the VONS, a human rights league. This trip was a partial failure, but at least it gave a little courage to Czechs fighting for human rights. The three speakers were arrested for one day and Thibaud was even somewhat brutalized. Dieudonné and Chéreau shared a miserable cell with no chairs. Chéreau recollects being rather frightened, which is normal enough. But Dieudonné threw one of his famous tantrums, banging on the wall and shouting "What kind of manners are these, I'm French, I'm an Academician and I want a chair!" The frightened guards brought a chair – but only one! During the night, they were brought in a rickety truck to the German frontier, near Bayreuth. Chéreau is actually related to the Wagner family, who came to pick him up. Dieudonné, who had never been in Bayreuth, came too and found himself with Chéreau in the house of the descendants of the famous musician and ended up declaring himself delighted by the conclusion to his adventures! They gave a press conference together in Paris. The appeals trial took place on December 20, 1979. Among the principal accused were Benda, Peter Uhl and Vaclav Havel. The sentences were heavy. Six representatives of French committees went to Prague; one of them was Marcel Berger, a geometer and professor at the University of Paris VII, who had been one of the ENS students in Nancy in 1948, and was the acting president of the French mathematical society. Obviously, they couldn't do much, apart from encouraging Czech resistants and making contacts. After intensive interrogation, there were released in a forest at night, near the German border. Berger was released near Nuremberg. Paul Thibaud was secretly held a day longer than the others and treated rather brutally. Only the British lawyer Platt-Mills was released with excuses! Before their expulsion, Berger and Thibaud gave a letter signed by twenty French intellectuals to the Czech Union of Writers. The evening of the trial, a thousand people demonstrated in front of the Czech Embassy in Paris. With incredible courage, seven hundred and seventy-three Czech citizens sent a letter to the president of the supreme court demanding the immediate liberation of the accused and all other political prisoners as well. The authenticity of the signatures was guaranteed by several well-known figures, for instance a son of Rudolph Slansky, who was hung in 1952. On December 4, as a gesture of support for Charter 77 and VONS, the Atelier Theater in Paris put on a very interesting play by Vaclav Havel called *Audience et vernissage*, contrasting the courage of some and the cowardice of others, in a Czech-style situation. The evening was organized by Ariane Mnouchkine, whom I met there. What a difference with the Moscow trials of August 1936. Then, at the age of twenty-one, I was isolated and alone in the middle of a mass of intellectuals who were completely indifferent or even admiring of Stalin's proceedings, considering the Moscow trials to be at worst an unimportant slip, whereas I already saw clearly that it was the prelude to a long litany of false trials and crimes. It took twenty-six years after the death of Stalin to open the eyes of French left-wing intellectuals.

## Refuseniks

Well before the end of the Massera affair, the problem of *refuseniks* arose in Russia. Starting around 1977, the big Soviet universities were seized by violent anti-Semitism. The names of Jewish mathematicians took to disappearing systematically from all publications, and no Jew was allowed to enter the Academy. Furthermore, no one was allowed to enter the university without passing an entrance examination which was full of questions not covered in the regular program of the final year of high school. The questions were already difficult, but the examiners were told to pose questions from a certain list of almost insolvable problems whenever the candidate had a Jewish name. Over a long period, almost no Jews entered the large universities. Gel'fand and Krein were excluded from the Academy, but mathematicians like Novikov and Shafarevitch were excluded as well because they opposed the regime. In the mathematical journal *Matematicheski Sbornik*, one counts eleven out of thirty-four articles written by Jewish mathematicians in 1970, seventeen out of thirty-four in 1974, and one out of thirty-two in 1977. Eighty-four out of four hundred and ten students admitted to Moscow University in 1969 were Jewish, four out of four hundred in 1977 and one out of five hundred in 1978 (and even that one was only admitted because Kolmogorov and Manin insisted). Mathematicians all over the world protested this open anti-Semitism. There has always been a rather strong popular anti-Semitism in the USSR. Lenin's government was strongly against it, but Stalin's government favored it and after his victory, the regime became more and more openly anti-Semitic. All the monuments to the dead of the Soviet Union (and Poland too, alas) give the names of the dead without ever mentioning a single Jew.

When we were invited to conferences in the East, there was never a single Jew among the speakers. When we invited Soviet mathematicians to France, they were not allowed to come if they were perceived as being the least bit "doubtful". We were told that they had no time, and they were replaced by other visitors. The Jewish mathematician Mityagin was once invited to a conference in Poland. Two other mathematicians appeared in his place and declared that he was too busy. Suddenly, right in the middle of the conversation, Mityagin suddenly appeared. Having been forbidden to attend as an invited mathematician, he had simply come with a tourist visa (which shows, at least, that confusion did sometimes reign within the Soviet police). It is pleasing to imagine the expressions of the two Soviet mathematicians who had just finished giving their little explanation. At any rate, it was Mityagin who gave a talk at the conference, as planned. Little by little, the Jewish mathematicians in the Soviet Union lost their jobs in the universities, particularly the ones who had asked for a visa to Israel. These people became known as "refuseniks", and were excluded from science. As they were not allowed to enter the university or to use the libraries, they set up a seminar amongst themselves. The Moscow refusenik seminar met regularly in the apartment of one or another, for meetings lasting several hours. They belonged to widely varying disciplines, but each one followed every lecture. They kept up a continuous, fruitful contact and learned many things about subjects different from theirs. The existence of this seminar circulated around the world, and soon, when foreign scientists were invited to visit Soviet universities, they took to visiting the clandestine seminars and giving talks there, too. This gave rise to many contacts and exchanges.

The "Scientific Committee of the National French Counsel for the protection of the rights of Jews in the USSR" organized a campaign to help the Jewish refuseniks and others who were forced, because of their religion or other reasons, to endure all kinds of segregation and even years of prison. The "three presidents" of the committee were Alfred Kastel, a Nobel Prize in Physics, André Lwoff, who had won the Nobel Prize in Medicine togetre with Jacques Monod and François Jacob, and myself. On the death of Alfred Kastler, in 1984, we added other people to this collective presidency. Hundreds of documents were regularly published on the subject of individual cases; the activity of the committee was enormous and large numbers of people were kept informed of the development of anti-Semitism in the USSR. We had five to ten meetings a year where fifty to one hundred members of the committee came to hear a French scientist who had visited the USSR and met refuseniks, and to keep abreast of the general situation. Ruth Fein from the Contemporary Jewish Library was a central figure in all our activities. The Library was in contact with the Embassy of Israel, but it was free to take any action it wished. Ruth Fein believed in peace between Israel and the Palestinians; she had a remarkably open mind. She was very dynamic and seemed to be everywhere at once. Without her, nothing of our activity would have functioned properly!

The refuseniks showed enormous courage. A certain number of them were arrested. This was the case of Iossip Begun and Victor Brailovsky, an excellent physicist who spent a year in prison. Ida Nudel, who helped many persecuted Jews (earning the name of "guardian angel of the Jews in the USSR") was exiled to the depths of the Soviet Union. I remember a particularly moving meeting about her, in a little room in the Mutuality. As was often the case, the Russian speeches were (remarkably well) translated by Vitia Hessel, the first wife of the ambassador Stephane Hessel. (We were all deeply saddened when she died of cancer. Stephane Hessel himself was very active in many of our struggles, and we met often – most of the meetings of the Sakharov committee were held in his house – and I admire him very much. I am now a member of the college of mediators in the affair of the clandestine Africans in Saint-Ambroise, and Stephane Hessel and his wife also play an important role.) The committee for the defense of Soviet Jews did not coincide with the Committee of Mathematicians. It was specifically designed to fight against the conditions suffered by Soviet Jews, whether mathematicians or not, imprisoned or not. Also, the Committee of Mathematicians only defended unjustly imprisoned mathematicians, anywhere in the world. At any rate, many other committees devoted to similar causes sprang up: committees of physicists, psychiatrists and so on. The principal organizer of the committee of physicists was Jean-Paul Mathieu, a professor at the University of Paris VII, who ceaselessly collected signatures. We struggled together against the Algerian and Vietnamese wars, and were close friends for almost forty years.

The effervescence of our activity generated analogous committees in neighboring subjects. A committee for the rights of scientists was formed in Great-

Britain, within the Royal Society, and another was formed in the USA, within the National Academy of Sciences. During a trip to France where he was received by our Academy, Lipman Bers suggested that we form a similar committee. After he left, several of my fellow members protested that he should mind his own business, but most of them agreed with him. We formed this committee and called it CODHOS; it was very active. It had many successive presidents, notably Guinier, Dausset, and now François Jacob.

The International Congress of Mathematicians of 1978 was held in Helsinki, but even though it was close to the USSR, almost no Soviet mathematicians attended. They were "too busy", as usual. The Soviet delegation was one of the smallest ones at the conference. Obviously, Russian scientists were kept under strict control whether they were Jewish or not. Thirty or so American mathematicians had prepared a very detailed and objective paper on the subject of anti-Semitism in the USSR. This text was available to everyone at the congress. It denounced by name – a rare event – the Soviet mathematicians who actually favored and organized anti-Semitic policies, particularly the academicians Vinogradov and Pontryagin. Vinogradov wrote several remarkable papers on prime numbers. For a long time, he was highly respected in the USSR and in the world at large, but he turned into a particularly virulent anti-Semite, displaying pathological aggressivity towards anything that could possibly seem Jewish. As for the blind mathematician Pontryagin, he had made several fundamental discoveries in different domains, and together with Vinogradov, he was the main force behind the organization of anti-Semitic practices in the universities. Other mathamaticians were also explicitly named. Everone at the congress, except the Soviets who carefully made a large detour on entering the main hall, procured the document. Among the Soviet delegates, there were as usual some spies from the KGB. Finally, one of them had the audacity to step up and get a paper. The same evening, the whole Soviet delegation shut themselves in one of their rooms to avidly read the paper. They were particularly interested in knowing exactly who had been named. Some of them probably had been names themselves, as those who were allowed to attend foreign congresses were usually strong supporters of the regime.

The general anti-Semitism which reigned in the USSR naturally caused a drastic lowering in the mathematical level. Russia was one of the best countries for mathematics during their glorious epoch of more than a century ago. Students at the Moscow University were recruited via incredibly stringent criteria. When the examiners were pressured to pose insoluble questions to Jews attempting the entrance exam, most high-level mathematicians refused to act as examiners. So the university turned to second or third-rate mathematicians, who were willing to accept the corrupt procedure. Obviously, these mathematicians weren't even capable of distinguishing the best candidates. So recruitment at Moscow University was essentially based on chance rather than on true value, and the level was noticeably lowered. I learned about this situation from the Soviet mathematician Gel'fand on one of his visits to Paris, and I talked about it at an international colloquium on Antisemitism in the USSR, organized by the Committee for the Protection of Jews in the USSR. Four reputed mathematicians, Alexandre Joffe, Naoum Meiman, Grigori Freimann and Gennady Hassin, had published a little book containing the problems posed at the oral exams for Jewish and non-Jewish candidates during the entrance examinations of 1978, together with the solutions which they had found themselves, in their seminar. The difference between the two types of question is quite edifying. The candidates had ten to twenty minutes to answer.

Questions posed to non-Jews:
1) write the law of cosines,
2) is the function $x - |x|^3$ differentiable at the origin?,
3) draw the graph of $y = 2|x|^{-1}$.

It's easy to see that these are simple questions.

Questions posed to Jewish candidates:
1) which of the following two numbers is larger, $(413)^{1/3}$ or $6+3^{1/3}$ (the first of these numbers is equal to $7.447\ldots$ and the second to $7.442\ldots$, but of course the candidate doesn't have a calculator);
2) find the integer solutions to the equations $x^y = y^x$,
3) Prove that a convex polygon of surface area equal to 1 contains a triangle of surface area equal to 1/4.

The Soviets had created an original system of exceptional high schools grouping the most talented students, who had extracurricular mathematics courses to allow them to think about various questions. Some of the best Soviet mathematicians taught in these high schools. During my trip to the USSR in 1964, I sat in on one of these courses, taught by Gel'fand to a group of tenth, eleventh and twelfth grade students; it was really fantastic. When I returned to France, I tried to set up a similar program. The only high school which accepted to try it was the Lycée Condorcet, where every Saturday from two to four a group of volunteers came to study all kinds of mathematics; it was a great success for a whole year. Unfortunately, anything which ressembles elitism is very looked-down on in France where for decades, the most absolute egalitarianism has been the rule, so that I had to stop. In the USSR, the experiment continued for a long time, and such programs also exist in the USA, in Viet-Nam, in Cuba and in Taiwan. The USSR eventually suppressed these special classes, because there were just too many students there coming from Jewish or liberal families. It was easy enough to reject Jews applying to university, but became a bit more difficult if they had been among the most brilliant students of these special high schools.

## The Shcharansky affair

The incident of Anatoly Shcharansky occurred right in the middle of all this period of persecution of Jews and other dissidents. He was unanimously considered to be a man of great importance. He was a friend of Andrei Sakharov, working closely with his secretariat, and he worked hard in favor of Jews excluded from science. These two groups tended to avoid each other so as not to interfere with each other's work, but Shcharansky represented a kind of link between the two. Personally, he would have preferred that the two groups work together. However, one can also consider that the separation constituted a reasonable protection. In the meantime, because of the rapid increase of cases, the little circle of our Committee of Mathematicians had gained three new members: Claude Bardos, Alain Chenciner and Jean-Louis Verdier (who died tragically, together with his wife, in a car accident).

On March 22, 1977, the Committee for the Revision of the Trial of Dr. Mikhail Stern had organized a meeting in the little room of the Mutuality. But Stern was liberated on that very day, and he had arrived in France where he was holding a press conference. An incredible article had also appeared in *Izvestiya* on that day, called "Stealers of Souls", directed against Shcharansky and some of his comrades, calling them traitors to the Soviet State, and publishing – incredibly – their names, addresses and photographs. This was obviously simply a preparation for something bigger. And indeed, shortly afterwards, Shcharansky was arrested and kept secretly for "observation" for six months. He had just religiously married the very young Natalia Stiglitz, but they had not yet had time to prepare the civil marriage. Thus, according to Soviet law, they were not married. She had asked for a visa to Israel, which was given to her the day after her religious marriage. She knew that if she refused it, she would not be able to leave the country for a long time. Considering the danger she was in, the couple decided together that it would be better for her to leave the country. They were separated for nine years and it was terrible for both of them. Shcharansky was kept absolutely hidden for the six months sentence of observation, which was extended another six months by decree of the Soviet regime. Even during the repressive trials, the State tried to observe some semblance of legality.

During this period, we had no news of him at all, and apart from the extension of his period of observation, we didn't even know whether he was dead or alive. He was interrogated almost ceaselessly, although not tortured. Both from the physical and the moral point of view, he survived magnificently. He was obviously a man of exceptionally iron will. Later he told us the following fascinating episode of his captivity. His guard had shown him a film which was supposed to prove that he was considered as a traitor everywhere. He wasn't particularly impressed, but suddenly he saw an image of his wife, who had taken the name Avital in Israel, demonstrating for his liberation, in London, before an enormous audience. He exulted. Not only had he the momentary pleasure of seeing her, but he knew that she was rushing all over the world in his defense. He asked to see this part of the film at least ten times over, without the imbecile guardian having any idea why. During the whole of this period, Avital Shcharansky traveled the world as though she were made of mercury. She constantly moved from capital to capital and telephoned everywhere on the planet. She laid siege to the most influential heads of state in France, in England and in the USA, and succeeded in meeting them. And the press followed her everywhere, relating her story, her problems, her difficulties, so that eventually the Shcharansky affair came onto the international stage, as the Pliushch affair had done.

I was in Israel a few days after Shcharansky's arrest, for a one-month scientific visit, and I had taken advantage of the occasion to see some Russian immigrants, former refuseniks, in particular the mathematician Voronel, who had founded one of the clandestine seminars in Moscow. I proposed to him that the Committee of Mathematicians work on Shcharansky's case since after all, he was a computer scientist. This could be advantageous, but perhaps there would be some disadvantages as well. At any rate, we immediately contacted his wife, who was in the US and who agreed with our proposal instantly. Since the committee is most active in France, she used to come to see Broué, Cartan and me at our house. The bulletin edited by Broué wrote up everything as it occurred. She was a strictly religious Jew, eating only kosher food and accepting only a glass of water or at most orange juice from us.

The conference in Helsinki was to be followed by a conference in Belgrade, and the third commission was to be on human rights. One day when Avital was at our house, we tried to contact the French delegate to the conference. I had no idea who he was or how to look him up, except maybe at the Ministry of Foreign Affairs, but they couldn't help me. I had to go out for half an hour, so I left the others talking about other things. I went out and did what I had to do, but then I ran into the ambassador Philippe Richer whom I had known well when he was the French ambassador to Viet-Nam in 1976. He was a very nice and quite exceptional man; probably the only ambassador ever to have succeeded the exploit of being a union member of the CFDT. He had helped us a great deal during our stay in Viet-Nam; he had organized Pham Van Dong's trip to Paris in 1977. So I stopped to chat with him for a few moments. "What are you doing now that you've left Viet-Nam?" I asked. "Oh, it's great – *I'm going to be the French delegate to the Belgrade conference*" he answered. After a moment's totally incredulous silence, I hastened to tell him all about the Shcharansky affair, which he only knew slightly. Returning home, I triumphantly announced "The French delegate to Belgrade will be here in a quarter of an hour." Avital was overjoyed, and turning to me she said "God guides our steps." I could hardly object! It's true that Richer ended up being replaced by someone else, but our task was accomplished. Indeed, Shcharansky was on the menu of both the Belgrade conference and the conference in Madrid. The Soviets were finding themselves in a more and more difficult position because of human rights problems. It's not for nothing – and it was really a brilliant idea since no one could stop them – that Soviet dissidents had decided to form groups "in defense of the Helsinki agreements". The USSR had signed these agreements, and they were quite strict. Shcharansky was one of the founders of these groups. At any rate, after twelve months, it was impossible from the legal point of view, which the Soviets always treated with great respect, to avoid a trial. The trial for treason opened on July 11, 1978. About two weeks earlier, preparations began for a grand demonstration to take place on that very day; more than ten thousand people marched in it. Only mathematicians were members of our Committee, but a Socialist municipal counselor called Thérèse Etner founded a Committee for the Defense of Anatoly Shcharansky which became very active. The demonstration was organized by her committee. There were not only militants, but deputies, mayors and many important people of all persuasions. The trial took place in the predictable manner. Only Ida Milgrom, Shcharansky's mother, was allowed to attend. The lawyer, appointed by the government, pleaded guilty. But Anatoly Shcharansky insisted on assuring his own defense, and he did it remarkably well. Nevertheless, he was condemned to ten years in the severe labor camp Perm. He even was made to spend weeks in the prison cell of the camp, but he was eventually liberated before the completion of his sentence.

During their nine years of separation, Anatoly and Avital wrote to each other regularly, but their letters rarely reached their destinations. Yet their relation continued. Avital obstinately persisted with her international activities. Since she (rather astutely) wanted a Communist lawyer, I asked Roland Rappaport to support and defend her. He did it admirably, even accompanying the Committee of Mathematicians to the Soviet Embassy. In vain, of course. On our advice, Avital also asked help from other lawyers, such as Daniel Jacoby, a member of the international federation for human rights, and the barrister Louis Pettiti, a member of the Association of Catholic Jurists. This trio of lawyers played an important role throughout the affair and later published a book about the trial, called *Un procès sans défense*, which analyzed the mechanism of false accusations and ended with a text by each of the authors describing his own personal evolution throughout the affair. The most interesting evolution was that of Roland Rappaport's, who abandoned every reference to Communism after the trial, and ended up leaving the party. This period led to close and warm relations within our group, in particular with the Rappaport family. Later, when Roland Rappaport defended the children of Izieux in the trial of Klaus Barbie, he asked me to be a moral witness against Barbie.

Ida Milgrom, Shcharansky's mother, was sometimes allowed to visit him in camp, but it also happened several times that after having taken the long trip from Moscow to Perm, she had to return again withou having been able to see him or give him the packages of food she had brought for him. She was very courageous and always remained in contact with us.

To reinforce our position, a certain number of mathematicians decided to boycott all official relations with the USSR until Shcharansky was freed. We refused to go to any official conference in the USSR and we didn't invite any Soviet mathematician who occupied a position in the hierarchy. This had some disadvantages – "real" Soviet mathematicians became even more isolated – but they were already dramatically isolated since they were forbidden from speaking in conferences and were replaced whenever they were invited abroad. Other mathematicians decided to go to the USSR whenever they were invited but to mention Shcharansky at every one of their talks, and to participate in the clandestine seminars. Globally, I think that the boycott made the masters of the Kremlin finally realize the failure of the anti-Semitic Soviet mathematicians, so it had a positive influence. Finally, Shcharansky was liberated on February 16, 1986, before the end of his sentence, and joined Avital in Israel. His visit to Paris was the occasion of a grand reunion. I later had the pleasure of seeing him again, in Israel, together with his wife and children. Unfortunately, I do not share his political opinions, but that does not diminish my esteem and friendship for him.

## Sakharov

The situation in the USSR remained extremely worrisome. It is impossible to overestimate the courage of those Soviets who braved the ferociously repressive regime. Tatyana Velikovana was one of the founders, in 1969, of the group "Initiative for the defense of human rights in the USSR". She was arrested on November 1, 1979. She was sent to Lefortovo prison, then to a severe labor camp. On November 17, 1981, Yuri Shikhanovitch found himself back in prison, eleven years after his first arrest. In spite of a vibrant call for his liberation by Sakharov, the Soviet psychiatrists declared him a case of "torpid schizophrenia"; this is the translation by a French psychiatrist of a Russian expression, which passed into our personal ironic language, so that sometimes I actually think I invented it! Once again, we sent a delegation to the Embassy, together with Amnesty International, but starting at that moment, the Embassy closed its doors to all delegations, and never received anyone again. In spite of the presence of many important people in the delegation, the entrance grille remained obstinately closed. We ended up being dispersed by the French police. One day, I remember one of the policemen asking Marie-Claire Mendès-France for her papers. Unfortunately, she had none on her right then. The police threatened to carry her off, and we had to negotiate with them and explain who she was, after which they calmed down. On November 6, 1984, Shikhanovitch was condemned to five years of camp plus five years of internal exile for anti-Soviet activities.

The deportation of Andrei Sakharov to Gorky and the imprisonment of the physicist Yuri Orlov mobilized us once again. Orlov was deported to a remote little village in Yakutia, a place which was glacial in winter and torrid and full of mosquitos in summer. At first, the local population thought he must be a criminal, and it took time before they accepted him. He was finally liberated, before Gorbachev came to power, and expelled in 1984 to the US. After his liberation, he was invited to a meeting of a Republican circle in Paris. I also invited him to our house with his wife, and we had a long and interesting conversation on the future of the USSR.

Sakharov was a world-famous physicist, and this fact afforded him protection. During the first part of his life, he had worked secretly – which was normal in his speciality – on making the Soviet H bomb, which was actually completed, as we now know, before the American H bomb. In collaboration with Zeldovitch, he also obtained controlled fusion (at several million degrees) during a very short period, concentrated in his torus-shaped "tokamak", thanks to a very intense magnetic field. One of his most interesting scientific discoveries was the spontaneous disintegration of the proton, over an average of $10^{32}$ years – a theoretical prediction which has not yet been verified, but which is apparently accepted by physicists. He also worked on cosmology, on the big bang and on antimatter. In addition, he was in contact with the most powerful and somber leaders of the regime, Beria in particular. His contempt for him grew as he came to understand more and more what his real nature was like. In spite of this, he actually wrote, on the day of Stalin's death, that a great man had disappeared! In his Memoirs, he says that he can't even begin to explain how he could possibly have thought and written such a thing. The atmospheric experiments with the H bomb, which he had proved would cause several thousand new cancers, worried him deeply. He began to militate against the tests, and call for underground testing. After hemming and hawing, Khrushchev allowed himself to be convinced. The USSR, the USA, France and England all adopted the universal practice of subterranean testing, and that was the beginning of the French tests in Mururoa.

Sakharov won the Nobel Prize in 1975. His life in Gorky was extremely difficult. His apartment was under constant surveillance, and almost nobody was allowed to visit him. However, the Soviet Academy of Sciences did protect him somewhat. The president, Alexandrov, accepted his deportation but did exercise a little pressure to make sure he was not harmed. He was always considered by many (though not all) members of the Academy as one of theirs. And Brezhnev needed the members of the Academy. To show that he was in perfect health, the Soviets put out a film about his life in Gorky. Perhaps his health was all right, but from the film he looked like a deeply humiliated man. He was surrounded by insignificant people and obliged to obey their commands; he was always followed, his house surrounded; he had no telephone. He was also made to submit to numerous vexations; a scientific manuscript of several hundred pages was stolen. Infinite sadness could be read in his face. It was of course impossible to write to him. His wife lived with him. Two members of the Academy were allowed to visit him, and although it was a small thing, he appreciated their visits as a sign of solidarity. The mathematician Olga Ladizhenskaya, known for her work on partial differential equations, decided to spoil the party and permanently take on the role of disturbing the apparatchiks. During one conference, she spent her entire talk writing on the board over and over "Free Andrei Sakharov". Behind her, an apparatchik kept on erasing it, but she was not in the least intimidated: nobody had the right to interfere with her talk. When I traveled to Moscow in 1964, she came to see me from Leningrad, where she lived. She had a lot of things to tell me. For four hours, she told me details about life in the USSR, about the "good" and "bad" people. For myself, I wasn't worried about our conversation being overheard, but for her? She had decided once and for all not to worry about it, and told me: "Of course, our conversation is being recorded by some microphone hidden somewhere, maybe in the chandelier, so I'm telling them how much I despise them right now." This just a few days after Khrushchev was replaced by Brezhnev!

A group of six French members of the Academy of Sciences stood out in the struggle for Sakharov: Jean Dieudonné and René Thom, mathematicians, Jean-Claude Pecker, an astrophysicist, Francis Perrin and Louis Michel, physicists, and Françoi Gros, a biologist. They asked to be allowed to visit him, and received no answer. The Six held a press conference, presided by Henri Cartan, which received the support of the CFDT, the FEN, FO, Amnesty International and the League for Human Rights. Then the Six signed a letter to Sakharov, which was published in *Le Monde* on June 7, 1981, and which received an indirect response from him. In between, he had been elected member of the French Academy of Moral and Political Sciences (on June 30, 1980) and the French Academy of Sciences for his work in physics (on February 16, 1981). The French members of the Academy of Sciences Louis Michel and Jean-Claude Pecker traveled to Moscow to try to see him. Alexandrov, the president of the Academy of Sciences in Moscow, received them with every appearance of courtesy, assuring them that he guaranteed Sakharov's security, but he did not give them permission to go to Gorky. They returned disappointed, but after all, they had expected no more. They gave a press conference on their visit to the Academy, presided by Cartan. They received thanks from Sakharov; it was the last communication we had with him until his liberation. Louis Michel returned to Moscow once again in December 1983, but again was not allowed to see him. All this was a kind of "summit", interacademic activity, conforming to the oldest established scientific customs in every country, and efficient for this very reason. On May 25, 1984, personal friends of Sakharov hidden near his house were able to catch sight of Sakharov's wife, Elena Bonner, on the balcony, but not of Sakharov. On June 21, François Mitterrand went to Moscow and spoke of Sakharov to Zamiatin, who told him that Sakharov was in perfect health, and that Mitterrand would do better to worry less about him and more about the French unemployed.

Sakharov was liberated only in 1987, after Gorbachev came to power. He returned to Moscow to take his position as a member of the Academy, and became very involved in the internal and international politics of the USSR, which changed drastically under Gorbachev. He was able to write in the newspapers and to exchange letters with foreigners as he wished, and no more cases like his occurred in the country. He died later, after his return to Moscow and his election as a senator. The years of harassment and a hunger strike of several weeks in Gorky had left him physically weakened. After his liberation, a delegation from the Institute, consisting of Henri Cartan, André Guinier and myself representing the Academy of Sciences (Guinier was our delegate for international relations) and Jerome Lejeune for the Academy of Moral and Political Sciences, traveled to Moscow to meet him. We asked to be allowed to organize a little conference in his honor; it's a perfectly normal thing among scientists. But it was refused. Gorbachev's audacity was not absolutely unlimited. For this reason, Louis Michel refused to be part of the delegation. But Sakharov was allowed to come to the French Embassy, accompanied by a delegation from the Academy of Sciences of the USSR and by the four of us, to receive his diploma and medal of associate member of our Academy. There were several speeches, by the ambassador, by Guinier, an answer by Sakharov and another by one of the vice-presidents of the USSR Academy. Sakharov did not mince his words. The ambassador had alluded to the solidarity of the entire world behind him, and in particular of every Academy. He responded that there was one Academy in the world which had not supported him, the Academy of his own country, and accused its president of having published a negative article about him in a Soviet newspaper. He added that the president ought to make a public self-criticism about it. In his answer, the president reproached Sakharov for washing dirty linen in public, arguing that the French Embassy was not the best place to make these statements. Perhaps he was a little bit right, but the question was secondary compared to the importance of Sakharov's accusations. There was less arbitrary injustice under Gorbachev, but Sakharov continued his combat just as before. During our visit, he invited us to his house one afternoon, and received us superbly. He shared our feelings about the Afghanistan war. Together with Sakharov, we were all received by a group of former refuseniks, in the home of I.L. Alpert.

## Other battles

Our activities did not only concern the USSR, Czechoslovakia and Uruguay. On February 23, 1972, the young Moroccan mathematician Sion Assidon was arrested for conspiring against the security of the state, and tortured. The proof of his guilt was a tract, printed out on a personal machine, which he had tried to spread around without much success. In December 1972, he went on a hunger strike of thirty-one days in order to obtain the right to be considered as a political prisoner, and to have a public trial. He was judged in April 1973 and condemned to 15 years of prison. In October 1979, finding himself in the hospital, he tried to escape, but he was caught, tortured and his sentence was extended by three years. He was, however, allowed to work in prison.

Sarah Assidon, the elder sister of Sion Assidon, had heard of the Committee of Mathematicians, and she came to see me in 1980 I believe, at the Institut Henri Poincaré in Paris. And we took on the case of Sion Assidon. We were snowed under with work! Assidon's mother was allowed to visit him. She brought him packages which he shared with his fellow-prisoners. We demonstrated several times for his liberation, and sent a delegation to the Moroccan Embassy in Paris. As usual, the ambassador received us cordially, and assured us that he would let his government know about our visit. I'm sure he did, but he also reproached us our concern for a man who had conspired against the security of his country. We tried to tell him about Shcharansky, and about the very different types of cases we had been led to defend, but he cried "You can't compare them! Sion Assidon is a criminal, whereas Shcharansky is an innocent man, who was unjustly accused and imprisoned!" This was rather amusing; indeed, just around the same time a delegation from the Embassy of the USSR (an aide, not the ambassador himself, who never deigned to receive us) reproached us with defending only the cases of enemies of the USSR. We tried to tell him about the cases of Massera and Sion Assidon, but he burst out "You can't compare Massera with Shcharansky! Shcharansky is a traitor to his country, and Massera is a hero defending liberty in Uruguay!" Sion Assidon was married and had a little girl, whom he barely met before she was ten years old. His younger sister, Elsa Assidon, who remained close to us, became the main connecting link between us and the Assidon family. We obtained a lot of help from Charles-André Julien, a retired professor of history at the Collège de France. He was very familiar with the Moroccan hierarchy, and he knew king Hassan II and his Prime Minister; we wrote to them several times and mentioned Julien, but they never answered. Yet surely these efforts protected Assidon from severe torture.

Broué and I decided to take a two-day trip to Morocco, to try to meet the king, the Prime Minister or at least the head of the cabinet, and to try to see Assidon. We did not succeed in a single one of these endeavors. We entered the country on a normal visa. We immediately contacted the family, who lived in Casablanca, whereas Assidon was imprisoned in Rabat, but we had to give up. The two people we wanted to see were both away that day, for reasons having nothing to do with our visit. "Our two zebras escaped and we won't be able to capture them," I said. I called them zebras so as not to mention any names in public, and Broué found it very funny. We asked for the authorization to accompany his mother on her usual visit, but we didn't receive it; his mother was at least able to let him know that we were there. Two months later, Broué went to Morocco alone, and calling himself a cousin of Assidon, he was able to obtain the authorization and visit Assidon with his mother; he had a long conversation with him. After his liberation, he took a trip to France to see his friends from the Committee of Mathematicians, and invited us to dinner with his sister Elsa.

We also worked on the cases of other mathematicians. I remember once sending a telegram in favor of three young Chilean mathematicians imprisoned under Pinochet; the telegram was to the president of the tribunal and to the rector, and I signed it Laurent Schwartz, member of the Academy of Sciences. When I was dictating it on the telephone, the young lady taking it down suddenly cried "Oh, Laurent Schwartz, mathematician, that's marvelous!" I really couldn't understand why. "Guess my name," she continued with enthusiasm. "Maybe Schwartz? how should I know?" "No, but it's still a name close to yours." I gave up and she told me her name was . . . Cauchy. It was very interesting. This young lady was no relation of the family of the famous mathematician Cauchy, but she knew that he had been one of the most important French mathematicians of the 19th century. Schwarz (without a t, as she quite correctly told me) was a great Austrian mathematician, Hermann Amandus Schwarz. The famous Cauchy-Schwarz inequalities are taught in every university. "You see," she added, "when people sign their telegrams with the name Schwartz, and I tell them my name is Cauchy, they don't get it, but if it's a Schwartz mathematician, I can finally tell my story!" I found this so amusing that I wrote about it in the *Gazette des mathématiciens*, a weekly bulletin addressed to the community of mathematicians, which published it. Two of the Chilean mathematicians were quickly liberated, but the third was kept under house arrest in a small village. His university colleagues organized a conference around his work, which they were allowed to hold in the village, and he was liberated soon after.

I certainly haven't finished with the activities of the Committee of Mathematicians. But I can already conclude that this committee was an experiment quite unique in history. For causes stemming from human rights abuses, the international community of mathematicians uprose, took action, contacted powerful heads of state, and ended every time, after great efforts, by obtaining what they were asking for. The activities of the committee were recounted in newspapers all over the world. All of that, and even the name of the committe, has been forgotten by now. The young generation of mathematicians probably knows very little about the whole story. I think it was worth telling it here so it doesn't just slide into oblivion. Why did mathematicians *specially* take action for such diverse reasons? There were antecedents, for instance the visa problems for the International Congress of Mathematicians in 1950, in the US, and previous combats over academic deontology. Mathematicians transport their rigorous reasoning into situations of daily life. Mathematical discovery is subversive and always ready to overthrow taboos, and it depends very little on established powers. Many people nowadays seem to consider scientists, mathematicians and others, like people uninterested in moral questions, locked away in their ivory towers and indifferent to the outside world. The Committee of Mathematicians is a brilliant illustration of the contrary.